50 YEARS *of* QUARKS

50 YEARS *of* QUARKS

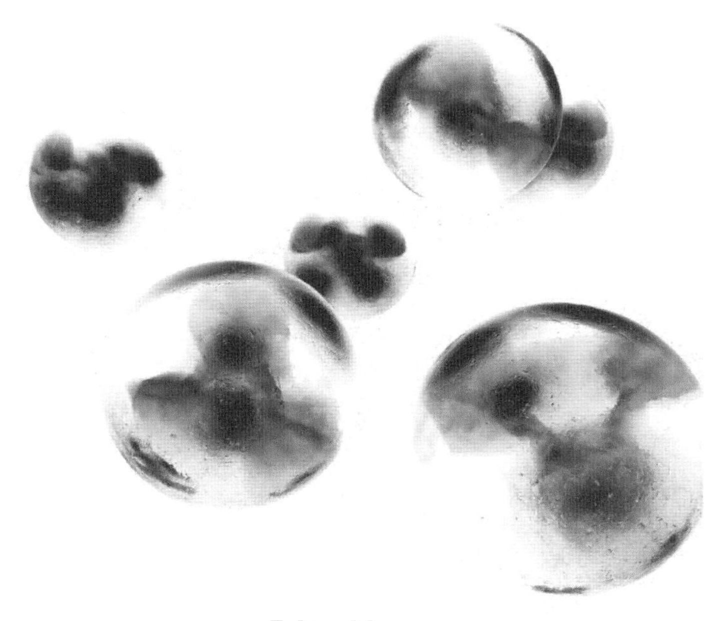

Edited by

Harald Fritzsch
Ludwig Maximilian University of Munich, Germany

Murray Gell-Mann
Santa Fe Institute, USA

NEW JERSEY • LONDON • SINGAPORE • BEIJING • SHANGHAI • HONG KONG • TAIPEI • CHENNAI

Published by

World Scientific Publishing Co. Pte. Ltd.
5 Toh Tuck Link, Singapore 596224
USA office: 27 Warren Street, Suite 401-402, Hackensack, NJ 07601
UK office: 57 Shelton Street, Covent Garden, London WC2H 9HE

British Library Cataloguing-in-Publication Data
A catalogue record for this book is available from the British Library.

50 YEARS OF QUARKS

ISBN 978-981-4618-09-0
ISBN 978-981-4618-10-6 (pbk)

Printed in Singapore

Preface

Harald Fritzsch

Physics Department, Ludwig-Maximilians-University,
D-80333 Munich, Germany

The year 1964 was a remarkable year in physics. In this year CP-violation was discovered, the Ω^--resonance was observed, the cosmic microwave background radiation was discovered and the quark model was introduced.

In 1947 the four K-mesons and the Λ-hyperon were discovered, followed by the discovery of the three Σ-hyperons and of the two Ξ-hyperons. In 1961 Murray Gell-Mann and Yuval Ne'eman suggested that these particles are described by a new symmetry, based on the group SU(3). The SU(3)-symmetry is an extension of the isospin symmetry, which was introduced in 1932 by Werner Heisenberg.

The observed hadrons are members of specific representations of SU(3). The lowest baryons and mesons are described by octet representations. The baryon octet contains the two nucleons, the Λ-hyperon, the three Σ-hyperons and the two Ξ-hyperons. The members of the meson octet are the three pions, the four K-mesons and the η-meson. The ten baryon resonances are members of a decuplet representation.

In the SU(3) scheme the observed hadrons are octets and decuplets. There are no hadrons, described by the triplet or sextet representation. In 1964 Gell-Mann introduced the triplets as constituents of the baryons and mesons, the "quarks". This name appears in the novel "Finnegans wake" of James Joyce on page 383: "Three quarks for Muster Mark", as mentioned in the contribution of Gell-Mann to this volume. The triplet model was also considered by the Caltech graduate student George Zweig, who was visiting CERN in 1964 (see the contribution of Zweig).

The three quarks are the "up", "down" and "strange" quarks: (u, d, s). The baryons are bound states of three quarks — the proton consists of two up-quarks and one down-quark: $p \sim (uud)$. The Λ-hyperon has one strange quark: $\Lambda \sim (uds)$. The Ω^--resonance is a bound state of three strange quarks: $\Omega^- \sim (sss)$. Thus the s-quark has the electric charge $(-1/3)$, as the d-quark. The charge of the u-quark is $(+2/3)$.

In the quark model it remained unclear, why there are only bound states of three quarks or three antiquarks, the baryons or antibaryons, or quark and antiquark, the mesons. Why are there no bound states, consisting of two quarks? Why are there no free quarks? Those would be stable particles with nonintegral electric charges. Thus the quark model had many problems.

Gell-Mann considered the quarks as mathematical objects, not as real particles. Zweig, however, assumed that the quarks are constituents of the mesons and baryons.

In 1968 the first experiments in the deep inelastic scattering of electrons and nuclei were carried out at the Stanford Linear Accelerator Center (SLAC). The measured cross sections depend on two variables, the energy transfer from the electron to the nucleus and the mass of the virtual photon. It was observed that the cross sections show a scaling behavior — they were only functions of the ratio of the two variables. This indicated that there are point-like constituents inside the nucleons with non-integral electric charges — the quarks were discovered. They were not only mathematical objects, but constituents of the atomic nuclei.

Murray Gell-Mann and I introduced in 1971 a new quantum number for the quarks, which we called "color". Each quark has three different colors: red, green or blue. The transformations among the three colors are described by the group SU(3). The quarks are color triplets, two quarks can either be in a color sextet or in a color anti-triplet, but three quarks can form a color singlet, likewise a quark and an anti-quark. The color symmetry was considered to be an exact symmetry of nature (see the contribution of Fritzsch).

The simplest color singlets were the bound states of a quark and an antiquark (meson) or three quarks (baryon). We assumed that all hadrons are color singlets, and all configurations with color, e.g. the color triplet quarks, are permanently confined inside the hadrons.

In 1972 Gell-Mann and I discussed a gauge theory for the description of the strong interactions. The color degree of freedom was gauged, as the electric charge in QED. The gauge bosons were mass-less gluons, which transformed as an octet of the color group. Later we called this theory "Quantum Chromodynamics".

The gluons interact with the quarks, but also with themselves. This self-interaction leads to the interesting property of asymptotic freedom. The gauge coupling constant of QCD decreases logarithmically, if the energy is increased. Experiments at SLAC, at DESY, at the Large Electron–Positron Collider in CERN and at the Tevatron in the Fermi National Laboratory have measured this decrease of the coupling-constant.

In QCD the scaling property of the cross sections, observed in deep inelastic scattering, is not an exact property, but it is violated by small logarithmic terms. The scaling violations were observed and are in good agreement with the theoretical predictions.

Today the theory of QCD is considered to be the true theory of the strong interactions and is an essential part of the Standard Theory of particle physics (see also the contributions of Heinrich Leutwyler and Finn Ravndal).

The birth of the quark model is described in the contributions of David Horn, Sidney Meshkov and Lev Okun. Details of the quark model are discussed in the contribution of Willy Plessas.

The color quantum number can be observed in electron–positron annihilation at high energy, but also in the electromagnetic decay of the neutral pion, as described in the contribution of Rodney Crewther. Although quarks are permanently confined, they can be observed indirectly as “quark jets”, as discussed in the contribution of Rick Field.

Also gluons can be observed as “gluon jets” — this is the main topic of the contributions of John Ellis and Sau Lan Wu. In QCD one can also make predictions about exclusive processes, as discussed in the contribution of Stan Brodsky.

Although quarks are permanently confined inside the hadrons, they do have masses, which determine in particular the symmetry breaking. The determination of the quark masses is described in the contribution of Cesareo Dominguez.

If hadrons are compressed to high density, e.g. in heavy ion collisions or in a big neutron star, the quarks and gluons form a “quark–gluon plasma”, as discussed in the contribution of Ulrich Heinz.

In reality there are not only three quarks, but today we know six quarks: u, d, c, s, t, b (see also the contribution of Shelley Glashow and of Marek Karliner). The six quarks had been predicted in 1972, since with six quarks one can understand CP-violation, as discussed by Makoto Kobayashi.

The six quarks form three electroweak doublets: u-d, c-s and t-b. These quarks are not mass eigenstates, but mixtures, and one obtains the flavor mixing, as discussed by Zhi-zhong Xing.

Also the leptons are forming three electroweak doublets, involving the electron and its neutrino, the muon and its neutrino and the tauon and its neutrino. Since neutrinos have masses, flavor mixing happens also for the leptons. This can be observed by the neutrino oscillations, as discussed by Rabindra Mohapatra.

The Standard Theory of particle physics is the theory of quantum chromodynamics and the gauge theory of the electroweak interactions. Many physicists speculate about a unified theory beyond the Standard Theory. There might be a unification of QCD and the electroweak theory at very high energy — the QCD coupling constant and the two coupling constants of the electroweak theory come together. An interesting possibility is the gauge group SO(10).

A unified theory might have new kinds of symmetries, e.g. supersymmetry (see the contributions of Stephen Adler, of Gordon Kane and Malcolm Perry, and of Mikhail Shifman and Alexei Yung).

Some physicists speculate that the leptons, quarks and the gauge bosons are manifestations of small one-dimensional objects, the superstrings. A theory, based on superstrings, might also be a theory of quantum gravity.

Contents

Preface v
H. Fritzsch

A Schematic Model of Baryons and Mesons 1
M. Gell-Mann

Quarks 5
M. Gell-Mann

Concrete Quarks 25
G. Zweig

On the Way from Sakatons to Quarks 57
L. B. Okun

My Life with Quarks 95
S. L. Glashow

Quarks and the Bootstrap Era 105
D. Horn

From Symmetries to Quarks and Beyond 115
S. Meshkov

How I Got to Work with Feynman on the Covariant Quark Model 127
F. Ravndal

What is a Quark? 149
G. L. Kane & M. J. Perry

Insights and Puzzles in Particle Physics 163
H. Leutwyler

Quarks and QCD 181
H. Fritzsch

The Discovery of Gluon 189
J. Ellis

Discovery of the Gluon 199
S. L. Wu

The Parton Model and Its Applications 227
T. M. Yan & S. D. Drell

From Old Symmetries to New Symmetries: Quark, Leptons and $B - L$ 245
R. N. Mohapatra

Quark Mass Hierarchy and Flavor Mixing Puzzles 265
Z.-Z. Xing

Analytical Determination of the QCD Quark Masses 287
C. Dominquez

CP Violation in Six Quark Scheme — Legacy of Sakata Model 315
M. Kobayashi

The Constituent Quark Model — Nowadays 325
W. Plessas

From Ω^- to Ω_b, Doubly Heavy Baryons and Exotics 345
M. Karliner

Quark Elastic Scattering as a Source of High Transverse Momentum Mesons 367
R. Field

Exclusive Processes and the Fundamental Structure of Hadrons 381
S. J. Brodsky

Quark-Gluon Soup — The Perfectly Liquid Phase of QCD 413
U. Heinz

Quarks and Anomalies 435
R. J. Crewther

Lessons from Supersymmetry: "Instead-of-Confinement" Mechanism 453
M. Shifman & A. Yung

Quarks and a Unified Theory of Nature Fundamental Forces 473
I. Antoniadis

SU(8) Family Unification with Boson–Fermion Balance 487
S. L. Adler

A Schematic Model of Baryons and Mesons*

M. Gell-Mann

California Institute of Technology, Pasadena, California, USA

If we assume that the strong interactions of baryons and mesons are correctly described in terms of the broken "eightfold way",[1–3] we are tempted to look for some fundamental explanation of the situation. A highly promised approach is the purely dynamical "boot trap" model for all the strongly interacting particles within which one may try to derive isotopic spin and strangeness conservation and broken eighfold symmetry from self-consistency alone.[4] Of course, with only strong interactions, the orientation of the asymmetry in the unitary space cannot be specified; one hopes that in some way the selection of specific components of the F-spin by electromagnetism and the weak interactions determines the choice of isotopic spin and hyper-charge directions. Even if we consider the scattering amplitudes of strongly interacting particles on the mass shell only and treat the matrix elements of the weak, electro-magnetic, and gravitational interactions by means of dispersion theory, there are still meaningful and important questions regarding the algebraic properties of these interactions that have so far been discussed only by abstracting the properties from a formal field theory model based on fundamental entities[3] from which the baryons and mesons are built up. If these entities were octets, we might expect the underlying symmetry group to be SU(8) instead of SU(3); it is therefore tempting to try to use unitary triplets as fundamental objects. A unitary triplet t consists of an isotopic singlet s of electric charge z (in units of e) and an isotopic doublet (u, d) with charges $z + 1$ and z respectively. The anti-triplet $\bar{t}$ has, of course, the opposite signs of the charges. Complete symmetry among the members of the triplet gives the exact eightfold way, while a mass difference, for example, between the isotopic doublet and singlet gives the first-order violation. For any value of z and of triplet spin, we can construct baryon octets from a basic neutral baryon singlet b by taking combinations $(bt\bar{t})$, $(btt\bar{t}\bar{t})$, etc.[a] From $(bt\bar{t})$, we get the representations **1** and **8**, while from $(btt\bar{t}\bar{t})$ we get **1**,

*Work is supported in part by the U.S. Atomic Energy Commission.

[a]This is similar to the treatment in Ref. 1. See also Ref. 5.

8, **10**, $\overline{\mathbf{10}}$, and 27. In a similar way, meson singlets and octets can be made out of $(t\bar{t})$, $(bt\bar{t}\bar{t})$, etc. The quantum number $n_t - n_{\bar{t}}$ would be zero for all known baryons and mesons. The most interesting example of such a model is one in which the triple has spin $\frac{1}{2}$ and $z = -1$, so that the four particles d^-, s^-, u^0 and b^0 exhibit a parallel with the leptons. A simpler and more elegant scheme can be constructed if we allow non-integral values for the charges. We can dispense entirely with the basic baryon b if we assign to the triplet t the following properties: spin $\frac{1}{2}$, $z = -\frac{1}{3}$, and baryon number $\frac{1}{3}$. We then refer to the members $u^{\frac{2}{3}}$, and $d^{-\frac{1}{3}}$, and $s^{-\frac{1}{3}}$ of the triplet as "quarks"[6] q and the members of the anti-triplet as anti-quarks $\bar{q}$. Baryons can now be constructed from quarks by using the combinations (qqq), $(qqqq\bar{q})$, etc., while mesons are made out of $(q\bar{q})$, $(qq\bar{q}\bar{q})$, etc. It is assuming that the lowest baryon configuration (qqq) gives just the representations **1**, **8**, and **10** that have been observed, while the lowest meson configuration $(q\bar{q})$ similarly gives just **1** and **8**. A formal mathematical model based on field theory can be built up for the quarks exactly as for p, n, Λ in the old Sakata model, for example[3] with all strong interactions ascribed to a neutral vector meson field interacting symmetrically with the three particles. Within such a framework, the electromagnetic current (in units of e) is just

$$i\left\{\frac{2}{3}\bar{u}\gamma_\alpha u - \frac{1}{3}\bar{d}\gamma_\alpha d - \frac{1}{3}\bar{s}\gamma_\alpha s\right\}$$

or $\mathscr{F}_{3\alpha} + \mathscr{F}_{8\alpha}/\sqrt{3}$ in the notation of Ref. 3. For the weak current, we can take over from the Sakata model the form suggested by Gell-Mann and Lévy,[7] namely $i\bar{p}\gamma_\alpha(1+\gamma_5)(n\cos\theta + \Lambda\sin\theta)$, which gives in the quark scheme the expression[b]

$$i\bar{u}\gamma_\alpha(1+\gamma_5)(d\cos\theta + s\sin\theta)$$

or, in the notation of Ref. 3,

$$\begin{aligned}&[\mathscr{F}_{1\alpha} + \mathscr{F}^5_{1\alpha} + i(\mathscr{F}_{2\alpha} + \mathscr{F}^5_{2\alpha})]\cos\theta \\ &\quad + [\mathscr{F}_{4\alpha} + \mathscr{F}^5_{4\alpha} + i(\mathscr{F}_{5\alpha} + \mathscr{F}^5_{5\alpha})]\sin\theta\,.\end{aligned}$$

We thus obtain all the features of Cabibbo's picture[8] of the weak current, namely the rules $|\Delta I| = 1$, $\Delta Y = 0$ and $|\Delta I| = \frac{1}{2}$, $\Delta Y/\Delta Q = +1$, the conserved $\Delta Y = 0$ current with coefficient $\cos\theta$, the vector current in general

[b] The parallel with $i\bar{\nu}_e\gamma_\alpha(1+\gamma_5)e$ and $i\bar{\nu}_\mu\gamma_\alpha(1+\gamma_5)\mu$ is obvious. Likewise, in the model with d^-, s^-, u^0, and b^0 discussed above, we would take the weak current to be $i(\bar{b}^0\cos\theta + \bar{u}^0\sin\theta)\gamma_\alpha(1+\gamma_5)s^- + i(\bar{\mu}^0\cos\theta - \bar{b}^0\sin\theta)\gamma_\alpha(1+\gamma_5)d^-$, The part with $\Delta(\eta_t - \eta_{\bar{t}}) = 0$ is just $i\bar{u}^0\gamma_\alpha(1+\gamma_5)(*d^-\cos\theta + s^-\sin\theta)$.

as a component of the current of the F-spin, and the axial vector current transforming under SU(3) as the same component of another octet. Furthermore, we have[3] the equal-time commutation rules for the fourth components of the currents:

$$[\mathscr{F}_{j4}(x)\pm\mathscr{F}^5_{j4}(x), \mathscr{F}_{k4}(x') \pm \mathscr{F}^5_{k4}(x')] = 2f_{jkl}[\mathscr{F}_{j4}(x) \pm \mathscr{F}^5_{l4}(x)]\delta(x - x'),$$
$$[\mathscr{F}_{j4(x)} \pm \mathscr{F}^5_{j4}(x), \mathscr{F}_{k4}(x') \mp \mathscr{F}^5_{k4}(x')] = 0\,,$$

$i = 1, \ldots 8$, yielding the group SU(3) $\times$ SU(3). We can also look at the behaviour of the energy density $\theta_{44}(x)$ (in the gravitational interaction) under equal-time commutation with the operators $\mathscr{F}_{j4}(x') \pm \mathscr{F}_{j4}5(x')$. That part which is non-invariant under the group will transform like particular representations of SU(3) $\times$ SU(3), for example like $(3, \bar{3})$ and $(\bar{3}, 3)$ if it comes just from the masses of the quarks. All these relations can now be abstracted from the field theory model and used in a dispersion theory treatment. The scattering amplitudes for strongly interacting particles on the mass shell are assumed known; there is then a system of a linear dispersion relations for the matrix elements of the weak currents (and also the electromagnetic and gravitational interactions) to lowest order in these interactions. These dispersion relations, unsubtracted and supplemented by the non-linear commutation rules abstracted from the field theory, may be powerful enough to determine all the matrix elements of the weak currents, including the effective strengths of the axial vector current matrix elements compared with those of the vector current. It is fun to speculate about the way quarks would behave if they were physical particles of finite mass (instead of purely mathematical entities as they would be in the limit of infinite mass). Since charge and baryon number of exactly conserved, one of the quarks (presumably $u^{\frac{2}{3}}$ or $d^{-\frac{1}{3}}$) would be absolutely stable*, while the other member of the doublet would go into the first member very slowly by β-decay or K-capture. The isotopic singlet quark would presumably decay into the doublet by weak interactions, much as Λ goes into N. Ordinary matter near the earth's surface would be contaminated by stable quarks as a result of high energy cosmic ray events throughout the earth's history, but the contamination is estimated to be so small that it would never have been detected. A search for stable quarks of charge $-\frac{1}{3}$ or $+\frac{2}{3}$ and/or stable di-quarks of charge $-\frac{2}{3}$ or $+\frac{1}{3}$ or $+\frac{4}{3}$ at the highest energy accelerators would help to reassure us of the non-existence of real quarks. These ideas were developed during a visit to Columbia University in March 1963; the author would like to thank Professor Robert Serber for stimulating them.

References

1. M. Gell-Mann, California Institute of Technology Synchrotron Laboratory Report CTSL-20 (1961).
2. Y. Ne'eman, *Nuclear Phys.* **26** (1961) 222.
3. M. Gell-Mann, *Phys. Rev.* **125** 1962) 1067.
4. E.g.: R. H. Capps, *Phys. Rev. Lett.* **10** (1963) 312; R. E. Cutkosky, J. Kalckar and P. Tarjanne, *Phys. Lett.* **1** (1962) 93; E. Abers, F. Zachariasen and A. C. Zemach, *Phys. Rev.* **132** (1963) 1831; S. Glashow, *Phys. Rev.* **130** (1963) 2132; R. E. Cutkosky and P. Tarjanne, *Phys. Rev.* **132** (1963) 1354.
5. P. Tarjanne and V. L. Teplitz, *Phys. Rev. Lett.* **11** (1963) 447.
6. James Joyce, *Finnegan's Wake* (Viking Press, New York, 1939) p. 383.
7. M. Gell-Mann and M. Lévy, *Nuovo Cimento* **16** (1960) 705.
8. N. Cabibbo, *Phys. Rev. Lett.* **10** (1963) 531.

Quarks*

M. Gell-Mann[†]

CERN – Geneva

In these lectures I want to speak about at least two interpretations of the concept of quarks for hadrons and the possible relations between them.

First I want to talk about quarks as "constituent quarks". These were used especially by G. Zweig (1964) who referred to them as aces. One has a sort of a simple model by which one gets elementary results about the low-lying bound and resonant states of mesons and baryons, and certain crude symmetry properties of these states, by saying that the hadrons act as if they were made up of subunits, the constituent quarks q. These quarks are arranged in an isotopic spin doublet u, d and an isotopic spin singlet s, which has the same charge as d and acts as if it had a slightly higher mass.

The antiquarks $\bar{q}$ of course, have the opposite behaviour. The low-lying bound and resonant states of baryons act like qqq and those of the mesons like $q\bar{q}$. Other configurations, e.g., $q\bar{q}q\bar{q}$, $qqqq\bar{q}$, etc., are called exotic, but they certainly exist in the continuum and may have resonances corresponding to them.

In this way one builds up the low-lying meson and baryon states and it is frequently useful to classify them in terms of an extremely crude symmetry group $U_6 \times U_6 \times O_3$, where one U_6 is for the quarks (three states of charge and two spin states) and one for the antiquarks, whereas O_3 represents a sort of angular momentum between them. This symmetry is however badly violated in the lack of degeneracy of the spectrum. The mesons then have as the lowest representation

$$(\underline{6}, \underline{\bar{6}}), \quad L^P = 0^-$$

which gives the pseudoscalar and vector mesons, nine of each, just the ones which have been observed. ($L = 0$ would normally have parity plus, but since we have a q and a $\bar{q}$ the intrinsic parity is minus.) The next pattern would

*Lecture given at XI. Internationale Universitätswochen für Kernphysik, Schladming, February 21 – March 4, 1972.

[†]On leave from CALTECH, Pasadena. John Simon Guggenheim Memorial Fellow.

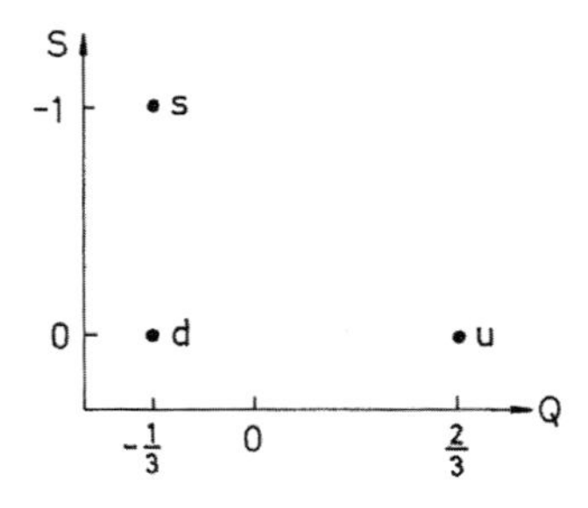

be

$$(\underline{6}, \underline{\bar{6}}), \quad L^P = 1^+$$

and this gives us the tensor mesons, axial vector mesons, another kind of axial vector mesons with opposite charge conjugation, and scalar mesons. All of these kinds have been seen, although not yet quite nine in every case. Then one can go up with $L = 2, 3, \ldots$ where there is just scattered experimental information. But whatever experimental information exists is at least compatible with this trivial picture. As is well known this group is not very well conserved in the spectrum and the states are badly split, e.g., $m_\pi = 140$ MeV and $m_{\eta'} = 960$ MeV.

For the baryons the lowest configuration is assigned to

$$(\underline{56}, \underline{1}), \quad L^P = 0^+$$

that is three quarks in a totally symmetric state of spin, isospin, etc., without antiquarks; the parity is plus by definition. This gives the baryon octet with spin $\frac{1}{2}$ and just above it the decimet with spin $\frac{3}{2}$, which agree with the lowest-lying and best-known states of the baryons. The next thing one expects would be an excitation of one unit of angular momentum which changes the symmetry to the mixed symmetry under permutations:

$$(\underline{70}, \underline{1}), \quad L^P = 1^-$$

and that seems to be a reasonable description of the low lying states of reversed parity. If one goes on to higher configurations things become more uncertain both experimentally and theoretically; presumably

$$(\underline{56}, \underline{1}), \quad L^P = 2^+$$

exists and contains the Regge excitations of the corresponding ground state, likewise

$$(\underline{70}, \underline{1}), \quad L^P = 3^- \quad \text{and so on}.$$

In doing this we have to assume something peculiar about the statistics obeyed by these particles, if we want the model to be simple. One expects the ground state to be totally symmetric in space. If the quarks obeyed the usual Fermi–Dirac statistics for spin $\frac{1}{2}$ particles, then there would be an over-all antisymmetry and one would obtain a totally antisymmetric wave function in spin, isospin and strangeness, whereas $(\underline{56}, \underline{1})$ is the totally symmetric configuration in these quantum numbers. What most people have assumed therefore from the beginning (1963) is that the quarks obey some unusual kind of statistics in which every set of three has to be symmetrized but all other bonds have to be made antisymmetric, so that, e.g., two baryons are antisymmetric with respect to each other. One version of this came up under the name of parastatistics, precisely "para-Fermi statistics of rank three", which gives a generalization of the result I just described. I will discuss it in a slightly different way, which is equivalent to para-Fermi statistics of rank three with the restriction that physical baryon states are all fermions and physical meson states are all bosons.

We take three different kinds of quarks, that is line altogether, and call the new variable distinguishing the sets "color", for example red, white and blue (R–W–B). The nine kinds of quarks are then individually Fermi–Dirac particles, but we require that all physical baryon and meson states be singlets under the SU_3 of "color". This means that for the meson $q\bar{q}$ configuration we now have

$$q_R\bar{q}_R + q_W\bar{q}_W + q_B\bar{q}_B$$

and for a baryon qqq we have

$$q_R q_W q_B - q_W q_R q_B + q_B q_R q_W - q_R q_B q_W + q_W q_B q_R - q_B q_W q_R\,,$$

which is totally antisymmetric in color and permits the baryon to be totally symmetric in the other variables space, spin, isospin and strangeness. This restriction to color singlet states for real physical situations gives back exactly the sort of statistics we want.

Now if this restriction is applied to all real baryons and mesons, then the quarks presumably cannot be real particles. Nowhere have I said up to now that quarks have to be real particles. There might be real quarks, but nowhere in the theoretical ideas that we are going to discuss is there any insistence that they be real. The whole idea is that hadrons act as if they are made up of quarks, but the quarks do not have to be real.

If we use the quark statistics described above, we see that it would be hard to make the quarks real, since the singlet restriction is not one that can be easily applied to real underlying objects; it is not one that factors: a singlet can be made up of two octets and these can be removed very far from each other such that the system over-all still is a singlet, but then we see the two pieces as octets because of the factoring property of the S matrix. If we adopt this point of view we are then faced with two alternatives: one is that there are three quarks, fictitious and obeying funny statistics; the other is that there are actually three triplets of real quarks, which is possible but unpleasant. In the latter case we would replace the singlet restriction with the assumption that the low lying states are singlets and one has to pay a large price in energy to get the colored SU_3 excited. I would prefer to adopt the first point of view, at least for these lectures.

Various crude symmetries and other related methods have been applied to these constituent quarks. First of all there is the famous subgroup of the classifying $U_6 \times U_6 \times U_3$, namely $[U_6]_w \times [O_2]_w$ which is applied to processes involving only one direction in space, like a vertex or forward and backward scattering (in general, collinear processes). $[O_2]_w$ has the generator L_z (assuming z is the chosen direction) and $[U_6]_w$

$$\frac{1}{2}\Big(\sum_i \lambda_i + \sum_j \lambda'_j\Big)\,,$$

$$\frac{1}{2}\Big(\sum_i \lambda_i\sigma_{iz} + \sum_j \lambda'_j\sigma'_{jz}\Big)\,,$$

$$\frac{1}{2}\Big(\sum_i \lambda_i\sigma_{ix} - \sum_j \lambda'_j\sigma'_{jx}\Big)\,,$$

$$\frac{1}{2}\Big(\sum_i \lambda_i\sigma_{iy} - \sum_j \lambda'_j\sigma'_{jy}\Big)\,,$$

where the sum over i extends over the constituent quarks and the primed sum over j extends over the constituent antiquarks; we have 36 operations. There is a very crude symmetry of collinear processes under this group.

Another thing that has been done is to draw simple diagrams following quark lines through the vertices and the scattering. These have been recently used by Harari and Rosner, who called them "twig" diagrams after Zweig, who introduced them in 1964.

The twig diagram, e.g., for a meson-meson-meson vertex, looks like

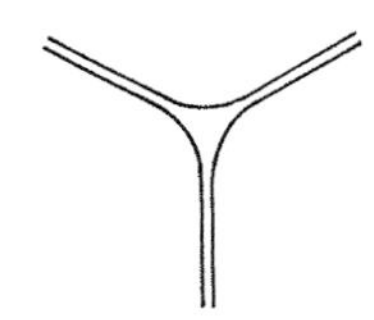

But another form

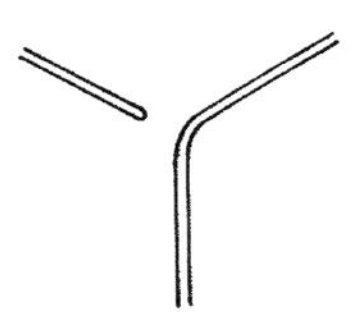

is forbidden by "Zweig's rule". This rule then leads to important experimental results, especially that the ϕ cannot decay appreciably into a ρ and π ($\phi \not\rightarrow \rho + \pi$), since the ϕ is composed of strange and antistrange quarks whereas ρ and π have only ordinary up and down quarks, and, therefore, the decay could take place only via the forbidden diagram. Similarly, we have the baryon-baryon-meson vertex

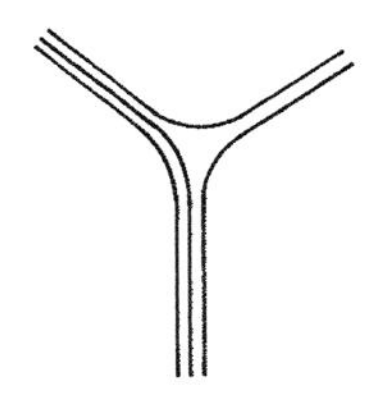

One can extend this concept to scattering processes and get a graphical picture of the so-called duality approach to scattering, e.g., for meson-meson scattering one can introduce the following diagram:

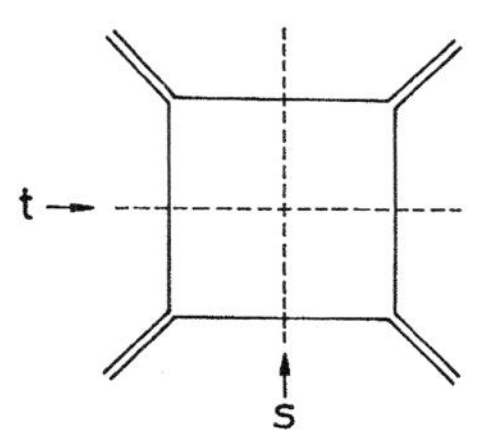

If we cut the diagram in the s and t channels we get $q\bar{q}$ in both cases: therefore, in meson-meson scattering we have ordinary non-exotic mesons in the intermediate states and exchange non-exotic mesons. We run into something of a trap, though, if we try to apply this to baryon-antibaryon scattering, because then we have a situation like

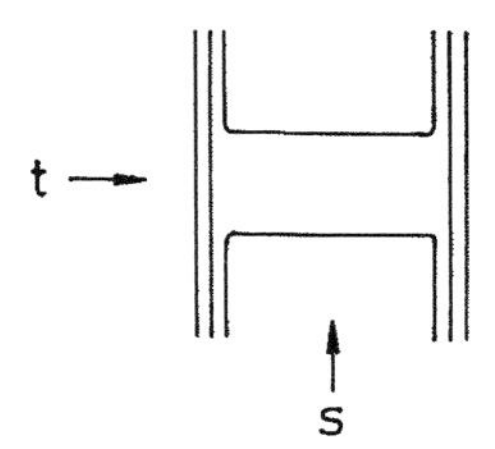

where the intermediate state is $q\bar{q}q\bar{q}$, which includes exotic configurations. In order to interpret this inconsistency different people have done different things.

The diagrams have been used in two different ways: one involves saying what the diagram means mathematically, and the other one involves not saying what it means mathematically. This is possible since here we do not have a priori a definite mathematical rule for computing the diagram, in contrast to a Feynman diagram for which a specific integral always exists. But we can obtain same results by never giving such a rule, only by noticing that we have zero when there is no diagram. Those so-called "null-relations" have been used by Schmid, by Harari and Rosner, by Zweig, Weyers and Mandula and by others for establishing a number of extremely useful sum rules. They give a correspondence between lack of resonances in the s channel, in places where the resonances would have to be exotic, and exchange degeneracy in the t channel. Exchange degeneracy is a noticeable feature of low energy hadron physics and a number of cases of agreement with experiments have been obtained.

All I want to say about the null-relation approach is that from the point of view of constituent quarks we are dealing here with a non-exotic approximation, because we are leaving out exotic exchanges, and that cannot be expected to be completely right. The simple null-equation duality approach is just another feature of the same kind of approximation we were talking about before, i.e., the classification under $U_6 \times U_6 \times O_3$ and the rough symmetry of collinear processes under $[U_6]_w \times [O_2]_w$, and when it fails that resembles a failure of such an approximation.

Another school of people consists of those who do the Veneziano duality kind of work and actually attempt to assign mathematical meaning to these diagrams. They go very far and construct almost complete theories of hadron scattering by means of extending these simple diagrams to ones with any number of quark pairs, but they run into trouble with negative probabilities or negative mass squares and the difficulty of introducing quark spin. There are also some difficulties with high energy diffraction scattering, etc. So that approach is not yet fully successful, while the much more modest null-relation approach has borne some fruit. However, if they overcome their difficulties, the members of the other school will have produced a full-blown hadron theory and advanced physics by a huge step.

There is a second use of quarks, as so-called "current quarks", which is quite different from their use as constituent quarks; we have to distinguish carefully between the two types in order to think about quarks in a reasonable manner. Unfortunately many authors including, I regret to say, me, have in past years written things that tended to confuse the two. In the following discussion of current quarks we attempt to write down properties that may be exact, at least to all orders in the strong interaction, with the weak, electromagnetic and gravitational interactions treated as perturbations. (It is necessary always to include gravity because the first order coupling to gravity is the stress-energy-momentum tensor and the integral over this tensor gives us the energy and momentum which we have to work with.)

When I say we attempt exact statements I do not mean that they are automatically true — there is also the incidental matter that they have to be confirmed by experiment, but the statements have a chance to be exact. Such statements which are supposed to be exact at least in certain limits or in certain well-defined approximations, or even generally exact, are to be contrasted with statements which are made in an ill-defined approximation or a special model whose domain of validity is not clearly specified. One frequently sees allegedly exact statements mixed up with these vague model statements and when experiments confirm or fail to confirm them it does not mean anything. Of course, we all have to work occasionally with these vague models because they give us some insight into the problem but we should carefully distinguish highly model-dependent statements from statements that have the possibility of being true either exactly or in a well defined limit.

The use of current quarks now is the following: we say that currents act as if they were bilinear forms in a relativistic quark field. We introduce a quark field, presumably one for the red, white and blue quarks and then we have for the vector currents in weak and electromagnetic interaction

$$F_{i\mu} \sim i\bar{q}_R \gamma_\mu \frac{\lambda_i}{2} q_R + i\bar{q}_W \gamma_\mu \frac{\lambda_i}{2} q_W + i\bar{q}_B \gamma_\mu \frac{\lambda_i}{2} q_B \,.$$

The symbol $\sim$ means the vector current "acts like" this bilinear combination. Likewise the axial vector current acts like

$$F^5_{i\mu} \sim i\bar{q}_R \gamma_\mu \gamma_5 \frac{\lambda_i}{2} q_R + i\bar{q}_W \gamma_\mu \gamma_5 \frac{\lambda_i}{2} q_W + i\bar{q}_B \gamma_\mu \gamma_5 \frac{\lambda_i}{2} q_B \,.$$

The reason why I want all these colors at this stage is that I would like to carry over the funny statistics for the current quarks, and eventually would like to suggest a transformation which takes one into the other, conserving the statistics but changing a lot of other things. An important feature of this discussion will be the following: is there any evidence for the current quarks that they obey the funny statistics? the answer is yes, and the evidence depends on a theoretical result due to many people but principally S. Adler.

The result is that in the PCAC limit one can compute exactly the decay rate of $\pi^0 \to 2\gamma$. The basis on which Adler derived it was a relativistic renormalized quark-gluon field theory treated in renormalized perturbation theory order by order, and there the lowest order triangle diagram gives the only surviving result in the PCAC limit:

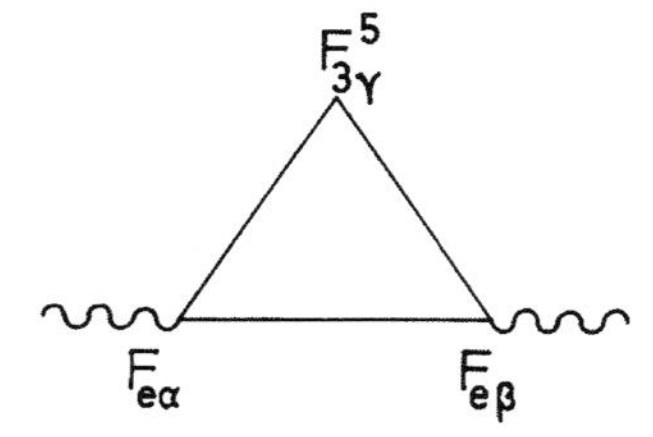

Here F_e are the electromagnetic currents, F_3^5 the third component of the axial current that is converted into a π^0 through PCAC. We reject this derivation because order by order evaluation of a renormalizable quark-gluon field theory does not lead to scaling in the deep inelastic limit. Experiments at SLAC up to the present time are incapable of proving or disproving such a thing as scaling in the Bjorken limit but they are certainly suggestive and we would like to accept the Bjorken scaling. So we must reject the basis of Adler's derivation but we can derive this result in other ways, consistent with scaling, as I shall describe briefly later on.

What is sometimes said about Adler's computation is that this result completely contradicts the quark model. What is true is that it completely contradicts a hypothetical quark model that practically nobody wants, with Fermi–Dirac statistics. It agrees beautifully on the other hand with the ancient model that nobody would conceivably believe today in which things are made up of neutrons and protons. If you take basic neutrons and protons you get for an over-all coefficient of that diagram the following: we have charges squared multiplied by the third component of I, which means +1 for up isotopic spin and −1 down isotopic spin, so for protons and neutrons as basic constituents we get

$$(p, n) : +1(1)^2 - 1(0)^2 = 1$$

and with this Adler obtained exactly (within experimental errors) the right experimental decay rate for π^0 and even the right sign for the decay amplitude. If we take quarks u, d, s we get $(u, d, s) : +1(2/3)^2 - 1(-1/3)^2 = 1/3$. So we obtain a decay rate which is wrong by a factor of 9. However, if we have the funny statistics — say in the easiest way with the red, white and blue color — we should put in

$$\begin{aligned}&u_R, d_R, s_R\\&u_W, d_W, s_W: \quad 3[+1(2/3)^2 - 1(-1/3)^2] = 1\,,\\&u_B, d_B, s_B\end{aligned}$$

remembering that the current is a singlet in R–W–B, but in the summation we obtain a loop for each color. So we get back the correct result. Thus, there is this, in my mind, very convincing piece of evidence from the current quarks too for the funny statistics such as the constituent quarks seem to obey. The transformation between them should preserve statistics and so the picture seems to be a consistent one.

The relation "acts like" ($\sim$) which we used to define the current quarks can be strengthened as we introduce more and more properties of the currents which are supposed to be like the properties of these expressions. In other words there will be a hierarchy of strength of abstraction from such a field theory to the properties that we suggest are the exact characteristics of the vector and axial vector currents. We have to be very careful then to abstract as much as we can so as to learn as much as we can from the current quark picture, but not to abstract too much, otherwise first of all experiments may prove us wrong, and secondly that it may involve us with the existence of actual quarks, maybe even free quarks — and that, of course, would be a disaster.

If quarks are only fictitious there are certain defects and virtues. The main defect would be that we will never experimentally discover real ones and thus will never have a quarkonics industry. The virtue is that then there are no basic constituents for hadrons — hadrons act as if they were made up of quarks but no quarks exist — and, therefore, there is no reason for a distinction between the quark and bootstrap picture: they can be just two different descriptions of the same system, like wave mechanics and matrix mechanics. In one case you talk about the bootstrap and when you solve the equations you get something that looks like a quark picture; in the other case you start out with quarks and discover that the dynamics is given by bootstrap dynamics. An important question about the bootstrap approach is whether the bootstrap gives everything or whether there are some symmetry considerations to be supplied aside from the bootstrap. I do not know the answer, but some people claim to have direct information from heaven about it.

Let us go back to the current quarks. Besides the V and A currents we might have well defined tensor (T), scalar (S) and pseudoscalar (P) currents, which would act like

$$T_{i\mu\nu} \sim \bar{q}\lambda_i\sigma_{\mu\nu}q$$
$$S_i \sim \bar{q}\lambda_i q$$
$$P_i \sim i\bar{q}\lambda_i\gamma_5 q\,.$$

I think these currents all can be physically defined: the S and P currents would be related to the divergences of the V and A currents and the tensor currents would arise when you commute the currents with their divergences.

The first of the most elementary abstractions was the introduction of the $SU_3 \times SU_3$ charges, that is

$$\int F_{io}d^3x = F_i$$

$$\int F_{io}^5 d^3x = F_i^5$$

with their equal time commutators. We do not have very good direct evidence that this is true, but the best evidence comes from the Adler–Weisberger relation which is in two forms: first the pure one, namely just these commutators

$$[F_i^5, F_j^5] = if_{ijk}F_k$$

which give sum rules for neutrino reactions, and a second form which involves the use of PCAC giving sum rules for pion reactions, and those have been verified. So an optimist would say that the commutation relations are okay and PCAC is okay; the pessimist might say that they are both wrong but they compensate. That will be checked relatively soon as neutrino experiments get sophisticated enough to test the pure form. In the meantime I will assume that the equal-time commutators (ETC) and PCAC are okay. In order to get the Adler–Weisberger relation it is necessary to apply the ETC of light-like charges for the special kinematic condition of infinite momentum (in the z direction). We make the additional assumption that between finite mass states we can saturate the sum over intermediate states by finite mass states. In the language of dispersion theory this amounts to an assumption of unsubtractedness; in the language of light-cone theory it amounts to smoothness on the light cone. In this way Adler and Weisberger derived their simple sum rule.

We are considering here the space integrals of the time components of the V and A currents, but not those of the x or y components at $P_z = \infty$. We are restricting ourselves, in order to have saturation by finite mass intermediate states, to "good" components of the currents, those with finite matrix elements at $P_z = \infty$. These are $F_{io} \approx F_{iz}$ and $F^5_{io} \approx F^5_{iz}$ at $P_z = \infty$. The "bad" components F_{ix}, F_{iy}, F^5_{ix}, F^5_{iy} have matrix elements going like P_z^{-1} at infinite P_z. The components $F_{iz} - F_{io}$ and $F^5_{iz} - F^5_{io}$ have matrix elements going like P_z^{-2} and are "terrible".

One generalization that we can make of the algebra of $SU_3 \times SU_3$ charges is to introduce the tensor currents $T_{i\mu\nu}$. In the case of the tensor currents the good components are

$$T_{ixo} \cong T_{ixz}$$

$$T_{iyo} \cong T_{iyz}$$

from which we construct the charges

$$\text{“}T_{ix}\text{”} = \int T_{ixo} d^3x$$

$$\text{“}T_{iy}\text{”} = \int T_{iyo} d^3x$$

at $P_z = \infty$ and adjoin them to F_i and F^5_i. Thus we get a system of 36 charges and that just gives us a $[U_6]_w$. One reason why I introduced these tensor currents is that it is simpler to work with $[U_6]_w$, which we have met before, rather than with one of its subgroups $SU_3 \times SU_3$. These charges then form generators of an algebra which we call

$$[U_6]_{w,\infty,\text{currents}}\,;$$

to be contrasted with

$$[U_6]_{w,\infty,\text{strong}}$$

which had to do with the constituent quarks. The contrast is between the $[U_6]_{w,\infty,\text{strong}}$ which is essentially <u>approximate</u> in its applications (collinear processes), and the $[U_6]_{w,\infty,\text{currents}}$ drawn from presumably exact commutation rules of the currents and having to do with current quarks. Although they are isomorphic they are not equal.

For those who cannot stand the idea of introducing the tensor currents, we can just reduce both groups to their subgroups $SU_3 \times SU_3$. In that case for $[U_6]_{w,\infty,\text{currents}}$ we are discussing only the vector and axial vector charges and for $[U_6]_{w,\infty,\text{strong}}$ only the so-called coplanar subgroup. Then we can make the same remark that these two are not equal, they are mathematically similar but their matrix elements are totally different. So one of them is the transform of the other in some sense.

The transformation between current and constituent quarks is then phrased here in a way which does not involve quarks; we discuss it as a transformation between $[U_6]_{w,\infty,\text{currents}}$ and $[U_6]_{w,\infty,\text{strong}}$ (or their respective subgroups). What would happen if they were equal? We know that for $[U_6]_{w,\infty,\text{strong}}$ the low-lying baryon and meson states belong approximately to simple irreducible representations $\underline{35}$, $\underline{1}$, $\underline{56}$, etc. If this were true also of $[U_6]_{w,\infty,\text{currents}}$ then we would have the following results:

$$-\frac{G_A}{G_V} \cong \frac{5}{3}$$

which we know is more like $5/3\sqrt{2}$; the anomalous magnetic moments of neutron and proton would be approximately zero, while they are certainly far from zero, and so on.

Many authors have in fact investigated the mixing under this group, and they found that there is an enormous amount of mixing, e.g., the baryon is partly $\underline{56}$, $L_z = 0$ and partly $\underline{70}$, $L_z = \pm 1$ and the admixture is of the order of 50%. There are higher configurations, too.

So these two algebras are not closely equal although they have the same algebraic structure. And there is some sort of a relation between them, which might be a unitary one, but we cannot prove that since they do not cover a complete set of quantum numbers. But we can certainly look for a unitary transformation connecting the two algebras and my student J. Melosh is pursuing that problem. He has found this transformation for free quarks where it is simple and leads to a conserved $[U_6]_{w,\text{strong}}$. But, of course, we are not dealing with free quarks and have to look at a more complicated situation. What I want to emphasize is that here we have the definition of the <u>search</u>; the search is for a transformation connecting the two algebras. In popular language we can refer to it as a search for the transformation connecting constituent quarks and current quarks.[a]

Let me mention here the work of another student of mine, Ken Young, who has cleaned up this past year at Caltech work that Dashen and I began about 7 years ago, and which we continued sporadically ever since.

[a]Buccella, Kleinert and Savoy have suggested a simple phenomenological form of such a transformation.

That is the attempt to represent the charges and also the transverse Fourier components of the charge densities at infinite momentum completely with non-exotic states, so as to make a non-exotic relativistic quark model as a representation of charge density algebra. We ran into all kinds of troubles, particularly with the existence of states with negative mass squares and the failure of the operators of different quarks to commute with each other. Young seems to have shown that these difficulties are a property of trying to represent the charge algebra at infinite momentum with non-exotics alone. Therefore, the transformation which connects the two algebras does not just mix up non-exotic states but also brings in higher representations that contain exotics. In simple lay language the transformation must bring in quark pair contributions and the constituent quark looks then like a current quark dressed up with current quark pairs.

We therefore must reject all the extensive literature, which I am proud not to have contributed to over the last few years, in which the constituent quarks are treated as current quarks and the electromagnetic current is made to interact through a simple current operator $F_{e\mu}$ with what are essentially constituent quarks. That is certainly wrong.

Another way of describing the infinite momentum and the smoothness assumption is to perform an alibi instead of an alias transformation, i.e., instead of letting everything go by you at infinite momentum you leave it alone at finite momentum and you run by it. These two are practically equivalent. In that case one is not talking about infinite momentum but about the behaviour in coordinate space as we go to a light-like plane and about the commutators of light-like charges which are integrated over a light-like plane instead of an equal-time plane. Leutwyler, Stern and a number of other people have especially emphasized this approach. From that again one can get the Adler–Weisberger relation, and so forth.

On the light-like plane (say $z+t=0$) we have the commutation rules not only for the charges, but also for the local densities of the good components of the currents, namely $F_{io}+F_{iz}$. and $F^5_{io}+F^5_{iz}$ for V and A, with the possible adjunction of the good components $T_{ixo}+T_{ixz}$ and $T_{iyo}+T_{iyz}$ of the tensor currents. Especially useful is the algebra of these quantities integrated over the variable z–t and Fourier-transformed with respect to the variables x and y. We obtain the operators $F_i(\vec{k}_\perp)$, $F^5_i(\vec{k}_\perp)$, $T_{ix}(\vec{k}_\perp)$, and $T_{iy}(\vec{k}_\perp)$ and they have commutation relations like

$$[F_i(\vec{k}_\perp), F_j(\vec{k}'_\perp)] = if_{ijk}F_k(\vec{k}_\perp + \vec{k}'_\perp)\,.$$

By the way, if we take this equation to first order in $\vec{k}_\perp$ and $\vec{k}'_\perp$, we get the so-called Cabibbo–Radicati sum rule for photon-nucleon collisions, which seems to work quite well.[b]

So far we have abstracted from quark field theory relations that were true not only in a free quark field model but also to all orders of strong interactions in a field model with interactions, say through a neutral "gluon" field coupled to the quarks. (We stick to a vector "gluon" picture so that we can use the scalar and pseudoscalar densities S_i and P_i to describe the divergences of the vector and axial vector currents respectively.) We shall continue to make use of abstractions limited in this way, since if we took all relations true in a free quark model we would soon be in trouble: we would be predicting free quarks! However, in what follows we consider the abstraction of propositions true only formally in the vector "gluon" model with interactions, not order by order in renormalized perturbation theory. We do this in order to get Bjorken scaling, which fails to each order of renormalized perturbation theory in a barely renormalizable model like the quark-vector "gluon" model but which, as mentioned before, we would like to assume true.

That leads us to light-cone current algebra, on which I have worked together with Harald Fritzsch. It has been studied by many other people as well, including Brandt and Preparata, Leutwyler *et al.*, Stern *et al.*, Frishman, and, of course, Wilson, who pioneered in this field although he disagrees with what we do nowadays. Those whose work is most similar to ours are Cornwall and Jackiw and Llewellyn-Smith.

The first assumption in light-cone current algebra is to abstract from free quark theory or from formal vector "gluon" theory the leading singularity on the light cone $(x-y)^2 \approx 0$ of the connected part of the commutator of two currents at space-time points x and y. For V and A currents we find

$$\begin{aligned}
[F_{i\mu}(x), F_{j\nu}(y)] &\hat{=} [F^5_{i\mu}(x), F^5_{j\nu}(y)] \\
&\hat{=} \frac{1}{4\pi}\partial_\rho\{\varepsilon(x_o - y_o)\delta(x-y)^2)\}\{(if_{ijk} - d_{ijk})(s_{\mu\nu\rho\sigma}F_{k\sigma}(y,x) \\
&\quad + i\varepsilon_{\mu\nu\rho\sigma}F^5_{k\sigma}(y,x)) + (if_{ijk} + d_{ijk})(s_{\mu\nu\rho\sigma}F_{k\sigma}(x,y) \\
&\quad - i\varepsilon_{\mu\nu\rho\sigma}F^5_{k\sigma}(x,y))\}\,,
\end{aligned}$$

[b]Strictly, the operators $F_i(\vec{k}_\perp)$ are not exactly the Fourier transforms of integrals of current densities but rather these Fourier transforms multiplied by

$$\exp\{i[k_x(\Lambda_x + J_y) + k_y(\Lambda_y - J_x)][P_o + P_z]^{-1}\}\,,$$

where $\vec{\Lambda}$ is the Lorentz-boost operator, $\vec{J}$ is the total angular momentum, and the component $P_o + P_z$ of the energy-momentum-four-vector is conserved by all the operators $F_i(\vec{k})$.

$$[F_{i\mu}(x), F^5_{j\nu}(y)] \hat{=} \frac{1}{4\pi}\partial_\rho\{\varepsilon(x_o - y_o)\delta(x-y)^2)\}\{(if_{ijk} - d_{ijk})(s_{\mu\nu\rho\sigma}F^5_{k\sigma}(y,x)$$
$$+ i\varepsilon_{\mu\nu\rho\sigma}F_{k\sigma}(y,x)) + (if_{ijk} + d_{ijk})(s_{\mu\nu\rho\sigma}F^5_{k\sigma}(x,y)$$
$$- i\varepsilon_{\mu\nu\rho\sigma}F_{k\sigma}(x,y))\}\,.$$

On the right-hand side we have the connected parts of bilocal operators $F_{k\sigma}(x,y)$ and $F^5_{k\sigma}(x,y)$; that reduce to the 1ocal currents $F_{k\sigma}(x)$ and $F^5_{k\sigma}(x)$ as $y \to x$. The bilocal operators are defined only in the vicinity of $(x-y)^2 = 0$. Here $s_{\mu\nu\rho\sigma} = \delta_{\mu\rho}\delta_{\nu\sigma} + \delta_{\nu\rho}\delta_{\mu\sigma} - \delta_{\mu\nu}\delta_{\rho\sigma}$.

The formulae give Bjorken scaling by virtue of the finite matrix elements assumed for $F_{k\sigma}(x,y)$ and $F^5_{k\sigma}(x,y)$; in fact the Fourier transform of the matrix element of $F_{k\sigma}(x,y)$ is just the Bjorken form factor. The fact that all charged fields in the model have spin 1/2 determines the algebraic structure of the formula and gives the prediction $\sigma_L/\sigma \underset{Tbj}{\to} 0$ for deep inelastic electron scattering, not in contradiction with experiment. The electrical and weak charges of the quarks in the model determine the coefficients in this formula, and give rise to numerous sum rules and inequalities for the SLAC-MIT experiments and for corresponding neutrino and antineutrino experiments in the Bjorken limit, none in contradiction with experiment, although the inequality $1/4 \leq F^{en}(\xi)/F^{ep}(\varepsilon) \leq 4$ appears to be tested fairly severely at the lower and near $\xi = 1$.

The formula for the leading light-cone singularity in the commutator contains, of course, the physical information that near the light-cone we have full symmetry with respect to $SU_3 \times SU_3$ and with respect to scale transformations in co-ordinate space. Thus there is conservation of dimension in the formula, with each current having $\ell = -3$ and the singular function of $x - y$ also having $\ell = -3$.

A simple generalization of the abstraction we have considered turns it into a closed system, called the basic light-cone algebra. Here we commute the bilocal operators as well, for instance $F_{i\mu}(x,u)$ with $F_{j\nu}(y,v)$, as all the six intervals among the four space-time points approach O, so that all four points tend to lie on a light-like straight line in Minkowski space. Abstraction from the model gives us on the right-hand side a singular function of one coordinate difference, say x–v, times a bilocal current $F_{k\sigma}$ or $F^5_{k\sigma}$ at the other two points, say y and u, plus an expression with (x,v) and (y,u) interchanged, and the system closes algebraically. The formulae are just like the ones for local currents.

We shall assume here the validity of the basic lightcone algebraic system, and discuss possible applications and generalizations.

First of all, we may consider what happens when the points x and u lie on a light-like plane with one value of $z+t$ and y and v lie on another light-like plane with a slightly different value of $z+t$, and we let these values approach each other.

For commutators of good components of currents, the limit is finite, and we get a generalization of the light-plane algebra of commutators of good densities $F_{io}+F_{iz}$ and $F_{io}^5+F_{iz}^5$. There is now a fourth argument in each density, namely the internal co-ordinate n, which runs only in the light-like direction $z-t$. As before, we get the most useful results by integrating over the average $z-t$ and Fourier-transforming with respect to the transverse average co-ordinates x and y, obtaining operators $F_i(\vec{k}_\perp,\eta)$, $F_i^5(\vec{k}_\perp,\eta)$, with commutation relations like

$$[F_i(\vec{k}_\perp,\eta),F_j(\vec{k}'_\perp,\eta')]=if_{ijk}F_k(\vec{k}_\perp+\vec{k}'_\perp,\eta+\eta')\,.$$

Remember that $\vec{k}_\perp$ is in momentum space and η in (relative) co-ordinate space.

The non-local operators $F_i(\vec{k},\eta)$ acting on the vacuum create strings of mesons with all values of the meson spin angular momentum J. In fact, a power series expansion in η of $F_i(\vec{k}_\perp,\eta)$ is just an expansion in η^{J-1}. At large η, we can reggeize and obtain a dominant term in $\eta^{\alpha(-k_\perp^2)-1}$, where $\alpha(-k_\perp^2)$ is the leading Regge trajectory in the relevant meson channel, for instance P or ρ. (In the work on these questions, Fritzsch has played a particularly important role.) The couplings of the Regge poles in the bilocals are proportional to the hadronic couplings of meson Regge poles.

If we commute bad components with bad components as the two light planes approach each other, then the leading singularity on the light-cone leads to a singular term that goes like a δ-function of the difference in coordinates $z+t$. This singular term, which is multiplied by a good component of a bilocal current on the right-hand side, gives the Bjorken scaling in deep inelastic scattering. Unlike the good-good commutators on the light plane, it involves a commutator of local quantities on the left giving a bilocal on the right, a bilocal of which the matrix elements give the Fourier transforms of the Bjorken scaling functions $F(\xi)$.

We may now generalize, if we keep abstracting from the vector gluon model, to a connected light-cone algebra involving V,S,T,A and P densities, where the divergences of the V and A currents are proportional to the S and P currents respectively, with coefficients that correspond in the model to the three bare quark masses, forming a diagonal 3×3 matrix M. The divergences of the axial vector currents, for example, are given by masses M multiplied by normalized pseudoscalar densities P_i.

The scalar and pseudoscalar densities are all bad, and do not contribute to the good-good algebra on the light plane, but we can commute two of these densities as the light planes approach each other and obtain the singular Bjorken term. In fact, the leading light-cone singularity in the commutator of two pseudoscalars or two scalars just involves the same vector bilocal densities as the leading singularity in the commutator of two vector densities

$$[P_i(x), P_j(y)] \hat{=} [S_i(x), S_j(y)] \hat{=} [F_{i\mu}(x), F_{j\mu}(y)]$$

so that the P's and S's give Bjorken functions that are not only finite but known from deep inelastic electron and neutrino experiments. The Bjorken limit of the commutator of two divergences of vector or axial vector currents is also measurable in deep inelastic neutrino experiments, albeit very difficult ones, since they involve polarization and also involve amplitudes that vanish when the lepton masses vanish. The important thing is that the shapes of the form factors in such experiments are predictable from known Bjorken functions and the overall strength is given by the "bare quark mass" matrix M, which is thus perfectly measurable, according to our ideas, even though the quarks themselves are presumably fictitious and have no real masses.

The next generalization we may consider is to abstract the behaviour of current products as well as commutators near the light-cone. Here we need only abstract the principle that scale invariance near the light-cone applies to products as well as commutators. The result is that products of operators, and even physical ordered products, are given, apart from subtraction terms that act like four-dimensional δ functions, by the same expressions as commutators, with $\varepsilon(x_0-y_0)\delta((x-y)^2)$ replaced by $1/[(x-y)^2-i(x_0-y_0)\varepsilon]$ for ordinary products and by $1/[(x-y)^2-i\varepsilon]$ for physical ordered products. The subtraction terms can often be determined from current conservation; sometimes they are zero and sometimes, for certain processes, they do not matter even when they are non-zero.

Using the current products, one can design experiments to test the bilocal-bilocal commutators, for example fourth order cross-sections like those for $e^- + p \to e^- + X + \mu^+ + \mu^-$, where X is any hadronic state, summed over X.

Using products, and employing consistency arguments, we can determine the form of the leading light-cone singularity in the disconnected part of the current commutator, i.e., the vacuum expected value of a current commutator, and it turns out to be the same as in free quark theory or formal quark "gluon" theory. The constant in front is not determined in this way, and we must abstract it from the model. It depends on the statistics. With our funny "quark statistics" or with nine real quarks, the constant is three times as large as for three Fermi–Dirac quarks.

We can then predict the asymptotic cross-section for $e^+ + e^- \to$ hadrons using single photon annihilation, namely

$$\frac{\sigma(e^+ + e^- \to \text{hadrons})}{\sigma(e^+ + e^- \to \mu^+ + \mu^-)} \to 3\left[\left(\frac{2}{3}\right)^2 + \left(-\frac{1}{3}\right)^2 + \left(-\frac{1}{3}\right)^2\right] = 2$$

where we would have obtained 2/3 with three Fermi–Dirac quarks.

We are now in a position to go back and rederive the Adler result for the rate of $\pi^0 \to 2\gamma$ in the PCAC approximation. Following the lead of Crewther, who first showed how such an alternative derivation could be given, Bardeen, Fritzsch and I use the connected light-cone algebra and the disconnected result just given to obtain the Adler result without invoking renormalized perturbation theory. The answer, as we indicated earlier, agrees with the experimental $\pi^0 \to 2\gamma$ amplitude in both sign and magnitude.

A final generalization, about which Fritzsch and I are not so convinced as we are of the others, involves a change in our approach from considering only quantities based on currents that couple to electromagnetism and the weak interaction to including quantities that are not physically determinable in that way. I have mentioned that bilocals like $F_{i\mu}(x, y)$, which are analogous to quantities in the model that involve one quark operator and one anti-quark operator, can be applied to the vacuum to create Regge sequences of non-exotic meson states. It might also be useful to define trilocals $B_{\alpha\beta\gamma,abc}(x, y, z)$ that are analogous to operators in the model involving three quark operators at x, y and z, when these points lie on the same straight light-like line, and to abstract their algebraic properties from the model, so that sequences of baryon states could be produced from the vacuum. We could, in fact, construct operators that would, between the vacuum and hadron states, give a partial Fock space for hadrons with any number of quarks and antiquarks lying on a straight light-like line. Whether this makes sense, and how many properties of hadrons we can calculate from such a partial Fock space of "wave functions" we do not know.

If we go too far in this direction, and try to construct a complete Fock space for quarks and antiquarks on a light-like plane, abstracting the algebraic properties from free quark theory, we are in danger of ending up with real quarks, and perhaps even with free real quarks, as mentioned before. In our work, we are always between Scylla and Charybdis; we may fail to abstract enough, and miss important physics, or we may abstract too much and end up with fictitious objects in our models turning into real monsters that devour us.

In connection with the written version of these lectures, I should like to thank Dr. Heimo Latal and his collaborators for the excellent lecture notes that they provided me. I should also like to thank Dr. Oscar Koralnik of Geneva for providing the beautiful table on which most of my writing was done. I acknowledge with thanks the hospitality of the Theoretical Study Division of CERN.

Concrete Quarks

George Zweig
MIT, Research Laboratory of Electronics,
26-169, 77 Massachusetts Ave, Cambridge, MA 02139, USA
zweig@mit.edu

A short history of the physics of strongly interacting particles is presented. Events leading to the discovery, and eventual acceptance, of concrete quarks are described.*

1. Introduction

This year is the fiftieth anniversary of the discovery of quarks, and last year was the fortieth anniversary of the birth of QCD (Quantum Chromodynamics). QCD developed in two phases, the first involving the discovery of quarks, the second specifying the nature of their interactions. These phases arose from two very different traditions, that of Rutherford and Bohr, and that of Einstein. The first was grounded in observation, and the startling interpretation of what was observed, the second in a triumph of the imagination made possible by the power of quantum field theory. Here the first phase, culminating in the discovery and acceptance of concrete quarks, is described.

2. How it Started

The story of QCD begins with the discovery of spontaneous radioactivity by Henri Becquerel in 1896 who thought that X-rays, discovered just a few months earlier, might be emitted by phosphorescent substances. He noticed, quite by accident, that crystalline crusts of uranium salts created silhouettes of great intensity when left in a drawer next to a photographic plate wrapped in black paper, even though the salt had not been exposed to sunlight (Fig. 1).

The nature of radioactivity was elucidated three years later by Ernest Rutherford who found that two types of particles, distinguished by their penetration power, were present in uranium radiation, particles he called α

*Other aspects of quark history are given in Refs. 1 and 2.

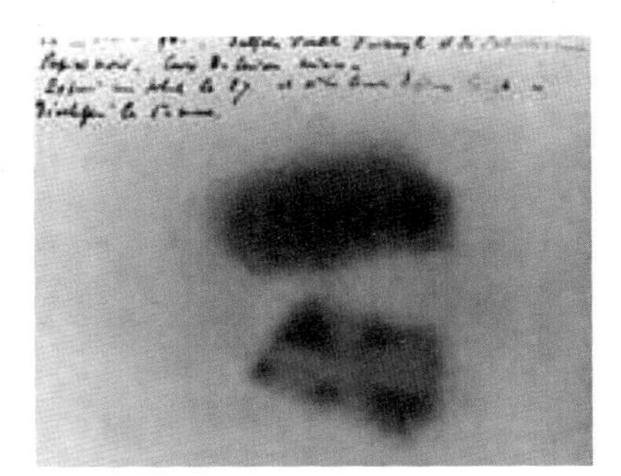

Fig. 1. One of Becquerel's photographic plates fogged by exposure to radiation from potassium uranyl sulfate. The shadow of a metal Maltese Cross placed between the plate and the uranium salt is visible.

and β. After three more years, Rutherford, with his young collaborator Frederick Soddy, interpreted the phenomena of radioactivity as "the spontaneous disintegration of [a] radio-element, whereby a part of the original atom was violently ejected as a radiant particle, and the remainder formed a totally new kind of atom with distinct chemical and physical character." Rutherford received the Nobel Prize in Chemistry in 1908. Soddy also received a Nobel Prize, but in 1921 for other work.[a] A photo of Rutherford with his group at Manchester University, taken two years after his prize, is shown in Fig. 2.

QCD speaks of protons, neutrons, and more remotely, the nuclei of atoms. Thompson's discovery of the electron in 1897 indicated that the atom was divisible. This meant that the charge on an electron in a neutral atom must be cancelled by a positive charge. The first indication that this positive charge is point-like was reported by Hans Geiger and Ernest Marsden in 1909.[3] Geiger was a postdoctoral fellow from Germany who came to Rutherford's lab to study the deflection of α-particles as they passed through thin metal plates. Marsden was a 20-year-old undergraduate from New Zealand, Rutherford's own country of origin. Geiger and Marsden reported that

> "conclusive evidence was found of the existence of a diffuse reflection of the α-particles. A small fraction of the α-particles falling upon a metal plate have their directions changed to such an extent that they emerge again at the side of incidence."

This innocuous sounding statement, when properly interpreted, was revolutionary, completely contradicting Newton's laws of mechanics, and Maxwell's

[a] All told, eleven of Rutherford's students, collaborators, and members of his laboratory went on to receive the Nobel Prize in Physics or Chemistry. Many more made remarkable but less recognized contributions. Rafi Muhammad Chaudhry went on to pioneer experimental nuclear physics in Pakistan, and with his student Mustafa Yar Khan, founded Pakistan's successful nuclear weapons program.

equations of electricity and magnetism as they had been formulated some 50 years earlier. Two years later in 1911 Rutherford published the proper interpretation,[4] which negated the common view that an atom consisted of negatively charged electrons embedded within a ball of positive charge. He showed that a large point-like concentration of charge was essential to account for the dramatic reflection of the positively charged α-particles back towards their source.[b] This meant that positive and negative charges were separated in an atom, but how? Not statically, for they would fall together. Not circling one another, for they would radiate energy like tiny antennae spiraling together. Rutherford had discovered the nucleus, an object that could not exist according to the existing laws of physics.

Fig. 2. Rutherford's group at Manchester University in 1912. Rutherford is seated second row, center. Also present: C. G. Darwin, J. M. Nuttall, J. Chadwick; H. Geiger, H. G. J. Moseley, and E. Marsden.

When Marsden in 1914 and 1915 presented the first experimental evidence that nuclei contain protons, a second contradiction appeared. Why didn't the positively charged nucleus explode? What was binding protons together?

Shortly after receiving his doctorate in 1911, Niels Bohr visited Rutherford's lab for several months, eventually settling there from 1914 to 1916. During his stay Bohr correctly combined two incomprehensible ideas, that of charge separation in the atom, and that of Planck's quantization of

[b]Rutherford concluded the nuclear charge could have either sign, was roughly proportional to the atomic number, and approximately $\pm 100e$ for gold.

radiation, into one incomprehensible idea: electrons exist in "stationary states" within the atom, emitting quanta of light as they jumped from one state to another. This view of the atom solved none of the contradictions with classical physics, but provided a conceptual framework within which the frequencies of spectral lines could be fruitfully organized, and their patterns contemplated.

The contradictions present in the Bohr atom, but not the nucleus,[c] were resolved with the introduction of quantum mechanics a decade later, which was formulated for atoms by Heisenberg solely in terms of the possible frequencies of light emitted as electrons changed their stationary states. His formulation consisted both of an equation involving a matrix of these possible frequencies, and *a philosophy of how theoretical physics should be practiced.* He emphasized that the contradictions inherent in Bohr's formulation of atomic physics would have been avoided if the equations of physics had been formulated entirely in terms of observables.

When experimental particle physics came to consist primarily of two-particle collisions, the observables were scattering amplitudes, like the observables in the Geiger–Marsden experiment. In three papers published during World War II Heisenberg argued that scattering amplitudes should also be organized into a matrix he called the S-matrix, and that the fundamental laws governing particles and their strong interactions should be formulated solely in terms of this matrix. Field theory, whose constructs and interactions were difficult or impossible to observe, was suspect. S-matrix theory in the form of the bootstrap[5] became the dominant school of thought in particle physics in the early 1960s.

My first exposure to nuclear physics came at the age of 10 in 1947, two years after the atomic bomb had been dropped on Hiroshima. One of my favorite after-school radio programs was the "Lone Ranger," sponsored by Kix breakfast cereal. Quite unexpectedly during a commercial break, the announcer asked his little listeners to send away for an "Atomic Bomb Ring."

---

[c]Rutherford and colleagues kept scattering α-particles off ever lighter nuclei. Rutherford and Chadwick, who later discovered the neutron, write:[7] "The study of the collisions of α-particles with hydrogen nuclei has shown that the force between the α-particle and the hydrogen nucleus obeys Columb's law for large distances of collision, but that it diverges very markedly from this law at close distances. The experiments of Chadwick and Bieler showed that for distances less than about 4×10^{-13} cm, the force between the two particles increased much more rapidly with decrease of distance than could be accounted for on an inverse square law of force. ... Possible explanations of the origin of these additional forces are discussed, and it is suggested tentatively that they may be due to magnetic fields in the nuclei." Here is the first direct observation of the nuclear force, the force holding protons together in the nucleus, but no new force is postulated to make the connection! The correct interpretation of truly novel phenomena can be difficult, even for giants like Rutherford and Chadwick.

I jumped at the chance. After mailing in my name, address, 15 cents, and a Kix box top, I received the ring (Fig. 3), took it into a dark closet full of winter coats, and waited till my eyes had adapted to the dark. Removing the red cap and peering down the long axis of the "bomb," I was rewarded with brilliant punctate flashes of light as one α-particle after another, emitted from a tiny piece of radioactive polonium, barreled into a zinc sulfide screen (a spinthariscope invented by William Crookes in 1903).[d] This was the same kind of screen used by Geiger and Marsden in their 1909 α-particle scattering experiment.[e]

In that same year of 1947, C. F. Powell discovered the pion in emulsions. Earlier in 1932 J. Chadwick had discovered the neutron. With these discoveries the cast of characters within the nucleus was complete. Both Chadwick and Powell had been Rutherford's students.

But also in 1947 a strange form of matter was discovered. G. D. Rochester and C. C. Butler published two cloud chamber photographs of cosmic-ray events, providing the first evidence of the existence of the K meson. Unlike the pion, whose existence had been predicted by Yukawa in 1935 to provide the force necessary to bind protons and neutrons together, the appearance of the K meson was entirely unexpected. As I. I. Rabi famously quipped when the muon was discovered, "Who ordered that?"[f].

This proliferation of elementary particles led E. Fermi and C. N. Yang to publish a paper in 1949 titled "Are Mesons Elementary Particles?" with the abstract:[6]

> "The hypothesis that π-mesons may be composite particles formed by the association of a nucleon with an antinucleon is discussed. From an extremely crude discussion of the model it appears that such a meson in most respects would have properties similar to those of the meson of the Yukawa theory."

However, in the body of the paper they caution,

> "Unfortunately we have not succeeded in working out a satisfactory relativistically invariant theory of nucleons among which such attractive forces act."

---

[d]Crookes was remarkable. He also discovered thallium in 1861, invented the radiometer, and developed the Crookes tube that was later used by W. C. Roentgen and J. J. Thomson in their discoveries of the X-ray and electron. As a young teenager wandering the stacks of the Detroit Public Library, I was fascinated by a book written by Crookes chronicling seances held in his house in the 1870s, not realizing his seminal contribution to the development of nuclear physics and the creation of the atomic bomb ring.

[e]I was so moved by the magic of the ring that I got a second one, not knowing that the half-life of polonium was only 138 days.

[f]We still don't know who ordered the muon.

Lone Ranger Atom Bomb Ring Spinthariscope (1947 - early 1950s)

This ring spinthariscope was known as the Lone Ranger Atom Bomb Ring and advertised as a "seething scientific creation." The Lone Ranger was more closely associated with silver bullets than atomic bombs but that's what it was called. When the red base (which served as a "secret message compartment") was taken off, and after a suitable period of time for dark adaptation, you could look through a small plastic lens at scintillations caused by polonium alpha particles striking a zinc sulfide screen.

Distributed by Kix Cereals (15 cents plus a boxtop), the instructions stated: "You'll see brilliant flashes of light in the inky darkness inside the atom chamber. These frenzied vivid flashes are caused by the released energy of atoms. PERFECTLY SAFE - We guarantee you can wear the KIX Atomic "Bomb" Ring with complete safety. The atomic materials inside the ring are harmless."

The following advertisement was appearing in newspapers in early 1947.

Fig. 3. The Atom Bomb Ring based on Crookes's 1903 spinthariscope.

This unsurmountable problem was to haunt Sakata's 1956 extension[8] of their work. He augmented the proton and neutron with the newly discovered Λ to construct K mesons out of pairs like the Λ and antineutron ($\Lambda\bar{n}$), but the binding mechanism was still obscure. Creating the light π meson from the other hadrons, all much heavier, would also be problematic for the bootstrap,[5] the fulfillment of Heisenberg's S-matrix theory. Advocates of the bootstrap ignored this problem.

By 1957 the list of "elementary particles" had grown to 19.[g] M. Gell-Mann and A. Rosenfeld summarized the situation in a paper for the Annual Review of Nuclear Sciences.[9] Working with E. P. Rosenbaum, an editor for Scientific American, Gell-Mann reworked this material for the general public in an article titled "Elementary Particles."[10] The 19 elementary particles they listed are shown in Fig. 4.

In the same issue of the Annual Review of Nuclear Science, S. Lindenbaum discussed a relatively new phenomenon, *resonances* in pion nucleon scattering, of which there were at least two (Fig. 5). Elementary particles and resonances coexisted side by side in journals and theoretician's minds, unconnected. Resonances, the "elephants in the room," were about to explode in number.

What distinguished resonances from the so-called elementary particles was their lifetimes. Resonances were created and decayed via the strong interactions. They lived for only 10^{-23} seconds, about the time it takes light to travel across a proton. Elementary particles lived much longer, either being stable, or decaying slowly through the electromagnetic or weak interactions. Elementary particles were grouped into two classes, point-like and extended in size. Only later was it realized that the long-lived extended elementary particles and the short-lived resonances were cut from the same cloth.

In the summer of 1957 I had just finished my sophomore year as a math major at the University of Michigan. I remember reading the Gell-Mann–Rosenbaum Scientific American article. My reaction can be summarized by lines from a Bob Dylan song of later years,

> "But something is happening here
> And you don't know what it is
> Do you, Mister Jones?"[h]

---

[g] Willis Lamb, in the first paragraph of his 1955 Nobel Prize Lecture, joked that he had "heard it said that 'the finder of a new elementary particle used to be rewarded by a Nobel Prize, but such a discovery now ought to be punished by a $10,000 fine.'"

[h] Bob Dylan, Ballad of a Thin Man, final track on Side One of Highway 61 Revisited, 1965.

Point particles

Spin 1/2 leptons	
Particle	Mass
e^-	1
μ^-	206.7
ν	0

Spin 1 photon	
Particle	Mass
γ	0

Extended particles (strongly interacting)

Spin 1/2 baryons		
Multiplet	Particle	Mass (m_e)
Ξ	Ξ^0	?
	Ξ^{-1}	2585
Σ	Σ^{-1}	2341
	Σ^+	2325
	Σ^0	2324
Λ	Λ	2182
N	n	1838.6
	p	1836.1

Spin 0 mesons		
Multiplet	Particle	Mass
π	π^+	273.2
	π^{-1}	273.2
	π^0	264.2
K	K^+	966.5
	K^-	966.5
	K_1^0	965
	K_2^0	965

Fig. 4. The elementary (long-lived) particles in 1957.[10]

3. Caltech

Two years later in 1959, 50 years after the Geiger–Marsden experiment, I started graduate school in physics at Caltech,[i] and shortly thereafter started my PhD thesis as an experimentalist in Alvin Tollestrup's group, piggy-

[i]The Caltech Physics Department was remarkable. Carl Anderson, chairman, discovered both the positron and muon. His department included six soon-to-be Nobel Prize winners: R. Feynman, W. Fowler, M. Gell-Mann, Shelly Glashow, R. Mössbauer, and K. Wilson; Sidney Coleman and R. Dashen were students, and Y. Ne'eman and J.J. Sakurai visitors. If that wasn't enough, I could always go across campus and talk with ex-particle-physicist Max Delbrück, who had invented molecular biology, or Linus Pauling, a phenomenologist par excellence. I got the impression that it was possible to do things. Discoveries were happening, just down the hall!

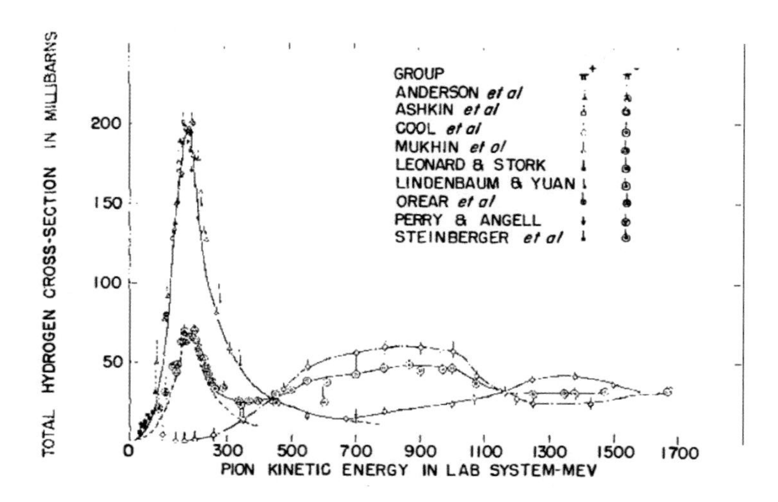

Fig. 5. $\pi^{\pm} + p$ total cross-sections as of 1957 showing evidence for the first pion–nucleon resonances.[11]

backing on their proposed study of $K^+ \to \pi^+ + \pi^0 + \gamma$, to measure the polarization of the μ, out of the plane of decay, in $K^+ \to \mu^+ + \pi^0 + \nu$. A non-zero value would imply a violation of time-reversal symmetry. This was the first "user's group" experiment, involving one faculty member, two research fellows, and two graduate students. The experimental equipment was built and tested at Caltech over a two-year period, but run at the Bevatron in Berkeley. After 21 half-days of running in the early spring of 1962, an adventure that should be chronicled elsewhere, I returned to Caltech with several hundred thousand spark-chamber photographs. A preliminary scan found no effect. I then faced an additional two years of tedious analysis determining the magnitude of errors, and establishing an accurate upper bound. With this unpleasant prospect, I embraced denial and went camping in the Yucatán.

On returning at the end of summer, I switched to theory. Screwing up my courage, I asked Murray Gell-Mann if he would be my thesis advisor. He seemed the natural choice: he supervised many graduate students, all of whom graduated quickly. I had spoken with him on several occasions after Alvin suggested that I talk to him about Alvin's K-decay experiment, meant to elucidate the mysterious $\Delta I = 1/2$ rule for weak nonleptonic decays. When asked, Murray immediately said "No," paused, and then said he was going to the East Coast on sabbatical, but would "talk to Dick." It wasn't until I returned from CERN two years later that I would speak with Murray once again.

"Dick" was Richard Feynman. I would never have had the courage to approach him. I had just watched him, in the second lecture of his gravity course, write down the Feynman rules for gravity for the first time, and compute the scattering of Mercury off the sun, getting the advance of the perihelion in 45 minutes, a result that I had seen H. P. Robertson, the grand old man of general relativity, obtain only after seven months into his course on general relativity.[j]

About a week later I asked Feynman if he would be my thesis advisor. He replied in a hoarse low-pitched voice, "Murray says you're okay, so you must be okay." Then he laid down the ground rules. I was to see him every Thursday afternoon from 1:30 to 4:15 when we would adjourn for tea, and then proceed to the physics colloquium. Each week I prepared frantically for our Thursday sessions, trying to pick some subject that would interest him, never touching any subject more than once, not even my thesis. My job was to make sure that Feynman was never bored. During the 1962–63 academic year we covered essentially all of particle physics. Here is some of what we talked about:

- Theory in the abstract:
  (a) Axiomatic field theory was championed by Arthur Wightman at Princeton. Feynman hated it, and I didn't see how it helped explain the wealth of observations being made.
  (b) Theory related to belief was championed by Geoffrey Chew at Berkeley. At a La Jolla conference in June 1961 in support of the bootstrap Chew said:

  > "I believe the conventional association of fields with strongly interacting particles to be empty. ... field theory..., like an old soldier, is destined not to die but just fade away."

  Feynman liked the idea of "nuclear democracy," but never tried to develop it.
- Theory related to experiment:
  (a) Particle classification (no dynamics): G_2 was first championed by Lee and Yang, and by Gell-Mann, with Gell-Mann eventually switching to SU_3, also Ne'eman's choice. Feynman, on the sidelines, had little

[j]Robertson, of the "Robertson–Walker metric," had the distinction of rejecting a paper submitted to the Physical Review by Einstein. Einstein claimed that an accelerating mass would *not* emit gravitational radiation. Robertson's review was longer than Einstein's paper. Einstein was furious, but Robertson's evaluation was correct (`http://scitation.aip.org/content/aip/magazine/physicstoday/article/58/9/10.1063/1.2117822`). Einstein never again submitted a paper to the Physical Review.

doubt that SU_3 was the correct symmetry group, and that the 1, 8, and 10 dimensional representations were useful for organizing the low-lying mesons and baryons. Glashow and Sakurai were using the 27-dimensional SU_3 representation to classify hadrons.

In the fall of 1962 I organized the weak and electromagnetic currents into representations of SU_3, an octet of currents for the familiar weak and electromagnetic interactions, but also included currents in a 27-dimensional representation to accommodate what was then thought to be experimental evidence for weak interactions with $\Delta S = -\Delta Q$ (the Barkas event, $\Sigma^+ \rightarrow n + \mu^+ + \nu$). Feynman said "Forget the experiment! Throw away the 27, and keep the 8," but I couldn't because the experiment was clean, and the chance that the Barkas event was background was only 2 or 3×10^{-5}. Feynman was right. The event turned out to be an unlikely statistical fluctuation.[k]

(b) Dynamics (no particle classification): Although the bootstrap had its problems, Fred Zachariasen was able to bootstrap the ρ meson from two pions (Fig. 6). Forces between particles are created by the exchange of particles. If the ρ exists, it will be exchanged between two pions, creating an attractive force strong enough to bind them into a state with the quantum numbers of the ρ. Equating the particle exchanged (ρ) with the state it creates leads to a determination of the ρ mass and $\rho\pi\pi$ coupling constant that have the right order of magnitude.[12] As simply described by Chew, Gell-Mann, and Rosenfeld, the bootstrap was the future.[13]

Feynman was sympathetic to the goals of S-matrix theory and the bootstrap, but never used it himself. I didn't see how to bootstrap the π.

- Experimental physics:

(a) More elementary particles and many more resonances were discovered:

i. Point particles: The 4th lepton (ν_μ),

ii. Extended elementary particles: An 8th spin 0 meson (η) and an 8th spin 1/2 baryon (Ξ^0),

iii. Resonances: 26 meson resonances (ρ, ω, $K^*, \ldots$), the first of which had been discovered only two years earlier in 1961.

[k] More than 30 years later I ran into Nicola Cabibbo and asked him if he hadn't worried about the Barkas event before he published his famous 1963 paper on the weak interactions postulating the existent of an octet of currents. "Yes," he said, "but I had an important advantage over you. I was at CERN and there they had observed an additional 10^4 Σ^+ decays, none with $\Delta S = -\Delta Q$, so I decided to publish."

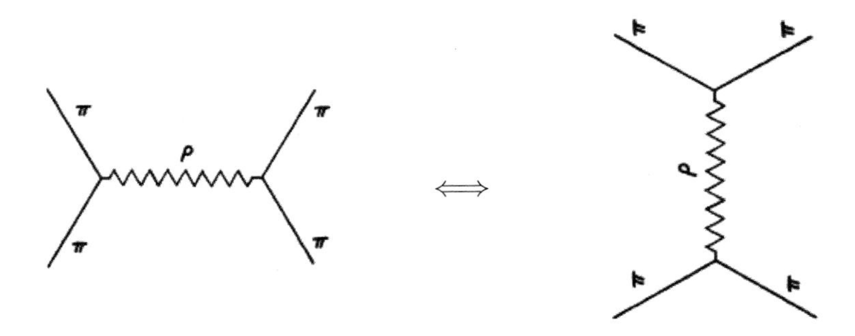

Fig. 6. The ρ bootstrap. Exchanging a ρ binds two pions into a ρ.[12]

4. Something Unbelievable

One Thursday afternoon I showed Feynman a paper in Physical Review Letters titled "Existence and Properties of the ϕ Meson."[14] Other papers had presented evidence for the existence of the ϕ, but this paper found no evidence where evidence was primarily expected. It was the nonexistence of a decay mode that should have broadened the width of the ϕ that fascinated me. Figure 7 shows the Dalitz plot for $K^- + p \to \Lambda + K + \bar{K}$. Note the long thin vertical pencil of $K\bar{K}$ events at the left edge of phase space clearly indicating the existence of the ϕ. So far, so good.

But now look at Fig. 8 where the 3-π mass distribution is displayed. There is no statistically significant evidence for a peak, so there is no evidence for the decay $\phi \to \pi^+ + \pi^- + \pi^0$, which I thought should be dominant. This decay channel is expected to proceed through the chain of reactions $\phi \to \rho + \pi$, followed by $\rho \to \pi + \pi$. The authors comment on this unexpected absence:

> "The observed rate [for $\phi \to \rho + \pi$] is lower than ... predicted values by one order of magnitude; however the above estimates are uncertain by at least this amount so that this discrepancy need not be disconcerting."

But I was very disconcerted. My calculations indicated that the $\rho\pi$ mode was suppressed by at least *two* orders of magnitude,[l] an *unprecedented* suppression for the strong interactions. I had learned from Feynman that in the

[l]The simple estimate was:

$$\begin{aligned} \frac{\Gamma_{K\bar{K}}}{\Gamma_{\rho\pi}} &\sim \left(\frac{p_{K\bar{K}}}{p_{\rho\pi}}\right)^3, \\ &= 1/4 \text{ (expected)}, \\ &\geq 35 \quad \text{(observed)}. \end{aligned}$$

Here p_{ij} is the momentum of either particle i or j in the rest frame of the ϕ.

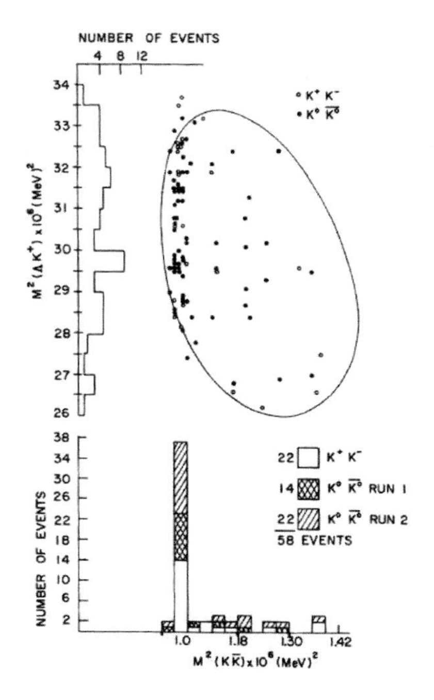

Fig. 7. Figure 1 from Ref. 14. "Dalitz plot for the reaction $K^- + p \to \Lambda + K + \bar{K}$. The effective-mass distributions for $K\bar{K}$ and ΛK^+ are projected on the abscissa and ordinate." The cluster of events at the edge of phase space indicate the existence of a meson called the ϕ decaying into $K + \bar{K}$.

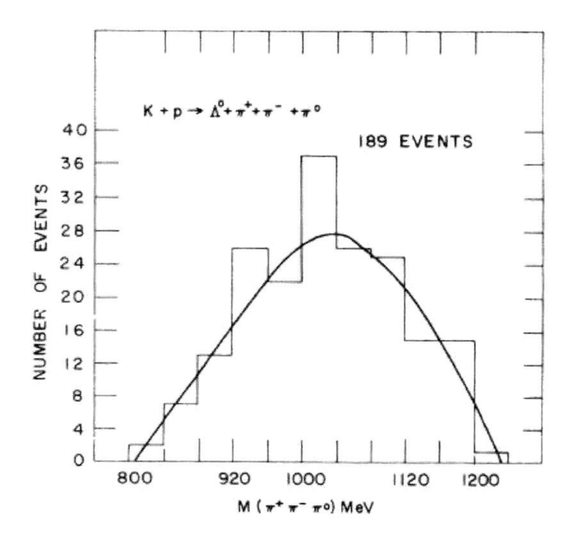

Fig. 8. Figure 4 from Ref. 14. "The $M(\pi^+\pi^-\pi^0)$ distribution from the reaction $K^- + p \to \Lambda + \pi^+ + \pi^- + \pi^0$." The absence of a large peak above phase space (solid line) indicates the suppression of the decay $\phi \to \rho + \pi$.

strong interactions everything that can possibly happen does, and with the maximum strength allowed by unitarity. Suppression implied symmetry, but no symmetry was present. Therefore the suppression must be dynamical, but no dynamical mechanism was available.

Feynman was not disconcerted. He launched into a tirade about how unreliable experiments were, and explained that at the time he proposed the V−A theory for the weak interactions all experiments were against him, and those experiments were wrong.[m]

5. The Explanation

I couldn't stop thinking about the suppression of ϕ decay, and finally realized that this paradox could be resolved by assuming that hadrons had constituents that obeyed a simple *dynamical* rule when they decayed.[15,16] Like Fermi and Yang, and later Sakata, I assumed that mesons are constructed out of fermion-antifermion pairs, but the fundamental fermions are not nucleons, but new fields called "aces," for reasons that will become apparent (they are now associated with constituent quarks, but unlike constituent quarks, aces are also the fundamental fields in electromagnetic and weak currents). With this assumption, the problems Fermi and Yang had in making the pion a nucleon-antinucleon bound state disappeared. Some unknown force, not the nuclear force, bound aces together. As in the Sakata model, there are three building blocks: a doublet $N_0 \equiv \{p_0, n_0\}$ with isotopic spin $I = 1/2$ and strangeness $S' = 0$, analogous to the nucleons $\{p, n\}$, and a singlet Λ_0 with isotopic spin $I = 0$ and strangeness $S' = -1$, analogous to the Λ. Following Dirac, aces are paired with antiaces, $\{\bar{p}_0, \bar{n}_0, \bar{\Lambda}_0\}$. The vector mesons with their constituents are shown in Fig. 12(d). The sawtooth line connecting an ace a (shaded circle) with an antiace $\bar{a}$ (open circle) represents the spring holding them together. The $a\bar{a}$ pair has angular momentum $L = 0$, and total spin $S = \frac{1}{2} + \frac{1}{2} = 1$. The pion, a pseudoscalar meson, has the same constituents as the ρ, but the spins of its constituents cancel, making $S = 0$. A meson decays when its $a\ \bar{a}$ constituents fly off in different directions. Their separation induces a polarization of the vacuum creating an $\bar{a}'a'$ pair that also splits, forming $a\bar{a}'$ and $a'\bar{a}$, the decay products (Fig. 9).

[m] The V−A theory (1957) was initially at variance with angular correlations measured in He^6 decay and the absence of the decay $\pi^- \to e^- + \bar{\nu}$, which was not seen in two independent experiments by Jack Steinberger in 1955, and Herb Anderson in 1957. In 1958, in CERN's first major discovery, Fazzini, Fidecaro, Merrison, Paul and Tollestrup observed this decay at the predicted rate, confirming V−A. As a testimony to the difficulty of measurement, both Steinberger and Anderson were outstanding experimentalists, students of Fermi. Steinberger later shared the Nobel Prize for demonstrating that the electron and muon each have their own neutrinos.

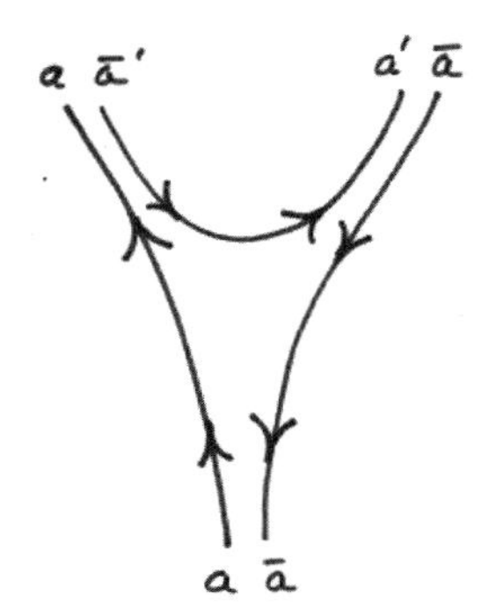

Fig. 9. The decay of meson $a\bar{a}$, a is an ace, $\bar{a}$ an antiace.

With this rule of ace conservation (a component of Zweig's rule), aces in the initial state must appear in the final state, i.e., "aces don't eat each other." Since the ϕ consists only of Λ_0 and $\bar{\Lambda}_0$, constituents not found in ρ or π, the decay ϕ to $\rho + \pi$ cannot take place, and empirically is strongly suppressed.[n] Remarkably, the constituent assignment for the mesons required to suppress $\phi \to \rho + \pi$ gave mass formulae for the vector mesons that worked exceptionally well (see below).[o]

In order to obtain the observed baryon spectrum it is necessary to build baryons out of three aces, giving them each baryon number $B = 1/3$. This distinguishes them from the $\{p, n, \Lambda\}$ of the Sakata model, which fails to group the low-lying baryons into the observed group of 8 (Fig. 4). The charge Q of an ace is then given by the NNG charge formula $Q = e[I_z + (B + S')/2]$, where I_z is the $z-$projection of the isotopic spin. With the I_z values of $\{1/2, -1/2, 0\}$ for $\{p_0, n_0, \Lambda_0\}$, their charges are $\{2/3, -1/3, -1/3\}$.

These ideas were extended to characterize additional properties of hadrons and their strong, electromagnetic, and weak interactions. The essence of the model as it appeared in two CERN preprints at the start of 1964, (Fig. 10[15,16]), is as follows:[p]

---

[n]Originally a graphically different, but functionally equivalent, visual representation of the decay was given, involving a "tinker toy" construction (see Fig. 13 of Ref. 16).

[o]Okubo obtained the suppression of ϕ decay, and mass formulae, with a different argument that didn't involve constituents.[17]

[p]Work on aces was almost finished by Thanksgiving 1963 when Ricardo Gomez, a Caltech Research Fellow I had worked with on the K-decay experiment at Berkeley, came to visit. When I told him what I was doing, he smiled and said I was "crazy." He wanted to go to the Bataclan, the one and only strip club in town (Calvin's Geneva). Much to my surprise, we skipped a very long line and were seated at a small table up front with a free bottle of champagne. The manager thought Ricardo was *Ricardo Gomez*, the famous bicycle racer, and Ricardo did nothing to disabuse him of this idea.

AN SU_3 MODEL FOR STRONG INTERACTION SYMMETRY AND ITS BREAKING

G. Zweig *)
CERN - Geneva

A B S T R A C T

Both mesons and baryons are constructed from a set of three fundamental particles called aces. The aces break up into an isospin doublet and singlet. Each ace carries baryon number $\frac{1}{3}$ and is consequently fractionally charged. SU_3 (but not the Eightfold Way) is adopted as a higher symmetry for the strong interactions. The breaking of this symmetry is assumed to be universal, being due to mass differences among the aces. Extensive space-time and group theoretic structure is then predicted for both mesons and baryons, in agreement with existing experimental information. An experimental search for the aces is suggested.

*) This work was supported by the Air Force Office of Scientific Research and the National Academy of Sciences - National Research Council, U.S.A.

8182/TH.401
17 January 1964

Fig. 10. Title page of Ref. 15. Only certain SU_3 representations, quantum numbers, and decays are allowed, constraints not found in the Eightfold Way.

(1) Hadrons have *point* fermion constituents with baryon number 1/3.
(2) There is a correspondence between leptons, the point particles of the weak interactions, and aces, the point particles of the strong interactions. In 1963 four leptons were known, hence the name aces.[q] This correspondence was just a hunch, in the Einstein tradition.
(3) The ultimate number of constituents was unknown. Representing them by $\clubsuit$, $\heartsuit$, $\spadesuit$ and $\diamondsuit$ was not general enough. To allow for the possibility of an expanding set, each constituent was represented by a regular polygon of ever increasing size, corresponding to increasing mass: a circle represented p_0, a triangle n_0, a square Λ_0, a pentagon the fourth ace, and so on.
(4) Each hadron is represented by a linear combination of $a\bar{a}'$ pairs (deuces) for mesons, and $aa'a''$ triplets (treys) for baryons. The deuces or treys are weighted by SU_3 coefficients to form meson and baryon wave functions, e.g., $\rho^0 = \frac{1}{\sqrt{2}}(p_0\bar{p}_0 - n_0\bar{n}_0)$, indicating that the ρ^0 is equally likely to be a $p_0\bar{p}_0$ or $n_0\bar{n}_0$, neglecting electromagnetic interactions.

Aces provide a rationale for the existence of SU_3 symmetry, but more restrictively, lead to unique predictions for the existence of only certain SU_3 representations and quantum numbers, and provide relations among masses, and coupling constants. Baryons only come in groups of 1, 8, and 10 ($3 \times 3 \times 3 = 1 + 8 + 8 + 10$), mesons only in groups of 1 and 8 ($3 \times \bar{3} = 1 + 8$), with these two SU_3 representations mixing strongly for vector mesons, leading to a group of 9. Keeping track of ace spins, and allowing aces to have angular momentum, connects the spin-parity of baryons with the dimension of their SU_3 representations, and creates higher mass excited states. Assuming aces in the lightest mesons and baryons have zero angular momentum leads to the observed baryon octet ($J^P = \frac{1}{2}^+$), decuplet ($\frac{3}{2}^+$), and singlet ($\frac{1}{2}^-$), as well as a pseudoscalar meson octet ($J^{PC} = 0^{-+}$) and vector meson nonet (1^{+-}). Mesons with certain quantum numbers are forbidden, i.e., those with $J^{PC} = 0^{--}$, and those in the series $0^{+-}, 1^{-+}, 2^{+-}, \ldots$. Such "exotic" states may still be absent, but weakly bound deuteron-like bound states with exotic quantum numbers may now exist.

[q] A more obvious reason for choosing the word ace is that it is derived from the frequently used Latin word *as*, designating a *unit*, *whole* or *one*, and also designating a *small copper coin*. In English, it meant the side of the die with only one mark before it meant the playing card. In High Energy Physics, an ace might now be taken to mean any one of the six faces of the die. Who says that "God doesn't play dice with the Universe?"

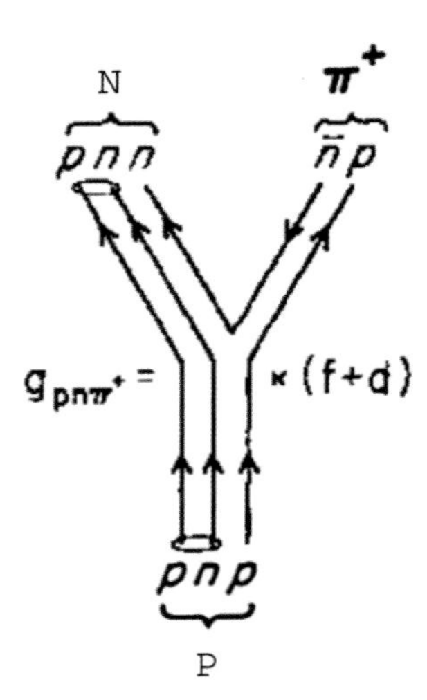

Fig. 11. Graphical representation of the "$f + d$" contribution to meson-baryon coupling. The "little loop" encloses anti-symmetrized aces. A different contribution to the coupling arises if the anti-symmetrized aces are separated in the decay, one ace terminating in the baryon, the other in the meson. The subscript "0" on aces is suppressed. Adapted from Ref. 18.

(5) *Aces, not hadrons, interact:*

(a) For the strong interactions, hadrons decay when their constituents separate and initiate the formation of decay products (Zweig's rule,[r] Figs. 9 and 11).

Graphically, when a meson decays into other mesons, it decays in all ways the deuces present can be connected, each connection having the same amplitude. The decay amplitude is proportional to the sum of these amplitudes, each amplitude weighted by the amplitude of the deuces in the meson wave functions (examples given in Figs. 10 and 11 of Ref. 16). This is a simple application of Feynman's "sum over all paths" formulation of quantum mechanics.

(b) The electromagnetic or weak interactions of hadrons occur through the interaction of their constituents with the photon or intermediate vector boson. For example, when the neutron n undergoes β-decay

[r] Gell-Mann enjoyed using Rosner's term, the "Twig rule" ("twig" is derived from the German word "zweig," which is conventionally — and more flatteringly! — translated as "branch"). Zweig's rule differs from that of Okubo's or Iizuki's because it not only says what is forbidden, but also what is allowed, *and by how much.* It involves enumerating, weighting, and summing all graphs available to a hadron for decay. An application of Zweig's rule that is not covered by the OZI rule is found in Ref. 16, and illustrated in Fig. 11.

($n \to p + e^- + \nu$), it is really its n_0 constituent that β-decays ($n_0 \to p_0 + e^- + \nu$), the lighter p_0 becoming a constituent of the final-state proton. In this respect aces behave like fields in a field theory, i.e., like "current quarks."[19]

(6) The mass of a hadron is the weighted average of the masses of its deuces or treys, the weights given by the relative probabilities of the hadron existing in its possible deuce or trey configurations. The mass of a deuce $D_a^{\bar{a}'}$ or trey $T_{a\,a'a''}$ is the sum of their constituent masses minus their *pairwise* binding energies. Subscripts refer to aces, superscripts to antiaces. Three-body forces are negligible. Ace mass differences Δm are assumed to be significantly greater than their binding energy differences ΔE, i.e., $|\Delta m| \gg |\Delta E|$, so symmetry breaking arises primarily from ace mass splittings:

   (a) For mesons: $m(D_a^{\bar{a}'}) = m(a) + m(\bar{a}') - E_a^{\bar{a}'}$. The binding energy $E_a^{\bar{a}'}$ depends on the SU_3 representation of the meson containing the deuce, and on the total spin $\vec{S}$ and angular momentum $\vec{L}$ of the $a\,\bar{a}$ system.

   (b) For baryons: $m(T_{a\,a'a''}) = m(a) + m(a') + m(a'') - E_{a\,a'\,*} - E_{a\,*\,a''} - E_{*\,a'\,a''}$. A trey $T_{a\,a'a''}$ in a baryon is represented by a triangle with the aces a, a' and a'' at its vertices, with $E_{a\,a'\,*} = E_{a'\,a\,*}$. The binding energies depend on the SU_3 representation of the baryon containing the trey, and on the spins and angular momentum of the aces.

The strong interaction symmetry SU_3 is broken by distinguishing the mass and binding energies of Λ_0 from N_0, (Fig. 12).

The electromagnetic symmetry SU_2 is broken by distinguishing the mass and binding energies of n_0 from p_0 (Fig. 13).

No potential function is assumed. This is not the naive quark model! The naive quark model missed the point. It made detailed assumptions about the potential that were surely incorrect, with very little to show for it. I had no idea how the potential varied with distance, just that it gave rise to a very strong two-body force.[s]

[s]When Finn Ravndal first came to Caltech as a graduate student in 1968 he was eager to work on the quark model, and was very surprised to find that Murray and I were not thinking about quarks. I had done what I could five years earlier, and didn't know how to find the force binding aces. Even if I guessed correctly, I thought the theory would be so nonlinear that I couldn't prove the guess correct. Murray wasn't trying to find the force because he didn't think quarks existed. Ken Young, one of Murray's brightest graduate students, laments that Murray didn't ask him to compute the force between two quarks when he asked Murray for a thesis problem in 1970, a problem that was assigned to Politzer and Wilczek by their thesis advisors two years later, after Gerard 't Hooft had found the answer, but didn't fully publish his result.

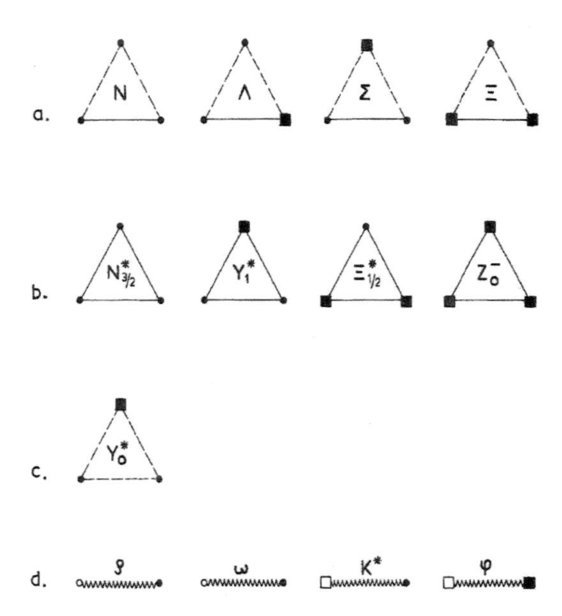

Fig. 12. Figure 2 of Ref. 15. "We view the particle representations with unitary symmetry broken. One of the three aces has now become distinguishable from the other two. It is pictured as a shaded square. ... The mass splittings within representations are induced by making the squares heavier than the circles. Since the same set of aces are used to construct all hadrons, mass relations connecting mesons and baryons may be obtained." The assignment of constituents to the vector mesons was suggested by the observation that the ρ and ω had the same mass, and the square of the K^* mass was the average of the mass squares of the ρ and ϕ.

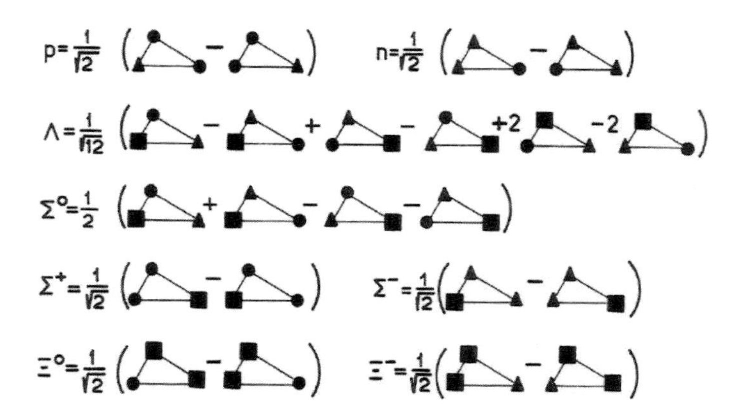

Fig. 13. Figure 3 of Ref. 16. "After SU3 has been broken by the strong and electromagnetic interactions the baryon octet looks like this. The three aces are now completely distinguishable from one another." A baryon wave function is a linear combination of treys. The trey coefficients come from SU3. Although aces are point-like, the mass of an ace is represented by the size of its symbol.

6. The Question

Would you, the reader, have adopted aces as constituents? To help answer this question, some results beyond SU_3 symmetry from the 80 page CERN report are outlined below. Except when in curly brackets, the masses in MeV are those known at the time.

(1) SU_3 symmetry breaking: $m(p_0) = m(n_0) < m(\Lambda_0)$, leading to:
A hierarchy of baryon relations, e.g., for the $J^P = \frac{1}{2}^+$ octet:

(a) If all binding energies are equal, then the stranger a baryon in a multiplet, the heavier it is:

i. $m(N) < m(\Lambda) \approx m(\Sigma) < m(\Xi)$,
939 1115 1193 1315

ii. $\frac{1}{2}[m(\Xi) + m(N)] \approx \frac{1}{2}[m(\Lambda) + m(\Sigma)]$.
1127 1154

(b) If averaging relations for the binding energies hold, i.e., $\frac{1}{2}(E_{\Lambda_0\,\Lambda_0\,.} + E_{\alpha\beta\,.}) \approx E_{\Lambda_0\,\alpha.} \approx E_{\Lambda_0\,\beta\,.}, \alpha, \beta = p_0, n_0$, then:

i. $\frac{1}{2}[m(\Xi) + m(N)] \approx \frac{1}{4}[3m(\Lambda) + m(\Sigma)]$.
1127 1134
(Gell-Mann–Okubo mass formula)

A hierarchy of meson relations,[t] e.g., for the $J^{PC} = 1^{--}$ nonet:

(a) If all binding energies are equal, then:

i. $m^2(\rho) \approx m^2(\omega) < m^2(K^*) < m^2(\phi)$.
750^2 784^2 888^2 1018^2

(b) If averaging relations for the binding energies hold, i.e., $\frac{1}{2}(E^{\bar{\Lambda}_0}_{\Lambda_0} + E^{\bar{\beta}}_{\alpha}) \approx E^{\bar{\beta}}_{\Lambda_0} \approx E^{\bar{\Lambda}_0}_{\alpha}$, $\alpha, \beta = p_0, n_0$, then:

i. $m^2(\phi) \approx 2m^2(K^*) - m^2(\rho)$.
1018^2 1007^2

(2) $SU(2)$ symmetry breaking: $m(p_0) < m(n_0)$, leading to:
A hierarchy of baryon relations, e.g., for the $J^P = \frac{1}{2}^+$ octet:

(a) If all binding energies are equal, then the more negative a baryon in a multiplet, the heavier it is:

[t] Mass relations were linear for baryons, and quadratic for mesons, in analogy with the linear Dirac equation for fermions, and the quadratic Klein-Gordon equation for mesons.

i. $m(p) < m(n)$,
938 939

ii. $m(\Sigma^+) < m(\Sigma^0) < m(\Sigma^-)$,
1190 1193 1196

iii. $m(\Xi^0) < m(\Xi^-)$.
1315 1321

(b) If averaging relations for the binding energies hold, i.e., $\frac{1}{2}(E_{p_0\,p_0\,.} + E_{n_0\,n_0\,.}) \approx E_{p_0\,n_0\,.}, \;\cdots$, then:

i. $\frac{1}{2}[m(\Sigma^+) + m(\Sigma^-)] \approx m(\Sigma^0)$.
1193.4 ± 0.3 1193.2 ± 0.7

(c) Even if all binding energies are different,

i. $m(n) - m(p) + m(\Xi^-) - m(\Xi^0) = m(\Sigma^-) - m(\Sigma^+)$.
7.3 ± 1.3 8.3 ± 0.5

2013 masses: $\{8.14 \pm 0.21\}$ $\{8.08 \pm 0.08\}$

2013 mass differences: $\{0.06 \pm 0.22\}$

This remarkable relation currently still holds to the known accuracy of the mass differences.

(3) Relations for the baryon decuplet and pseudoscalar meson octet.

(4) Additional mesons with orbital excitations ($L = 1, 2, \cdots$) and $\vec{L}\cdot\vec{S}$ mass splittings.

(5) Cross multiplet relations:

(a) $m(N) < m(\Lambda) \Longrightarrow m^2(\rho) < m^2(K^*)$,

(b) $m(\Xi^*) - m(\Sigma^*) \approx m(\Xi) - m(\Sigma)$,
145 122

(c) $m^2(K^*) - m^2(\rho) \approx m^2(K) - m^2(\pi)$.
0.22 Gev^2 0.22 Gev^2

I thought the empirical evidence for the existence of aces was overwhelming. Were these many successful relations a result of over-fitting?[u] Probably not, because over the past 50 years the accuracy of these relations has either remained essentially the same, or improved.

7. The Difficulties

All this was not as easy as it looks. Although aces resolved the problem of ϕ decay, explained many regularities, and made predictions that could easily be tested, some predictions contradicted experiment, and even violated basic theoretical principles.

Matt Roos's "definitive" compilation of resonances, published in the 1963 Reviews of Modern Physics, lists 26 meson resonances.[20] *We now know that 19 of these resonances do not exist, and of those 19, 7 are exotic, i.e., cannot be made from deuces.* As recounted previously:[1]

> "Matt Roos's ... compilation of particles and their properties referred to several hundred experimental papers. I read essentially all of them, taking care to understand how each measurement was made. Then an accurate appraisal of the results of each experiment was possible. Rational choices between conflicting experiments usually could be made."

My training both in experimental and theoretical physics made it possible to separate "wheat from chaff." Others did not have the training, interest, or patience.

A major theoretical problem was the existence of the celebrated Ω^- with spin $S = 3/2$ constructed from three identical aces all in the same state, thereby *violating Pauli's spin-statistics theorem.* I didn't have a resolution to this problem, but thought it eventually would be resolved, much as the contradictions with Rutherford's atom and Bohr's orbits were eventually resolved by quantum mechanics.[v]

Equally problematic, aces had no place in current dogma. They were incompatible with nuclear democracy, the basis of the bootstrap, and much prevailing thinking. In addition, because aces were not observed as free particles, formulating a theory of strong interactions in terms of them was

[u]Certain simple relations like $m(\Xi) = [3m(\Sigma) - m(N)]/2$, which were correct to electromagnetic mass splittings, were considered accidental.

[v]Feynman's 1970 solution to the spin-statistics problem was to make quarks bosons! The abstract to a paper written by Feynman, Kislinger, and Ravndal titled "The $\Delta I = 1/2$ Rule from the Symmetric Quark Model" reads "The $\Delta I = 1/2$ rule for the weak non-leptonic hyperon decays will result from quark currents interacting at a point, if the quarks obey Bose statistics." The paper was withdrawn before publication when Feynman learned that this idea had been previously proposed (Ravndal, private communication).

incompatible with Heisenberg's requirement that theory be based solely on observables. Like Copernicus's view of the solar system, simplicity was no excuse for challenging what was known to be true.[w]

Finally, working with aces made me uncomfortable. I knew what real theories looked like. As a student I had read and understood Schwinger's papers on Quantum Elecrodynamics. I had written "professional" papers.[21] Nothing I now did looked like that. Talking about aces in public was embarrassing, because I understood how theorists thought, and what many thought of me.[x] In spite of this, a good part of me believed that aces were real particles, and that the beginnings of a solution to the proliferation problem of hadrons had finally been found. Aces were real particles because they had *dynamics*. They moved from one hadron to another, avoiding annihilation, either forwards or backwards in time. They spun and rotated around one another giving rise to $\vec{L} \cdot \vec{S}$ mass splittings. In addition, they interacted with light and underwent β decay. I applied the "duck" test: If it walks like a duck, swims like a duck, and quacks like a duck, it's probably a duck. Others disagreed.

8. The Reaction

After returning from CERN in the early fall of 1964, I went into Murray's office and told him all about aces. Sometimes Murray would close his eyes when someone talked to him, but this time his eyes stayed open. After I finished at the blackboard he exclaimed from behind his desk "Oh, the *concrete* quark model. That's for blockheads!" As late as five years after the deep inelastic scattering experiments at the Stanford Linear Accelerator provided conclusive evidence for electrons scattering off point particles in nucleons, Murray still did not accept the existence of constituent quarks inside of hadrons. He writes:[22]

> "In these lectures I want to speak about at least two interpretations of the concept of quarks for hadrons and, the possible relations between them.
>
> First I want to talk about quarks as 'constituent quarks'. *These were used especially by G. Zweig (1964)* [italics added] who referred to them as aces. ..."

[w]Wegener's theory of continental drift provides another example of discovery contradicting dogma. Evidence from geology and paleontology clearly showed that Africa and South America were connected sometime in the past, but this idea was not accepted because the Earth's crust was thought to be "frozen." A dynamical mechanism for driving continents apart was not yet known.

[x]Having worked at the Bevatron in the early 1960s, I appreciated the excitement and beauty of Berkeley and the San Francisco area. I wanted a job at UC Berkeley. Gerson Goldhaber presented my application package at a physics faculty meeting, but a senior theorist blocked the appointment, passionately arguing that the ace model was the work of a "charlatan" (Goldhaber, private communication).

It is more precise to say: "*These were introduced by G. Zweig* (1964)..." After all, they did not exist at the time as a tool in a toolbox for anyone to use.

Murray then goes on to say:

> The whole idea is that hadrons act as if they are made up of quarks, but the quarks do not have to be real. ..."

That's a mischaracterization of my idea. I believed that they *were* made up of quarks.

> "There is a second use of quarks, as so-called 'current quarks' which is quite different from their use as constituent quarks ...
>
> If quarks are only fictitious there are certain defects and virtues. The main defect would be that we never experimentally discover real ones and thus will never have a quarkonics industry. The virtue is that then there are no basic constituents for hadrons — hadrons act as if they were made up of quarks but no quarks exist — and, therefore, *there is no reason for a distinction between the quark and bootstrap picture: they can be just two different descriptions of the same system, like wave mechanics and matrix mechanics.* [italics added]"

This was Murray's vision. Aces, which had aspects of both current and constituent quarks, are not mentioned.

Although Feynman had no quarrel with "current quarks,"[19] or the current-quark aspect of aces, he disliked Zweig's rule. Every time I would "explain" it to him he became angry, and said that it didn't make sense. Unitarity mixed all states with the same quantum numbers, so the suppression of the ϕ to $\rho\pi$ was not possible. He also believed that the correct theory of the strong interactions should not allow one to say which particles are elementary, a key element of the bootstrap.

And what about Heisenberg? In an interview in the early 1970s for broadcast as part of a CBC radio documentary series entitled Physics and Beyond he said:[24]

> "Even if quarks should be found (and I do not believe that they will be), they will not be more elementary than other particles, since a quark could be considered as consisting of two quarks and one anti-quark, and so on. I think we have learned from experiments that by getting to smaller and smaller units, we do not come to fundamental units, or indivisible units, but we do come to a point where division has no meaning. This is a result of the experiments of the last twenty years, and I am afraid that some physicists simply ignore this experimental fact."

No, I'm afraid not. Quarks certainly are "more elementary than other particles," and a quark is not "considered as consisting of two quarks and one anti-quark." Heisenberg's variant of the Sakata model, but at a smaller scale, simply doesn't work.

9. Acceptance

Late in May of 1968 I bumped into Feynman as we both were walking to the Greasy (the Caltech cafeteria) for lunch. He was very excited about the High Energy Physics course he had just finished teaching, and was back in full swing after a lull in research following his unsuccessful attempt to renormalize gravity. After listing the many areas he had covered in his course he stopped, turned to me and asked, "Did I miss anything Zweig?" Patiently, once again, I told him about aces. This time he was quiet, intent, and listened. After I finished, he hitched up his pants with both thumbs, looked straight into my eyes, and in his most official voice replied: "All right, I'll look into it!" Shortly thereafter he created the parton model, and the following fall Feynman, with Finn Ravndal and Mark Kislinger, two graduate students, looked for evidence of concrete quarks in pion photoproduction, a place where I, and others, had never looked.

Three years later, as I was walking down the corridor on the fourth floor of Lauritsen where we both had offices, I noticed Feynman in the distance with an enormous grin, swaggering like a sailor, thumbs hooked inside his belt, fingers splayed apart. When he was almost in my face, no more than a foot away, he extended his right hand and said "Congratulations Zweig! You got it right." By that time I had switched to neurobiology and didn't know that his work with Finn and Mark had finally been completed.

9.1. *Bayes' theorem*

In retrospect, Bayes' theorem could have been used to elucidate some of the issues involved in deciding if aces should have been accepted as constituents of hadrons. With some massaging, Bayes' theorem says that the probability that hadrons have ace constituents, given the experimental evidence ($\phi \not\to \rho + \pi$, mass relations, etc.), is

$$P(A|E) = \frac{1}{1+\lambda}, \text{ where } \lambda = \frac{P(E|\bar{A})}{P(E|A)} \frac{P(\bar{A})}{P(A)}.$$

Here $P(E|\bar{A})$ is the probability of the evidence given that aces do not exist, $P(A)$ is the a priori probability that they do exist, and $P(\bar{A})$ that they don't.

Since

$$P(\bar{A}) \approx P(E|A) \approx 1,$$

the expression for $P(A|E)$ simplifies,

$$P(A|E) \approx \frac{1}{1 + P(E|\bar{A})/P(A)}.$$

Whether or not a rational person would have believed in the existence of aces depends only on the ratio λ of two small numbers. *They would have believed if and only if the likelihood of obtaining the evidence, assuming that aces didn't exist, was much lower than the a prior probability of aces existing* i.e., if and only if

$$P(E|\bar{A}) \ll P(A).$$

When this inequality holds, even though initially $P(\bar{A}) \approx 1$, when the experimental evidence is taken into account, just the opposite holds, i.e., $P(A|E) \approx 1$.

Because it is not possible to empirically determine these probabilities, how should this argument be interpreted? It suggests that in deciding whether hadrons have ace constituents, consider how easy it would be to find an alternative explanation for all the relations that follow assuming their existence, and weigh that against how much you like the idea of ace constituents in the abstract, without any experimental evidence for or against their existence. I suspect that those working in the Rutherford–Bohr tradition found acceptance easier than those trying to follow in Einstein's footsteps.[y]

9.2. *Different priors, different times*

The early adopters were Linus Pauling and Dick Dalitz, with Pauling coming first. Shortly after the CERN preprints were circulated, I received galley proofs of the third edition of Linus Pauling's "College Chemistry,"[23] where Pauling presented aces to undergraduates. He asked for my comments and corrections. Essentially none were necessary. Pauling recognized a good thing when he saw it. Aces didn't have anything to do with college chemistry, but I suspect he thought aces were so beautiful, and so right, that he couldn't resist ($P(A)$ was not *that* small for him).

Dalitz's conversion is less surprising. He came from nuclear physics where nucleon constituents, and the forces between them, were his daily bread and butter. He just changed constituents.

[y]A Baysian inspired evaluation of current speculative theories is also possible, even if $P(E|Theory)$ is unknown. If a theory has not yet advanced to the stage where it can compute observables, the evidence E should be replaced by the Standard Model.

For Feynman acceptance of concrete quarks came when he discovered them for himself in photoproduction amplitudes, where no one else had looked. Others were convinced by deep inelastic scattering experiments at SLAC. The holdouts finally folded with the discovery of the ψ(J), whose forbidden decay modes and narrow width were just like ϕ with the fourth ace replacing Λ_0. It was "déjà vu all over again."

10. Invention or Discovery

Last summer Murray and I were present at a talk given by Chris Llewellyn Smith at the Santa Fe Institute. In his introduction, Geoffrey West said "Murray and George invented quarks, which were later discovered at SLAC." I interrupted to suggest a different ending to Geoffrey's sentence: "quarks, whose existence was later confirmed at SLAC." But what was invented, and what discovered? According to Merriam-Webster, discovery is "the act of finding or learning something for the first time," and invention is "a product of the imagination." Current quarks were invented in the tradition of Einstein. Aces were discovered in the Rutherford–Bohr tradition, buried in the data, obscured by the contradictions they implied. Bohr would have loved them.

11. Final Thoughts

The CERN report concludes:[15]

> "There are, however, many unanswered questions. Are aces particles? If so, what are their interactions? Do aces bind to form only deuces and treys? What is the particle (or particles) that is responsible for binding the aces? Why must one work with masses for the baryons and mass squares for the mesons? And more generally, why does so simple a model yield such a good approximation to nature?"

After dutifully listing possible interpretations of the results, the report concludes with:

> "there is also the outside chance that the model is a closer approximation to nature than we may think, and that fractionally charged aces abound within us."

This then was the beginning of QCD, the theory that will eventually give us a real understanding of how it came to be that Becquerel's photographic plate clouded in his drawer.

12. Epilogue

In modern terms, an ace, or equivalently a concrete quark, was used as a point particle in a field theory to construct the electromagnetic and weak currents, and as a much heavier particle (the point particle dressed in glue) to construct the hadrons.[z] I thought observation required a dual description.

Although Bohr was able to calculate the spectral lines of hydrogen, his model could not accurately account for the spectrum of helium, and was conceptually incomplete. For example, the wave nature of particles had not yet been discovered by de Broglie, and was absent from Bohr's thinking. There was no Lagrangian. There was no theory!

The concrete quark model provided predictions for the particle spectrum, and enabled calculations of hadronic masses and couplings that were accurate to varying degree, but it was conceptually incomplete, lacking a specification of the interaction between quarks. Like the wave–particle duality of quantum mechanics, concrete quarks were chimeric: sometimes acting as fields in a field theory for weak decay, sometimes as convenient objects for the calculation of masses and coupling constants. And like the Bohr atom, the Ω^-, made from three quarks in the same state, could not exist according to the laws that were known. There was no Lagrangian. There was no theory!

Details aside, the important observation was that hadrons had fermion constituents of baryon number 1/3, with dynamics that suggested they were real, and therefore should have corresponding fields. It was generally believed that hadrons had constituents, but those constituents were other hadrons. Some even thought that field theory was irrelevant. No! Concrete quarks said that there is a deeper level of reality to be described with field theory, and channelled thinking into more productive directions, eventually leading to QCD. It was this deeper level of reality, with fractional charges, that made the acceptance of concrete quarks so difficult, and truly revolutionary.

The regularities in the spectral lines of the hydrogen atom responsible for the creation of the Bohr model, that eventually led to the invention of quantum mechanics, were immediately derived from quantum mechanics. Although ever more accurate numerical calculations of masses of the low-lying hadrons are now possible with QCD, the *regularities* among these masses that led to the discovery of the concrete quark model have not been derived. Their derivation would provide a satisfying test of that fledgling theory.

[z]By constructing the weak currents from aces I hoped that it would be possible to eliminate the unintelligible Cabibbo angle. The suppression of strangeness-changing decays would then be dynamical.

How can such a complicated nonlinear theory as QCD give rise to the simple low-energy relations of the concrete quark model? There should be some way to approximate QCD so that the existence and hierarchy of relations among masses become apparent. It's not like the physics of water where the greater the nonlinearity, the more complex the flow. Quite the contrary. In QCD, when energies are small and nonlinearities large, the particle spectrum and its couplings are simple.

Recall the history of classical mechanics. Newton's laws were reformulated again and again over centuries, each time with a different purpose. In the following centuries QCD will be reformulated again and again, and with one of those reformulations the concrete quark model may appear, a fortiori.

Acknowledgments

I thank Erica Jen and Jeffrey Mandula for their insightful comments, and David Donoho, Stanislaw Mrowczynski, Finn Ravndal, and Harvey Shepard for improving the accuracy and content of the text.

References

1. G. Zweig, Origins of the Quark Model, in *Proc. Fourth Int. Conf. Baryon Resonances*, ed. N. Isgur, University of Toronto, Canada, pp. 439–479 (1980), `http://www-hep2.fzu.cz/~chyla/talks/others/zweig80.pdf`.
2. G. Zweig, Memories of Murray and the Quark Model, *Proc. Conf. in Honor of Murray Gell-Mann's 80th Birthday*, eds. H. Fritzsch, K. K. Phua and B. E. Baaquie, World Scientific, Singapore, pp. 7–20 (2010), `http://arxiv.org/abs/1007.0494`.
3. H. Geiger and E. Marsden, *Proc. R. Soc. Lond. A* **82**, No. 557, 495 (1909).
4. E. Rutherford, *Phil. Mag.*, Series 6, **21**, 669 (1911).
5. G. F. Chew and S. Frautschi, *Phys. Rev. Lett.* **7**, 394 (1961).
6. E. Fermi and C. N. Yang, *Phys. Rev.* **76**, 1739 (1949).
7. E. Rutherford and J. Chadwick, *Phil. Mag.* Series 7 **4:22**, 605 (1927).
8. S. Sakata, *Progr. Theor. Phys.* **16**, 686 (1956).
9. M. Gell-Mann, and A. H. Rosenfeld, *Annu. Rev. Nucl. Sci.* **7**, 407 (1957).
10. M. Gell-Mann and E. P. Rosenbaum, *Scientific American*, 72 (July 1957).
11. S. J. Lindenbaum, *Annu. Rev. Nucl. Sci.* **7**, 317 (1957).
12. F. Zachariasen, *Phys. Rev. Lett.* **7**, 112 (1961); *Erratum* **7**, 268 (1961).
13. G. F. Chew, M. Gell-Mann, and A. H. Rosenfeld, *Scientific American*, **74** (February 1964).
14. P.L. Connolly *et al.*, *Phys. Rev. Lett.* **10**, 371 (1963).
15. G. Zweig, An SU_3 Model for Strong Interaction Symmetry and its Breaking, *CERN Report 8419/TH.401* (January 17, 1964), `http://cdsweb.cern.ch/record/352337?ln=en`.

16. G. Zweig, An SU_3 Model for Strong Interaction Symmetry and its Breaking II, *CERN Report 8419/TH.412* (February 21, 1964), in *Developments in the Quark Theory of Hadrons, A Reprint Collection*, Vol. I: 1964–1978, eds. D. B. Lichtenberg and S. P. Rosen, Hadronic Press, Inc., Nonantum, MA, pp. 22–101 (1980), `http://cdsweb.cern.ch/record/570209?ln=en`.
17. S. Okubo, *Phys. Lett.* **5**(2), 165 (1963).
18. J. Mandula, J. Weyers, and G. Zweig, *Annu. Rev. Nucl. Sci.* **20**, 289 (1970).
19. M. Gell-Mann, *Phys. Lett.* **8**, 214 (1964).
20. M. Roos, *Rev. Mod. Phys.* **35**, 314 (1963).
21. G. Zweig, *Il Nuovo Cimento* **XXXII**, No. 5, 689 (1964).
22. M. Gell-Mann, *Acta Physica Austriaca, Suppl. IX*, 733 (1972).
23. L. Pauling, *College Chemistry*, 3rd edn. (W. H. Freeman and Co., San Francisco, 1964).
24. W. Heisenberg, *Glimpsing Reality: Ideas in Physics and the Link to Biology*, P. Buckley and F. D. Peat eds., University of Toronto Press, Canada, 15 (1996).

On the Way from Sakatons to Quarks

L. B. Okun

ITEP, Moscow, Russia

okun@itep.ru, levokun@gmail.com

This is my contribution to "50 Years of Quarks" written on the invitation from Prof. K. K. Phua and Prof. Harald Fritzsch to the 50th Jubilee of the discovery of quarks. It consists of two parts. In the first part I describe chronologically personal recollections from the years 1956–1980. The second part contains excerpts from my book "Particle Physics: The Quest for the Substance of Substance" published in 1985.

1956. Moscow

The book[1] contains theses of talks presented at my first conference May 14–22, 1956. I presented four talks on phenomenology of elementary particles dealing with their isotopic properties and met the most active Soviet and foreign physicists who played a major role in fostering the concepts of particle physics.

Shoichi Sakata was the first foreigner who visited the ITEP theory division. He came in the spring of 1956 and compiled a list of the ITEP theorists — I. Ya. Pomeranchuk, V. B. Berestetsky, A. D. Galanin, A. P. Rudik, B. L. Ioffe, V. V. Sudakov, I. Yu. Kobzarev and myself. Sakata also took a photo of those who were present. (It would be interesting to find this picture in his archives.) I still have the three pages of thin rice paper with the Sakata model which he left with us. They correspond to his paper.[2] These three pages were crucial for all my life in physics.

Sakata[2] considered 7 mesons ($3\,\pi$'s, $4\,K$'s) and 8 baryons ($2\,N$'s, Λ, $3\,\Sigma$'s, $2\,\Xi$'s) known at that time. He postulated that 3 baryons — p, n, Λ — are more fundamental than the other 5 baryons and 7 mesons and demonstrated that these 12 particles could be composed from p, n, Λ and $\bar{p}$, $\bar{n}$, $\bar{\Lambda}$. The paper had a philosophical flavor and contained no experimental predictions. In 1956 particle physicists were discussing the $\tau\theta$-puzzle and parity violation (see Ref. 3 for further details). Therefore the paper[2] as well as three accompanying papers of Sakata's students[4–6] had no immediate response. (S. Tanaka[5] discussed $\tau\theta$-parity degeneracy in the Sakata model, Z. Maki[6] attempted to calculate bound states of baryons and antibaryons, while K. Matumoto[4] suggested a semi-empirical formula for masses of composite particles.)

1957. Moscow

My first review[7] "Strange Particles (The Scheme of Isotopic Multiplets)" was published in Russian in April 1957, it presented a detailed explanation of the masterly articles concerning the concepts of isotopic spin and strangeness by M. Gell-Mann in 1953–1955.[8]

1957. Padua — Venice

In the summer of 1957 I suddenly "reinvented" the Sakata model and realized its beauty and its potential. Then I recalled the three rice pages and reread them.

My first paper[9] on the Sakata model was presented by I. I. Gurevich at the conference in Padua — Venice, September 1957. A slightly different text[10] was published in a Russian journal. In these publications the three "sakatons" were not physical p, n, Λ, but some primary particles denoted by the same letters, so "we can assume that for the primary particles $m_\Lambda = m_N$."[10] Strong and weak interactions of sakatons were considered and for the latter a number of selection rules were deduced, in particular, those which are known as $|\Delta S| = 1$, $\Delta T = 1/2$ for nonleptonic decays of strange particles via the $\bar{n}\Lambda$ transition, while for the leptonic (or semileptonic) ones $|\Delta S| = 1$, $\Delta Q = \Delta S$ and $\Delta T = 1/2$ via $\bar{p}\Lambda$ current.

As for the strong interactions, the existence of η- and η'-mesons was predicted in Refs. 9 and 10; I denoted them ρ_1^0 and ρ_2^0:

> "In the framework of this scheme there is a possibility of two additional neutral mesons which have not so far been observed:
>
> $$\rho_1^0 = \Lambda\bar{\Lambda}\,, \quad \rho_2^0 = (p\bar{p} - n\bar{n})/\sqrt{2}\,.$$
>
> The isotopic spin of the ρ-mesons is zero."[9]

(The unconventional minus sign in the definition of ρ_2^0 was in accord with the not less unconventional definition $\pi^0 = (p\bar{p} + n\bar{n})/\sqrt{2}$.)

1957. Stanford and Berkeley

In December 1957, four Soviet particle physicists (D. I. Blokhintsev, V. P. Dzhelepov, S. Ya. Nikitin and myself) visited Palo Alto, Berkeley, Boston, New York, Brookhaven. For me it was my first trip abroad and the first flight in my life. (The next time the Soviet authorities allowed me to visit the USA was only in 1988 for the Neutrino '88 conference.)

During the 1957 trip, I talked with M. Baker, H. Bethe, S. Drell, R. Feynman, R. Gatto, M. Gell-Mann, S. Goldhaber, C. Sommerfield, F. Zachariasen, C. Zemach and many others, gave a seminar at Berkeley. As a result of this seminar, E. Segre invited me to write an article for the Annual Review of Nuclear Science. It appeared in 1959 (see below).

1958. Geneva

My second paper on the Sakata model "Mass reversal and compound model of elementary particles" was published in June 1958 as a Dubna preprint[11] and I had it with me at the 1958 Rochester Conference at CERN. On the initiative of J. R. Oppenheimer and R.E. Marshak a special seminar was arranged at which I presented my paper at the start of the conference and then was asked to present it also at Session 7, "Special theoretical topics," see Ref. 12. (Note that selection rules for weak interactions in Secs. 14, 15 and Refs. 24–28 of the Dubna preprint[11] were deleted by the editors of the Proceedings;[12] see the Appendix for the deleted pages.)

In Refs. 11 and 12, ρ_1^0 and ρ_2^0 became mixtures of the states discussed above. What is more important, all interactions were assumed to be γ_5-invariant following papers[13–15] and especially.[16] The conservation of the vector non-strange current, postulated in Refs. 13 and 17, was shown in Refs. 11 and 12 to be inevitable in the Sakata model. Unfortunately the strong interaction was written as an ugly four-fermion interaction of sakatons.

The discussion of my talk involved R. Gatto, G. Lüders, R. Adair, G. Wentzel, T. D. Lee, Y. Yamaguchi (see page 228 of the Proceedings). The discussion with Yoshio Yamaguchi continued during the lunch in the CERN canteen. In the afternoon of the same day J. Oppenheimer commented on my argument that in the Sakata model conservation of the weak non-strange vector current is inevitable (see page 257). He again at length commented on the subject in his "Concluding Remarks" at the Conference (see page 293). R. Marshak stressed the novelty of chiral invariance for strong interactions (see page 257). In his talk "K_{e3} and $K_{\mu 3}$ decays and related subjects" Marshak repeatedly underlined that for these decays "$\Delta I = 1/2$ in Okun's model" (see Ref. 18, pp. 284 and 285).

In the discussion[19] I described an upper limit on $\Delta S = 2$ transitions which had been derived by B. Pontecorvo and myself.[20]

On the basis of the selection rules for weak interactions which follow from the Sakata model the lifetime of K_2^0 and its branching ratios were predicted[21] by I. Yu. Kobzarev and myself. This prediction was cited by me in December 1957 at Stanford and as Ref. 28 in the Dubna preprint[11] and was soon confirmed experimentally.[22]

1959. Kiev Symmetry

In 1959 my paper[23] appeared as well as its Russian twin.[24] I received a hundred requests for reprints, many of them — from Japan. Strangely enough, rereading now this paper, I do not see in it the prediction of η and η' and any statement that p, n, Λ are not physical baryons, but some more fundamental particles. Both the prediction and the statement were in Refs. 9–11. I cannot understand now their irrational omission in Refs. 23 and 24.

In 1959 other authors started to publish papers on the Sakata model. A. Gamba, R. Marshak and S. Okubo[25] pointed out the symmetry between the three leptons (μ, e, ν) and three baryons (Λ, n, p) "in models of Sakata[2] and Okun.[10]"[a] This symmetry has been emphasized by Marshak (in his rapporteur talk[26] at the 1959 Rochester conference in Kiev) and became known as the Kiev symmetry. (I served as a scientific secretary of R. Marshak and participated in preparation of his report.)

At the Kiev conference M. Gell-Mann told me: "If I were you, I would introduce in the Λnp model the linear superposition $(n \cos\theta + \Lambda \sin\theta)$." I do not understand why I did not follow his advice. The angle θ is known now as the Cabibbo angle. The weak current $\bar{p}(n + \varepsilon\Lambda)/(1 + \varepsilon^2)^{1/2}$ first appeared next year in the paper by M. Gell-Mann and M. Levy.[27]

In 1959 the symmetry which is now called SU(3) was introduced into Sakata model. Y. Yamaguchi[28] with reference to Ref. 12 stressed the existence of 9 pseudoscalar mesons $(9 = \bar{3} \times 3)$. O. Klein[29] and S. Ogawa[30] discussed generalizations of isotopic symmetry. In particular, S. Ogawa with reference to Ref. 28 considered 3 doublets (pn), $(n\Lambda)$, (Λp) and 3 meson triplets. M. Ikeda, S. Ogawa, Y. Ohnuki[31] with reference to Ref. 30 developed some mathematical constructs of the symmetry to which they referred as U(3). O. Klein[29] discussed the interaction between the triplet of sakatons and the octet of pseudoscalar mesons and stressed the symmetry between Λnp and $\mu e\nu$.

[a] Here and in other quotations the reference numbers correspond to my list of references.

1960. Rochester

In 1960 I was invited to give a rapporteur talk at the Rochester Conference in Rochester. I worked on this talk for half a year, but was not allowed Soviet authorities to attend the conference.

The talk "Weak Interactions (Theoretical)" was given by M. L. Goldberger who started with the words:[32] "Since the subject of this Session is weak interactions, it is perhaps appropriate to have a rapporteur who is only weakly attached to the subject. The more strongly interacting Professor L. B. Okun was unfortunately unable to be here and as member of the organizing committee, I was thrown into the breach at a rather late date. I was fortunate enough to have at my disposal Okun's preliminary draft of his own report. I have also received considerable help from experts, in particular Gell-Mann. Many of the contributions which I shall not speak of will appear in Okun's summary."

As the appendix to this Session, my summary "Certain problems of weak interaction theory" was published with the footnote: "This appendix has been prepared for publication by S. Weinberg from a preliminary version of a paper prepared by L. B. Okun for presentation as a rapporteur at this session."[33]

R. Feynman[34] spoke on the conserved vector current. He said that in the model of Fermi and Yang "as has been pointed out in much more detail by Okun, in any complex structure, the coupling of the beta decay is proportional to the total isotopic spin." M. Gell-Mann[35] spoke on the conserved and partially conserved currents. He said: "... there is the scheme mentioned by Feynman and favored by Okun, Marshak, and others, based on just n, p, and Λ. Of course, if that is right we do not need the elaborate machinery I just described. We simply draw an analogy." But as it is clear from their talks both Feynman and Gell-Mann at that time preferred to use the composite model only as a tool to formulate more general phenomenological approaches. Among the talks at Rochester 1960 was that by Y. Ohnuki[36] who with a reference to Ref. 10 assumed $m_\Lambda = m_N$ and the three-dimensional unitary symmetry.

An important paper of 1960 was that by J. Sakurai.[37] With a reference to Ref. 12 he mentioned that instead of N, Λ one can use as "elementary" Ξ, Λ. He considered the absence of η-meson as a serious problem: "... within the framework of Fermi–Yang–Sakata–Okun model it may be difficult to explain why the η does not exist" (see pp. 32–36).

The η-meson was discovered within a year.[38]

Further progress in SU(3) symmetric Sakata model was achieved by M. Ikeda, Y. Miyachi, S. Ogawa,[39] who applied this symmetry to weak decays. Z. Maki, M. Nakagava, Y. Ohnuki, S. Sakata published a paper on Sakata model.[40] They wrote: "... it has recently become clear that Feynman–Gell-Mann current derived from the Sakata model is quite sufficient to account for the experimental facts concerning the weak processes.[10,41]" They postulated the existence of a so-called B^+ matter. The bound state eB^+ had been identified with n, bound state μB^+ — with Λ, while νB^+ — with p.

In 1960–61 I was giving lectures[42,43] based on the Sakata model. My major mistake at that time was that I did not consider seriously eight spin 1/2 baryons as an SU(3) octet in spite of the "eightfold way" papers by M. Gell-Mann[44,45] and Y. Ne'eman.[46] (The former referred to papers Refs. 31, 28 and 23.)

1962. Geneva Again

The conference was held in Geneva, July 4–11, 1962. I. Ya. Pomeranchuk and V. N. Gribov attended this conference and gave the leading talks on strong interactions and Regge poles. I gave the rapporteur talk on the theory of weak interaction. I said in it,[47]

> Notwithstanding the fact that this report deals with weak interactions, we shall frequently have to speak of strongly interacting particles. These particles pose not only numerous scientific problems, but also a terminological problem. The point is that "strongly interacting particles" is a very clumsy term which does not yield itself to the formation of an adjective. For that reason, to take one instance, decays into strongly interacting particles are called non-leptonic. This definition is not exact because "non-leptonic" may also signify "photonic." In this report, I shall call strongly interacting particles *hadrons* and the corresponding decays *hadronic* (the Greek hadros signifies large, massive, in contrast to leptos which means small, light). I hope that this terminology will prove to be convenient.
>
> When reasoning about the universality of the weak interaction, one usually says: "Let us presume that the strong interaction is switched off..." The first step in this direction was apparently taken by Gell-Mann when he postulated that if the strong interaction is "switched off," the electromagnetic interaction of the particles will be completely described by their charges (i.e. principle of minimal electromagnetic interaction).

In 1962 the Sakata model was "falsified" for a short time by experiments,[48,49] which discovered decays $\Sigma^+ \to n\mu^+\nu$ and $K^0 \to e^+\nu\pi^-$ forbidden by $\Delta S = \Delta Q$ rule. At the 1962 Geneva conference I tried to find a mistake in the results[48,49] but failed. Pomeranchuk who witnessed the argument commented later that my "feathers were flying." I do not remember now how the mistake was found subsequently by experimentalists. Maybe it was a statistical fluctuation.

The authors of articles[48,49] referred to the paper by Feynman and Gell-Mann.[13] While in my papers[9,10] the forbidden decays were simply listed in Ref. 13 the notations ΔQ and ΔS were used and the currents with $\Delta Q = \Delta S$ and $\Delta Q = -\Delta S$ ($\bar{p}\Lambda$ and $\bar{n}\Sigma^+$-currents) were phenomenologically considered on the same footing. The product of these currents gives transitions with $\Delta S = 2$. The limit on these transitions from the absence of decays $\Xi^- \to n\pi^-$ was not reliable because "so few Ξ particles have been seen that this is not really conclusive."[13] (The paper[20] (published in June 1957) had put a much better limit on $\Delta S = 2$ processes from $K^0 \leftrightarrow \bar{K}^0$ transitions. But it was not known to Feynman and Gell-Mann when they wrote.[13])

In 1962 M. Gell-Mann predicted the existence of Ω-hyperon.[50] I. Yu. Kobzarev and myself[51] derived the SU(3) relations between semileptonic decays of π and K-mesons. Together with relations for the decays of baryons they were later derived by N. Cabibbo.[52]

1962. From 3 to 4 Sakatons

The discovery of ν_μ prompted attempts to reconcile the existence of two neutrinos with the lepton–sakaton symmetry. In order to preserve the Kiev symmetry Z. Maki, M. Makagawa, S. Sakata[53] modified the B^+ matter model.[40] They assumed that $p = \nu_1$, B^+, where ν_1 is one of the two orthogonal superpositions of ν_e and ν_μ. The other superposition ν_2 was assumed either not to form at all a bound state with B^+ or to form a baryon with a very large mass. On the basis of this model the paper introduced $\nu_e - \nu_\mu$ oscillations. Another way to lepton–sakaton symmetry was suggested in the paper by Y. Katayama, K. Motumoto, S. Tanaka, E. Yamada,[54] where the fourth sakaton was explicitly introduced.

1963. My First Book

In 1963, my stenographed lectures[42,43] were published as a book "Weak Interaction of Elementary Particles"[55] in Russian, consisting of 19 chapters.

1964. Quarks

In 1964 η'-meson and Ω-hyperon were discovered.[56,57] Earlier that year G. Zweig[58] and M. Gell-Mann[59] replaced the integer charged sakatons by fractionally charged particles (aces — Zweig; quarks — Gell-Mann). This allowed them to construct not only the octet and singlet of mesons, but also the octet and decuplet of baryons. When establishing the electromagnetic and weak currents in the quark model M. Gell-Mann[59] referred to similar expressions in the Sakata model.

1965. Moscow

In 1965, my book was translated into English.[60] For the English edition, I have added Chapter 20 "Weak Interaction and Unitary Symmetry," where I compared the model of three-integer charged sakatons with the model of three fractionally charged fundamental particles.

The joint paper with Ya. B. Zeldovich and S. B. Pikelner[61] presented one of the many attempts organized by Zeldovich to consider fractionally charged quarks as "realistic," ordinary particles and to calculate their abundance in various media. The abundance turned out to be too large.

1967. 14th Conference on Physics at the University of Brussels

I was invited to this Solvay meeting, but made no comments at it. I had a long and heated discussion with E. P. Wigner during which he asked me to speak only while I was holding his pen.

At the conclusion of this meeting, "Some concluding remarks and reminiscences" were presented by Leon Rosenfeld (who, by the way, in 1948 introduced the term "leptons"). He briefly recalled the previous Solvay meetings and spoke about Pauli and Heisenberg, about Einstein and Bohr, about Serber and Teller, and about Chew and Gell-Mann.

During my stay at Brussels, I worked with Carlo Rubbia on our joint contribution on CP-violation to the Heidelberg Conference.[62]

1968. Moscow

I organized in Moscow an International Seminar on the Problems of Violation of CP Invariance. The program of the seminar included the talks by C. Rubbia, L. Wolfenstein, G. A. Leksin, V. V. Anisovich, B. Aubert, M. Veltman, B. A. Arbuzov, A. T. Filipov, G. Finocchiaro, C. Baglin, B. G. Erozolimskii, F. L. Shapiro, I. S. Shapiro, P. Miller, I. I. Gurevich, B. A. Nikolskii,

N. A. Burgov, L. I. Lapidus, V. Ya. Fainberg, S. M. Bilenkii, R. M. Ryndin, and myself. The talks were published in 1968 in *Uspekhi Fiz Nauk* and in *Yadernaya Fizika.*

Also in that year the Russian edition of the Volume IV of L. D. Landau and E. M. Lifshitz, "Theoretical Physics" appeared in two parts.[63,64] The Preface to Part 2 ended with the statement: "In the exposition of the weak interaction we benefited from the well known book by L. B. Okun."

1980. Moscow

Mr. Martin B. Gordon, who visited Moscow two months before the 20th International Conference on High Energy Physics (Madison, USA, July 1980) suggested that an enlarged text of the summary talk that I was preparing for the conference to be included in a series of books he was publishing. I was grateful for his suggestion, at that time I did not even suspect how much time this work would take.

I addressed the book to those who work at different sites of construction of the Babel tower of high energy physics and do not always understand the language spoken at the adjacent sites. As a result, the book contains two "strata": a science-popularizing layer and a professional layer.

What follows are some excerpts from the book.

1985. "Particle Physics: The Quest for the Substance of Substance"

Hadrons and quarks

In contrast to leptons, hadrons can only be called elementary particles with certain reservations. All the numerous hadrons are indeed elementary, in the sense that none of them can be broken into constituents. At the same time, it has been reliably established that hadrons have an internal structure: they consist of quarks. Like leptons, quarks appear at our present level of knowledge as structureless, truly elementary particles. For this reason, leptons and quarks are often called "fundamental particles."

The paradoxical properties of quarks have no precedent in the history of physics, itself so rich in paradoxes. Experimentalists, using beams of high-energy particles, definitely observe these constituents inside hadrons and measure their spins, masses, and electric charges. But no one has succeeded — nor ever will succeed, if today's theoretical concepts are correct — in knocking a quark out of a hadron. Quarks in hadrons are imprisoned for life. Physicists use a milder term for this imprisonment — confinement.

Theoretical concepts of the mechanism of confinement will be discussed later. Now we shall have a better look at different sorts of quarks.

It is convenient to begin our discussion of the properties of quarks with a nonrelativistic quark model that operates with the so-called constituent quarks. Hadrons are built out of the constituent quarks as if of building blocks. A constituent quark of a given type is a complex object with the same electric charge and the same spin as the "bare" quark present in the Lagrangian (these Lagrangian quarks are usually referred to as current quarks). The complex structure of the constituent quark appears because the current quark is surrounded by a cloud of virtual particles generated by strong interactions. As a result, the mass of a constituent quark exceeds that of the corresponding current quark by about 300 MeV. Hereafter, when we mention the masses of quarks, we shall always mean the masses of current quarks.

Protons and neutron are built up of the lightest quarks, u (up) and d (down). As with all other quarks, their spin is $1/2$. The charge of the u quark is $+2/3$, and that of the d quark is $-1/3$. The mass of the u quark is approximately 5 MeV and that of the d quark is 7 MeV. The proton is built of two u quarks and one d quark: $\mathrm{p} = \mathrm{uud}$. The neutron consists of two d quarks and one u quark: $\mathrm{n} = \mathrm{ddu}$.

According to the nonrelativistic quark model, quarks have zero angular orbital momenta within nucleons. The total spin of two u quarks in the proton is unity. This unity, added geometrically to the spin of the d quark, gives the proton a spin of $1/2$. The neutron is constructed similarly, by interchanging u and d quarks.

A whole sequence of other hadrons can be constructed of quarks as of toy cubes. Thus, for example if the spins of three quarks are parallel, they form the quartet of spin $3/2$ Δ baryons:

$$\Delta^{++} = \mathrm{uuu}\,, \quad \Delta^{+} = \mathrm{uud}\,, \quad \Delta^{0} = \mathrm{udd}\,, \quad \Delta^{0} = \mathrm{ddd}\,.$$

The orbital angular momenta of quarks in Δ-baryons, according to the nonrelativistic quark model, are zero. The attentive reader has probably noticed that this structure of Δ-baryons and nucleons seems to contradict the Pauli principle: indeed, two or even three quarks of the same sort seem to occupy the same quantum state. In fact, the Pauli principle is not in danger. We shall see below that these quarks differ from each other in color.

The Δ baryons are the lightest of baryon resonances. They decay into nucleons and π mesons in about 10^{-23} s: $\Delta \to N\pi$. A large number of heavier baryon resonances, also composed of u and d quarks, are known. In these resonances, quarks are in states with orbital and/or radial excitations. In this respect, resonances resemble excited states of atoms.

Baryons thus consist of three quarks. Another type of hadron — mesons — consists of a quark and an antiquark. For example, the lightest of mesons — π mesons — have the following structure:

$$\pi^+ = \mathrm{u\bar{d}}\,, \quad \pi^0 = \frac{1}{\sqrt{2}}(\mathrm{u\bar{u}} - \mathrm{d\bar{d}})\,, \quad \pi^- = \mathrm{d\bar{u}}\,.$$

(the meaning of the minus sign in a quantum-mechanical superposition of states forming the π^0 meson will be explained later). The quark and antiquark in the π meson are in the state with zero orbital momentum and oppositely directed spins, so that the total spin of the π meson is zero.

If a quark and an antiquark with zero orbital momentum have parallel spins, then they form mesons with spin 1: ρ^+, ρ^0, ρ^-. These mesons are resonances and decay into two π mesons over a time of the order of 10^{-23} s: $\rho \to 2\pi$. The ρ mesons are the lightest among meson resonances. A large number of heavier resonances, in which the quark–antiquark pairs are in excited states, are known.

Decays of Δ and ρ resonances can be illustrated by the following quark diagrams. In Figs. 1 and 2, an arrow directed backward in time represents an antiquark. Do not overlook the difference between conventional Feynman diagrams and quark diagrams: The quarks that fly to infinity are not free but confined within hadrons. Furthermore, as a rule, strong interactions between quarks are not indicated on quark diagrams. Thus, the interaction that results in the creation of a quark–antiquark pair, which appears as a "hairpin" on quark diagrams, is not shown.

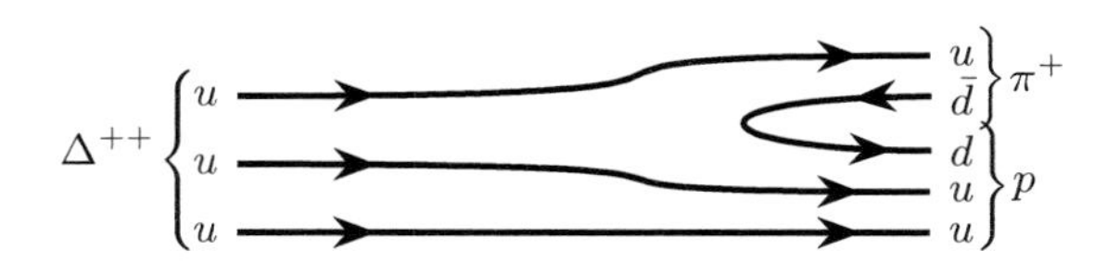

Fig. 1. Quark diagram for the $\Delta^{++} \to p\pi^+$ decay.

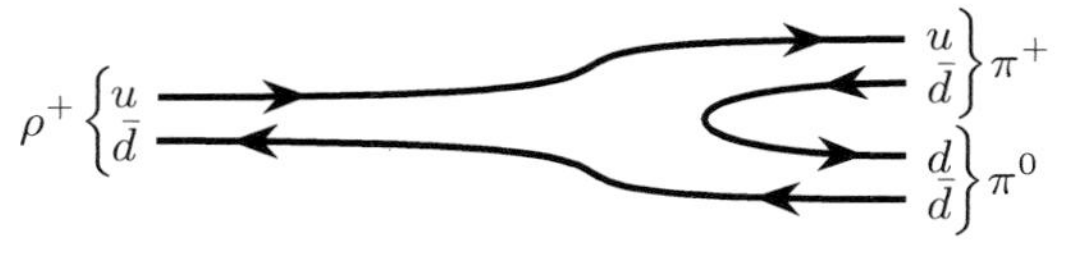

Fig. 2. One of the two quark diagrams for the $\rho^+ \to \pi^+\pi^o$ decay.

Isotopic Spin. SU(2) Group

The difference between the masses of the u and d quarks is much smaller than the mass of a hadron composed of these quarks. Therefore, it is reasonable to consider an approximation in which the masses of the u and d quarks are equal. In the theory of the strong interaction that we shall describe shortly, strong interactions of the u and d quarks are identical. If the mass difference of the u and d quarks and the difference between their electric charges are neglected, then the quark Lagrangian has an additional symmetry called the isotopic symmetry.

In the framework of the isotopic symmetry, the u and d quarks are treated as upper and lower states of a spinor in the so-called isotopic space. The u quark corresponds to the projection of isotopic spin equal to $+1/2$ and the d quark to the projection of spin equal to $-1/2$. (The projection is onto some axis in the isospin space; it is usually referred to as the z axis.) Transformations of the isotopic spinor under which the Lagrangian is invariant are realized by complex 2×2 matrices (reads "two by two") satisfying the conditions of unitarity ($U^+U = I$, where U^+ is the Hermitian conjugate matrix, and I is the 2×2 identity matrix) and unimodularity ($\det U = 1$). These 2×2 matrices give the simplest representation of the group SU(2) (reads "ess you two"). Here the letter S signifies that the transformations are special (unimodular in the present case) and the letter U indicates that they are unitary, while the numeral 2 denotes that the simplest representation of the group is formed by 2×2 matrices and that the simplest representation space is formed by a two-component spinor.

The group SU(2) — and the more complicated unitary unimodular groups SU(N), where $N > 2$ — play an important role in the physics of elementary particles. This calls for a more detailed discussion of the properties of 2×2 matrices U. Higher representations of SU(2) and representations of higher-order groups than SU(2) have much in common with these matrices.

In the general case, a 2×2 matrix U is determined by three real parameters α_k ($k = 1, 2, 3$) and can be written in the form

$$U = e^{i\alpha_k \tau_k/2} = 1 + i\alpha_k \frac{\tau_k}{2} + \frac{1}{2} i \left(\frac{\alpha_k \tau_k}{2} \right)^2 + \cdots ,$$

where summation is implied over the subscript k and τ_k are three Pauli matrices

$$\tau_1 = \begin{pmatrix} 0 & 1 \\ 1 & 0 \end{pmatrix}, \quad \tau_2 = \begin{pmatrix} 0 & -i \\ i & 0 \end{pmatrix}, \quad \text{and} \quad \tau_3 = \begin{pmatrix} 1 & 0 \\ 0 & -1 \end{pmatrix}.$$

When acting on a spinor, the matrix $\tau_+ = \frac{1}{2}(\tau_1 + i\tau_2)$ raises its lower component upward, while the matrix $\tau_- = \frac{1}{2}(\tau_1 - i\tau_2)$ lowers the upper component downward and the matrix $\frac{1}{2}\tau_3$ gives the eigenvalues of the projection of the isotopic spin on the z axis in the isotopic space. Pauli matrices do not commute:

$$[\tau_i, \tau_k] = \tau_i\tau_k - \tau_k\tau_i = i2\epsilon_{ikl}\tau_l \quad (i, k, l = 1, 2, 3)\,,$$

where ϵ_{ikl} is a completely antisymmetric tensor: $\epsilon_{123} = \epsilon_{231} = \epsilon_{312} = 1$ and $\epsilon_{213} = \epsilon_{132} = \epsilon_{321} = -1$.

Components of the tensor ϵ_{ikl} with at least two identical subscripts equal zero. A group whose consecutive transformations do not commute with each other are called non-Abelian. The group SU(2) is one of the simplest non-Abelian groups.

SU(2) can be used to illustrate another important concept. If the parameters of group transformations (in this case, α_1, α_2, α_3) do not depend on space–time coordinates, the symmetry is called global. But if they are functions of space–time coordinates, the symmetry is called local. In the second half of this chapter, we shall see that the isotopic symmetry caused by the resemblance of the properties of the u and d quarks is global; we shall consider a very interesting example of local symmetry involving the concept of color. The foregoing mathematical definitions are presented for future use. They will help the reader understand the more complicated physical symmetries which will come up later both in this book and in other books on elementary particles. As for "constructing" baryons of three quarks or mesons out of a quark and an antiquark, elementary operations with such a "quark erector set" are intelligible to even the youngest schoolchildren. This is also true of a number of aspects of isotopic symmetry. For instance, for an arbitrary isotopic multiplet with isotopic spin I, the number of particles in the multiplet is given by the simple formula

$$n = 2I + 1\,,$$

which is readily derivable from the observations that the maximum isospin projection on the third axis is I, the minimal projection is $-I$, and the "step" ΔI is unity. Correspondingly, the isospin of nucleons is 1/2, that of π mesons is 1, and that of Δ isobars is 3/2. Note that this quark-based outline of isotopic symmetry is anti-historical. Historically, the concept of isotopic spin was introduced into physics by Heisenberg at the beginning of the 1930s, immediately after the discovery of the neutron, and was applied to nucleons and the nuclear forces between them. It was soon extended to the then-hypothetical π mesons, whose existence was predicted by Yukawa. Multiplets of real π mesons and Δ isobars were discovered approximately 20 years later. The quark hypothesis was suggested only in 1964. The path to this hypothesis lay through a study of the properties of the so-called strange particles and SU(3) symmetry.

Strange particles

The family of strange hadrons is more numerous than that of non-strange hadrons. In comparison to nucleons and π mesons, strange hadrons play a rather minor role in nuclear physics because strange hadrons are unstable (the most long-lived of them, the K_L^0 meson, lives for 5×10^{-8} s) and heavy, so that they are produced only at relatively high energies of colliding particles. The first strange particles were discovered in cosmic rays in the 1940s. In the 1950s they were already being mass-produced by specially designed accelerators. What seemed paradoxical or strange in their behavior (and gave rise to the name) was that these particles are born copiously (strongly) (at sufficiently high energies of colliding hadrons), but decay into non-strange hadrons slowly (weakly). (On the nuclear scale of time, 10^{-8} s is a very long time, the characteristic time of the strong interaction being 10^{-23} s. Thus, the K_L^0 meson lives for about 10^{16} "nuclear days — compare it to the age of the Earth: about 10^{12} terrestrial days.) The solution to this paradox is that strange particles are produced via the strong interaction and decay individually via the weak interaction. At the present time we know that this happens because each strange particle contains at least one strange quark: the s quark. Like the d quark, the strange quark's charge is 1/3. But is much heavier than the d quark; its mass is about 150 MeV. Decays of s quarks will be discussed in the section devoted to the weak interaction. Here we shall look into their strong interaction. Strong interaction produce quark–antiquark pairs: $\mathrm{s} + \bar{\mathrm{s}}$. Figure 3 shows a quark diagram of the process $\pi^- p \to K^0 \Lambda^0$. Notice that the creation of a pair of strange particles results in a "strange hairpin" ($\mathrm{s}\bar{\mathrm{s}}$) on the quark diagram. In this particular case, one end of the hairpin ($\bar{\mathrm{s}}$) belongs to a K meson and the other (s) to the Λ hyperon.

SU(3) symmetry

The K meson is the lightest of the strange mesons and the Λ hyperon is the lightest of the strange baryons (strange baryons were given the name hyperons). Strange and non-strange hadrons together form common families:

Fig. 3. Quark diagram for the $\pi^- p \to K^0 \Lambda^0$ reaction.

meson octets and singlets and baryon octets and decuplets (a singlet contains 1, an octet 8, and a decuplet 10 particles). The structure of these families is easily understood in terms of the SU(3) symmetry. In terms of quarks, the SU(3) symmetry is reduced to a symmetry (degeneracy) among u, d, and s quarks. This SU(3) symmetry is a generalization of the isotopic SU(2) symmetry. The SU(3) symmetry is much more strongly broken in nature than SU(2) because the s quark is much heavier than non-strange quarks:

$$m_{\mathrm{s}} - m_{\mathrm{u}} \approx m_{\mathrm{s}} - m_{\mathrm{d}} \gg m_{\mathrm{d}}, m_{\mathrm{u}} \,.$$

As a consequence, there are large mass splittings among hadrons within each SU(3) multiplet. It was not easy to deduce the existence of the SU(3) symmetry by studying hadrons. The decisive contribution to the understanding of symmetry properties of hadrons was made by Gell-Mann. At the beginning of the 1950s, he extended the notion of isotopic spin to strange particles. At the beginning of the 1960s, he gave the present formulation of the SU(3) symmetry of mesons and baryons. And finally, in 1964, he advanced the quark hypothesis (three independent contributions by Nishijima, Ne'eman, and Zweig should be mentioned in connection with the isotopic spin, SU(3), and quarks, respectively). SU(3) multiplets are conveniently plotted on the plane I_3Y where I_3 is the third projection of isotopic spin, and Y is the hypercharge (by definition, hypercharge equals twice the mean charge of an isotopic multiplet). Figure 4 shows the octet of pseudoscalar mesons ($J^P = 0^-$, where J is the spin of the particles and P is their parity; parity will be discussed in detail in the chapter devoted to the weak interaction). Figure 5 shows the octet of vector mesons ($J^P = 1^-$). The quark structure of these SU(3) multiplets is given in Fig. 6. The structure of the particles at the vertices of the hexagon is obvious, but the combinations at its center need clarification.

Nine different combinations can be constructed out of three quarks and three antiquarks. Three of them are truly neutral: $\mathrm{u\bar{u}d\bar{d}}$, and $\mathrm{s\bar{s}}$. As a result of strong interactions, these three quark–antiquark states can transform into each other so that definite mass values exist for three quantum-mechanical superpositions of these states. Were the SU(3) symmetry strict, an SU(3)-

invariant superposition,

$$\frac{1}{\sqrt{3}}(\mathrm{u\bar{u}} + \mathrm{d\bar{d}} + \mathrm{s\bar{s}})$$

would split off. In the case of pseudoscalar mesons, this would correspond to the SU(3) singlet η' meson; for vector mesons, it would correspond to the SU(3) singlet ω meson. Among the two remaining superpositions, one has an isotopic spin equal to 1 (this is π^0 for pseudoscalars and ρ^0 for vectors),

$$\frac{1}{\sqrt{2}}(\mathrm{u\bar{u}} - \mathrm{d\bar{d}})$$

(it is constructed of quark wave functions by means of the matrix τ_3), and the other superposition has zero isospin:

$$\frac{1}{\sqrt{6}}(\mathrm{u\bar{u}} + \mathrm{d\bar{d}} - 2\mathrm{s\bar{s}})\,.$$

Its form is determined by requiring orthogonality to the first two superpositions. It corresponds to the η meson for pseudoscalars and the ϕ meson for vectors (note that the coefficients with all three superpositions stem from the normalization of the quantum-mechanical state to unity).

Like the three Pauli matrices τ in the case of SU(2) symmetry, the matrices important in SU(3) are the eight Gell-Mann matrices λ:

$$\lambda_1 = \begin{pmatrix} 0 & 1 & 0 \\ 1 & 0 & 0 \\ 0 & 0 & 0 \end{pmatrix}, \quad \lambda_2 = \begin{pmatrix} 0 & -i & 0 \\ i & 0 & 0 \\ 0 & 0 & 0 \end{pmatrix}, \quad \lambda_3 = \begin{pmatrix} 1 & 0 & 0 \\ 0 & -1 & 0 \\ 0 & 0 & 0 \end{pmatrix}, \quad \lambda_4 = \begin{pmatrix} 0 & 0 & 1 \\ 0 & 0 & 0 \\ 1 & 0 & 0 \end{pmatrix},$$

$$\lambda_5 = \begin{pmatrix} 0 & 0 & -i \\ 0 & 0 & 0 \\ i & 0 & 0 \end{pmatrix}, \quad \lambda_6 = \begin{pmatrix} 0 & 0 & 0 \\ 0 & 0 & 1 \\ 0 & 1 & 0 \end{pmatrix}, \quad \lambda_7 = \begin{pmatrix} 0 & 0 & 0 \\ 0 & 0 & -i \\ 0 & i & 0 \end{pmatrix}, \quad \lambda_8 = \frac{1}{\sqrt{3}}\begin{pmatrix} 1 & 0 & 0 \\ 0 & 1 & 0 \\ 0 & 0 & -2 \end{pmatrix}.$$

One readily notices a relationship between the quark structure of the η meson and the matrix λ_8.

Since the SU(3) symmetry is broken in nature, SU(3) singlet mesons and the eighth components of SU(3) octets are partly mixed. This phenomenon is called mixing. Mixing is stronger for vector mesons than for pseudoscalar mesons. The physical states produced by this mixing are

$$\omega \approx \frac{1}{\sqrt{2}}(\mathrm{u\bar{u}} + \mathrm{d\bar{d}})\,, \quad m = 783\ \mathrm{MeV}\,,$$

$$\phi \approx \mathrm{s\bar{s}}\,, \quad m = 1020\ \mathrm{MeV}\,.$$

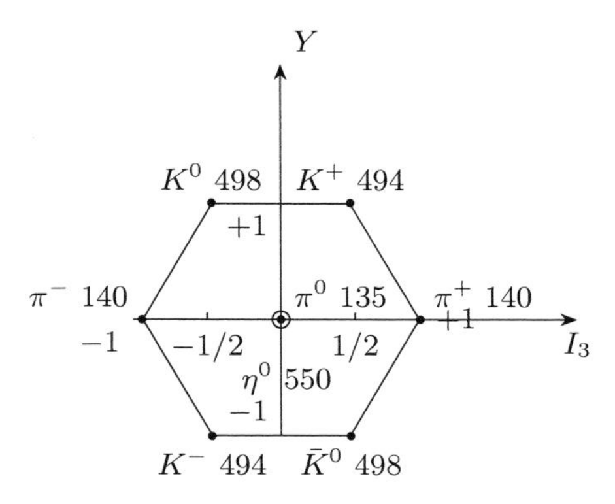

Fig. 4. I_3Y-diagram for the octet of the lightest pseudoscalar mesons.

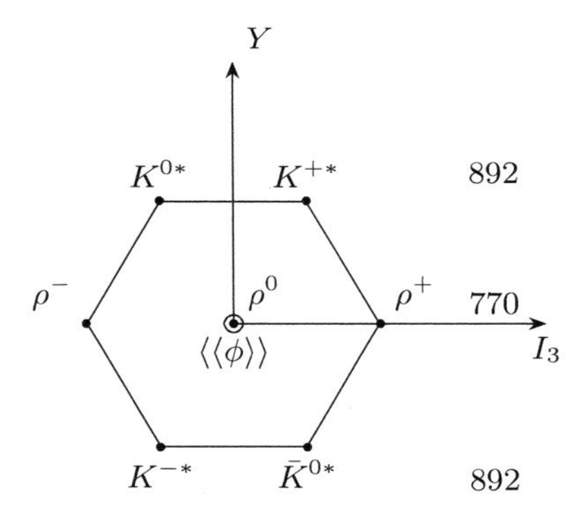

Fig. 5. I_3Y-diagram for the octet of the lightest vector mesons.

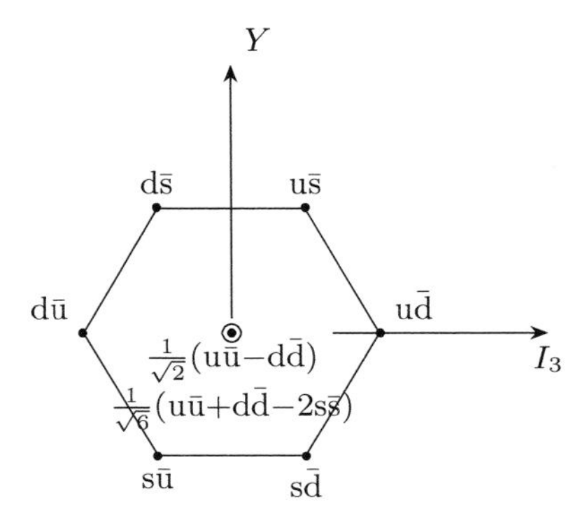

Fig. 6. Quark structure of an octet of mesons.

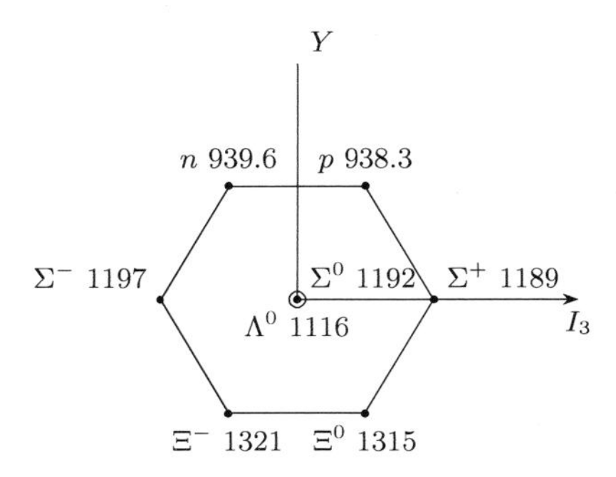

Fig. 7. I_3Y-diagram for the octet of the lightest baryon with $J^P = 1/2^+$.

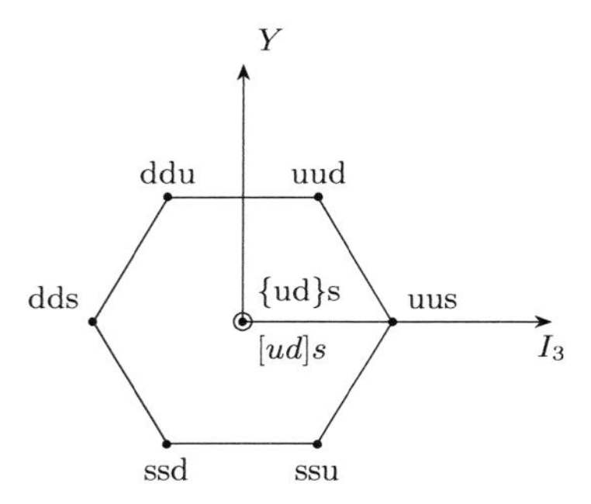

Fig. 8. Quark structure of an octet of baryons.

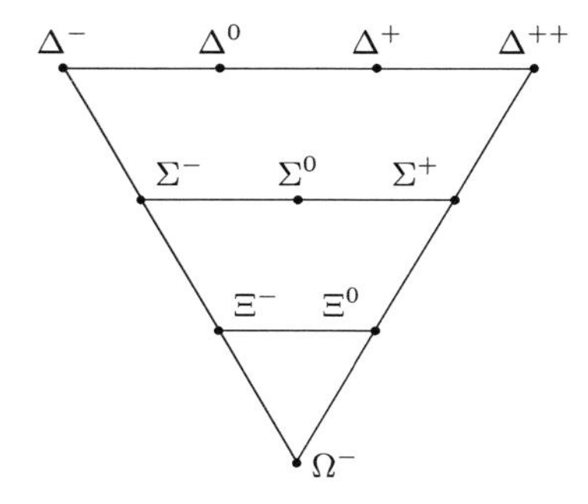

Fig. 9. Decuplet of the lightest baryons with $J^P = 3/2^+$.

Figure 7 shows the baryon octet with $J^P = 1/2^+$. The quark structure is given in simplified form in Fig. 8. The combination {ud}s at the center of Fig. 8, symmetric with respect to the substitution u $\leftrightarrow$ d, has isospin 1 and describes the Σ^0 hyperon; the combination [ud]s, antisymmetric with respect to u $\leftrightarrow$ d, has $I = 0$ and describes the isoscalar Λ^0 hyperon. Figures 9 and 10 show the baryon decuplet with $J^P = 3/2^+$ and its quark structure. A

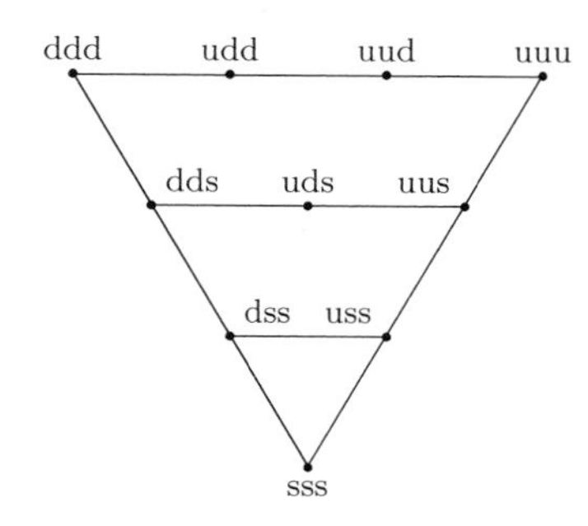

Fig. 10. Quark structure of a decuplet of baryons.

number of SU(3) multiplets with other values of spin and parity are known. But the decisive role in establishing the SU(3) symmetry and the quark structure of hadrons was played by the octets shown in Figs. 4 and 7 and the decuplet shown in Fig. 9.

The charmed quark

The quark theory of hadrons was conclusively confirmed by the discovery of charm made by Richter and Ting and their collaborators.

Charmed particles comprise quarks of the fourth sort, the so-called charmed quarks denoted by the letter c. First came the discovery in the autumn of 1974 of the J/ψ meson, a vector particle with "hidden charm" consisting of a pair $c\bar{c}$ in the $^3S_1$ state. A number of other levels of the $c\bar{c}$ system (dubbed charmonium) were soon discovered. The diagram of the charmonium levels currently known is shown in Fig. 11. The masses of the levels are given in GeV (the diagram is purely schematic, with no mass scale). Primed particles represent radial excitations of lower states. S-states correspond to zero orbital momentum of $c\bar{c}$; P-states to an orbital momentum equal to unity. The right-hand lower suffix indicates the spin of the meson and the upper left-hand suffix indicates the total spin of the quark and antiquark.

Some particles with explicit charm were also found: the mesons $D^0[c\bar{u}](1.863)$, $D^+[c\bar{d}](1.868)$, and $F^+[c\bar{s}](2.02)$ and the baryon $\Lambda_c^+[cdu](2.72)$ (the symbols in brackets are the quark compositions of the particles; the numbers in parentheses are the masses of the particles expressed in GeV).

The study of the properties of these particles made it possible to determine not only the charge but also the mass of the c quark. The charge of the c quark is $+2/3$ and its mass is approximately 1.4 GeV. Therefore, the c quark is a very heavy analogue of the u quark.

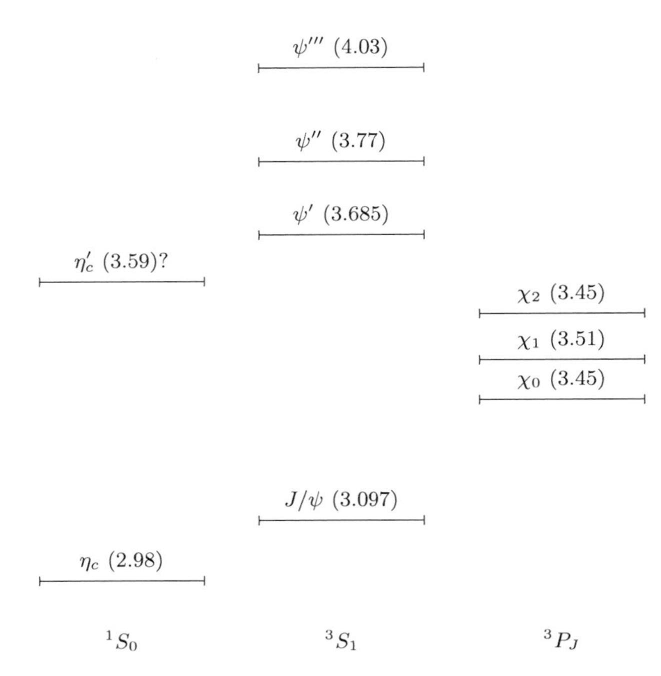

Fig. 11. Spectroscopy of charmonium states. The masses of the states (energy levels) are given in GeV in parentheses; vertical spacings between levels in the figure are arbitrary.

The b quark and others

In 1976, Lederman and his collaborators at FNAL discovered a new particle: the Υ meson (read "upsilon"), consisting of quarks of the fifth sort, the so-called b quarks. The charge of the b quark was found to be $-1/3$. The b quark is a heavy analogue of the "lower" d and s quarks; hence, the term "bottom" and the symbol b. (Some physicists prefer to derive the letter b from beauty.) The b quark is much heavier than the c quark: its mass is approximately 4.8 GeV. The Υ meson, whose mass is 9.46 GeV, is the lowest $^3S_1$ state of the hidden-beauty pair $b\bar{b}$. Three radially excited $^3S_1$ levels of this system, sometimes called upsilonium (and sometimes bottomonium), have been found: Υ' (10.02), Υ'' (10.40), and Υ''' (10.55).[b] Mesons with a single b quark have also been found: $B^+ = \bar{b}u$, $B^0 = \bar{b}d$, $B^- = b\bar{u}$, $\bar{B}^0 = b\bar{d}$. The masses of these naked-beauty mesons (~ 5.274 GeV) are such that

$$m_{\Upsilon''} < 2m_B < m_{\Upsilon'''} .$$

The existence has thus been established of three quarks of the "lower type": d, s, b; and two quarks of the "upper type": u, c. It is expected, owing

[b] Author's note (autumn 1983). P-levels of upsilonium have recently been discovered.

to quite persuasive arguments (see below), that a third "upper" quark also exists. It is called the t quark (from "top"). The search for the t quark has so far been unsuccessful. The search for "toponium," the $\bar{t}t$ pair, in colliding electron–positron beams at PETRA at an energy of 18 GeV in each beam has been especially thorough. It has been concluded from these experiments that if the t quark exists, its mass must be greater than 18 GeV. At present, there are no serious arguments in favor of quarks still heavier than t.

Flavors and generations

Quarks of different types are often said to differ by their flavors. These quark flavors have nothing in common with the ordinary notion of flavors. The word "flavor" is used here as a synonym of the words "type" or "sort," adorning by its unexpected appearance "dry" texts in physics. The term "flavor" is also convenient because it is semantically in contrast to the term "color" that we shall start discussing in the next section.

There seems to exist some profound symmetry between leptons and quarks of different flavors. The following table points to the existence of such a symmetry:

$$\begin{array}{ccc} \nu_e & \nu_\mu & \nu_\tau \\ e & \mu & \tau \\ u & c & t(?) \\ d & s & b \end{array}$$

It was the quark–lepton symmetry that allowed the existence of the c quark to be predicted as early as 1964 (four leptons and only three quarks were known at that time). After the τ lepton was discovered in 1975, the same symmetry was used to predict the existence of b and t quarks.

We shall soon see that the lepton–quark symmetry is especially well pronounced in weak interactions. Of course, this symmetry is not perfect: although the differences between the charges of the neutrino and charged leptons equal those between the charges of the upper and lower quarks, the charges themselves are different for leptons and quarks.

The twelve leptons and quarks are naturally classified into three groups, or three generations of fundamental fermions. Each generation consists of four particles forming a column in the table: an "upper" and a "lower" lepton and "upper" and "lower" quark. The first generation is formed of the lightest particles. In each subsequent generation charged particles are heavier than in the preceding one.

Together with photons, fermions of the first generation form the matter of which the universe is built at present. Atomic nuclei consist of u and d quarks, atomic shells consist of electrons, and nuclear fusion reactions inside the sun and stars would be impossible without the emission of electron neutrinos. As for the fermions of the second and third generations, their role in the world around us appears to be negligible. At first glance, the world would not seem to be any worse off if these particles never existed. These particles resemble draft versions that the Creator has thrown out as unsuccessful, but that we, using sophisticated instruments, have retrieved from his waste basket.

Now we have begun to understand that these particles were very important at the first instants of the Big Bang. Thus, the number of sorts (flavors) of neutrinos determined the ratio between the hydrogen and helium abundances in the universe. Cosmological calculations of helium abundance indicate that the number of neutrino flavors does not exceed four. In the framework of the scheme of lepton–quark generations, this means that the total number of quark flavors is not greater than eight.

The importance of the subsequent generations seems to lie in that it is because of them that the particles of the first generation have precisely the values of mass that we observe. The relation between the masses of the u and d quarks and the electron is essential for our very existence. Indeed, the difference between the masses of the neutron and proton is primarily due to the difference between the masses of the u and d quarks. And if the inequality $m_{\rm p} - m_{\rm n} + m_{\rm e} > 0$ were correct, hydrogen would be unstable.

We thus begin to guess that higher order generations are not as insignificant as we thought. To find their profound significance, and the nature of the quark–lepton symmetry itself, is one of the most important problems in physics. These remark conclude the discussion of quark flavors and we shall now begin a new topic: quark colors.

Color and gluons

So far we have carefully avoided the question of how the forces between quarks are realized. It is time to answer at this juncture the following questions: (i) What charges are the sources of these forces? and (ii) What particles mediate these forces? Short answers to these two questions are: (i) color charges and (ii) gluons.

It has been established that quarks of each flavor exist in three strictly degenerate species. These species are said to differ only in color. The following three colors are normally used: yellow, blue, and red. These quark colors obviously have no bearing on the usual optical colors. "Color" for quarks is simply a convenient term to denote a quantum number that characterizes quarks. Color charge of antiquarks are different from those of quarks. Sometimes they are referred to as antiyellow antiblue, and antired, and sometimes as violet, orange, and green, in correspondence with the well-known sequence of complementary colors in the optical spectrum (recall the convenient mnemonic phrase "Richard Of York Gained Battles (In) Vain").

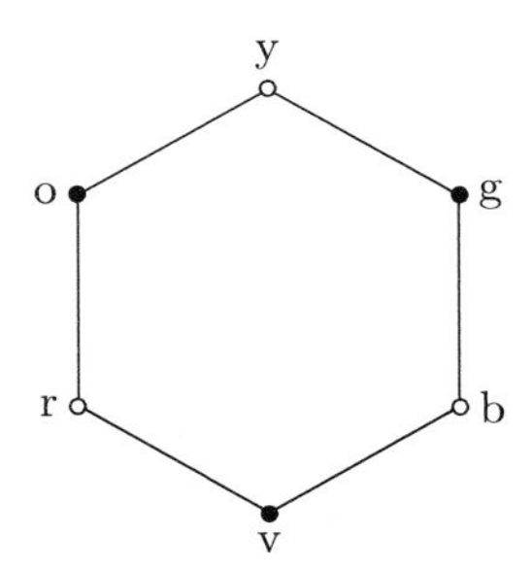

Fig. 12. Three colors of quarks (red, blue, yellow) and three colors of antiquarks (green, orange, violet).

With such a selection of quark colors it is natural to regard hadrons as colorless, or white, particles. Baryons are colorless because they consist of three quarks of three mutually complementary colors.[c] Mesons are colorless superpositions of quarks and antiquarks.

Mathematically, color degeneracy of quarks implies the color SU(3) symmetry: $\mathrm{SU}(3)_c$ (subscript c for color). A color triplet of quarks, q_α, $\alpha = 1, 2, 3$, for yellow, blue, and red forms the so-called fundamental representation of the group SU(3). A triplet of antiquarks, $\bar{\mathrm{q}}_\alpha$, forms a conjugate representation (this is an antitriplet). Mesons (M) and baryons (B) are $\mathrm{SU}(3)_c$ singlets:

$$\mathrm{M} = \frac{1}{\sqrt{3}}\bar{\mathrm{q}}_\alpha \mathrm{q}^\alpha = \frac{1}{\sqrt{3}}(\bar{\mathrm{q}}_1\mathrm{q}^1 + \bar{\mathrm{q}}_2\mathrm{q}^2 + \bar{\mathrm{q}}_3\mathrm{q}^3)\,,$$

$$\mathrm{B} = \frac{1}{\sqrt{6}}\mathrm{q}^\alpha \mathrm{q}^\beta \mathrm{q}^\gamma \epsilon_{\alpha\beta\gamma}\,,$$

where $\epsilon_{\alpha\beta\gamma}$ is a completely antisymmetric tensor that we have encountered already when discussing the properties of the Pauli matrices. It is because of the antisymmetrization with respect to color that three quarks in a baryon do not violate the Pauli principle and thus behave as ordinary fermions.

The color charges of quarks play the same part in the strong interaction that the electric charges of particles play in the electromagnetic interaction;

[c]Gell-Mann has chosen the colors of quarks as colors of the national flag of USA: Red, White and Blue. However such choice does not allow one to call the usual colorless baryons as white, so I proposed the above choice.

and the photon's part is played by electrically neutral vector particles christened gluons. Quarks exchanging gluons are thus "glued together" to form hadrons.

The main difference between gluons and photons is that there is only one photon and it is electrically neutral, but there are 8 gluons and they all carry color charges. Because of their color charges, gluons strongly interact with one another and emit one another. As a result of this nonlinear interaction, the propagation of gluons through vacuum is quite unlike the propagation of photons, and color forces do not resemble electromagnetic forces.

When a beam of light traverses a dark room it is visible to a side observer only because light is scattered by suspended dust. Otherwise it would be invisible since photons are neutral and therefore do not emit photons. Gluons do emit gluons, and one would expect that a beam of gluons, in a "dark" room, would behave as a "shining light." As we shall shortly see, this is not so because of a phenomenon called confinement.

Quantum chromodynamics (QCD)

The theory of the interaction between quarks and gluons is called quantum chromodynamics (from the Greek word $\chi\rho\omega\mu\alpha$, color). QCD is based on the postulate that the SU(3) color symmetry is a local, i.e. gauge symmetry. The requirement of local invariance (gauge invariance) leads to the inevitable conclusion of the existence of the octet of gluon fields with their specific self-coupling. Thus, the symmetry determines the entire dynamics of strong interactions. In this respect, the local color symmetry $\mathrm{SU}(3)_c$ is much deeper than the global flavor symmetry, $\mathrm{SU}(3)_f$ (subscript f for flavor), which appears because the masses of the u, d, and s quarks are approximately degenerate.

The Lagrangian of QCD is very much like the QED Lagrangian (see Chapter II). The difference lies in that (i) the electromagnetic coupling constant, that is, electric charge e, is replaced by the strong coupling constant g, (ii) quark spinors, in contrast to the electron spinor, have color suffixes over which summation is carried out, and (iii) the gluon vector potential A_μ in the Lagrangian, in contrast to the photon Lagrangian, is a matrix in color space:

$$A_\mu = A^i_\mu \lambda_i/2\,, \quad i = 1, 2, \ldots, 8\,.$$

Here A^i_μ are vector potentials of eight gluon fields and λ_i are eight Gell-Mann matrices. Note that, in QCD, the covariant derivative takes the following (matrix) form:

$$D_\mu = \partial_\mu - igA_\mu .$$

The matrix of the gluon field strength is

$$F_{\mu\nu} = F^i_{\mu\nu}\lambda_i/2 .$$

In the case of gluons, the field strength $F_{\mu\nu}$ is expressed through A_μ by a more complex formula than in the case of photons:

$$F_{\mu\nu} = \partial_\mu A_\nu - \partial_\nu A_\mu - ig[A_\mu A_\nu - A_\nu A_\mu] .$$

Here g is the strong interaction constant. For photons, A_μ is a number, not a matrix, and the commutator in the expression for $F_{\mu\nu}$ vanishes. But, for non-Abelian gauge fields, such as gluons, this commutator is not equal to zero. Consequently, it determines the character of the nonlinear self-interaction of gluons peculiar to gluon forces.

This form of the QCD Lagrangian, and of the tensor $F_{\mu\nu}$ is dictated by the requirement that the Lagrangian be invariant under gauge transformation:

$$\mathrm{q} \to S\mathrm{q} , \quad \bar{\mathrm{q}} \to \bar{\mathrm{q}}S^+ , \quad A_\mu \to SA_\mu S^+ - \frac{i}{g}(\partial_\mu S)S^+ ,$$

where $S = \exp[-i\alpha_i(x)\lambda_i/2]$ and α_i are eight parameters that depend on the coordinates of the world point x.

Asymptotic freedom and confinement

If we take into account the contribution of the nonlinear coupling of gluons to the polarization of the gluon vacuum by a gluon (see Fig. 13), we find that, in contrast to the polarization of the quark vacuum by a gluon (Fig. 14), this

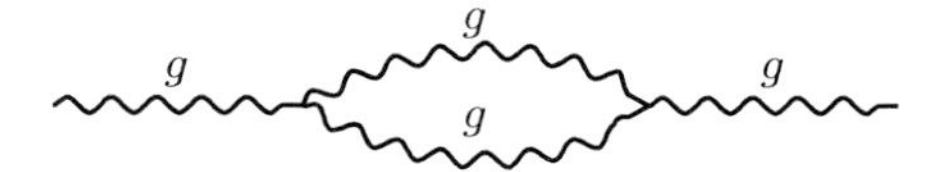

Fig. 13. Gluon loop: the contribution to the polarization of vacuum by a gluon.

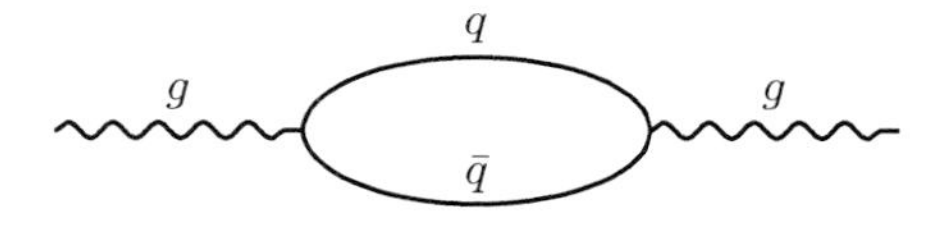

Fig. 14. Quark loop: the contribution to the polarization of vacuum by a gluon.

polarization results not in a screening but in an antiscreening of the color charge. The color charge of a quark diminishes as we go deeper into the gluon cloud surrounding the quark. This means that, in the limit of infinitely small distances between quarks, the color interaction between them "switches off." This phenomenon was given the name "asymptotic freedom." Quarks are nearly free at short distances: the color potential between them is a Coulomb-type potential, α_s/r, where the running constant $\alpha_s(r) = g^2(r)/4\pi$ decreases logarithmically as r diminishes or as the momentum transfer q increases. At sufficiently large q,

$$\alpha_s(q) \approx \frac{2\pi}{b\ln(q/\Lambda)}\,.$$

Here the dimensionless coefficient b is found theoretically by calculating the diagrams in Figs. 13 and 14; $b = 11 - (2/3)n_f$, where n_f is the number of quark flavors ($b = 7$, if $n_f = 6$).

As for the constant Λ, which has the dimension of momentum, its value is extracted from experimental data (using the widths and masses of energy levels of heavy quarkonia, the properties of hadronic jets produced in e^+e^- annihilations at high energies, and the properties of cross section of deep inelastic scattering) and is found to be of the order of 0.1 GeV. The constant Λ (sometimes denoted $\Lambda_{\rm QCD}$) plays a fundamental role in QCD.

The reverse side of asymptotic freedom is the growth of the color charges as the distance between quarks increases. At distances $r \sim 1/\Lambda \sim 10^{-13}$ cm, color interaction becomes truly strong. In this region, perturbation theory fails and no reliable calculations can be made. Nevertheless, qualitative arguments show that the strengthening of this interaction with distance can be expected to result in quark confinement, that is, the impossibility of getting single free quarks.

In order to clarify the anticipated picture of confinement, imagine first a world completely devoid of light quarks. Consider a heavy quark and a heavy antiquark ($m \gg \Lambda$). At short distances ($r \ll 1/\Lambda$), the color potential between the quarks resembles the Coulomb law ($\sim 1/r$), and the force decreases with distance as $\sim 1/r^2$. Lines of color force diverge from the charge isotropically, so that their flux across a unit surface area is inversely proportional to the whole surface area (Fig. 15). At large distances between quarks ($r \gg 1/\Lambda$), the strong nonlinear interaction between gluons makes the surrounding vacuum "squeeze" the lines of force into a tube with the radius $1/\Lambda$. This produces a "gluonguide" resembling a conventional light-guide. In this situation, the flux across unit surface area is constant, the force between quarks is independent of the distance between them, and the

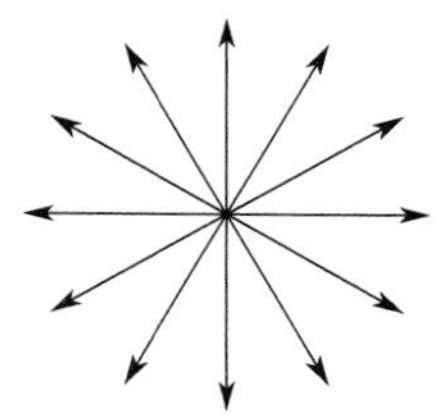

Fig. 15. Lines of force of the Coulomb field.

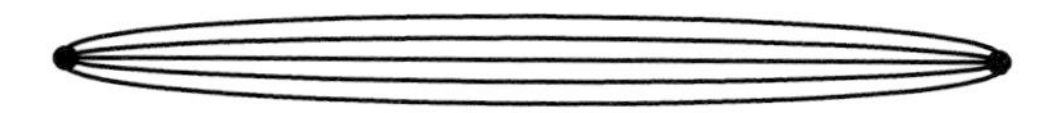

Fig. 16. Lines of force of the gluon field between a quark and an antiquark.

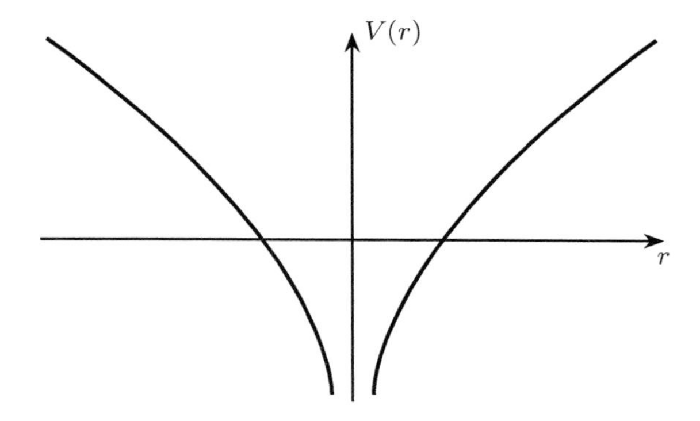

Fig. 17. Funnel-like potential between a quark and an antiquark.

potential is proportional to this distance. As a result, the color potential resembles a funnel (Fig. 17).

The levels of charmonium and upsilonium calculated on the basis of the funnel-type phenomenological potential are in good agreement with experimental data. However, it has not been possible thus far to construct an analytic theory of the gluon string based on a solution of the equations obtained from the QCD Lagrangian. But a large number of numerical computations on powerful computers, with QCD equations simplified by converting the space–time continuum to four-dimensional lattices with a finite number of cells (up to 10^4), point to the existence of such strings.

In principle, a gluon string between a heavy quark and a heavy antiquark could be infinitely extended. The energy spent on pulling the quarks apart

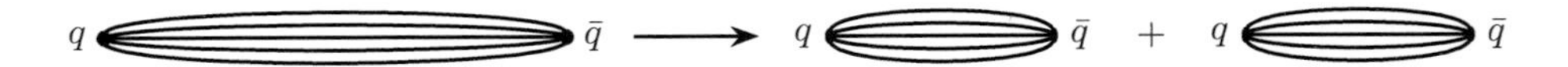

Fig. 18. Gluon strong broken because of the production of a quark–antiquark pair. As a result, one meson transforms into two.

would thereby be transformed into the mass of the string. However, in the real world where light quarks exist ($m \ll \Lambda$), this does not occur. The string breaks into segments about $1/\Lambda \sim 10^{-13}$ cm long, constituting new mesons. A light quark–antiquark pair appears at the breaking point (see Fig. 18).

An attempt to "break" a meson into a quark and an antiquark would be similar to trying to break a magnetic needle into the south and north poles. We would always get two dipoles.

Chiral symmetry

Since quarks are dressed with dense virtual gluon and quark–antiquark clouds, one cannot specify the mass of the quark without qualifying the distance at which this mass is measured. The shorter this distance, the smaller the mass is.

The masses given in the preceding sections refer to distances of the order of 10^{-14} cm, at which the clouds are rarefied owing to asymptotic freedom. It has already been mentioned that these "half-bare" quarks devoid of their heavy gluon coat are usually called current quarks; quarks completely enveloped by gluon clouds are called block or constituent quarks. Assuming that the mass of a nucleon is the sum of the masses of the three non-relativistic constituent quarks, we conclude that the masses of the gluon clouds around the u and d quarks are approximately 300 MeV.

It will be interesting to consider an imaginary world in which the current masses of light quarks are zero. It can be expected that this massless limiting case would fit the real world quite closely since the current masses of the u and d quarks are small: $m_u \approx 5$ MeV and $m_d \approx 7$ MeV. Indeed, a theoretical analysis shows that the masses of all baryons and nearly all mesons remain practically unaltered. The only exceptions are the lightest among mesons: π mesons whose squared masses are proportional to $(m_u + m_d)\Lambda_{\rm QCD}$. This special position of the π mesons stems from the key role they play in the spontaneous breaking of chiral symmetry. We shall explain now what stands behind the terms "chiral symmetry" and "spontaneous breaking" of a symmetry.

By analyzing the QCD Lagrangian with massless u and d quarks, it is easy to notice that it has not only the isotopic symmetry SU(2) but also a higher global symmetry $SU(2)_L \times SU(2)_R$. The point is that massless particles possess a specific conserved characteristic that cannot be defined in a Lorentz-invariant manner for particles with nonzero mass. This quantity is the so-called helicity, that is, the projection of the spin of a particle on its momentum. Helicity is said to be left-handed (L) if the directions of spin and momentum are opposite and right-handed (R) if they are identically oriented. Massless particles move at the velocity of light; thus, it is impossible to change the helicity of a massless particle by any motion of a reference frame, whose velocity is always less than c. However, this is easy for a particle with nonzero mass.

Emission and absorption of vector gluons by color charges of quarks do not change quark helicities. Hence, the QCD Lagrangian of massless quarks naturally decomposes into two symmetric terms, one of which contains left-handed quarks u_L, d_L and the other, right-handed quarks u_R, d_R. Each of these terms is isotopically symmetric, so that the total Lagrangian is invariant under the product $SU(2)_L \times SU(2)_R$.

At the level of the Lagrangian, both symmetries $SU(2)_L$ and $SU(2)_R$ are on the same footing as the ordinary isotopic symmetry SU(2). However, the difference becomes obvious when we look into how these symmetries are realized in the world of hadrons.

The ordinary isotopic symmetry is realized linearly: rotations of the isospinor of quarks and the isospinor of composite nucleons occur synchronously. This is not the case for chiral symmetry because, in contrast to massless quarks, nucleons are massive and have no definite helicity. Here, for the first time, we encounter a case in which the Lagrangian has a definite symmetry but the physical states have not. In such situations, we speak of a spontaneous breaking of symmetry. In this particular case, we face an example of spontaneous breaking of global chiral symmetry.

It was established that spontaneous symmetry breaking is always accompanied by the appearance of massless zero-spin bosons, the so-called Goldstone bosons. The three massless π mesons in our imaginary world in which the u and d quarks are massless would represent such Goldstone bosons. In the real world, where the masses of the u and d quarks are small but distinct from zero the chiral symmetry of the Lagrangian is approximate, the π mesons are called pseudo-Goldstone bosons: although their masses are not zero, they are small compared to the masses of other hadrons. In principle, in the chiral limit, all the masses of hadrons consisting of light quarks should be expressible through a single dimensional parameter Λ_{QCD} that enters the expression for the running constant α_s. This problem has not yet been solved.

QCD on the march

The advent of quantum chromodynamics drastically changed the situation in elementary particles physics. QCD explained the fundamentals of such already familiar symmetries as isotopic invariance SU(2) and its generalization, the flavor SU(3) symmetry of strong interactions, and the chiral symmetries $\mathrm{SU}(2)_L \times \mathrm{SU}(2)_R$ and $\mathrm{SU}(3)_L \times \mathrm{SU}(3)_R$. New light was shed on such phenomenological models as the model of nonrelativistic quarks, the bag model, and the parton model. A number of new physical objects and phenomena were predicted on the basis of quantum chromodynamics: quark and gluon jets and glueballs (hadrons consisting only of gluons, without quarks).

No competitors contest the role of quantum chromodynamics as the ultimate theory of strong interactions. The main mountain pass on the way to a total understanding of hadrons is behind us: the Lagrangian has been written. Nevertheless, we are still far from the final goal because we are unable to solve the QCD equations as soon as the color interaction becomes strong. The problem of confinement is a challenge to the theorists. The mathematical structure of the theory and, in particular, the properties of the chromodynamic vacuum, with its quark and gluon condensates and intricate topological relief whose simplest elements are the so-called instantons, remains largely unprobed.

Further experimental study of hadrons may prove invaluable to further progress of the theory. It is remarkable that not only experiments at the highest possible energies but also experiments at low energies will be valuable for the theory. The latter will make it possible to put in order the spectroscopy of hadrons, including the spectroscopy of exotic (not of the types qq and qqq) and cryptoexotic mesons and baryons, baryonium, two-baryon resonances, and glueballs. (We advise the reader frightened by the avalanche of new terms to look them up in the Glossary.)

When Yang and Mills published, in 1954, a paper that pioneered the analysis of the local non-Abelian theory (SU(2)), it was difficult to discern in it the prototype of the future theory of strong interactions. Indeed, the theory operated with massless gauge fields that seemed to lead inevitably to long-range forces that do not exist in nature. Many a theorist considered the Yang–Mills theory an exciting mathematical toy. A long development that culminated in quarks proved necessary before Nambu could introduce, in 1965, a hypothesis of gauge fields coupled to the degree of freedom that Gell-Mann later (at the beginning of the 1970s) named color.

But QCD is not the only descendant of the Yang–Mills theory. The reader will see later that the current theory of the electroweak interaction and the models of grand unification of the strong, weak, and electromagnetic interactions are also realized as non-Abelian gauge theories.

On the reliability of experimental data

The physics of elementary particles is done by people. It is characteristic of man to err: mistakes are made both by experimenters and by theorists. Examples of such mistakes were mentioned earlier in the book. Thus, the current structure of the β-decay interaction was initially determined incorrectly and erroneous figures for the probabilities of the decays $\pi \to e\nu$, $K^0 \to \mu^+\mu^-$, and $K^0 \to 2\pi^0$ were regarded as correct for a long time. More recent examples could also be given.

Why then do physicists regard a multitude of phenomena as experimentally established, despite such mistakes? Could similar mistakes crop up in the experiments on which we unreservedly rely today? How can it be guaranteed that these experiments are correct if so many incorrect results occurred in the past?

The only guarantee is to accept a result as reliable only if it is obtained independently by several different groups employing different experimental methods. This condition is absolutely necessary but may not be sufficient, and does not provide a 100 percent guarantee. The 100 percent guarantee appears when the phenomenon recedes from the frontline of the science, when it is reproduced routinely, with the statistics of events exceeding by thousands or millions that with which the discovery was made, and when the quantities characterizing the phenomenon become known to an accuracy of several decimal places. Another way is not so much quantitative as it is qualitative: the search and discovery of a number of related phenomena that often follow the original discovery. Both these paths are well traced in the discoveries of P violation, CP violation, charm, and so on.

One of the difficulties of work in the field of high energy physics is the fact that very preliminary (not necessarily correct) results often provoke a premature general discussion. The guilt partially lies with theoretical physicists who snatch the hot experimental data "right off the griddle." This frequently leads to an expenditure of very considerable efforts to give an explanation of a "result" that in a year or two bursts like a soap bubble. Of course, competition between experimental groups also plays an important role. But, somehow or other, in several years the truth usually comes to the surface and turbulence evolves to a quiet clarity.

Appendix. Four Pages from the Dubna Preprint[11]

- 19 -

where $\mathcal{L}^{s}_{int}$ is the Lagrangian of strong interaction in the form (10). Taking advantage of (22), we can easily prove that for the nucleon current of β-decay interaction $j^{V}_{\mu} = G^{1}_{V} \bar{\chi} \tau \gamma_{\mu} \chi$ the following relation is fulfilled:

$$\frac{\partial j^{V}_{\mu}}{\partial x_{\mu}} = 0 \qquad (23)$$

It was shown in /21,23/ that relation (23) results in the conclusion that the value G^{1}_{V}, like an electrical charge, does not change when corrections connected with the strong interaction are taken into account. A similar proof can be given for G^{2}_{V}. The above-obtained result is true only in an approximation which makes no allowance for virtual slow and electromagnetic processes.

Unlike the leptonic interaction of nucleons, the leptonic interaction of Λ-hyperons (expressions (17) and (18)), as can easily be seen, does not possess the property of non--renormalizability of the vector coupling constant.

14. From expressions (17) and (18) it follows that if the strangeness of strongly interacting particles changes in leptonic decays of strange particles (as is clear from the foregoing, we attribute no strangeness to leptons), this change can only be $\Delta S = \pm 1$. Here $\Delta S = -1$ corresponds to emission of positively charged leptons, while $\Delta S = +1$ corresponds to the emission of negatively charged leptons.

Now let us enumerate the decays allowed by (15), (16), (17) and (18) and those forbidden by these interactions:

A l l o w e d :

$$\Lambda \to p + e^-(\mu^-) + \tilde{\nu}, \quad \Sigma^- \to n + e^-(\mu^-) + \tilde{\nu}, \quad \Sigma^- \to \Lambda + e^- + \tilde{\nu}$$
$$\Sigma^+ \to \Lambda + e^+ + \nu, \quad \Xi^- \to \Lambda + e^-(\mu^-) + \tilde{\nu}, \quad \Xi^- \to \Sigma^0 + e^-(\mu^-) + \tilde{\nu}$$
$$\Xi^0 \to \Sigma^+ + e^-(\mu^-) + \tilde{\nu}, \quad K^0 \to \pi^- + e^+(\mu^+) + \nu, \quad \bar{K}^0 \to \pi^+ + e^-(\mu^-) + \tilde{\nu}$$

F o r b i d d e n :

$$\Sigma^+ \to n + e^+(\mu^+) + \nu, \quad \Xi^- \to n + e^-(\mu^-) + \tilde{\nu}, \quad \Xi^0 \to p + e^-(\mu^-) + \tilde{\nu}$$
$$\Xi^0 \to \Sigma^- + e^+(\mu^+) + \nu, \quad K^0 \to \pi^+ + e^-(\mu^-) + \tilde{\nu}, \quad \bar{K}^0 \to \pi^- + e^+(\mu^+) + \nu$$

We do not claim this list to be exhaustive. The forbiddances arising for K^0_{e3} and $K^0_{\mu 3}$-decays result in a known relationship between the number of e^+ (μ^+)-decays and the number of e^- (μ^-) in a beam of neutral K-mesons (see /24, 25, 26, 16/ on this question):

$$\frac{n_{\mu^+}}{n_{\mu^-}} = \frac{n_{e^+}}{n_{e^-}} = \frac{e^{-\frac{t}{\tau_1}} + e^{-\frac{t}{\tau_2}} + 2e^{-\frac{t}{2\tau_1} - \frac{t}{2\tau_2}} \cos \Delta m t}{e^{-\frac{t}{\tau_1}} + e^{-\frac{t}{\tau_2}} - 2e^{-\frac{t}{2\tau_1} - \frac{t}{2\tau_2}} \cos \Delta m t} \quad (24)$$

Here n is the number of corresponding decays per unit time, τ_1 and τ_2 are the lifetimes of K_1^0 and K_2^0 - m e s o n s , the first of which has a combined parity of +1 and the second of -1. Δm is the difference of masses of K_1^0 and K_2^0-mesons.

From (17) and (18) it follows that in leptonic decays of strange particles involving strangeness changes the isotopic spin of strongly interacting particles changes by $\Delta T = 1/2$ /27, 26, 16/. This makes it possible to relate the probabilities of decay of K^+ and K^0-mesons /26, 16/:

$$w(K^0 \to e^+(\mu^+) + \nu + \pi^-) = 2w(K^+ \to e^+(\mu^+) + \nu + \pi^0) \quad (25)$$

which in its turn results in the relationships /28/:

$$w(K_2^0 \to e^+(\mu^+) + \nu + \pi^-) = w(K_2^0 \to e^-(\mu^-) + \tilde{\nu} + \pi^+) = w(K^+ \to e^+(\mu^+) + \nu + \pi^0) \qquad (26)$$

In combination with the rule $\Delta T = 1/2$ for non-leptonic decays relation (26) makes it possible to relate the lifetime of the K_2^0 with the lifetime of the K^+-meson. Taking advantage of data on the lifetime of the K^+-meson and on the frequency of various K^+-meson decays we can estimate the lifetime of the K_2^0-meson /23/:

$$\tau_{K_2^0} \sim 4 \times 10^{-8} \text{ sec.}$$

15. It can easily be seen that interaction (19), which is responsible for the non-leptonic decays of strange particles, allows a change of strangeness of $|\Delta S| = 1$ and forbids a change of strangeness of $|\Delta S| \geq 2$. Non-leptonic decays with $\Delta T > 3/2$ are forbidden by interaction (19). It is not clear, however, how the selection rule $\Delta T = 1/2$ can be obtained from (19), if this interaction is considered as an elementary one.

16. The author wishes to acknowledge the interest taken in his work, useful discussions and critical remarks by I.Ya.Pomeranchuk, Ya.B.Zeldovich, V.B.Berestetsky and A.P.Rudik. The author is also greatly indebted to Dr. R.Gatto for an interesting discussion.

References

1. J.Tiomno. Nuovo Cimento 1, 226)(1955)
2. S.Watanabe. Progr.Theor.Phys. 15, 81 (1956)
3. S.Watanabe. Phys.Rev. 106, 1306 (1957)
4. R.Marshak, E.Sudarshan. Phys.Rev. 109, 1860 (1958)
5. J.J.Sakurai. Nuovo Cimento 7, 649 (1958)
6. W.Pauli. Nuovo Cimento, 6, 204 (1957)
7. W.Heisenberg, W.Pauli (preprint)
8. M.Gell-Mann. Suppl.Nuovo Cim. 4, N 2, 848 (1956)
9. E.Fermi, C.N.Yang. Phys.Rev. 76, 1739 (1949)
10. M.A.Markov. Doklady Ac.Sci.101, 54 (1955)
11. S.Hori, A.Wakasa.Nuovo Cimento 6, 304 (1957)
12. S.Sakata. Progr.Theor.Phys. 16, 686 (1956)
13. S.Tanaka. Progr.Theor.Phys. 16, 625, 631 (1956)
14. Z.Maki. Progr.Theor.Phys. 16, 667 (1956)
15. R.W.King, D.C.Peaslee. Phys.Rev.106,360 (1957)
16. L.B.Okun. Zhurn.Exp.Teor.Fiz. 34, 469 (1958)
17. H.Fierz. Zeits.f.Phys. 104, 553 (1937)
18. A.Baldin. Zhurn.Exp.Teor.Fiz. (in print).
19. T.D.Lee, C.N.Yang. Nuovo Cim. 3, 749 (1956)
20. M.Gell-Mann. Phys.Rev. 106, 1296 (1956)
21. M.Gell-Mann, R.Feynman. Phys.Rev.109, 193 (1958)
22. S.S.Gerstein,Ya.B.Zeldovich,Zhurn.Exp.TeorFiz.29,698(1955)
23. B.L.Ioffe. Zhurn.Exp.Teor.Fiz. (in print)
24. Ya.B.Zeldovich. Zhurn.Exp.Teor.Fiz. 30, 1168 (1956)
25. S.B.Treiman, R.G.Sachs. Phys.Rev. 103, 1545 (1956)
26. L.B.Okun. Zhurn.Exp.Teor.Fiz. 32, 400 (1957)
27. M.Gell-Mann. Proc Rochester Conference, 1956
28. I.Yu.Kobzarev, L.B.Okun. Zhurn. Exp.Teor.Fiz.34,764 (1958)

References

1. Theses of talks at the all-union conference on high energy physics, 14–22 May, 1956 (USSR Academy of Science Publishing).
2. S. Sakata, *Progr. Theor. Phys.* **16**, 686 (1956).
3. L. B. Okun, Mirror particles and mirror matter: 50 years of speculation and search, arXiv:hep-ph/0606202v2.
4. K. Matumoto, *Progr. Theor. Phys.* **16**, 583 (1956).
5. S. Tanaka, *Progr. Theor. Phys.* **16**, 625 (1956).
6. Z. Maki, *Progr. Theor. Phys.* **16**, 667 (1956).
7. L. Okun, *Strange Particles (The Scheme of Isotopic Multiplets)*, UFN 1957 April, 535–559.
8. M. Gell-Mann, *Phys. Rev.* **92**, 833 (1953); M. Gell-Mann, talk at Pisa Conference 1955.
9. L. B. Okun, Some remarks concerning the compound model of fundamental particles, in *Proc. Int. Conf. on Mesons and Recently Discovered Particles*, Padova, Venezia, 22–28 September 1957, p. V-55.
10. L. B. Okun, *Sov. Phys. JETP* **7**, 322 (1958) [*ZhETF* **34**, 469 (1958) (in Russian)].
11. L. B. Okun, Mass reversal and compound model of elementary particles, Dubna preprint P-203 (1958).
12. L. B. Okun, Mass reversal and compound model of elementary particles, in *Proc. 1958 Annual Int. Conf. High Energy Physics at CERN*, 30th June–5th July, 1958, p. 223.
13. R. P. Feynman and M. Gell-Mann, *Phys. Rev.* **109**, 193 (1958).
14. E. C. G. Sudarshan and R. E. Marshak, *Phys. Rev.* **109**, 1860 (1958).
15. E. C. G. Sudarshan and R. E. Marshak, The nature of the four fermion interaction, in *Proc. Int. Conf. Mesons and Recently Discovered Particles*, Padova, Venezia, 22–28 September 1957, p. V-14.
16. J. J. Sakurai, *Nuovo Cimento* **7**, 649 (1958).
17. S. S. Gershtein and Ya. B. Zeldovich, *JETP* **2**, 576 (1956) [*ZhETF* **29**, 698 (1955) (in Russian)].
18. R. E. Marshak, K_e3 and $K_\mu 3$ decays and related subjects, in *Proc. 1958 Annual Int. Conf. High Energy Physics at CERN*, 30th June–5th July 1958, p. 284.
19. L. Okun, *Proc. 1958 Annual Int. Conf. High Energy Physics at CERN*, 30th June–5th July 1958, p. 201.
20. L. Okun and B. Pontecorvo, *Sov. Phys. JETP* **5**, 1297 (1957) [*ZhETF* **32**, 1587 (1957) (in Russian)].
21. I. Yu. Kobzarev and L. B. Okun, *Sov. Phys. JETP* **7**, 524 (1958) [*ZhETF* **34**, 763 (1958) (in Russian)].
22. M. Bardon, M. Fuchs, K. Lande, L. M. Lederman, W. Chinowsky and J. Tinlot, *Phys. Rev.* **110**, 780 (1958).
23. L. Okun, *Annu. Rev. Nucl. Sci.* **9**, 61 (1959).
24. L. Okun, *Usp. Fiz. Nauk* **68**, 449 (1959) (in Russian).
25. A. Gamba, R. Marshak and S. Okubo, *Proc. Natl. Acad. Sci.* **45**, 881 (1959).

26. R. Marshak, Theoretical status of weak interactions, in *Ninth Int. Annual Conf. High Energy Physics*, Vol. 2, Moscow, 1960, p. 269.
27. M. Gell-Mann and M. Levy, *Nuovo Cimento* **16**, 705 (1960).
28. Y. Yamaguchi, *Suppl. Progr. Theor. Phys.* **11**, 1 (1959).
29. O. Klein, *Arkiv för Fysik Swed. Acad. Sci.* **16**, 191 (1959).
30. S. Ogawa, *Progr. Theor. Phys.* **21**, 209 (1959).
31. M. Ikeda, S. Ogawa and Y. Ohnuki, *Progr. Theor. Phys.* **22**, 715 (1959).
32. *Proceedings of the 1960 Annual Int. Conf. on High Energy Physics*, p. 732.
33. *Proceedings of the 1960 Annual Int. Conf. on High Energy Physics*, p. 743.
34. R. P. Feynman, The status of the conserved vector current hypothesis, in *Proc. Rochester Conf.*, August 25–September 1 1960, p. 502.
35. M. Gell-Mann, Conserved and partially conserved currents in the theory of weak interactions, in *Proc. Rochester Conf.*, August 25–September 1 1960, p. 508.
36. Y. Ohnuki, Composite model of elementary particles, in *Proc. Rochester Conf.*, August 25–September 1 1960, p. 843.
37. J. J. Sakurai, *Annals of Physics* **11**, 1 (1960).
38. A. Pevsner *et al.*, *Phys. Rev. Lett.* **7**, 421 (1961).
39. M. Ikeda, Y. Miyachi and S. Ogawa, *Progr. Theor. Phys.* **24**, 569 (1960).
40. Z. Maki, M. Nakagava, Y. Ohnuki and S. Sakata, *Progr. Theor. Phys.* **23**, 1174 (1960).
41. S. Okubo, R. E. Marshak and E. C. G. Sudarshan, *Phys. Rev.* **113**, 944 (1959).
42. L. B. Okun, Lectures on the theory of weak interactions of elementary particles, 17 ITEP preprints, 1960-1961 (in Russian) [English translation: Theory of weak interactions: Thirteen lectures, AEC-tr-5226. US Atomic Energy Commission. Oak Ridge, Tenn. (For the translation of lectures 14–16 and of contents see NP-10254, 10842, 10845, 10840.)]
43. L. B. Okun, Lectures on the theory of weak interactions of elementary particles, JINR preprint P-833 (1961) (in Russian).
44. M. Gell-Mann, The eightfold way: a theory of strong interaction symmetry, Report CTSL-20 (1961).
45. M. Gell-Mann, *Phys. Rev.* **125**, 1067 (1962).
46. Y. Ne'eman, *Nucl. Phys.* **26**, 222 (1961).
47. *Proceedings of the 1962 Int. Conf. on High-Energy Physics at CERN*, p. 845.
48. R. Ely *et al.*, *Phys. Rev. Lett.* **8**, 132 (1962).
49. G. Alexander *et al.*, *Phys. Rev. Lett.* **9**, 69 (1962).
50. M. Gell-Mann, Strange particle physics. Strong interactions, in *Proc. Int. Conf. High Energy Phys.* (CERN, 1962), p. 805.
51. I. Yu. Kobzarev and L. B. Okun, *JETP* **15**, 970 (1962) [*ZhETF* **42**, 1400 (1962) (in Russian)].
52. N. Cabibbo, *Phys. Rev. Lett.* **10**, 531 (1963).
53. Z. Maki, M. Nakagawa and S. Sakata, *Progr. Theor. Phys.* **28**, 870 (1962).
54. Y. Katayama, K. Matumoto, S. Tanaka and E. Yamada, *Progr. Theor. Phys.* **28**, 675 (1962).
55. L. B. Okun, *Weak Interaction of Elementary Particles* (Moscow, 1963).
56. P. M. Dauber *et al.*, *Phys. Rev. Lett.* **13**, 449 (1964).
57. V. E. Barnes *et al.*, *Phys. Rev. Lett.* **12**, 204 (1964).

58. G. Zweig, An SU(3) model for strong interaction symmetry and its breaking, Preprints CERN-TH-401, 412.
59. M. Gell-Mann, *Phys. Lett.* **8**, 214 (1964).
60. L. B. Okun, *Weak Interaction of Elementary Particles* (Pergamon Press), translated from the Russian by S. Nikolic and M. Nikolic, translation edited by J. Bernstein.
61. Ya. B. Zeldovich, L. B. Okun and S. B. Pikelner, *Usp. Fiz. Nauk* **87**, 113 (1965).
62. L. B. Okun and C. Rubbia, CP violation, in *Proc. Heidelberg International Conference on Elementary Particles* (1968), pp. 301–345.
63. V. B. Berestetskii, E. M. Lifshitz and L. P. Pitaevskii, *Relativistic Quantum Theory*, Part 1 (Nauka, Moscow, 1968).
64. E. M. Lifshitz and L. P. Pitaevskii, *Relativistic Quantum Theory*, Part 2 (Nauka, Moscow, 1968).

My Life with Quarks

Sheldon Lee Glashow

Department of Physics, Boston University,
Boston, MA 02215, USA

This is a personal, anecdotal and autobiographical account of my early endeavors in particle physics, emphasizing how they interwove with the conception and eventual acceptance of the quark hypothesis. I focus on the years from 1958, when my doctoral work at Harvard was completed, to 1970, when John Iliopoulos, Luciano Maiani and I introduced the GIM mechanism, thereby extending the electroweak model to include all known particles, and some that were not then known. I have not described the profound advances in quantum field theory and the many difficult and ingenious experimental efforts that undergird my story which is not intended to be an inclusive record of this exciting decade of my discipline. My tale begins almost two years before I met Murray and over five years before the invention of quarks.

My thesis,[1] "The Vector Boson in Elementary Particle Decays," began with an appropriate remark by Galileo:

> *E forze dire che gl'ingegni poetici sieno di due spezie: alcuni destri ed atti ad inventar le favole; ed altri, disposti ed accomodati e crederle,*

which, loosely translated, says that the poetic imagination appears in two ways: as those who invent fables and as those inclined to believe them. Schwinger provided my personal fable: the quest for a unified theory of weak and electromagnetic interactions such as was suggested by their shared universal and vectorial natures. I found two other favorable hints: that gauge theories make sense in the zero mass limit[2] and (as my high-school buddy Gary Feinberg showed) that their one-loop contributions to charged lepton magnetic moments are finite.[3] More relevant to my thesis was the incompatibility revealed by the absence of radiative muon decay of the one-neutrino and intermediate vector boson hypotheses.

After examining and excluding various models based on the $SU(2)$-based Yang–Mills scheme, I nonetheless concluded: "It is of little interest to have a renormalizable theory of beta processes without the possibility of a renormalizable electrodynamics. We should care to suggest that a fully renormalizable theory of these interactions may only be achieved if they are treated together, in accordance with an identification of the neutral [boson] as the photon.... There is a mechanism [unspecified!] that produces a large mass splitting between the charged bosons and the massless photon... without sacrifice of the conservation laws." My high-school chum Steven Weinberg would identify the mechanism almost a decade afterward!

My final oral examination took place at the University of Wisconsin with a committee consisting of Paul Martin, Robert Sachs, Chen Ning Yang and my thesis advisor at Harvard, Julian Schwinger. All went well until I suggested, as I had been taught, that the muon neutrino could differ from the electron neutrino.[a] At this point Yang interrupted, claiming (falsely) that such a distinction could have no observable consequence. My examination ended as Schwinger patiently explained the matter to Yang. A year and a half later, Lee and Yang published a Physical Review Letter pointing out the importance of testing whether electron and muon neutrinos were identical. They cited Feinberg[3] and Schwinger[4] but neither my oral exam nor Schwinger's tutorial.

Murray's first passing mention of a second neutrino appears in a 1960 collaborative paper with Maurice Lévy that was mostly about the weak axial current. They wrote: "The two neutrinos involved in muon decay have been denoted by different symbols because we are not certain that they are identical." Furthermore, in a note added in proof, they were the first to propose the correct form of the hadronic weak current in the context of the Sakata model wherein the proton, neutron and lambda hyperon are the only 'fundamental' hadrons: "Such a situation $[G_V < G_\mu]$ is consistent with universality if we consider the weak vector current for $\Delta S = 0$ and $\Delta S = 1$ to be:

$$G_\mu \bar{p}\gamma_\mu(n + \epsilon\Lambda)/\sqrt{1+\epsilon^2}$$

and likewise for the axial current. If $(1+\epsilon^2)^{-1/2} \approx 0.97$, then $\epsilon^2 = 0.06$, which is of the right order of magnitude for explaining the low rate of β decay of the Λ particle. There is, of course, a renormalization ... so that we cannot be sure

[a]Schwinger regarded μ^+, ν and e^- as a particle triplet bearing "leptonic number" +1 under a rotation group whose third component is electric charge.[4] He argued that a sensible quantum number must distinguish between electrons and muons with the same charge, thus implying electron neutrinos to be distinct from muon neutrinos. Marshak (and later Gell-Mann) took the opposite view, endorsing a kind of lepton–hadron symmetry with ν, e^- and μ^- forming a triplet analogous to p, n and Λ, the three basic baryons of Sakata's scheme.

that the low rate really fits in with such a picture." Curiously, Murray failed to mention this brilliant forerunner of the Cabibbo current in his subsequent formulation of the Eightfold Way of strong-interaction symmetry.

Geneva and Copenhagen: I spent the academic years 1958–1960 as an NSF Postdoctoral Fellow at what would become known as the Niels Bohr Institute and partially at CERN. Finally, realizing that a sensible electroweak theory required more than three gauge bosons, I devised the $SU(2) \times U(1)$-based electroweak model in a paper[6] published in 1961 but written and circulated in the Spring of 1960. I lectured about it in Paris at Gell-Mann's invitation, where I explained how the universality of weak interactions would result if the charged weak currents generate a rotation group. (In particular, they generate the $SU(2)$ of my $SU(2) \times U(1)$ model, where the photon is a linear combination of the two neutral gauge bosons.) If this model could be extended to include hadrons, the observed universality of beta decay, muon decay and muon capture — the Puppi triangle — would find a natural explanation. Murray rephrased my exotic (Schwingeresque) notation and cited my ideas (duly giving me credit) at the 1960 International Conference on High Energy Physics at Rochester.[7] He introduced hadronic weak currents via the Sakata model, as he had done previously with Lévy, but he failed to mention the threat of strangeness-changing neutral currents, a problem which I had recognized[6] in 1960 when I wrote: "The modes $K \to \pi + \nu + \bar{\nu}$ and $K \to \pi + e^+ + e^-$ have never been seen. Since the coupling strengths of $W^\pm$ and Z^0 to their neutral currents are limited by the requirement of partial symmetry, the absence of neutral leptonic decay modes must be attributed to a mass splitting between $W^\pm$ and Z^0." [Here I use today's notation.]

That summer I was a student at the first Scottish Universities Summer School in Physics, held at Newbattle Abbey and organized by Peter Higgs as one of his first academic duties. Nicola Cabibbo and Tini Veltman (as well as Peter when he could escape his duties) were also students: three future Nobel Laureates and one plausible contender! Higgs, soon after he received the Nobel Prize in 2013, expressed himself somewhat wistfully: "There were a group of students... who stood up half the night discussing things like weak and electromagnetic interactions[b] but I wasn't part of that. [Peter had other responsibilities, such as providing us with adequate wine.] So I didn't learn of Glashow's theory when I could have." If only he had!

[b]Derek Robinson was another participant in our nightly adventures. He would achieve fame as an axiomatic field theorist.

California: Soon after my talk in Paris (where Murray treated me to an exquisite dinner at a two-star restaurant) I was offered — and accepted — a post-doctoral position at CalTech. I stayed for just one academic year. It began when I met Sidney Coleman, becoming his lifelong friend and frequent collaborator; it ended with the unexpected death of my father. In between Murray invented the eightfold way[8] and he and I published our one collaborative, prophetic but rarely cited paper.[9] Sidney and I became ardent disciples of the 8-fold way, advocating it (and betting on it) throughout the world. I remain proud of our discovery of the Coleman–Glashow formula which correctly relates the several electromagnetic mass differences of the baryon octet.[10]

Gell-Mann first introduced an $SU(3)$ group acting on a triplet of "leptons" ν, e^-, μ^- which "do not have anything to do with real leptons." These fictitious particles were precursors to the quarks Murray was gestating. He defined λ_i and F_i as the 3×3 and 8×8 matrix generators of $SU(3)$. His baryonic weak vector current was analogous to the leptonic current in a one-neutrino model:

$$\bar{\nu}\gamma_\mu(e_L + \mu_L) \leftrightarrow \bar{p}\gamma_\mu(n_L + \Lambda_L)\,,$$

which, in the context of the 8-fold way, became

$$\bar{\ell}\gamma_\mu([\lambda_1 + i\lambda_2] + [\lambda_4 + i\lambda_5])\ell_L \leftrightarrow \bar{\Psi}\gamma_\mu([F_1 + iF_2] + [F_4 + iF_5])\Psi_L$$

where ℓ denotes the lepton triplet and Ψ the baryon octet. Several features of this construct bear mention:

- Murray's "leptons" follow a one-neutrino scenario. Once again, he hedges with an *en passant* remark ("there may exist two neutrinos, one for electrons and one for muons,") which he proceeds to ignore.
- His weak vector current contains strangeness preserving and changing components "of equal strength, like the $e\nu$ and $\mu\nu$ currents .... The experimental evidence on the decay $K \to \pi +$leptons then indicates a renormalization factor, in the square of the amplitude, of the order of 1/20." His fascination with Marshak's lepton–hadron analogy had led Murray astray from his earlier thoughts with Lévy and me, thereby leaving an opening for Cabibbo's later triumph.
- His follow-up article on unitary symmetry,[11] submitted in September 1961, introduces the chiral $SU(3)$ current algebra but maintains the flawed lepton analogy. See his weak current in the Sakata model (Eq. 4.3) and its 8-fold way analog (Eq. 5.22), where the $\Delta S = 0$ and $\Delta S = 1$ terms are of equal strength! Both of Murray's 1961 eightfold way papers cite our joint

paper (in which what would become the Cabibbo angle appears in the context of the Sakata model) but here he sticks to the one-neutrino analogy where the $\Delta S = 1$ current must suffer dramatic renormalization. And I found it disconcerting that he would claim full credit (in footnote 29) for originating the notion of algebraic universality.

My Paper with Murray (hereafter GG): I begin by quoting from a recent article by John Iliopoulos:[12] "After looking at gauge theories for higher groups... the authors [of GG] try to apply the non-Abelian gauge theories to particle physics. They study both strong interactions, for which they attempt to identify the gauge bosons with the vector resonances that had just been discovered, as well as weak interactions. The currents were written in the Sakata model... Notice also that Gell-Mann had just written the paper on *The Eightfold Way*, but here they do not... exploit the property of the currents to belong to an octet. The paper is remarkable in many aspects.... For the weak interactions it considers the Glashow $SU(2) \times U(1)$ model and it correctly identifies the problems related to the absence of strangeness-changing neutral currents and the small value of the K_1–K_2 mass difference [which would be solved in a decade via the GIM mechanism]. The question of universality is addressed in a footnote [quoted below]:"

> Observe that the sum of the squares of the coupling strengths to strangeness-saving and to strangeness-changing charged currents is just the square of the universal coupling strength. Should the gauge principle be extended to leptons... the equality between G_V and G_μ is no longer the proper statement of universality, for in this theory $G_V^2 + G_\Lambda^2 = G_\mu^2$ (G_Λ is the unrenormalized coupling strength for the β-decay of Λ).

John doesn't mention another prescient footnote appearing in GG:

> The remarkable universality of the electric charge would be better understood if the photon were not merely a singlet, but a member of a family of vector mesons comprising a simple partially gauge invariant theory, such as in grand unified theories based on simple groups which would emerge a decade or so later.[13]

GG concluded on a pessimistic note, suggesting that we are "missing some important ingredient of the theory" because, "in general the 'weak' and 'strong' gauge theories will not be mutually compatible." Lots of missing ingredients would soon show up, such as quarks with charm and color. John's discussion of our paper ended with the following remark: "I do not know why this paper has not received the attention it deserves, but this is partly due to the authors themselves, especially Gell-Mann, who rarely referred to it."

It took another two years before my dear friend Nicola Cabibbo could put it all together.[14] Here is how Iliopoulos[12] described his great 1963 paper: "As Glashow and Gell-Mann, [Cabibbo] assigned the weak current to an octet of $SU(3)$, but he was in the 8-fold way scheme and not in the Sakata model. This allowed him to compare strangeness conserving and strangeness changing semileptonic decays and show that the scheme agreed with experiment, thus putting the final stone in the $SU(3)$ edifice. There are very few articles in the scientific literature in which one does not feel the need to change a single word and Cabibbo's is definitely one of them. With this work he established himself as one of the leading theorists of weak interactions."

The Birth of the Quark: Murray had been playing with triplets of hypothetical particles long before he dared to take them (a bit) more seriously. In a brief paper,[15] Murray introduced both the concept and the name. He offered a choice of triplets: His first version used an $SU(3)$ triplet t of integer-charged spin-$\frac{1}{2}$ fermions (u, d, s) supplemented by a neutral singlet b. Baryons had the composition $bt\bar{t}$ or $btt\bar{t}\bar{t}$. "The most interesting example of such a model," he wrote, "is one in which the four particles d^-, s^-, u^0 and b^0 exhibit a parallel with the leptons" (of which there were now four).

He reserved the word 'quark' for a "simpler and more elegant scheme [which] can be constructed if we allow non-integral values for the charges. We can dispense entirely with the basic baryon b .... We can then refer to the members $u^{2/3}$, $d^{-1/3}$ and $s^{-1/3}$ of the triplet as 'quarks' q .... It is amusing that the lowest baryon configuration qqq gives just the [$SU(3)$] representations 1, 8, 10 that have been observed ...."

For years Murray (and many others) remained uncertain whether quarks were real or merely useful fictions: "It is fun to speculate about the way quarks would behave if they were physical particles of finite mass (instead of purely mathematical entities as they would be in the limit of infinite mass) .... A search for stable quarks ... or diquarks ... at the highest energy would help to reassure us of the nonexistence of real quarks."[c]

A Senior Moment at 32? I spent the academic year 1963–1964 on leave from Berkeley: the fall semester at Harvard, where I collaborated with Sidney on tadpole diagrams. I spent that spring in Copenhagen, where I worked with James Bjorken. Our joint paper[17] introduced a fourth *charmed* quark so as to establish "A fundamental similarity between the weak and electromagnetic interactions of the leptons and the [quarks]." We imagined weak

[c]Murray is enjoying himself. Note his choices of the words 'fun' and 'amusing'. His jocular reference to Joyce has led many to rhyme quark falsely with lark.

interactions to involve two quark pairs (u, d_θ) and (c, s_θ) just as they involve two lepton pairs (ν_e, e^-) and (ν_μ, μ^-). We pointed out that the commutator of the weak charges was flavor conserving (as we would say today), a relation that "suggests an intimate connection between weak and electromagnetic interactions," but *we failed to notice how relevant this was to my work on the electroweak model* which, incredibly, *I had neglected to cite*!

Six months earlier (indeed, two weeks before Murray had submitted his quark paper!) Yasuo Hara[16] also had proposed a fourth quark and $SU(4)$, but he failed to note the algebraic properties of the weak current. Neither of our papers adopted fractional quark charges: baryons (ours and Hara's) were to be made of $qq\bar{q}$ or even $qqq\bar{q}\bar{q}$.

It is now the golden anniversary of a truly a remarkable year in fundamental physics. 1964 saw the discoveries of CP-violation, the Ω^- baryon and the cosmic microwave background radiation, the invention of the Higgs mechanism and the debuts of quarks and charm. It was also the year in which Wally Greenberg[18] introduced paraquarks of order 3, which would soon become known as quarks with color.

Weinberg's Masterstroke and the GIM Mechanism: Let me not explain the ways in which the works of Anderson, Brout, Englert, Goldstone, Guralnik, Hagen, Higgs, Kibble, Nambu, Salam and others may have inspired Steven Weinberg in 1967[19] to inject the notion of spontaneous symmetry breaking into my $SU(2) \times U(1)$ model, thereby explaining the origin of many particle masses, transforming my model into a genuine electroweak theory of leptons, posing a theoretical challenge that would be met by Veltman and 't Hooft in four years and an experimental challenge that would take another forty.

Weinberg's brilliant solution to the problem of W and Z masses remained academic and rarely cited until the problem of strangeness-changing neutral currents was addressed. The final portion of my tale concerns my work with John Iliopoulos and Luciano Maiani at Harvard in the Spring of 1970.[19] We began by identifying the problem: "With a [three-]quark model, we immediately encounter strangeness-violating couplings of neutral lepton currents and contributions to the neutral kaon mass splitting of order $G(G\Lambda^2)$ [where Λ is an appropriate cutoff]. For this reason, it appears necessary to depart from the original phenomenological model of weak interactions... by introducing four quark fields.... The extra quark completes the symmetry between quarks and the four leptons." Upon constructing the now-standard form of the hadronic current, we wrote: "This is just the form... suggested [earlier].[17] What is new is the observation that this model is consistent with the phenomenological selection rules and with universality even when all divergent first order terms [i.e., $G(G\Lambda^2)^n$] are considered."

Later in the paper we "briefly consider a more daring speculation. It has long been suspected [Refs. 4 and 6 herein] that there may be a fundamental unity between weak and electromagnetic interactions. For this reason it may have been wrong for us to introduce a gauge symmetry for the weak interactions not shared by electromagnetism ...." We then describe my 1960 electroweak model, noting that it "could be correct only if the weak bosons are very massive (100 GeV)." The paper goes on to stress the importance of searching for neutrino-induced neutral-current effects, concluding that "Evidence for the existence [of the Z^0] could come from colliding beam experiments." At no point did our paper cite Weinberg's seminal work[19] of which we were then quite unaware (even though we had discussed our paper with Steve prior to its publication).

Epilog: It took some doing to convince the particle-physics community to launch a dedicated search for particles containing one or more charmed quarks. Many physicists, such as Schwinger and Maiani, failed at first to recognize the J/Ψ particle as charmonium. Charmed particles were not decisively seen until 1976. Yet, immediately upon the discovery of the tau lepton, everyone became convinced that there had to be two more quarks, although it did take a rather long time and considerable effort for experimenters to capture the top. I would have preferred just two families of fundamental fermions: all the members of the first family but none of the second are essential to life on earth, indicating a perfect symmetry between the relevant and the irrelevant. But Nature did not see things my way.

References

1. S. L. Glashow, The Vector Meson in Elementary Particle Decays, Harvard University Thesis (July 1958).
2. K.-H. Tzou, *Comptes Rendus* **245**, 289 (1957).
3. G. Feinberg, Decay of the mu meson in the intermediate-meson theory, *Phys. Rev. Lett.* **112**, 1482 (1958).
4. J. Schwinger, A theory of the fundamental interactions, *Ann. Phys. (N.Y.)* **2**, 407 (1957).
5. M. Gell-Mann and M. Lévy, The axial vector current in beta decay, *Nuovo Cimento* **16**, 705 (1960).
6. S. L. Glashow, Partial symmetries of weak interactions, *Nucl. Phys.* **22**, 579 (1961) (received 9 Sept. 1960) [See also: S. L. Glashow, *Nucl. Phys.* **10**, 109 (1959)].

7. M. Gell-Mann, *Proc. Int. Conf. High Energy Physics* (1960), pp. 509–510.
8. M. Gell-Mann, The eightfold way: A theory of strong interaction symmetry, CalTech Report CTSL-20 (received 24 Apr. 1961) [See also: Y. Neeman, Gauges, groups & an invariant theory of the strong interactions (unpublished, Aug. 1961)].
9. S. L. Glashow and M. Gell-Mann, Gauge theories and vector particles, *Ann. Phys.* (*N.Y.*) **15**, 437 (1961).
10. S. R. Coleman and S. L. Glashow, Electrodynamic properties of baryons in the unitary symmetry scheme, *Phys. Rev. Lett.* **6**, 423 (1961).
11. M. Gell-Mann, Symmetries of baryons and mesons, *Phys. Rev.* **125**, 1067 (1962) (received 20 Sept. 1961).
12. J. Iliopoulos, Symmetries and the weak interactions, arXiv:1101.3442 (submitted 18 Jan 2011).
13. H. Georgi and S. L. Glashow, Unity of all elementary particle forces, *Phys. Rev. Lett.* **32**, 438 (1974).
14. N. Cabibbo, Unitary symmetry and leptonic decays, *Phys. Rev. Lett.* **10**, 531 (1963).
15. M. Gell-Mann, A schematic model of baryons and mesons, *Phys. Lett.* **8**, 214 (1964) (received 6 Jan. 1964) [See also: G. Zweig, An SU(3) model for strong interaction symmetry, CERN preprint 8182/TH-401 (17 Jan 1964)].
16. Y. Hara, Unitary triplets and the eightfold way, *Phys. Rev.* **134**, B701 (1964) (received 23 Dec 1963).
17. B. J. Bjørken and S. L. Glashow, Elementary particles and SU(4), *Phys. Lett.* **15**, 255 (1964) (received 19 June 1964).
18. O. W. Greenberg, A paraquark model of baryons and leptons, *Phys. Rev. Lett.* **13**, 598 (1964).
19. S. Weinberg, A theory of leptons, *Phys. Rev. Lett.* **19**, 1264 (1967).
20. S. L. Glashow, J. Iliopoulos and L. Maiani, Weak interactions with lepton hadron symmetry, *Phys. Rev. D* **2**, 1285 (1970).

Quarks in the Bootstrap Era

D. Horn

School of Physics and Astronomy,
Tel Aviv University, Tel Aviv, 69978, Israel

The quark model emerged from the Gell-Mann–Ne'eman flavor SU(3) symmetry. Its development, in the context of strong interactions, took place in a heuristic theoretical framework, referred to as the Bootstrap Era. Setting the background for the dominant ideas in strong interaction of the early 1960s, we outline some aspects of the constituent quark model. An independent theoretical development was the emergence of hadron duality in 1967, leading to a realization of the Bootstrap idea by relating hadron resonances (in the s-channel) with Regge pole trajectories (in t- and u-channels). The synthesis of duality with the quark-model has been achieved by duality diagrams, serving as a conceptual framework for discussing many aspects of hadron dynamics toward the end of the 1960s.

1. Introduction

In this short memoir I wish to present a personal perspective concerning the early history of the quark model. During the early 60s I was a graduate student of Yuval Ne'eman, spending several formative years at the Weizmann Institute, and in the late 60s I have been a postdoc at Caltech, witnessing the bustling research activity influenced by Murray Gell-Mann. The theoretical framework, within which Yuval and Murray and all their students worked at that time, was based on fundamental notions and principles which can be derived from Quantum Field Theory (QFT). Particle interactions were divided into four categories: strong, electromagnetic, weak and gravitational interactions. It seemed clear that, given the strength of the strong interactions, and lacking any analog of the electromagnetic fine structure constant α, there existed no hope of formulating a fundamental field theory of strong interactions.

Studying strong interactions one tried to get the utmost out of basic principles, like analyticity and unitarity of the S-matrix, and PCT symmetry principles, all of which did have their origin in QFT. Other than that, one tried to come up with principles that seemed to fit the observed experimental phenomena. I refer to this period as the bootstrap era, whose basic leading ideas will be described in short in the next paragraph. It is within this background that we, as young practitioners of novel particle symmetry ideas, tried to make critical observations and work out the consequences of the ideas that we have promoted or criticized.

2. The Bootstrap Era

S-matrix models of two-to-two particle-scattering were described in terms of the s, t, u Mandelstam variables[1] $s = (p_1 + p_2)^2 = (p_3 + p_4)^2$, $t = (p_1 - p_3)^2 = (p_2 - p_4)^2$ and $u = (p_1 - p_4)^2 = (p_3 - p_2)^2$, obeying the over-all constraint $s + t + u = \sum_i m_i^2$. The physical region for the process $1 + 2 \to 3 + 4$ was characterized by positive s and negative t. Physical regions in the crossed channels described processes involving the anti-particles, according to conventional associations in Feynman diagrams. This is exemplified in Fig. 1, which is taken from Mandelstam's paper, describing πN scattering.

Analyticity of the S-matrix implied the appearance of poles below the scattering threshold of the s-channel, and cuts above the threshold. The same types of structures appear in physical regions of the S-matrix in the t-channel

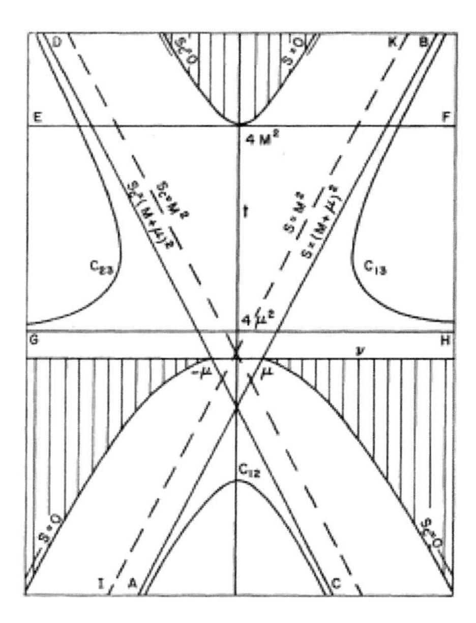

Fig. 1. The Mandelstam plot for πN scattering, taken from the original paper.[1] Masses of π and N are denoted by μ and M, respectively. The physical region of s is the shaded region on the bottom right.

and in the u-channel, where t or u obtain positive values. Analyticity was expressed in terms of dispersion relations, relating real and imaginary parts of the scattering amplitude to each other. These analytic structures, whose analogs can be defined for general m to n particle scattering amplitudes, were further supplemented by unitarity constraints. The simplest of these relationships is the optical theorem, relating the imaginary part of the elastic scattering amplitude $a + b \to a + b$ to the total cross-section $\sigma_T(ab)$. The rich set of analytic and unitarity constraints has led Chew and Frautschi[2] to propose the Nuclear Bootstrap idea, stating that these constraints may suffice to determine a unique set of poles (i.e. particles and resonances) in all channels, thus providing the basis of a theory of the strong interactions. An example of a detailed summary of all these ideas is the set of lectures delivered by Chew in the 1965 Les Houches Summer School.[3]

Another important element of the dynamics of strong interactions was the use of Regge poles.[4,5] The asymptotic behavior of a scattering amplitude at high energies (large s and negative t) has been composed of a set of terms of the type $s^{\alpha(t)}$ where $\alpha(t)$, for negative t values, was a (linear) extrapolation of angular momenta of particles (or strong interaction resonances) of masses m observed for a given set of quantum numbers in the crossed channel, i.e. for positive $t = m^2$ values.[6] Their presumed linear behavior is depicted in the Chew–Frautschi plot reproduced in Fig. 2. The Froissart bound,[7] stating that σ_T cannot increase asymptotically faster than $\log^2(s)$, meant that all

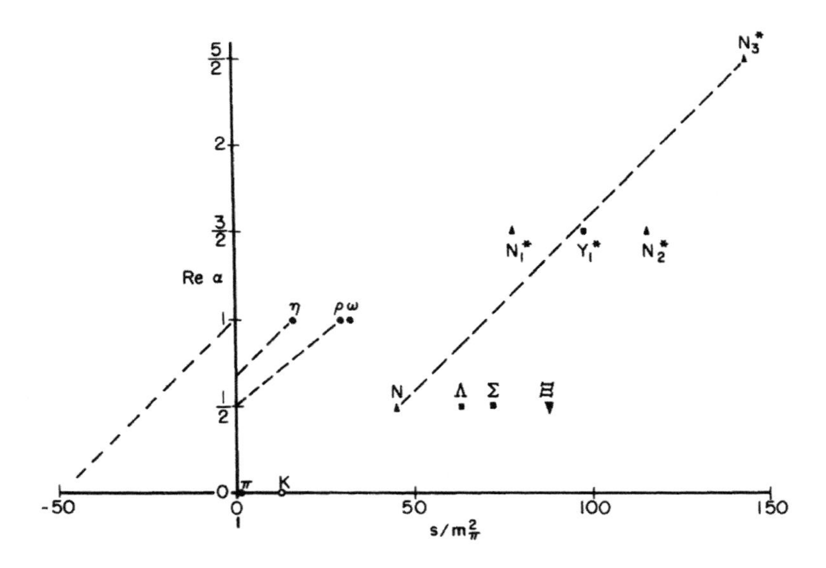

Fig. 2. Chew–Frautschi plot of Regge pole trajectories, taken from Ref. 6.

$\alpha(0)$ had to be smaller than 1. The Regge pole with $\alpha(0) = 1$ in Fig. 2 was called the Pomeron, leading to constant total cross-section and implementing the Pomeranchuk theorem, which stated that the asymptotic total cross-sections of particles and of their antiparticles were identical. The Pomeron was exceptional, because no particles were identified for $t > 0$ which were fit to lie on its trajectory. All other Regge poles were associated with known particles and resonances with relevant quantum numbers, accounting for many phenomenological observations at large s and negative t. Corrections to this picture were presumed to be due to lower-lying cuts in complex angular momentum. For positive t, one expected to find further resonances with ever-increasing angular momenta.

3. SU(3) and the Quark Model

The late 50s and early 60s have witnessed the uncovering of an increasing zoo of particles and resonances. Some of them can be seen in Fig. 2, where they served as the basis of deciding which Regge trajectories should influence high-energy scattering amplitudes. By that time there existed already many sets of particles with different spin (J) and parity (P) quantum numbers, and the hope was that a symmetry higher than the spin symmetry SU(2) will be found to account for ordering spectra of the same J^P, but different isospin and strangeness, into the same representation. The SU(3) symmetry (which today should be referred to as flavor-SU(3)) proposed by Gell-Mann[8] and Ne'eman[9] was one of the suggested symmetry models. It has accommodated pseudo-scalar and vector mesons in singlet and octet representations, and the lowest baryons of spin-$\frac{1}{2}$ were also nicely accounted for by an octet representation. There existed 9 spin-$\frac{3}{2}$ resonances which could fit into a decuplet, if a tenth strangeness -3 particle could be found.[10] With the experimental discovery of the stable Ω^- particle,[11] which served to complete the spin-$\frac{3}{2}$ decuplet, the Gell-Mann and Ne'eman SU(3) model has won overall recognition as the correct symmetry model of strong interactions. This symmetry model provided also a framework for mass formulas, as well as their electromagnetic corrections, and it also served as a basis for postulating the structure of weak interaction currents, etc. The symmetry considerations became so popular that, within a few years, various extensions of SU(3) into higher symmetries have been proposed, like SU(6) (including spin degrees of freedom) and more, but all of them looked quite speculative and the interest in them diminished throughout the years while flavor-SU(3) stood its ground. The early period of flavor-SU(3) has been summarized in "The Eightfold Way."[12]

In the meantime, Gell-Mann[13] has proposed the quark model, and Zweig[14] has independently proposed his ace model. The quark model built the isospin degrees of freedom from u and d quarks, and associated the strangeness quantum number with the s quark. Thus all mesons were accounted for by quark–antiquark combinations and all baryons could be viewed as three-quark structures. Most physicists have regarded this viewpoint as a mnemonic for SU(3) symmetry considerations, rather than viewing quarks as real physical objects. The strong belief that all true physical variables should be experimentally measurable was at the heart of this refusal to accept quarks as physical building blocks, because of their fractional electric charges and wrong spin-statistics relations. Nonetheless it won popularity because the quark model seemed to be the natural way to explain SU(3) representations, i.e. why representations other than 1, 8 and 10 have not been observed in hadron physics. Clearly, most of the dynamic consequences of quark-based descriptions like SU(3) breaking (mass differences) and electromagnetic mass-shifts, etc., could just as well be stated in terms of operators with specific SU(3) characteristics. There were however a few indications of experimental phenomena that required a quark-based rule. One example was the Zweig rule,[14] explaining the amazing dominance of the decay mode $\varphi \to K^+ + K^-$ in terms of the assumption that φ is a bound state of an s quark and antiquark. Note that this quark structure meant a particular mixing of SU(3) singlet and octet states. The striking argument was that if strong decay can proceed only in terms of Feynman diagrams in which the quark–antiquark pair does not annihilate ("rule of ace conservation" in Zweig's language), it accounts for the dominance of this particular decay mode.

Zweig's approach was an early example of what became known as the constituent-quark point of view; regarding quarks within particles the same way one considers nucleons within nuclei, without trying to explain the unresolved conceptual problems. An early review of hadron physics as accounted for by this naïve quark approach was given by Dalitz.[15] A detailed account of the development of the quark model has been presented by Lipkin.[16]

4. Duality of the Strong Interactions

By the mid-sixties there existed then two important observations regarding hadron spectra. One was that their states fit well into quark model constructs, and the other was that towers of resonances with ever-increasing masses and angular momenta are expected to exist in conjunction with the

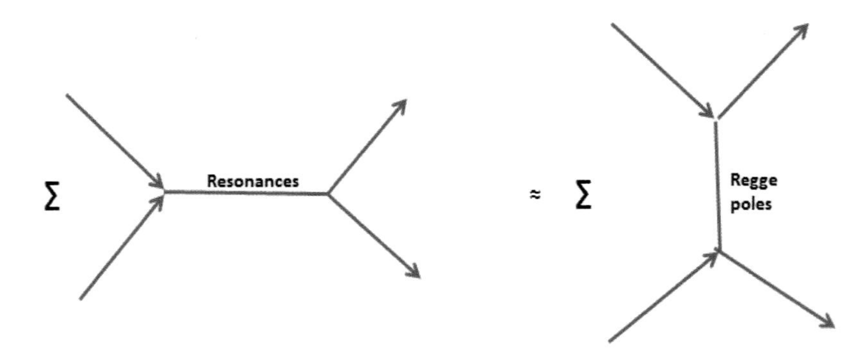

Fig. 3. Duality as implemented by the FESR.

Regge model. The question still remained how all this can fit into the bootstrap approach. It turned out that what was missing was a formulation of the bootstrap idea that can unite the two.

Such a formulation has been proposed by the Finite Energy Sum Rules (FESR).[17–21] Employing analyticity of the S-matrix, the FESR approach incorporated experimental scattering amplitudes (which was dominated by resonances) as input, up to some finite energy, and Regge pole parametrization from that point on. This allowed relating these two contributions using a Cauchy integral approach. Applying it to inelastic scattering, such as $\pi^- p \to \pi^0 n$, one ended up with a relationship of the type expressed in Fig. 3, implying that the Regge pole amplitude, when continued to low-energies, can approximate on average the resonance contributions. Or, vice versa, the resonances can be used to account for Regge pole properties. Inelastic scattering channels were used in order to avoid the Pomeron issue: the elastic channel was assumed to be dominated by Pomeron exchange, with $\alpha(0) = 1$ (see Fig. 2), and its s-channel amplitude was evidently not dominated by resonances.

The FESR approach was quite revolutionary on two counts. First, it went against the then common trend to sum both types of contributions, resonances and Regge exchanges, to the scattering amplitude. This practice has followed the common experience from the use of Feynman diagrams, which the FESR have shown to involve double-counting. Second, it has led to a concrete realization of the bootstrap idea: the resonances in the s-channel have now successfully been related to Regge poles that are the analytic continuation of resonances in the t-channel. This has also become to be known as hadron duality.[22]

The dual relationship of Fig. 3 can also be captured within a mathematical model, as demonstrated by Veneziano.[23] Using linear Regge trajectories he expressed his model in terms of the Euler Beta function. This mathematical realization has led to a flurry of theoretical activity studying dual resonance models in the following years.

5. Duality Diagrams

With hadrons appearing as quark-model resonances in both the s-channel and the t- and u-channels, it seemed natural to try and account for both in one diagrammatic description. A merger of the duality principle and the quark model came about through the duality diagrams which have been proposed by Harari[24] and by Rosner.[25] The diagrams, such as the examples shown in Fig. 4, display scattering amplitudes in terms of two- or three-quark components, as befitting mesons and baryons, but their s- and t-structure is that of a dual amplitude. Although they did not account for Pomeron exchanges, duality diagrams presented a theoretical framework which has described the understanding of strong interaction dynamics at that time. They have also extended the thinking underlying Zweig's rule into the realm of scattering amplitudes, thus encapsulating all allowed and forbidden resonances and Regge exchanges in their various channels.

One should however admit that, although duality diagrams served as an underlying conceptual framework, explaining allowed and forbidden reaction channels, they fell short of providing a model which can explain the observed dynamical features of scattering phenomena and multi-particle production. The discussion of these phenomena continued to make use of concepts

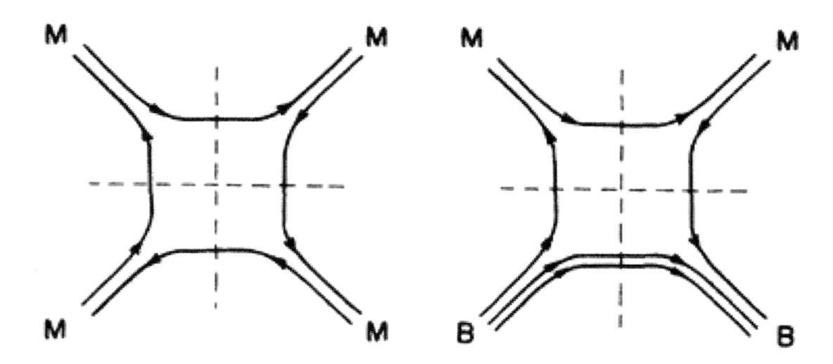

Fig. 4. Duality diagrams, taken from Harari,[24] representing meson–meson and meson–baryon scattering amplitudes. They may be realized as representing resonances in the s-channel and Regge exchanges in the t-channel, both obeying quark-model assignments.

derived from various field-theoretical or statistical models, which have dominated the relevant literature.[26] One important reason was that high-energy reactions consisted mostly of multi-pion production, in flagrant violation of what might be expected from a system which is symmetric under flavor-SU(3). Therefore one still needed theoretical tools which could deal with pion dominated phenomena.

6. Epilog

Now we know that the quark model is more fundamental than the flavor-SU(3) symmetry which gave birth to it in the 60s. After the discoveries of additional and heavier quarks it became clear that flavor SU(3) symmetry reflected the coincidence that the quarks u, d and s possess light masses. Now we also know that strong interactions can be formulated as a quantum field theory, QCD. Why should one then look back at the 1960s?

Other than just the simple answer, that it is nice to reminisce about the 1960s, these stories may serve as an account of a very active era, in which theoretical developments and experimental discoveries followed rapidly one another. It was a great achievement of the HEP community that, even in the absence of an underlying theory, it could proceed very far in establishing an adequate framework to provide consistent understanding of observed particles and their interactions.

Aficionados of field theory have criticized the bootstrap approach at the time, arguing that having everything emanate from self-consistency is too much to be asking for. It turned out that the underlying quark model was necessary in order to provide the duality framework, which was the manifestation of the bootstrap approach, with its particle-flavor foundation. This was sufficient to produce a conceptual framework for many aspects of hadron physics at the end of the 1960s.

References

1. S. Mandelstam, *Phys. Rev.* **112**(4), 1344 (1958).
2. G. F. Chew and S. C. Frautschi, *Phys. Rev. Lett.* **7**, 394 (1961).
3. G. F. Chew, The analytic S-matrix: A theory of strong interactions, in *High Energy Physics, 1965 Les Houches Summer School*, eds. C. DeWitt and M. Jacob (Gordon and Breach, New York, 1965), pp. 189–250.
4. T. Regge, *Nuovo Cimento* **14**, 951 (1959).
5. T. Regge, *Nuovo Cimento* **18**, 947 (1960).
6. G. F. Chew and S. C. Frautschi, *Phys. Rev. Lett.* **8**, 41 (1962).
7. M. Froissart, *Phys. Rev.* **123**, 1053 (1961).

8. M. Gell-Mann, The eightfold way: A theory of strong interaction symmetry, Unpublished report CTSL-20, 1961; *Phys. Rev.* **125**, 1067 (1962).
9. Y. Ne'eman, *Nucl. Phys.* **26**, 222 (1961).
10. M. Gell-Mann, *Proc. Int. Conf. on High-Energy Nuclear Physics*, Geneva, 1962, p. 805.
11. V. E. Barnes *et al.*, *Phys. Rev. Lett.* **12**, 204 (1964).
12. M. Gell-Mann and Y. Ne'eman, *The Eightfold Way* (W.A. Benjamin, New York, 1964).
13. M. Gell-Mann, *Phys. Lett.* **8**, 214 (1964).
14. G. Zweig, An SU3 model for strong interaction symmetry and its breaking, CERN Report 8419/TH.401 (January 17, 1964); An SU3 model for strong interaction symmetry and its breaking II, CERN Report 8419/TH.412 (February 21, 1964).
15. R. H. Dalitz, Quark models for the "elementary particles", in *High Energy Physics, 1965 Les Houches Summer School*, eds. C. DeWitt and M. Jacob (Gordon and Breach, New York, 1965), pp. 253–323.
16. H. J. Lipkin, *Phys. Rep.* **8**, 173 (1973).
17. D. Horn and C. Schmid, Finite energy sum rules, Unpublished report CALT-68-127, April 1967.
18. A. Logunov, L. D. Soloviev and A. N. Tavkhelidze, *Phys. Lett. B* **24**, 181 (1967).
19. K. Igi and S. Matsuda, *Phys. Rev. Lett.* **18**, 625 (1967).
20. R. Dolen, D. Horn and C. Schmid, *Phys. Rev. Lett.* **19**, 402 (1967).
21. R. Dolen, D. Horn and C. Schmid, *Phys. Rev.* **166**, 1768 (1968).
22. G. F. Chew, *Comments on Nuclear and Particle Physics*, 74–76 (1968).
23. G. Veneziano, *Nuovo Cimento A* **57**, 190 (1968).
24. H. Harari, *Phys. Rev. Lett.* **22**, 562 (1969).
25. J. L. Rosner, *Phys. Rev. Lett.* **22**, 689 (1969).
26. D. Horn and F. Zachariasen, *Hadron Physics at Very High Energies* (W. A. Benjamin, Reading, MA, 1973).

From Symmetries to Quarks and Beyond

Sydney Meshkov

LIGO Laboratory, California Institute of Technology,
Pasadena, California 91125, USA

Attempts to understand the plethora of meson baryon and meson resonances by the introduction of symmetries, which led to the invention of quarks and the quark model, and finally to the formulation of QCD, are described.

1. Introduction

In this paper[a] I would like to look at the sequence of events that led to the quark model, how it evolved, and some of its consequences. As always, these events did not follow a simple linear path. This journey went on for about 25 years, from the late 1950s to the mid-1970s. During this exciting period, there was a happy confluence of lots of data to be explained and some imaginative theoretical constructs. Avoiding some dead ends, elementary particle physics progressed from the Sakata model,[1] to the symmetry era culminating in the Eightfold Way of Gell-Mann[2] and Ne'eman,[3] to quarks and the simple quark model, to the study of SU(6), the introduction of color, and eventually to Quantum Chromodynamics (QCD). The interplay between experiment and theory was crucial to the progression of our understanding of each of these topics.

2. Personal Perspective

I was fortunate to be at the Weizmann Institute in the fall of 1961. Carl Levinson and I had been using the group SU(3) for dynamical nuclear physics calculations, first at Princeton University in 1960 and then at Weizmann during the winter and spring of 1961. SU(3) is the group of unitary 3×3 matrices with determinant 1. We had returned from giving talks about our nuclear structure work at conferences in Europe in the fall of 1961, Carl at the Varenna summer school, and I, at a conference in Manchester. We arrived just in time to hear a seminar by Yuval Ne'eman, in which he described his

[a]LIGO document number: P1400170.

PhD thesis. Working at Imperial College with Abdus Salam as his advisor, Yuval had produced work that paralleled that which Murray Gell-Mann had produced at Caltech. At no point in the seminar, did Ne'eman use the term SU(3), although it was evident that he had made great use of it in the transformations presented in his talk. When the seminar was over, Carl and I rushed up to Yuval, and asked him, "Isn't this just SU(3)?" Yuval agreed that it was. We said, "We know all about SU(3), but we do not know any particle physics. Please give us something to read. Yuval did, and we were on our merry way to learning about a whole new field. That very first night we were able to write down the mass breaking formula, but we did not know that the breaker had to transform like an $I = 0$, $Y = 0$ member of an octet. It took Murray to point out this crucial fact at the CERN conference in the summer of 1962.[4] Okubo,[5] independently, wrote down the mass formula

$$M = m_0\{1 + aY + b[I(I+1) - Y^2/4]\}\,. \tag{1}$$

I is the isospin and Y is the hypercharge.

3. G2 versus SU(3)

During the period 1960–1961, a big question was whether the proper classification group for the ever-growing list of mesons and baryons was G2 or SU(3). Ralph Behrends[6] was a proponent of G2, whereas Gell-Mann and Ne'eman advocated SU(3). The deciding factor for which extension of SU(2) isospin symmetry to use was the prediction for the number of pseudoscalar mesons. At the time, there existed seven pseudoscalars, namely three pions and four kaons. G2 predicted that there should be seven pseudoscalars, whereas SU(3) predicted that there should be an additional $I = 0$, $Y = 0$ meson. The issue was settled with the discovery of the pseudoscalar $\eta(548)$ meson, so G2 was ruled out and SU(3) was deemed the correct choice.

4. Sakata Model and its Demise

An early model (1956) to describe the baryons and mesons was the Sakata model. It was based on a fundamental triplet, composed of the physical proton, neutron and $\Lambda(1115)$ particles, called B. In this model, mesons were formed as $B\bar{B}$ ($3 \times \bar{3}$) composites, giving the now familiar octets and singlet. The pseudoscalar mesons could be accommodated in an octet of SU(3) as well. However, it was not clear in which representations the eight spin-$\frac{1}{2}$ baryons should be in the Sakata Model, whereas in the Eightfold Way they could be accommodated in a single octet. Fortunately, we (C. A. Levinson,

H. J. Lipkin, S. Meshkov, A. Salam and R. Munir),[7] were able to eliminate the Sakata model by looking at the prediction for proton anti-proton annihilation going into $K_L K_S$ compared to $K_L K_L$. The Sakata model forbids annihilation into $K_L K_S$ pairs, whereas it is allowed in the Eightfold Way. Experimentally, these decays are produced at a macroscopic rate, so the Sakata model was ruled out.

An aspect of the Sakata model that turned out to be very useful was that to describe meson decays to two other mesons, the couplings for $BB\bar{B}\bar{B}$ were needed. Fortunately, Ikeda, Ogawa and Ohnuki,[8] and Sawada and Yonezawa[9] had produced tables of these. From their work, I was able to abstract a complete set of 8×8 Clebsch–Gordan coefficients for SU(3), which Levinson, I and Harry Lipkin, who joined us a few weeks after we heard Ne'eman's seminar, exploited in our subsequent work.

5. Breaking SU(3)

Once it was established that SU(3) was the correct symmetry model of the strong interaction, there were some obvious problems. Just looking at the wide range of masses of the mesons and baryons, in their respective multiplets, it was clear that there was a large symmetry breaking going on. This was explained by Gell-Mann[2] and Okubo,[3] in 1962. They assumed that the symmetry breaker transformed like an $I = 0$, $Y = 0$ member of an octet. This neatly explained and correlated the observed splittings.

From the fall of 1961 through 1963 there was a lot of activity on the symmetry front. Using what we now call flavor SU(3) was not the favored area in which to work for most theorists. Many were busy in the complex plane and did not take kindly to the idea of using symmetries. However, there gradually developed a number of physicists in the United States, Israel and Europe who were interested in exploring the use and properties of SU(3) symmetry.

I enjoyed working on SU(3) with Carl Levinson and Harry Lipkin at the Weizmann Institute, and with Gaurang Yodh and George Snow at the University of Maryland. In our early work Levinson, Lipkin and I made copious use of Weyl reflections and applied them to decay widths, scattering amplitudes in hadronic processes, photoproduction and other electromagnetic processes.[10–12] E. C. G. Sudarshan and his group at Syracuse did analogous work.[13–16]

Later, we invented the U-spin and V-spin subgroups of SU(3)[17] and observed that the photon is a U-spin scalar,[12] useful in dealing with electro-

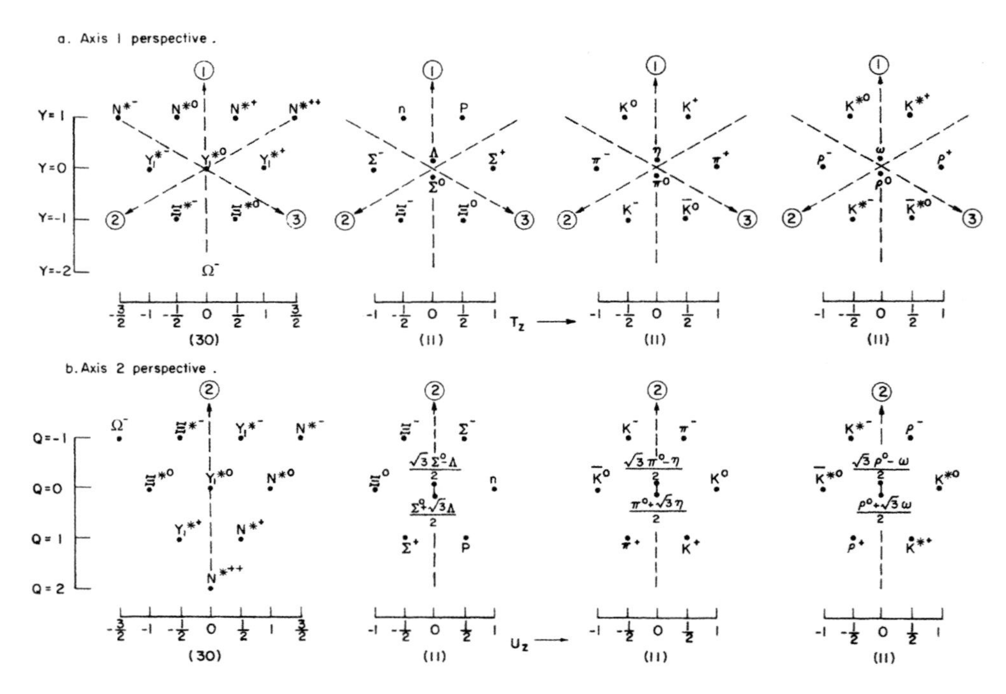

Fig. 1. Representation diagrams of the group SU_3. The (30) baryon-resonance representation diagram and the (11) baryon, meson and vector-meson representation diagrams are displayed.

magnetic processes. The classification of the decuplet, baryon octet, and the pseudoscalar and vector meson octets according to I-spin and U-spin assignments are illustrated in the figure above. The upper part of Fig. 1 shows the usual I-spin display. The lower part shows the U-spin display of these particles. Just as I-spin transformations are perpendicular to axis 1, so U-spin transformations are perpendicular to axis 2. U-spin multiplets all have the same electric charge. V-spin transformations are perpendicular to axis 3.

Yodh, Snow and I made the first successful comparison of SU(3) predictions with experiment for scattering processes.[18] In the modern era, Jonathan Rosner and Michael Gronau continue to make extensive use of U-spin in studying CP asymmetry in the weak decays of B mesons.[19]

As mentioned earlier, there was lots of data — some good, some not — that asked for explanation. Both meson and baryon resonances were being produced at a great rate. Knowing which new resonance was real was difficult, especially for a theorist. My favorite way of determining which result to believe was to consult Nick Samios at Brookhaven National Laboratory. He was never wrong!

6. SU(3) Wins

In SU(3) the product:

$$8 \times 8 = 27 + 10 + \overline{10} + 8 + 8 + 1\,. \tag{2}$$

For baryons and baryon resonances, only 10, 8 and 1 multiplets existed. Why there were no 27 or $\overline{10}$ representations, no one knew. The spin-$\frac{3}{2}$ positive parity decuplet, 10, was supposed to have a linear mass spacing, according to the Gell-Mann Okubo mass formula. The mass splitting formula, Eq. (1), simplifies for the decuplet to:

$$M = M_0(1 + a'Y)\,, \tag{3}$$

the familiar linear mass relation. This ordering is illustrated in the figure above. The non-strange N^* (1236) was the lowest state. The strangeness -1 Σ (1385) was next. The strangeness -2 state of the decuplet, the Ξ^*, was predicted to be at 1533 MeV. Once it was found at its predicted mass, the general belief that SU(3) was a good symmetry grew. To confirm this belief, it was necessary to find the last member of the decuplet, the strangeness -3 Ω^-. In 1963, Samios and his group at Brookhaven,[20] found the Ω^-, exactly where it was supposed to be, with a mass of 1672 MeV, once again in accordance with the SU(3) prediction. Joy reigned.

The mesons were even more restricted in representation content than the baryons, with only 8*s* and 1*s* occurring. The question at the time was how to explain the lack of 27 and $\overline{10}$ representations for both baryons and mesons as well as the lack of 10s for mesons.

7. Quarks and Aces

In one of the cleverest and simplest inventions in physics history, this mystery was solved. In 1964, Murray Gell-Mann[21] and, independently, George Zweig,[22] proposed a simple mechanism to get only 10, 8 and 1 for baryons, and 8 and 1 for mesons. They introduced a fundamental triplet of fractionally charged objects, (u, d, s) which Gell-Mann called quarks and which Zweig called aces. The u quark has charge 2/3 while the d and s quarks each had charge $-1/3$.

The simple SU(3) multiplications

$$3 \times 3 \times 3 = 10 + 8 + 8 + 1 \quad \text{for baryons, and} \tag{4}$$

$$3 \times \bar{3} = 8 + 1 \quad \text{for mesons} \tag{5}$$

produced the representations that occurred in nature, for baryons and mesons of various spins and parities. The quarks in the triplets, with fractional charge, were peculiar objects. Whether they could be physically detected was a matter of hot debate at the time. Were they an index symmetry or real physical particles? Many exotic experiments were carried out, including looking for them in oyster shells, with no success. Whatever they were, their introduction solved a longstanding mystery. For several years, the group SU(3) had been successful in many aspects, but there had been no real understanding of what the 3 was. Now we knew — quarks! Later, the fact that no free quarks were observed was embedded into the modern view of confinement, described by the theory of asymptotic freedom and QCD.

A proposal that lent additional credence to the validity of the quark model was the Zweig rule,[22] proposed also by Okubo,[23] that the decay of the ϕ meson into K^+K^- and not into $\rho\pi$ was due to the fact that the ϕ meson is an $s\bar{s}$ composite. This accounts for the narrowness of the decay width.

Once we had the quark model it was easy to understand the structure of the baryon octet and decuplet. The neutron, proton and N^* (1236) were made of u and d quarks, the Λ, Σ and Y_1^* were made of u, d and one s quark. The Ξ^* was made of one non-strange quark and two strange quarks, and the Ω^- contained three strange quarks.

The quark model gave a simple interpretation of Cabibbo's observation[24] in the prequark SU(3) era that the weak current mixed non-strange and strange weak decay amplitudes. It could now be interpreted as a mixture of d and s currents. He found that strangeness non-changing decays dominated, with a coupling $G\cos\theta_c$, compared to the weaker strangeness changing decay strength $G\sin\theta_c$. G is a universal weak coupling strength and θ_c is the Cabibbo angle, 13 degrees.

8. Gluons and Nambu

After fractionally charged quarks and aces were proposed, Han and Nambu[25] introduced an alternative scheme which included three triplets of integrally charged fundamental particles, held together by the exchange of vector gauge bosons, that we now call gluons. Although the integrally charged particles are no longer viable candidates, the concept of the gluon introduced by Fritzsch and Gell-Mann, eventually was experimentally verified, flourished and is now part of our field theory of strong interactions, QCD.

9. SU(6) and Color

Almost immediately after quarks were proposed they were given spin by Beg, Lee, Pais,[26] Pais,[27] Radicati and Gursey[28] and slightly later by Sakita and Wali.[29] When quarks are given a spin, however, there is a spin statistics problem. The SU(6) multiplets come from combining three quarks, each with spin-$\frac{1}{2}$, written as 6.

Baryons are the composites:

$$6 \times 6 \times 6 = 56 + 70 + 70 + 20\,, \qquad (6)$$

where $56 = 10^{3/2} + 8^{1/2}$. (SU(6) may be decomposed as SU(3) flavor $\times$ SU(2) spin). The problem that arose was that 56 is a symmetric combination, and the proton must be an antisymmetric state under the exchange of the constituents. What to do?

O. W. Greenberg,[30] on leave from the University of Maryland at the Institute for Advanced Study (IAS), solved that problem. He invoked parastatistics, now called color, and explicitly wrote down all of the baryon states in an SU(6) $\times$ O(3) model, though he did not call it that. In modern parlance, the symmetry problem for the 56 was solved by combining it with an antisymmetric color singlet giving a totally antisymmetric state.

Only color singlets are allowed in this scheme. (Periodic attempts have been made, at times, to invoke colored states, but to no avail.) Note that Sudarshan and Mahanthappa,[31] in a paper entitled, "SU(6) $\times$ O(3) Structure of Strongly Interacting Particles," also examined this problem, as did Richard Dalitz.[32]

In addition to solving the symmetry problem, the success of the approach described above, also cleared up a then extant problem of exactly how to describe the large catalog of resonances. Were baryon resonances described as composites of 4 quarks and an antiquark or as qqq in an L wave, and were mesons to be regarded, analogously, as $qq\bar{q}\bar{q}$ or $q\bar{q}$ in an L wave? The simplicity of the SU(6) $\times$ O(3) model, which included color, was almost universally accepted, and answered this question. A remarkable transition had taken place over a relatively short time span. Originally, we developed a symmetry description, based on the flavor group SU(3), which, while yielding many useful results, did not have a theoretical underpinning. The quark model of Gell-Mann and Zweig provided the basis for these successes. We have several earlier examples in physics and chemistry, where much was accomplished by developing and exploiting ad hoc models that "worked" and eventually led to the formulation of the true underlying principles. One example is the success in building the periodic chart long before we understood atomic structure, that the nucleus of the atom contained neutrons and protons, and that quantum mechanics was the fundamental theory that led to our true understanding of atomic physics. Another example, a bit later, was the development of the Bohr atom.

10. Combining Internal and Space-Time Symmetries

It was clearly interesting to try to combine internal and space-time symmetries. This effort took place all over the world through 1964 and 1965. My memory — a bit hazy since it was 50 years ago — was of going to the second Coral Gables Conference in January 1965 and hearing presentations by Salam, and several other groups claiming to have solved the problem by invoking the ill-fated symmetry $U(\widetilde{12})$. Shortly thereafter, I went to visit at Weizmann Institute and Harry Lipkin and I began to look at various subgroups of the symmetry. We found that with a subgroup decomposition into a particular SU(6) $\times$ SU(2) we could, within the SU(6), combine internal symmetries with a restricted version of the Lorentz Transformation. We were able to do this for collinear processes such as two body decays (3-point functions), but not for scattering amplitudes (4 point functions). We named the relevant SU(2) space-time symmetry W-spin and, combining it with flavor SU(3), called the combined symmetry $\mathrm{SU}(6)_W$.[33,34] W stood for Weizmann Institute. We did this with constituent quarks and learned that Dashen and Gell-Mann[35] had done similar work but with current quarks. In fact, at the then annual Washington APS meeting in the spring of 1965, Murray rushed up to me and said, excitedly, "Don't worry. Your work is OK." By that time, it had been accepted that $U(\widetilde{12})$ was not a good symmetry, but Murray was pointing out that our $\mathrm{SU}(6)_W$ subgroup symmetry was fine. Barnes, Carruthers and Von Hippel[36] also did analogous work.

The W-spin operators are invariant under Lorentz Transformations in the z direction, so it is a collinear symmetry. The W-spin classification for a particle with arbitrary momentum in the z direction is the same as the classification at rest. The generators of $\mathrm{SU}(2)_W$ are:

$$W_z = \sigma_z/2\,, \tag{7}$$

$$W_x = \beta\sigma_x/2\,, \tag{8}$$

$$W_y = \beta\sigma_y/2\,, \tag{9}$$

where β is the intrinsic parity of spin-$\frac{1}{2}$ particles in the rest frame.

A virtue of this symmetry is that it correctly describes decays that are forbidden in the standard SU(6) approach. For example the decay, $\rho \to \pi\pi$, was forbidden in the usual SU(6) but was allowed in $SU(6)_W$. Later, in 1973, together with Fred Gilman and Moshe Kugler,[37,38] and making use of the Melosh transformation[39] between constituent quarks and current quarks, we successfully analyzed the decay amplitudes of a myriad of meson and baryon resonances that had been produced in a SLAC partial wave analysis of $\pi N \to \pi\pi N$ and $\gamma N \to \pi\pi N$ experiments.[40]

11. Higher Mass Quarks

There had been predictions for the existence of a fourth heavier quark by Bjorken and Glashow[41] in 1964, shortly after quarks were invented and somewhat later in 1970, by Glashow, Illiopoulos and Maiani.[42] These predictions were fulfilled in 1974 by the discovery of the J/Ψ resonance at 3.10 GeV by a BNL group headed by Sam Ting,[43] and a SLAC group headed by Burton Richter.[44] The J/Ψ is a very narrow $c\bar{c}$ resonance. The charmed quark c with a charge of 2/3 has a mass of 1.275 GeV and charge +2/3. Shortly thereafter, the b quark with mass 4.18 GeV and a charge of −1/3 was discovered. The last quark to be found was the top quark, t, with a huge mass of 173 GeV and charge of +2/3. Searches for higher mass quarks have not yielded evidence for any new quarks. With these three heavy quarks, the Cabibbo model for the light quark transitions has been expanded to give the Cabibbo, Kobayashi, Maskawa 3×3 transition matrix.[45] The spectroscopy related to the c and b quarks is vast. In fact, it is much more extensive than that of the original u, d, s system. A prescient paper by Appelquist and Politzer,[46] written before the discovery of the J/Ψ resonance, is a guide to the study of this fertile heavy quark spectroscopy.

12. The Path to Quantum Chromodynamics (QCD)

Just as the path from symmetries led to the invention of the concept of quarks, so the sparkling success of the quark model in so many areas culminated in leading Harald Fritzsch and Murray Gell-Mann to the formulation of Quantum Chromodynamics.[47–49] Quantum Chromodynamics is a non-Abelian quantum field theory of strong interactions. In the QCD Lagrangian, which is of a Yang–Mills type, quarks, which come in three colors, are coupled to an octet of colored gluons. All physically observable systems are SU(3) color singlets. This quark–gluon field theory incorporates confinement of all colored states such as quarks and gluons. It is a major component of the Standard Model of Elementary Particle Physics.

13. Final Comments

Writing this historical review has reminded me of the travails and joys that accompanied our progress over the quarter century involved. We proceeded for a long time without a fundamental theory, piecing together an array of disparate experimental clues, interspersed with occasional clever theoretical constructs. The process marked the importance of invoking new mathematical techniques, in this case, group theory. The gradual acceptance of the role of unitary groups, in particular, marked a big change in the attitudes of physicists, many of whom preferred more analytic approaches. Fortunately, Harald Fritzsch and Murray Gell-Mann were clever enough to produce a grand synthesis of the earlier endeavors that culminated in the formulation of Quantum Chromodynamics.

References

1. S. Sakata, *Prog. Theor. Phys.* **16**, 686 (1956).
2. M. Gell-Mann, The eightfold way: A theory of strong interaction symmetry, California Institute of Technology Report CTSL-20 (1961), unpublished; *Phys. Rev.* **125**, 1067 (1962).
3. Y. Ne'eman, *Nucl. Phys.* **26**, 222 (1961).
4. M. Gell-Mann, Strange particle physics. Strong interactions, in *Proc. Int. Conf. High Energy Phys.*, CERN, 1962, p. 805.
5. S. Okubo, *Prog. Theor. Phys.* **27**, 949 (1962).
6. R. E. Behrends and A. Sirlin, *Phys. Rev.* **121**, 324 (1961).
7. C. A. Levinson, H. J. Lipkin, S. Meshkov, A. Salam and R. Munir, *Phys. Lett.* **1**, 125 (1962).
8. M. Ikeda, S. Ogawa and Y. Ohnuki, *Prog. Theor. Phys.* **22**, 715 (1959); **23**, 1073 (1960).
9. S. Sawada and M. Yonezawa, *Prog. Theor. Phys.* **23**, 662 (1960).
10. C. A. Levinson, H. J. Lipkin and S. Meshkov, *Nuovo Cimento* **23**, 236 (1962).
11. C. A. Levinson, H. J. Lipkin and S. Meshkov, *Phys. Lett.* **1**, 44 (1962).
12. C. A. Levinson, H. J. Lipkin and S. Meshkov, *Phys. Lett.* **7**, 81 (1963).
13. E. C. G. Sudarshan, Consequences of SU(3) invariance, in *Proc. Athens Conf. on Newly Discovered Resonant Particles*, Athens, 1963.
14. C. Dullemond, A. J. MacFarlane and G. Sudarshan, *Nuovo Cimento* **30**, 845 (1963).
15. C. Dullemond, A. J. MacFarlane and G. Sudarshan, *Phys. Rev. Lett.* **10**, 423 (1963).
16. A. J. Macfarlane and E. C. G. Sudarshan, *Nuovo Cimento* **31**, 1176 (1964).
17. S. Meshkov, C. A. Levinson and and H. J. Lipkin, *Phys. Rev. Lett.* **10**, 361 (1963).
18. S. Meshkov, G. A. Snow and G. B. Yodh, *Phys. Rev. Lett.* **12**, 87 (1964).
19. M. Gronau and J. L. Rosner, *Phys. Lett. B* **500**, 247 (2001).

20. V. E. Barnes *et al.*, *Phys. Rev. Lett.* **12**, 204 (1964).
21. M. Gell-Mann, *Phys. Lett.* **8**, 214 (1964).
22. G. Zweig, An SU(3) model for strong interaction symmetry and its breaking, CERN-8419-TH-412 (1964).
23. S. Okubo, *Phys. Lett.* **5**, 1975 (1963).
24. N. Cabibbo, *Phys. Rev. Lett.* **10**, 531 (1963).
25. M. Y. Han and Y. Nambu, *Phys. Rev. B* **139**, 1006 (1965).
26. M. A. B. Beg, B. W. Lee and A. Pais, *Phys. Rev. Lett.* **13**, 514 (1964).
27. A. Pais, *Phys. Rev. Lett.* **13**, 175 (1964).
28. F. Gursey and L. A. Radicati, *Phys. Rev. Lett.* **13**, 173 (1964).
29. B. Sakita and K. C. Wali, *Phys. Rev. Lett.* **14**, 404 (1965).
30. O. W. Greenberg, *Phys. Rev. Lett.* **13**, 598 (1964).
31. K. T. Mahanthappa and E. C. G. Sudarshan, *Phys. Rev. Lett.* **14**, 163 (1964).
32. R. H. Dalitz, *Proc. Oxford Int. Conf. on Elementary Particles*, Oxford, England, 1965.
33. H. J. Lipkin and S. Meshkov, *Phys. Rev. Lett.* **14**, 670 (1965).
34. H. J. Lipkin and S. Meshkov, *Phys. Rev.* **143**, 1269 (1966).
35. R. Dashen and M. Gell-Mann, *Phys. Lett.* **17**, 142 (1965).
36. K. J. Barnes, P. Carruthers and F. von Hippel, *Phys. Rev. Lett.* **14**, 82 (1965).
37. F. J. Gilman, M. Kugler and S. Meshkov, *Phys. Lett. B* **45**, 481 (1973).
38. F. J. Gilman, M. Kugler and S. Meshkov, *Phys. Rev. D* **9**, 715 (1974).
39. H. J. Melosh, *Phys. Rev. D* **9**, 1095 (1974).
40. D. Herndon *et al.*, Partial wave analysis of the reaction $\pi N \to \pi\pi N$ in the CM range 1300–2000 MeV LBL-1065 (1972), SLAC – PUB-1108, unpublished; R. Cashmore, private communication.
41. J. D. Bjorken and S. L. Glashow, *Phys. Lett.* **11**, 255 (1964).
42. S. L. Glashow, J. Illiopoulos and L. Maiani, *Phys. Rev. D* **2**, 1285 (1970).
43. J. J. Aubert *et al.*, *Phys. Rev. Lett.* **33**, 1404 (1974).
44. J. E. Augustin *et al.*, *Phys. Rev. Lett.* **33**, 1406 (1974).
45. M. Kobayashi and T. Maskawa, *Prog. Theor. Phys.* **49**, 652 (1973).
46. T. Appelquist and H. D. Politzer, *Phys. Rev. Lett.* **34**, 43 (1975).
47. H. Fritzsch and M. Gell-Mann, *Proc. Int. Conf. on Duality and Symmetry*, Tel Aviv, 1971.
48. H. Fritzsch and M. Gell-Mann, *Proc. Int. Conf. on High Energy Physics*, Chicago, 1972.
49. H. Fritzsch, M. Gell-Mann and H. Leutwyler, *Phys. Lett. B* **47**, 365 (1973).

How I Got to Work with Feynman on the Covariant Quark Model

Finn Ravndal

Department of Physics, University of Oslo,
Blindern, N-0316 Oslo, Norway

In the period 1968–1974 I was a graduate student and then a postdoc at Caltech and was involved with the developments of the quark and parton models. Most of this time I worked in close contact with Richard Feynman and thus was present from the parton model was proposed until QCD was formulated. A personal account is presented how the collaboration took place and how the various stages of this development looked like from the inside until QCD was established as a theory for strong interactions with the partons being quarks and gluons.

Already in the high school I was interested in elementary particle physics. Most of my knowledge stemmed from popular articles in Norwegian papers and magazines. Buzzwords like parity violation and strangeness were explained, while more abstract ideas like S-matrix or unified spinor field theory didn't make much sense except for the promise that there might be something deeper underneath. For me a particle was a track in a photographic emulsion or in the newly invented bubble chamber.

A great inspiration was the beautiful book *The World of Science* where I for the first time felt that I got acquainted with modern physics and the people working in the field.[1] In a chapter about elementary particles there was a picture of professors Feynman and Gell-Mann standing in front of a blackboard filled with diagrams and new particle symbols. They seemed very happy with what they were doing and to me it looked like an ideal occupation. This book also set Caltech on my map since it originated there.

A few years later in 1964 at the university I came across the article by Gell-Mann *et al.* in Scientific American about the Eightfold Way and Lie groups.[2] It represented a real revolution — and beautiful so! The walls in my dorm room were soon filled with pictures of octets and decuplets. Here was also the prediction of the Ω^- which had just been discovered with the right properties. Finally there was some order in the ever increasing number of new particles which didn't seem so elementary longer.

Around the same time came the news that this new symmetry could be explained by assuming that the particles were built up of quarks. These simple ideas explaining so much was something I could follow. From then on I knew that the structure of elementary particles was what I wanted to study more. There were no courses in nuclear or particle physics at the technical university where I studied. But from the elegant book by Okun on weak interactions I learned a lot and it showed me how particle physics was done in practice.[3] The copy I had was in Russian. That was fine since I was at that time into learning a new language. That particular book meant very much to me and set me on the track I wanted to follow.

In the summer of 1965 I was hired to work at CERN as a summer student in the Track Chamber division in an experimental group headed by B. French. My real task was to scan bubble chamber pictures for tracks coming from the decay of new particles, but soon I was asked to work more as a group theoretician than a scanner. That meant that I could spend long hours in the library reading about $SU(6)$ symmetries and the quark model. During this time I also found the book by Sakurai very useful.[4] After another year I got my engineering degree and was hired as an assistant in the physics department.

The following year I learned a lot about Lie groups and algebras from a Japanese guest professor. Knowing that the symmetry group of the 2-dimensional harmonic oscillator was $SU(2)$ and the one for the Kepler motion was $SO(4)$, I concluded from the relation $SO(4) \simeq SU(2) \times SU(2)$ that the physics of the hydrogen atom ought to be solvable in terms of the creation and annihilation operators of two 2-dimensional oscillators. Using parabolic coordinates I actually managed to show that and thus got my first publication.[5]

But it was clear that any future work on quarks and particle physics would be difficult to pursue in Norway. The summer of 1967 I spent at a school on particle physics organized at the Bohr Institute in Copenhagen. It turned out to be more of an advanced conference with current algebra as the hottest topic. To my disappointment I found most to be way over my head. But meeting other students in the same situation was a great inspiration. The following year I had to do my military service and happened to be stationed at the same research institute as Per Osland. He was in the process of applying to American universities for graduate studies. It didn't take much time to get me convinced to do the same thing.

In the spring of 1968 I was admitted to Caltech and left for California in August. The only scientific paper I brought with me, was a preprint from

Dubna on the quark model. It presented a relativistic calculation of the magnetic moment of a proton made up of three massless quarks in a spherical cavity.[6] What I found so nice and fascinating, in addition to being in Russian, was the use of the Dirac equation in describing quarks. But it also showed that assigning them a new quantum number taking three different values, one could solve the spin-statistics problem. I hoped I would be able to make use of this in the coming years.

1. Graduate Student at Caltech

During the first week at Caltech in the fall I went through two gruelling days with placement exams. As a result I also got a teachning assistantship and was assigned to a course on intermediate quantum mechanics taught by John Bahcall. One of the first problems he gave was to estimate the flux of neutrinos from the Sun. I soon realized that physics at Caltech meant getting out measureable numbers from theoretical formulas. I got a desk in a laboratory in the Kellogg building for nuclear physics together with David Schramm.

Initially I heard very little about quarks. But at one of the first institute colloquia the invited speaker was J. Bjorken from SLAC who talked about deep inelastic scattering and the parton model. The proton contained an infinite number of these new particles which seemed to be quarks. The atmosphere was electric. Feynman in the front row was smiling and Gell-Mann beside him was visibly excited. The old quark model seemed no longer to be relevant in this context and replaced by a more fundamental description. This was like a shock to me.

Everyone seemed to be fascinated with these new ideas but none of the graduate students or faculty I met were actively working neither on the quark nor on the parton model. In the theory group the research activity seemed to be concentrated around hadronic duality and in particular the recently proposed Veneziano scattering amplitude.

Most of my time went to take courses. One was an introductory, theoretical course on particle physics given by Feynman himself. He started out at the simplest, phenomenological level. From data in the Particle Tables he was trying to sort out all the known particles according to quantum numbers, masses and interactions. It was obvious that he searched for some underlying order and systematics. Quarks were mentioned, but at the most rudimentary level which I already knew well. They were called A, B and C. Some of the hadronic resonances could be grouped along approximately straight Regge trajectories so that the squared masses of the particles increased linearly with their spins. This was significant, hinting at internal dynamics of constituents. Deep inelastic scattering was discussed in a lecture or two and his parton model was briefly presented. No mention of partons being quarks. The partons were just unknown, fundamental particles which he had proposed earlier in the year trying to understand the outcome of future experiments to be done at the ISR accelerator under construction at CERN.

At the end of the fall semester I realized I needed a thesis advisor and found Steve Frautschi who tried to put me on the path of Regge theory. One of the first warm-up problems he gave me, was to generalize Yang's theorem forbidding the decay of the ρ-meson into two photons, to higher resonances on the ρ-trajectory. That was very useful because it forced me to get acquainted with the helicity formalism. Regge physics itself I did not find so tempting. But the next term I started reading more about scattering theory, Glauber description of nuclear scattering and the eikonal approximation.

During this term I followed a more advanced course on particle theory, again given by Feynman. Instead of quantum field theory which I had expected, it was again a more phenomenological course on current commutators, infinite momentum frames and dispersion relations. Late in the spring he finished with three lectures on deep inelastic electron scattering and the parton model. A possible connection between quarks and partons was not discussed. Instead he seemed to convey the idea that the parton model went beyond known physics and would give a completely new description of hadrons. While he had surveyed the experimental situation in particle physics during the previous term, he now seemed to go more into theoretical aspects. The lectures were hard to follow and I could make little use of the material he presented.

Instead I started to read up on the non-relativistic quark model on my own. In a recent paper by Faiman and Hendry they had extended the model to also describe hadron resonances by letting the quarks become orbitally excited.[7] They were assumed to be confined within a harmonic oscillator potential which had the attractive feature that explicit calculations involving the quark wave functions could easily be done analytically. From previous work I soon realized that this could be done even simpler in this symmetric quark model by using algebraic methods based on raising and lowering operators. Together with Jon Mathew who that term gave a course on advanced quantum mechanics, I reformulated the model along these lines and could soon calculate strong transition rates in a straightforward way.

During this time I became aware of the paper by Copley, Karl and Obryk from the Dalitz group in England. They used the same quark model to calculate helicity amplitudes for radiative decays rates of nucleon resonances.[8] Not only did most of the rates agree with experiments, but even the absolute signs of the different amplitudes came out right. This represented really a new, great success for the quark model. Previously it had been used primarily to explain the magnetic moments of the ground state baryons and vector mesons and the radiative transition moment for the $J = 3/2$ nucleon delta resonance. The agreement with measurements was now so significant that the quark model could no longer be just laughed away.

At the end of the semester Clem Heusch, who had given a very good course on experimental particle physics, asked me if I wanted to work at the Caltech electron synchrotron in the basement during the summer. He was heading a project within Bob Walker's experimental group where mesons were photoproduced on nuclear targets. They could measure cross-sections in different channels and were thus able to isolate the production amplitudes for different nucleon resonances. My job was to analyze them within the quark model framework established by Copley, Karl and Obryk. Finally I was in an environment where the quark model was at the center of most discussions!

During the summer it became clear to us that the measured photoproduction amplitudes for the nucleon resonance with mass 1720 MeV couldn't be explained if it was assigned the the known **56** multiplets of the symmetry group $SU(6)$. We found instead that it could be assigned to a **70** multiplet with total quark angular momentum $L = 0$. It was the first candidate established for this more exotic multiplet. I gave a talk on these calculations at an APS meeting in Boulder late in the summer. A more general report was presented by C. Prescott at the Electron-Photon Conference at Daresbury in September. There I had my first encounter with members of the enthusiastic Dalitz group who had opened up these new applications of the quark model. Heusch and I wrote up the results about the 1720 MeV resonance in a short paper which was published in the PRL a few months later.

2. Feynman and the Covariant Quark Model

By now I had moved in on the top floor of the new Downs–Lauritsen building with the rest of the high energy theory group. There I shared an office with a fellow graduate student, Chris Hamer from Australia. The office was just a few doors down the corridor from Feynman and Gell-Mann who bracketed the office of our beloved secretary Helen Tuck. Opposite my office resided George Zweig, the co-inventor of the quark model. Together with his foreign visitors he was drawing beautiful quark duality diagrams which could be pushed and pulled so to give both a s- and t-channel description of a hadronic reaction from one and the same underlying diagram. That was neat and seemed to be phenomenologically useful, but this approach seemed to be far away from the quark model I felt most comfortable with.

It was great to have followed Feynman through two advanced courses, but it didn't set me on my own line of investigations. His parton model and its applications to deep inelastic scattering which were made at SLAC, didn't seem to be much on his mind. I saw little purpose in asking him to be my thesis advisor. I had had so far no discussions with him. More exciting were the weekly Wednesday seminars directed by Gell-Mann. These were concentrated around scale invariance in quantum field theory in order to understand the new experimental results coming out of SLAC. My fellow student Rod Crewther felt at home here. But for me the seminars were often too theoretical and abstract. Feynman was usually not there and his partons were dismissed as 'putons'. Frautschi instead suggested to think about the physics which soon would come out from the new accelerator which was getting ready at Serpukhov in Russia. Together with Chris Hamer I therefore started to consider available models for such processes which were all based on Regge theory. This approach was the best we had to describe high energy collisions and had been pursued at Caltech for many years. Several other graduate students were working on related problems under the guidance of Frautschi. Among them were Bob Carlitz and Mark Kislinger who were both one year ahead of me.

One day early in the fall Feynman was suddenly standing in the door to my office, smiling broadly as he often did. He wanted to know everything about the quark model I had used and the calculations of strong and electromagnetic decay amplitudes. It sounded like he never had heard about quarks before. I didn't know then that he had been the advisor to George Zweig when he first started to think in terms of what he then called aces as fundamental constituents of hadrons. Someone in the synchrotron group must have told Feynman of the successes of the model. From that first encounter on we met more or less every day in the next five years. Quarks and the quark model was mostly on the agenda. Mark Kislinger was also drawn automatically into these discussions since Feynman already had established similar contacts with him.

Parallel with this very encouraging development, my work with Chris Hamer continued during the fall of 1969. Soon we realized that we could use the eikonal method to generate higher corrections or cuts to lowest-order predictions from Regge theory. The idea was simple and elegant. It would predict cross-sections starting to increase slowly at the highest energies. Frautschi liked it and we started soon to churn out numbers for different reactions. As soon as we received the first results from Serpukhov, we saw that our model worked quite well. When I told Feynman about these calculations, he didn't show much interest. He looked more forward to the results which soon would come out from the ISR accelerator at CERN in Geneva and where he hoped to see the $1/x$ scaling in the final state as a central prediction of the parton model.

Late in the fall it was announced that Gell-Mann had received the Nobel Prize in physics for his contributions to theoretical elementary particle physics, in particular the $SU(3)$ classification scheme. No mention of George Zweig and his much more detailed quark model. It was not fair. But we had a big celebration in the new conference room. Also a Nobel Prize went to Max Delbrück in biology. A great day for Caltech and us students.

In my discussions with Feynman we had become convinced that the quark model had to include a coupling to mesons which we called the 'Mitra term' after the Indian physicist who had previously studied it.[10] It was necessary in order to describe strong decays where the momentum of the final-state meson is small. This new term depended on the internal quark momentum and would therefore not be suppressed in these cases. From my earlier quark investigations I knew that it would automatically appear if a meson coupled to another hadron through the divergence of the four-dimensional axial vector current. Feynman found this so significant that he suddenly showed up one day with the first ideas for what later would be called the covariant quark model. While Mark had previously shown little interest in the old quark model, he was now suddenly all excited. We now had a Lorentz-invariant Hamiltonian giving the squared masses of hadrons in terms of the Dirac four-momenta of the quarks and the squares of their invariant separation. It was thus a relativistic generalization of the non-relativistic, harmonic oscillator quark model. By minimal coupling to external vector and axial vector fields, we could then isolate the hadronic vector and axial vector currents expressed in terms of quark variables. It was compact in its description and we found it to be very elegant. One nice feature Feynman stressed, was that it didn't assume any explicit values for the quark masses. For a ground state baryon they came automatically out to be one third of the baryon mass. We could early see that it would automatically reproduce many of the features of the non-relativistic model we needed to have in place. But all its consequences had now to be worked out in detail and checked against measurements.

Early in 1970 Feynman had gone to a meeting on the East Coast and reported on my work with Chris on the eikonalized Regge model for the Serphukov total cross-sections without telling me in advance. On the first day back at Caltech he came to my office, very excited. He told me that on the plane back to LA he had been thinking about these collisions in terms of his parton picture and had come up with a very simple model which predicted that the total cross-sections had to increase as the squared logarithm of the collision energy when this went into the asymptotic region. This simple behavior seemed to be consistent with the observations. All hadronic total cross-sections should approach the same asymptotic value. He sketched the derivation on the blackboard. In the center-of-mass frame for the collision, it is the partons with the smallest momenta and therefore with a known distribution, which will dominate the process. So one can express the total cross-section in terms of parton–parton cross-sections. But each parton is again made up of other partons so that the energy dependence of a parton–parton scattering must be the same as for the hadronic process we started with. This bootstrap idea is similar to what later could be formulated in terms of the renormalization group. As a result he could derive an integral equation for the cross-section which can then be shown to have the above-mentioned analytical solution. The meaning was that we should study this model more carefully later. But we never came back to it. We were too busy with the covariant quark model. Some twenty years later I presented this derivation at a meeting at Yale.[11]

During the spring of 1970 together with Hamer and Frautschi we finished two papers on the eikonalization of Regge amplitudes and sent them off to the Physical Review.[12] Late in the spring I was invited by G. Chew up to Berkeley to give an institute colloquium about these calculations. All the local big shots were in the audience and I felt greatly honored. But it was not the physics which my heart was beating for. I received $50 to cover my expenses and took the Pacific Coast Highway back to Los Angeles and Pasadena.

Together with Kislinger and Feynman I had during the same time used our covariant quark model to calculate every known strong and electromagnetic amplitude which had been measured or could be of interest. We had discovered that the model had a serious flaw related to the removal of unphysical degrees of freedom. It resulted in a form factor which was just wrong and showing up in all transitions. We had to replace it with a more *ad hoc* or phenomenological choice which we just called the 'fudge factor'. There were so many other dramatic assumptions in the model, that we didn't worry too

~~to be submitted to Physical Review Letters.~~

CALT-68-279

AEC RESEARCH AND
DEVELOPMENT REPORT

The $\Delta I = 1/2$ Rule from the Symmetric Quark Model*

R. P. FEYNMAN, M. KISLINGER, and F. RAVNDAL

California Institute of Technology, Pasadena, California 91109

(Received October , 1970)

ABSTRACT

The $\Delta I = 1/2$ rule for the weak non-leptonic hyperon decays will result from quark currents interacting at a point, if the quarks obey Bose statistics.

*Work supported in part by the U. S. Atomic Energy Commission, under Contract AT(11-1)-68 of the San Francisco Operations Office.

Fig. 1. Front page of preprint where quarks are suggested to be bosons.

much about this particular point. What we liked and gave us confidence in this rough construction, was the following properties:

1. It was covariant so that calculations could be done in any Lorentz frame.
2. No explicit quark masses appeared in the model.

3. Excited hadrons were on straight Regge trajectories.
4. The vector current was conserved in the symmetric limit and the axial current satisfied PCAC.

No other quark model had these desirable properties. Needless to say, it also reproduced all the good results of the non-relativistic model. We had nothing new to say about the spin-statistics problem. Feynman liked to say that the Pauli theorem relating spin and statistics only applied to free particles and therefore not to permanently confined quarks. Thus he joked that they could just as well be bosons and there would be no problem. Gell-Mann in his seminars had already discussed both parastatistics and the color quantum number of Han and Nambu. We felt this was a much deeper problem which would be solved at a later time. But one day we realized that if quark really had symmetric statistics as for bosons, the $I = 1/2$ selection rule in non-leptonic, weak decays would be explained. Feynman became very excited, wrote together a short letter and submitted it to PRL.[13] But a couple of weeks later we decided to withdraw it since we had been informed that the same observation had been made previously by others. The arguments were later included in our main article about the covariant quark model.

In the fall it became clear that we had enough results and Feynman wanted to write it all up. We were all three happy with the results. He suggested to write the first version himself. I checked the formulas and added in the references. In order to check the numerical calculations, Feynman had gotten hold of a small Wang office computer in the Downs building. He had learned to program it in some cryptic language and loved to see all the numbers come out. The final paper was sent off to the Physical Review just before Christmas 1970.[14]

Although very few else around us took much interest in what we had done, this two-year long, concentrated effort by Feynman had been necessary for him to become convinced about the physical existence of quarks. We realized certainly that our version of the quark model was just a better crutch for that purpose and would not have a lasting impact. But for Feynman the quarks themselves were from then on real and he was now open to identify them with his proposed partons.

At the start of the following term I was told by Frautschi that I would be able to graduate in the spring with a thesis based on the quark model. In order to finish off with something new, I extended the covariant quark model to calculate radiative transitions with off-shell photons. The resulting amplitudes could be measured in the electroproduction of nucleon resonances.[16]

On February 9, 1971 Southern California was rocked by the strong San Fernando earthquake. That afternoon Feynman was scheduled to give his first, regular seminar about the parton model at Caltech. The building was shaking from aftershocks which didn't seem to worry him while the rest of us were pretty nervous. This was the first time that his partons were officially identified with quarks. From now on it would be the quark–parton model as already explored at SLAC for more than two years. The same day Harald Fritzsch had also arrived in Pasadena to work with Gell-Mann.

Some weeks later Feynman asked me if I wanted to stay on for another two years in a postdoc position starting in the fall. I was elated, hoping that I could then devote myself to the parton model instead. It was obvious that it represented the future. In addition, this new position meant that I didn't have to go back to Norway or find a new job some other place.

3. Postdoc at Caltech

The next Electron-Photon Conference took place at Cornell in August, 1971. Feynman went there and wanted me to come with him. During the meeting he was highly sought out by people who wanted to see him and discuss with him. Sometimes I felt that the main reason for me being there was to act like a 'bodyguard' for him. From the talks we became more and more convinced that quarks and partons were behind everything that was presented. Feynman was very happy. During a talk on what was then called the dual parton model which we never had understood, Feynman pushed his elbow into my side when he saw on a slide that the model could even explain a small bump around 3 GeV in the $\mu^+\mu^-$ invariant mass in new Drell–Yan cross-section data. He didn't believe it. Two years later this bump became the J/ψ particle.

On the plane back to Los Angeles, we encountered bad weather and Feynman became visibly nervous. This was in stark contrast to his careless appearance during the earthquake half a year earlier. As a distraction from this bumpy ride he wanted that we should plan our activities at Caltech in the fall. They would be concentrated around a new lecture course he would give on the evidence for quarks and partons. For every lecture he would prepare notes which I then should correct, elaborate and make publishable in a joint book. In addition he wished to terminate a project with a graduate student who was working on a QED problem.

It didn't take many weeks before I concluded that I couldn't continue on the book project. The first reason was that I felt that I already was much too close to Feynman and identified with him in almost everything I did.

To a large degree that was the actual situation and I had to establish some kind of independence, even if he had asked me to work with him. Secondly, when I read his notes with his own words and explanations, I also felt it like a sacrilege to the rest of the physics community if I should start to modify his own formulations. He accepted that and the job was given to another graduate student, Arturo Cisneros. He also let the notes essentially intact and the book was published the following year.[15] I myself skipped to a large extent the lectures since I felt that I knew it all. Instead I finished a paper on electroproduction of nucleon resonances in the quark model[16] and started to consider new applications to diffractive and weak production in neutrino scattering.[17]

At this time the light-cone current algebra of Fritzsch and Gell-Mann seemed to give a better theoretical understanding of the scaling seen in deep inelastic electron scattering.[18] Feynman himself didn't think so and showed little interest. The main reason was the fundamental mass scale which was implicit in his model and manifested itself for instance in the transverse momentum cut-off of the hadrons in the final state. The current commutators on the light cone were abstracted from free quark theory and did not contain such a quantity. From the over-all structure of this description applied to inclusive scattering it was clear that it could reproduce many of the parton model results. This was in particular demonstrated by my colleagues Tony Hey and Jeff Mandula.[19]

In my own discussions with Feynman we tried instead to extend our quark model so to give a more symmetric description of hadronic couplings. Previously this had been done via PCAC where we could replace a meson field with the divergence of the axial current. We had in mind the dual diagrams which George Zweig and collaborators had considered. John Schwarz had by that time joined the theory group. In his string theory such couplings could easily be constructed. Together with the straight Regge trajectories in that description we thought that there could be some connection to our harmonic quark model, but we didn't pursue this much further.

When Feynman continued his lectures in the spring of 1972 he extended the parton model to describe the final hadronic state in terms of quark fragmentation functions. This opened up a new chapter of applications and we started to talk about jets in the final state. Together with two postdocs Mike Gronau and Yair Zarmi I started discussions about how these ideas could be tested in deep inelastic scattering experiments. Later in the year this was written up in joint paper.[20]

During that term Feynman went to Hungary to take part in a neutrino conference at Balatonfüred. He came back fired up with the first quantitative experimental confirmation of the parton model and the fractional quark charges. This was the measurement of the famous factor of 5/18 which related the deep inelastic electron scattering cross-section to the corresponding cross-section with neutrinos. Around this time he had started to think that the charge of a single quark could be measured directly by summing up the charges from the jet fragments in the final state. This possibility caught the attention of Glennys Farrar and Jon Rosner.[21] In the fall I travelled to Fermilab with Feynman where we among others met Stan Brodsky who also had considered this proposal in his collaboration with Farrar. He argued convincingly that such a measurement would not be possible. Feynman realized that he had been wrong and was visibly shaken afterwards.

In the meantime Murray Gell-Mann and Harald Fritzsch had returned from CERN where they had spent the last year. The quarks in the light-cone algebra had now three colors. In addition to solving the spin-statistics problem in hadronic spectroscopy, Rod Crewther had shown from the Adler anomaly of the axial current that the decay $\pi^0 \to \gamma\gamma$ got the right rate when multiplied by this factor of three. It also meant that the cross section for electron–positron annihilation into hadrons at high energies would be three times larger than what followed from just adding up the three squared quark charges which were expected to contribute. But both the light-cone approach and the parton model said that the ratio of this cross section to that of annihilation into a muon pair, should be constant with increasing energy. Instead the available data showed that the ratio continued to rise with increasing energy. These results seemed to undermine our whole understanding.

Feynman didn't seem to be so worried. His parton model was an attempt to give a framework for high energy hadronic reactions built upon as few assumptions as possible. There was no Lagrangian or particular field theory behind, but just the most basic properties of special relativity and quantum mechanics. And with this extra, built-in mass scale the underlying theory had to represent really new and different physics not contained in conventional quantum field theory. This became especially clear in a discussion between Feynman and Fritzsch about the final state in hadronic electron–positron annihilation. While Feynman talked about and drew pictures of two hadronic jets coming out in the directions of the produced quarks, Fritzsch argued instead for having hadrons in all directions.

At the Batavia conference in September 1972 Gell-Mann presented his new understanding of strong interactions worked out together with Fritzsch and J. Bardeen during the year at CERN. The three new color charges represented an invariance under local gauge transformations governed by the group $SU(3)$. Observed hadrons should be color singlets described by a Yang–Mills field theory with eight colored gluons. Such a field theory had many times earlier come up as a possibility in Gell-Mann's seminars, most often in connection with chiral invariance. But this time it was much more convincing. It could be the definite theory of strong interactions with a structure very similar to QED. It was already christened QHD for Quantum Hadrodynamics but was renamed QCD the year after.

The question was now what one could do with it. It was a beautiful proposal, but in many ways too good to be true. How could it be compatible with the parton model and in particular with Feynman's description of quark fragmentation and jets which the now standard light-cone formulation said very little about? Feynman himself found it interesting but remained sceptical. For instance, in a field theory with gluons it would be difficult or impossible to recover the simple separation of quark distribution functions in the initial state and quark fragmentation functions in the final state which was so important for phenomenological applications of the parton model. Similar problems would arise in Drell–Yan production and the final state in e^+e^- annihilation. We stuck or clung to the belief that the parton model was something else than standard quantum field theory.

I continued with the quark model and finished also a couple of papers on the parton model. In one I showed that there should be no azimuthal dependence in the hadronic distribution of deep inelastic scattering resulting from quark fragmentation in a jet,[22] a result which survived the later and more accurate treatment in QCD. But even if no one around me started to work on Yang–Mills theories, it was certainly discussed. And the use of them in electroweak unification got more attention for the first time. Feynman had never shown any particular interest in these developments. When he once was asked why, he answered that one could not discover new physics in the electroweak sector from the requirement of making a theory renormalizable. It sounded like he thought that also in this sector one would need a more radical approach. After these comments there would be nothing more said about unified theories. On my own I tried to orient myself and wanted to consider the possibility of seeing charmed quarks in current-induced reactions at high energies. Together with a graduate student I wrote a paper about this within the parton model.[23] When I told Feynman about our results, he got almost upset. He saw no reason to introduce new degrees of freedom in a situation where there still were so much uncertainty.

My two years as a postdoc were now running out. In the meantime I had already secured a permanent job at my old university in Norway. But

Feynman meant that I could find a better job at a more exciting place. And to my great surprise he could a short time afterwards tell me that he had secured money for me to stay another year in my postdoc position. That was a relief, but it also meant that I didn't have a permanent job to return to the year after.

Late in the spring 1973 we heard that the beta-function for QCD had been calculated on the East Coast and found to be negative, implying asymptotic freedom. The implications for deep inelastic scattering we already knew from the lecture notes of Sidney Coleman. My first reaction was great exhilaration since suddenly the treatment of partons as free particles could be understood. But Feynman showed little or no interest in this result. That was surprising since his parton model now had a field-theoretic formulation. One reason was the unsettling situation with the total e^+e^- cross section for which the latest experiments at the Cambridge accelerator still gave values much larger than expected. The factor of three due to the new colors didn't seem to be the solution.

Back at Caltech in the fall we in the younger generation realized that renormalization and the calculation of Feynman diagrams would be necessary in order to participate in the exploration of the new QCD. But this was a direction of particle physics for which we were not prepared, in spite of having Feynman and Gell-Mann around us on a daily basis. There had been very little or no quantum field theory in standard courses with the weight instead on more phenomenological aspects. In one of his Wednesday seminars Gell-Mann wanted to discuss the renormalization group and its use in QCD. It was not of much help and we felt disappointed, expecting more from one of the originators of this fundamental method.

Instead of going into all the new and detailed calculations having to do with applications of QCD, I stuck with the original parton model and all its predictions which seemed to be confirmed in an increasing number of new experiments. I got engaged in a new project together with Thom Curtright and Jeff Mandula which we called the covariant operator parton model.[24] This new formulation enabled us to derive Feynman's results more systematically based on the properties of the quarks fields entering the current matrix elements for the different processes. It gave us a certain satisfaction although we knew that it was most likely doomed. In particular, there were no gluon degrees of freedom and therefore also no gluon jets.

Feynman didn't take much interest in what we were doing, but Gell-Mann asked me to give a seminar about it. What instead caught Feynman's attention was 2-dimensional QED and its use as an illustration of the parton model.[25] It gave exact scaling and could also describe the fragmentation in the final state. In addition, the partons were permanently confined by a linear potential. We had earlier been reminded about the attractiveness of this potential for the quark model by Ken Kauffmann who was graduate student at the time when we worked with harmonic confinement. He had pointed out that if the potential was due to some underlying field theory, then it ought to be linear since only in that case does it have a simple Fourier transform, namely $1/Q^4$. Some years later this idea would be a central part of the dual quark model as formulated by Fred Zachariassen and his collaborators.[26]

In December that year we drove down in Feynman's car to the particle physics meeting at Irvine. The first experimental results were presented confirming the small scaling violations in deep inelastic electron scattering which should follow from QCD. The original parton model seemed to some of us to be defeated and we worried even more for the implied lack of factorization between initial and final states which were so central to its successes. But Feynman was unperturbed, partly because the e^+e^- cross section was still not understood. At the meeting Abdus Salam even proposed that it showed the existence of an electromagnetic Pomeron, an idea Gell-Mann in the audience found laughable. We headed back to Pasadena and Xmas vacation, confused and not knowing that this great bewilderment would be gone in less than a year.

My last term at Caltech in the spring 1974 went primarily into finishing up older projects and preparing for the future. I got a position by Nordita in Copenhagen and looked forward to going back to Europe after six wonderful years at Caltech. In this period I had been so fortunate to be present when QCD was established as a possible fundamental theory for strong interactions. But in order to verify it and extract its experimental consequences one would need methods of quantum field theory. Many of us felt we were ill equipped in this area with the emphasis on phenomenology that we had had. New results and insights were soon coming out from groups on the East Coast where they were better prepared in pure theory. A certain gloom settled in the last months I was there. Some of us felt that our group had lost the leading rôle we previously had felt we had. Half jokingly it was said that the only contribution from Caltech that would remain, would be the u, d and s quark names and not the script p, n and λ used until then by these new QCD theoreticians.

On one of my last days in Pasadena I was sitting in the conference room where so much of the new physics had been presented and discussed in the previous years. With me was Steve Frautschi who also had felt that the group had lost some of its momentum because of this new direction in particle physics opening up. And that was perhaps a paradox. Feynman and Gell-Mann were to a large degree the founders of modern particle physics, but now it seemed that a new breed of younger people with a different background would be needed there. It was time for me to leave Caltech.

4. Feynman Seen a Little from the Inside

I never understood why I should be the one of the many students and post-docs at Caltech that Feynman chose to engage himself with in such a close way for five years in this period. It started obviously out from his desire to get a fast start on the evidence for quarks and the quest to understand their relation to his own partons. But starting out as an ordinary student–professor collaboration, it changed soon also into a personal friendship which was to a large extent independent of a scientific collaboration. In some ways it could have been similar to an old-fashioned assistantship where the assistant fulfils a personal need of having someone around and always available for more practical chores. I would think that several earlier and later collaborators of Feynman have had similar experiences.

One of the tasks which I soon realized was to some extent my responsibility, was to keep him informed about what was happening, what kind of ideas were discussed and what was new in preprints and other publications. In his office there were usually very few papers and journals to see and essentially no books. That surprised me and certainly gave the impression that he was independent of the works of others. And to a large extent that was probably also true. He loved to work out every calculation himself, to understand every problem in his own way. This inclination was also reflected in the way he prepared his lectures. Usually he was then sitting in his office an hour or two before each lecture and writing out in detail every calculation and argument he needed even if he had given the same course the year before or earlier. Everything had to be fresh in his mind and worked out from scratch. In spite of this unusual effort, he never seemed to get stressed when it had do with teaching.

That he didn't read the books and works of others, was to some degree just a show. Early on in our collaboration I was asked to 'babysit' his two kids Carl and Michelle. I was certainly curious about how the great man lived and used the occasion to sneak around in his house. From his large living room there was a stair down to a cozy room in the basement, filled with books and scientific literature around the walls. That was a relief to see and made him a little less godlike. He later told me that this room was his 'Cave'.

At the same time he probably worked out much on his own what already was known in the literature. One morning he came directly to my office and was all excited about a new derivation he had discovered to explain the so-called 'twin paradox' in the theory of relativity. It involved two observers with identical radio emitters exchanging signals as one travels away from Earth and later returns. Using the standard, relativistic Doppler-effect he could then easily illustrate how time evolves differently for the two observers. What Feynman had found was nothing other than the K-calculus of H. Bondi.[27] Later I was inclined to think that he had arrived at this delightful insight through possible consulting work at JPL and one of their planetary satellite programs.

He was a very kind person, and almost often very happy and content. I never heard an angry word from him or a disparaging remark. He never made you feel inferior in any way although he was way above me and most others. Instead he could care about others and helped out when he could. Probably the first time I saw this side of him, was when he got me X-rayed. We went almost daily to lunch together, usually to the 'Greasy' student cafeteria. One day we passed one of these buses that travel around and offer examinations of peoples lungs. With a short account of how he had lost his first wife to tuberculosis and that he didn't want to loose me the same way, he almost pushed me into the bus and the X-ray machine. I couldn't protest but my lungs had probably preferred to avoid all this radiation.

Sometimes when he went away from Pasadena, he wanted me to go with him. On one of our first visits to Fermilab he was asked on a short notice by our experimental colleague Barry Barish to come and check out a potential discovery of a wrong-sign lepton in one of the first Caltech-run neutrino experiments there. We arrived late in the afternoon and drove directly to the primitive quonset hut where the data was analyzed. It was dark, rainy and muddy. In all respects very different from the clean facilities available today. After a short look at the measurements Feynman concluded that the signal was just a fluke and nothing to care about. At that time there was much talk of possible heavy leptons which in some theories should show up in these new experiments.

At a later visit to Fermilab we were invited to dinner by the director Bob Wilson. In the afternoon we drove over to his house that lay on a small hill out in the countryside. It was in the winter and the narrow road up there was icy and slippery. The car spun and Wilson came out and helped me push

Fig. 2. On the beach at Feynman's summer house in Mexico.

it to the house with Feynman at the wheel. This set the tone for the rest of the evening which took place in the kitchen with the three of us. Much of the talk was about all times and colleagues. When it touched upon the lives of Bob Serber and Kitty Oppenheimer, I felt like I was back in war-time Los Alamos.

It was also at Fermilab I for the first time saw a weaker and more vulnerable side of Feynman. This was in the evening after he had been told that he had made a mistake in his argument about measuring the quark charges from summing up the hadronic charges in a jet. He had afterwards become unusual quiet and withdrawn. After we had gone to bed in the guest apartment we disposed, he called me into his room and wanted to tell me about what was important in life and how one should avoid wasting time on meaningless endeavors. It was like a father-to-son talk for which I was not prepared and didn't feel had some much to do with me.

It was probably this other side I also got a glimpse of on the flight back to Los Angeles after the Cornell meeting in 1971. I had not expected to see him nervous in the plane because of bad weather outside. He, the most rational man on Earth! But sometimes he could show such contradictory sides of himself.

One weekend he had invited me and my wife to his summerhouse at the beach south of Tijuana in Mexico. We went in his car and he was at the wheel. Down there the roads were pretty narrow and curvy which he didn't seem to notice. Instead he drove rather fast and totally careless, as we saw it. This time I was the one who was scared while he was just happily grinning and talking. But as soon as we had safely arrived, everything around this scary experience was soon forgotten and we had a wonderful weekend with the family, including his sister and mother.

Many years later I came across an interview with Freeman Dyson.[28] There he told about a similar experience with Feynman already in 1947 when they drove from Cornell to a seminar in Rochester. That car ride also frightened Dyson a lot and shows that this streak in Feynman's character was already there when we experienced it on the road in Mexico. What it showed, I will probably never understand.

Through Feynman I was also so happy to meet many other famous physicists. Once I was with my wife invited to his house in Altadena in connection with a visit to Caltech by his colleague and PhD advisor John Wheeler. It was clear that Feynman had deep respect for Wheeler who in turn at that setting was dominated by an unusually assertive wife which reduced my own impression of the great physicist. But Wheeler made it up again the following day in a beautiful lecture on general relativity.

Most likely it was also Feynman who in 1971 set me in contact with Max Delbrück who wanted a house sitter for the summer months he was in Cold Spring Harbor or in Europe. Delbrück was very interested in what we could explain with the quark model. But the result was that he wanted me later to transfer to a postdoc position in his own group to work on the fungus *Phycomyces*. I'm happy I didn't accept the offer. Even so we were allowed to stay in his beautiful house the next three summers. It was also by the Delbrücks that W. Heisenberg stayed when he came to visit Caltech in 1974. While Gell-Mann kept away that day, Feynman was the considerate host. But at dinner in the evening Feynman became very direct in his criticism of the talk Heisenberg had given earlier in the day.[29] This side of Feynman I never saw myself, but I knew how important it was for him to sort out bad from good physics.

In the spring of 1974 he told me that he would give the commencement speech at the graduation ceremony at Caltech some weeks later. He already had some ideas around honesty in science that he wanted to emphasize and showed me a preliminary manuscript to read. Some of the arguments I had heard before, but didn't realize how much attention this talk about *Cargo Cult Science* would later generate. I didn't even attend the ceremony and was instead preparing my departure from Caltech after six wonderful years there. Looking back, there is much I would have done differently had I again been given the opportunity to be close to this great physicist and human being.

5. Epilogue

I came back on short visits to California several times in the following years. Already in 1975 I met him at a meeting at Stanford. We then had lunch together with L. Alvarez who was very happy for just having debunked a possible discovery of a magnetic monopole. Feynman was diagnosed with cancer around this time. During a talk at a meeting at Caltech on QCD in 1978 it seemed to me that he had lost some of the strength and confidence he earlier usually showed. On the last day the final talk was given by Gell-Mann who pointed out that Feynman just had turned 60 years. He ended by thanking him explicitly for his many contributions which had made the success of this new physics possible. I was sitting next to Feynman and know that this open acknowledgement from his colleague was much appreciated.

When I met Feynman the following year, he was just out of the hospital and I had to visit him at home in his own bedroom. As ten years earlier I was surprised to see that also here he had books all around the room. Next to his bed lay a copy of Newton's *Principia*. That we didn't discuss. Instead he was very excited about his new, personal computer. It was a PET which stood on a nearby desk. He was very proud of a couple of programs he ran for me and which recursively generated some very nice patterns. He always had loved computing and computers.

During all these years I stayed in contact with Helen Tuck who was Feynman's secretary and kept me informed about what happened. When I was on a visit to Santa Barbara in 1984, she called me and said that Feynman wanted to see me. Again in some way I couldn't accept or believe it, and declined with the excuse of not having enough time. Four years later I was at the University of Illinois in Urbana-Champaign and heard Feynman's voice out in the corridor happily talking physics. He was so alive! But what I heard was just some tapes the students played of his old lectures. They had just been informed that he had died the day before.

A few years ago I browsed through the book Feynman's daughter Michelle wrote about his many letters.[30] One is to a friend on the East Coast who is asked if he could come up with some money for a Norwegian postdoc who planned to go back for a job which he didn't find so attractive. Apparently Feynman got the money and I thereby my sixth year there. My postdoc arrangement had been against department traditions and regulations and came out just because of Feynman's expressed wish. After two years it should be over, but again he reached out a helping hand. I never understood what really happened.

References

1. J. W. Watson, *The World of Science* (Golden Press, New York, 1958).
2. G. F. Chew, M. Gell-Mann and A. H. Rosenfeld, *Sci. Am.* **120**(2), 74 (1964).
3. L. B. Okun, *Weak Interactions of Fundamental Particles* (Pergamon Press, London, 1965).
4. J. Sakurai, *Invariance Principles and Elementary Particles* (Princeton University Press, New Jersey, 1964).
5. F. Ravndal and T. Toyoda, *Nucl. Phys. B* **3**, 312 (1967).
6. P. N. Bogoliubov, JINR preprint P-2569, Dubna, USSR (1966).
7. D. Faiman and A. W. Hendry, *Phys. Rev.* **173**, 1720 (1968).
8. L. A. Copley, G. Karl and E. Obryk, *Phys. Lett. B* **29**, 117 (1969).
9. C. A. Heusch and F. Ravndal, *Phys. Rev. Lett.* **25**, 253 (1970).
10. A. N. Mitra and M. Ross, *Phys. Rev.* **158**, 1630 (1967).
11. F. Ravndal, Feynman's parton model for diffraction scattering, in *Baryons '92*, New Haven, Connecticut, June 1992, ed. M. Gai (World Scientific, Singapore, 1993).
12. S. C. Frautschi, C. J. Hamer and F. Ravndal, *Phys. Rev. D* **2**, 2681 (1970); C. J. Hamer and F. Ravndal, *Phys. Rev. D* **2**, 2687 (1970).
13. R. P. Feynman, M. Kislinger and F. Ravndal, Caltech preprint CALT-68-279 (1970).
14. R. P. Feynman, M. Kislinger and F. Ravndal, *Phys. Rev. D* **3**, 2706 (1971).
15. R. P. Feynman, *Photon-Hadron Interactions* (W. A. Benjamin, Inc., Reading, Massachusetts, 1972).
16. F. Ravndal, *Phys. Rev. D* **4**, 1466 (1971).
17. F. Ravndal, *Nuovo Cimento Lett.* **3**, 631 (1972).
18. H. Fritzsch and M. Gell-Mann, Caltech preprint CALT-68-297 (1971).
19. A. J. G. Hey and J. Mandula, *Phys. Rev. D* **5**, 2610 (1971).
20. M. Gronau, F. Ravndal and Y. Zarmi, *Nucl. Phys. B* **51**, 611 (1973).
21. G. R. Farrar and J. L. Rosner, *Phys. Rev. D* **7**, 2747 (1973).
22. F. Ravndal, *Phys. Lett. B* **43**, 301 (1973).
23. D. Novoseller and F. Ravndal, *Nucl. Phys. B* **79**, 333 (1974).
24. T. L. Curtright, J. E. Mandula and F. Ravndal, *Phys. Lett. B* **55**, 397 (1975).
25. A. Casher, J. Kogut and L. Susskind, *Phys. Rev. D* **10**, 732 (1974).
26. M. Baker, J. S. Ball and F. Zachariassen, *Nucl. Phys. B* **186**, 531 (1981).
27. H. Bondi, *Relativity and Common Sense* (Doubleday & Company, Inc., New York, 1964).
28. F. Dyson, http://www.webofstories.com/play/freeman.dyson/58.
29. H. Fritzsch, private communication.
30. M. Feynman, *Perfectly Reasonable Deviations from the Beaten Track: The Letters of Richard P. Feynman* (Basic Books, 2005).

What is a Quark?

Gordon L. Kane

Michigan Center for Theoretical Physics, Department of Physics, University of Michigan, Ann Arbor, MI 48109, USA

Malcolm J. Perry

DAMTP, Centre for Mathematical Sciences, Wilberforce Road, University of Cambridge, Cambridge, CB3 0WA, England

We are used to thinking of quarks as fundamental particles in the same way we think of the electron, or gauge bosons, neutrinos, leptons. In strong theory, these objects are unified with gravitation and the physics of space-time into what is hoped to be an ultimate theory, string/M theory. The string/M theory paradigm completely changes the way we think of the so-called elementary particles in quantum field theory.

We know how to describe an electron and its behavior. Since the discovery of the Dirac equation, we have had a satisfactory theory of electrons. Quantum electrodynamics was developed so that the electron interacting with photons could be described and this led to a theory that has been verified to an accuracy of around one part in 10^9. Let us examine the g factor for the magnetic dipole moment of the electron. The Dirac equation predicts that $g = 2$ but once virtual photons are taken into account in quantum electrodynamics, we find that $g - 2 = 115\,965\,218\,17.8 \times 10^{-13}$ theoretically, compared to $115\,965\,218\,07.3 \times 10^{-13}$ experimentally. This makes quantum electrodynamics extraordinarily accurate. In quantum electrodynamics, the electron is just a quantized fluctuation in the electron field. This explains why all electrons are identical. Why should one look further than this picture? The answer is because there is more to the world than electrodynamics. There are the other interactions, the strong and weak nuclear forces and gravitation. It is possible to include the strong and weak interactions into a straightforward generalization of quantum electrodynamics.

From the spacetime point of view, free quarks are treated in the same way, as solutions of the Dirac equation. But quarks carry an additional quantum number, a non-Abelian one, called "color" charge. The theory of the color force is called quantum chromodynamics. In quantum field theory the strong, or color, force is more complicated to describe, but basically under control.

Gravity however cannot be treated in the same way. For this reason, we need to move on to string theory to see how to describe electrons and quarks in a more fundamental way. Before doing that first let's look at the question of parity violation in the weak interactions and see how it is accommodated in the theory, and how its description originates in string theory.

Electrons and quarks are spin 1/2 fermions. We can think of them as massless when formulating the theory, and later allow them to gain mass by the Higgs mechanism. All massless fermions can be split up into their right-handed part and their left-handed part without violating Lorentz invariance. A right-handed fermion f has its spin pointing in its direction of motion and is written f_R whilst a left-handed fermion has its spin pointing in a direction opposite to its direction of motion and is written f_L. From the point of view of the $SU(2)$ electroweak symmetry of the Standard Model, amazingly the left-handed fermions are taken to transform as a doublet of the $SU(2)$ while the right-handed ones are taken to be singlets. For the up and down quarks, usually written as u and d, of which the proton and neutron are composed, the doublet is made from $(\mathrm{u}_L, \mathrm{d}_L)$, whereas the two $SU(2)$ singlets are u_R, and d_R. Under the parity operation $L \longleftrightarrow R$, so this classification violates parity conservation. This *ad hoc* procedure is how the Standard Model accommodates parity violation. Fermions that are just either left-handed or right-handed are called chiral fermions.

Thus to understand how quarks can emerge from string theory we need to see how chiral fermions emerge, and how the color $SU(3)$ emerges. Quarks also have non-integer electric charges (in units of the electron or proton charge), which has to emerge too.

What is string theory and why should one study it? These questions are those that one is always asked by those outside the string theory community. It does not matter who is doing the asking; it could be a condensed matter theorist or an immigration inspector or someone you meet at a cocktail party. In each case, the question is always the same. What follows is an attempt to provide some answers on the route to explaining what a quark is in string theory.

The Standard Model (SM) of particle physics is perhaps the most successful theory ever invented. It provides us with explanations of most of the phenomena in nature. It is a construction that assumes that point particles are properly described by a renormalizable relativistic quantum field theory.

Contained in it are the basic interactions of nature. It contains three generations of leptons and colored quarks together with gauge bosons that carry the fundamental forces of nature, massless colored gluons that mediate the strong force, the massless photons that are responsible for the electromagnetic interaction and the massive $W^{\pm}$ and Z bosons that mediate the weak interaction. Color is rather like electric charge but is carried only by the quarks and gluons. In addition, there are Higgs scalars which act to provide the mass to those particles that are massive. The whole theory is based on the gauge group $SU(3) \otimes SU(2) \otimes U(1)$. Each of the strong, electromagnetic and weak interactions are associated with a coupling strength.

What the Standard Model does not provide is any insight into dark matter or dark energy which make up around 27% and 68% of the Universe respectively. Also, what goes into the Standard Model does seem rather arbitrary. For example, there is no obvious reason why the gauge group should be what it is, and no obvious reason why there should be three generations. There is however a hint, from a knowledge of the magnitudes of the three coupling strengths, that there should be some deeper picture underlying the standard model. Couplings depend on energy in a way specified by the renormalization group equations. If one extrapolates the three couplings to an energy scale of around 10^{15} GeV, they become numerically similar. If one makes a modest extension of the standard model to include supersymmetry, then this numerical coincidence becomes even stronger. We take this to mean that there is some kind of unification of these three interactions at that scale, and that supersymmetry is an essential ingredient of any more fundamental theory than the standard model.

Missing from the Standard Model, and indeed from any conceivable extension of it is any description of the interaction that controls the behavior of the entire Universe; gravitation. Also missing from the standard model is any understanding of CP-violation or neutrino masses. To make progress, a radical new idea is needed.

It is often said that the gravitational force is a special case and unlike the other forces. Whilst it is very familiar, at the level of individual subatomic particles it is vastly weaker than the other forces, by about a factor of 10^{39}. This makes it difficult to study, and so the microscopic mechanism behind gravitation is yet to be fully understood. The only reason we notice gravitation is that unlike the other forces, it is purely attractive in nature. The other forces can either be attractive or repulsive; that is to say they have charges of both signs. In electromagnetism, like charges attract and unlike charges repel. That picture holds for all interactions except gravity. Gravity only has one sign of charge. So although it is weak on the atomic scale, its effects build up and become easily observable for large objects. It is for that reason that gravity is most easily observed on astronomical distance scales. There is an excellent theory of gravitation that works at the classical level and that is Einstein's general theory of relativity.

Yet, the unification scale of 10^{15} GeV is so close to the scale of gravitation, the Planck scale of 10^{19} GeV, that to ignore gravitation is surely misconceived. At these energy scales, the strengths of the other couplings become similar to that of gravitation. And since there can hardly be one set of laws for gravitation and a different set for everything else we are driven to try to find a picture in which all of the interactions can be accounted for at once. Superstring theory is the only known picture that includes gravitation that does not suffer from some kind of fatal difficulty. Trying to incorporate gravitation into a quantum field theory of particles, a method that has been astonishingly successful for the other forces, encounters numerous problems.

To do so, one postulates the existence of a graviton that transmits the gravitational force in a way that is similar to how a photon transmits the electromagnetic force. A graviton viewed as an elementary particle must be massless and have a spin of 2 in contrast to the photon which is massless and has a spin of 1. It is the fact that the gravity is universally attractive that requires the graviton to have a spin of 2. Perhaps the most serious difficulty is that nobody has ever found a way around is the ultraviolet divergence problem in a quantum field theory of gravitation. In relativistic quantum field theory, one always encounters infinite quantities. One can think of this as being associated with the fact that the self-energy of a point particle is infinite. In theories without gravitation, it is possible to control these infinities through a process known as renormalization. If one tries to construct a quantum theory of gravity by following the same route, it is impossible to get rid of these infinities without violating some essential physical principle. The principles that one comes into conflict with are either causality, the idea that no information can propagate faster than light or unitarity, the idea that the probability for any event lies in the range zero to one inclusively. These two principles as so basic to our understanding of the universe that it impossible for us to conceive of any theory that does not satisfy these conditions. A theory that has uncontrollable infinities is termed unrenormalizable and as a fundamental physical theory makes no sense whatsoever. Indeed, there are many that regard the process of renormalization as a defect itself, and would be happy to do away with the controllable infinities in relativistic quantum field theory.

In fact, it is easy to see why gravity is different. The idea of a point particle in gravitation does not really exist. To see why, consider a spherical body of mass M and radius R. To escape from it, one needs to achieve a the escape velocity. If a body of fixed mass gets smaller, its escape velocity increases. If this velocity becomes equal to the speed of light, then nothing can escape from the body. Such an object is a black hole. The idea of black holes was first discussed by John Michell in 1783 but in Newtonian theory. He knew that the velocity of light c was finite as was first demonstrated by Ole Rømer as far back as 1676. He showed that the escape velocity v_e in Newtonian theory was given by

$$v_e = \sqrt{\frac{2GM}{R}} \tag{1}$$

and realized that stars could be so condensed that v_e might exceed c rendering such objects invisible. In general relativity, the same formula holds but its significance is greater as since nothing can exceed the speed of light, such regions of spacetime cannot influence anything outside. In general relativity, this is the closest one can get to a point particle. Black holes have finite size. For example, if the Sun were to be shrunk to form a black hole, then it would have a radius of around 3 km. One might think that this would require a density that is impossible to obtain, but in fact for the Sun the density would be about $10^{15}\ \mathrm{g \cdot cm^{-3}}$ similar to the nucleus or the nuclear matter found in neutron stars. Since the density scales as M^{-2}, the density of black holes in the center of galaxies is about that of water.

The simplest string theory is the bosonic string. It is termed bosonic because it results in objects in spacetime that are all bosons. The bosonic string cannot describe fermions and therefore at best is a toy model for realistic physics. One replaces the "worldline" of a particle, a one-dimensional time-like line in spacetime generated as the particle moves, with a two-dimensional surface with one time direction and one spatial direction. The spatial direction can either be a line segment — the open string — or a circle — the closed string. Thus, a string is a surface in space time. An open string is one that has spatial endpoints that sweep out lines in spacetime. A closed string is a cylinder in spacetime. One can regard a string as being a generalization of a particle.

In classical mechanics, to describe the behavior of a massive particle, one constructs an action and uses its Euler–Lagrange equations as the particle equations of motion. Let $X(\tau)$ be the position of a particle in spacetime with metric tensor η_{ab} and where τ labels proper time along the particle worldline, A suitable action I is then

$$I = m \int d\tau \sqrt{-\eta_{ab} \frac{dX^a}{d\tau} \frac{dX^b}{d\tau}} \,. \tag{2}$$

The action is proportional to the proper length of the worldline. m has the dimensions of mass and is put in to give the action the correct dimensions. One can take m to be the mass of the particle. Extremizing the action with fixed initial and final positions yields the particle equations of motion. Interactions between particles can now happen but the interactions must be concentrated at points and in practice are restricted by conservation laws that constrain how elementary particles are observed to interact. At the classical level, these interactions must be introduced by hand. In quantum field theory, these interactions are introduced naturally as cubic or higher-order terms involving the fields corresponding to particular particles. Their diagrammatic treatment is as if the particles had classical interactions involving three or higher point vertices. The nature of these interactions is rather *ad hoc* and is governed by what one has observed experimentally together with some consistency requirements such as gauge invariance, hermiticity and renormalizability.

To describe the behavior of a string, a similar action can be constructed. The string is a two-dimensional surface with one time direction and one space direction. Let the time co-ordinate by τ and the space co-ordinate σ. Then the action for the string can be taken to be

$$I = \frac{1}{4\pi\alpha'} \int d\tau \, d\sigma \, \eta_{ab} \left(\frac{\partial X^a}{\partial \tau} \frac{\partial X^b}{\partial \tau} - \frac{\partial X^a}{\partial \sigma} \frac{\partial X^b}{\partial \sigma} \right) . \tag{3}$$

The action is proportional to the proper area of the string worldsheet. The factor α' has dimensions of the inverse string tension and is the only dimensionful parameter entering string theory. Unlike point particle theory, all string interactions are contained in this action. Whilst free strings are described by tubes or planar worldsheets, there is no restriction on the topology of the worldsheet. The surface that describes the string can have holes or junctions where string meet. The string action can describe how strings interact without having to introduce any assumptions about how interactions take place. A string can in principle live in d spacetime dimensions. A string is described by whether it is open or closed, by its center-of-mass motion, and how it is vibrating. If one thinks about a point on the surface of the string, it can move in two directions in the plane of the string, or in $d-2$ directions perpendicular to the surface of the string. Motion in the direction of the string is just moving one point on the worldsheet into another and does not correspond to any physical change in the string, just how it is described. Since the surface of the string has one timelike and one spacelike direction, the physical degrees of freedom all correspond to spacelike displacements and so it is only these $d-2$ directions that are physical string motions. These vibrations in $d-2$ directions transverse to the string should really be thought of as waves traveling along the string rather in the same way that vibrations of a string in a musical instrument, such as a guitar involve waves causing displacement of the string perpendicular to the string itself.

Let us first consider the closed bosonic string. The spectrum of states of the string consists of a collection of excitations that are classified by their mass and other quantum numbers. States can either be massless or have masses on the order of the string scale. Since the string scale is of the order of the Planck scale, roughly 10^{19} GeV, the only excitations directly observable are those that are massless. These massless states can be described using the same language one uses for elementary particles. That is really because on any distance scale we can measure directly, the strings are invisibly small, of the order of the Planck length, roughly 10^{-33} cm. One finds that the spectrum of massless states contains three types of object: something with the quantum numbers of a scalar particle, something that is a bit like a photon and an object with the quantum numbers expected for a graviton, the object that transmits the gravitational force. These massless excitations are universal features of all string theories. It is this graviton-like object that is most intriguing as it is something which comes out of the theory automatically rather than something that has to be put in by hand. It is a hint that string theory really does contain gravitation.

An open bosonic string is a little different. Here the waves can be reflected at the endpoints of the string and set up standing waves. An open string has massless excitations which look rather like a photon. A simple modification of the open bosonic string is to place a label, the so-called Chan–Paton factor, at each end of the string which can take integer values from 1 to N. suppose that we restrict the surfaces describing the string to be orientable. Then the result is that instead of an Abelian gauge theory like electromagnetism, one finds excitations that behave like Yang–Mills particles with gauge group $U(N)$. In a similar way, one can also realize gauge theories of the orthogonal groups $SO(N)$ and the symplectic groups $USp(N)$, but this can only be done by demanding that the strings are surfaces in spacetime that are non-orientable. Thus strings are represented in low-energy physics, the physics that we see, by fields in quantum field theory and can carry quantum numbers such a spin and charge. One way to think of Yang–Mills particles is that they are generalizations of the photon in which the gauge transformations have become rather more complicated and controlled by a component of a Lie algebra rather than just numbers as would be the case for electromagnetism. Just as photons can couple to objects with electric charge, Yang–Mills particles can couple to objects with generalizations of electric charges.

One can then ask if the string theory is quantum mechanically consistent. This is a rather nontrivial step which is forced on one because string theory has a hidden classical symmetry called conformal symmetry. The conformal symmetry is essential for the string to work properly. Yet quantum effects can cause conformal symmetry to be broken. This would be an undesirable state of affairs. It turns out that the fields that yield string theories with unbroken conformal symmetry are precisely those that obey the classical Einstein equations or Yang–Mills equations or for scalars a version of the wave equation. This indicates a deep relationship between the known laws of physics and string theory. In fact, one could say that these laws of physics instead of being postulated, have in fact been derived from just the symmetry of string theory.

There are some difficulties in promoting bosonic string theories into fully sensible models of fundamental physics. The first is that quantum effects result in the string making sense only if $d = 26$ which is rather a long way from the $d = 4$ we observe. A second problem is that they do not contain any fermions. A final problem is that all bosonic string theories contain a fatal flaw in the form of a tachyon. For the most part, we do not need to worry about massive string states simply because their mass is so high that we cannot observe them. An exception to this is what happens in bosonic string theory, and that is its spectrum contains a tachyon, which is a particle that has to move faster than light. Such objects as well as being objectionable due to their conflict with causality, also tend to indicate instabilities. Since their masses are of the order of the Planck scale, whatever instability they signal will be governed by a timescale of order of the Planck time of 10^{-43} seconds, in gross conflict with the observed age of the universe.

However, we have learnt some interesting facts. The first is that one finds a quantum field theory of massless particles as a low energy version of string theory. All other string excitations are not directly observable at energies scales that we can access. In amongst these string excitations, there is the graviton and so we have the potential to describe gravity using string theory. One might wonder if such a theory has the same kind of problems with infinities as gravity on its own does. The answer is no. String theory does not have the kind of divergences encountered in quantum field theory. The reason is that when one tries to calculate the types of quantities that diverge in quantum field theory, we find they do not diverge in string theory. Mathematically, this comes about because of the effect of all the massive excitations of the string is to kill off those divergences. We have made progress.

But now we need to fix up the difficulties found in bosonic string theory. The remedy is supersymmetry and the result of incorporating supersymmetry this way yields the superstring. The basic idea of supersymmetry is to introduce a symmetry that exchanges bosons with fermions. For every boson, there will be a fermion. The supersymmetry we are interested in is going to be a new kind of structure on the string worldsheet. The physical degrees of freedom of the string are the spacetime coordinates transverse to the worldsheet. These degrees of freedom behave like bosons living on the string worldsheet. Now, we introduce partners to some or all of these degrees of freedom. These are fermionic variables that pair up with bosonic variables. Remarkably, it turns out there are five different ways of doing this in a way consistent with quantum mechanics.

In each case, the first thing we discover is that instead of being consistent in only 26 spacetime dimensions, the superstring is only consistent in 10 spacetime dimensions. The second thing we discover is that there are no tachyons in the spectrum of the superstring. Lastly, although there are massive string states as before, there are still massless string states, but now as well as representing bosons, there are fermions too. For historical reasons, all these massless string states contain gravity as well as a collection of other fields. The five different string theories are usually referred to as type IIA string theory, type IIB string theory, $SO(32)$ string theory and two heterotic string theories, $SO(32)$ and $E_8 \otimes E_8$. They contain different spectra of massless particles. In each case, the fields are the same as those found in various ten-dimensional supergravity theories. Supergravity is an extension of general relativity that is supersymmetric. The two type II string theories are related to $N = 2$ pure supergravity and the other theories are related to $N = 1$ supergravity theories coupled to supersymmetric Yang–Mills theories with either the $SO(32)$ or $E_8 \otimes E_8$ gauge groups. Any of these theories has the potential to turn into realistic models of physics.

There are connections which enable one to translate any problem in any one of the string theories into any of the other string theories. These are known as duality symmetries. The complete picture is of the five string theories and an eleven-dimensional theory, in which the connections between all of them are realized, is known as M theory. Whilst quite a lot is known about M theory, it is still a work in progress. What is certain about M theory is that it contains various limiting cases, of which the five string theories are examples. These five string theories are all ten-dimensional theories. Another quite different limit is ordinary $N = 1$ supergravity in eleven dimensions. In many ways this seems to be more fundamental than the string theories.

M theory therefore is our candidate theory of everything. It does however seem a long way from something we recognize. What would we want from a theory that describes all the phenomena we observe? Firstly, it must be a theory with four spacetime dimensions not ten or eleven. Secondly at low energies, it must contain the Standard Model as well as gravitation. One thing we can be pretty certain of, and that is at around the GUT scale of roughly 10^{16} GeV, our theory must be described by a four-dimensional theory that has $N = 1$ supersymmetry. So assuming we start in eleven spacetime dimensions, we need to be able to get rid of seven of them. This process is known as compactification. One assumes that spacetime is described by the product of a compact seven-dimensional space with Minkowski spacetime. This compact space must have physical dimensions that are very small, typically the Planck length of around 10^{-33} cm. This compact space must have holonomy group G_2 in order for supersymmetry to exist at the GUT scale.

It is by this kind of method that we get rid of seven space time dimensions. We assume that they are so small, that on the scale on which we exist, they just do not get directly noticed. In a sense, this is derivable in both string and M theories. However, the precise shape and size of these extra dimensions is reflected in low energy physics by the existence of light scalars or some new fields called moduli fields and axions that describe relations among the small dimensions. If the compact manifold is smooth, then one only finds Abelian symmetry groups at low energies. These wrapped up spatial directions need not be smooth. They can have certain types of singularities that turn out to be interesting. If the space is singular, it can result in Yang–Mills fields appearing in our part of spacetime together with their fermionic superpartners. It can also turn out to produce chiral fermions. The nature of the singularities determines what kind of gauge group these Yang–Mills fields have and what kind of chiral matter result.

The types of allowed singularities are those that give rise to the gauge groups $SU(N)$, $SO(2N)$, $E6$, $E7$ and $E8$. There is however a geometric version of the Higgs mechanism. One can start with one of these gauge groups and break it down into its subgroups. This comes about because it is possible to partially resolve the singularities. As an example, if one starts with an $E8$ singularity and partially resolves it into $E7 \otimes U(1)$, then one finds gauge fields with the group $E7 \otimes U(1)$. However, it might be that the $E8$ singularity is resolved this way everywhere in the compact space except at a single point where it remains as $E8$. In this case, as well as the $E7 \otimes U(1)$ gauge fields we find that chiral fermions that transform under the action of $E7$ and have a $U(1)$ charge. this symmetry breaking scheme in many ways resembles that found in GUTs, except that we can now get chiral fermions for free. One can ask if this method can give rise to any kind of realistic grand unified theory based on the gauge group $SU(5)$. Remarkably, the answer is yes. Suppose one starts from an $E8$ singularity, and breaks it successively down to $E6 \otimes SU(3)$, then the $E6$ down to $SO(10) \otimes U(1)$ and then $SO(10)$ into $SU(5) \otimes U(1)$. Then one finds a fairly realistic $SU(5)$ grand unified theory that contains three families. It cannot contain more than three families or less than three families. From this stage as one goes to lower energies, the standard model can be recovered by more or less the standard rules of grand unification. In this way, we can see how quarks possess the $SU(3)$ of color. Furthermore, the appearance of various $U(1)$ charges in this process can account for the charge on the electron and the charges on the quarks being what they are. Whilst a fully realistic theory has yet to be obtained by these methods, we regard the whole approach as very promising.

One wonders if there is a ten-dimensional string theory description instead of an eleven-dimensional M theory one. The answer is yes. Suppose one looks at the $E8 \otimes E8$ heterotic string. There one finds a ten-dimensional theory that contains Yang–Mills fields with gauge group $E8 \otimes E8$. One of these $E8$'s can be broken down into $E6 \otimes SU(3)$ in a natural way. This gives a gauge theory that in many ways is similar to our geometric way of obtaining gauge groups and chiral fermions. However, it also leads to a second $E8$ gauge group which is apparently hidden from observation. So this seems like an alternative way of arriving at our destination. Of course, in the M theory approach, we did not have an analogous large hidden sector. However, we could easily manufacture a theory that had such a hidden sector simply by postulating that the compact seven-dimensional manifold started off with two singularities of the $E8$ type. There is nothing forbidding this, but equally there is nothing compelling it either, whereas in heterotic string theory, a pair of $E8$'s is inevitable.

How is it possible that there are two different descriptions of the same thing? One answer is that there is a curious collection of symmetries, known as string dualities, which relate one picture in string or M theory to another. There is a translation between phenomena in the heterotic string and phenomena in M theory. It just so happens that what is described by singularities in hidden parts of space in the M theory picture are just string excitations in the the heterotic string. It does not make sense to ask which one is right. They are both legitimate explanations of what is happening in nature. In this sense, the phenomenon is exactly like wave–particle duality in quantum mechanics. Whether you decide to think of an electron or a wave is up to you. Its behavior is accurately described by the quantum mechanical theory, however you care to describe it. M theory has its dualities which allow one to describe a quark or electron as one thing in the eleven dimensional picture and a different thing in the heterotic picture.

A somewhat different viewpoint is that M theory is more fundamental. After all, we only know about dualities between various limits of M theory. To support this viewpoint, it suffices to note that the $E8 \otimes E8$ heterotic string can be derived from M theory by the compactification of one dimension on a projective line. We thus find we are looking at ten-dimensional theory. $N = 1$, $d = 11$ supergravity contains solitons that are like membranes with two spatial dimensions and one timelike dimension. These are higher dimensional analogs of a string. In the compactified ten-dimensional spacetime, these look exactly like the heterotic strings.

One can also imagine asking what happens, if instead of campactifying on a projective line one compactifed on a circle. Then these solitons would look like the strings in type IIA string theory. However, in this picture it is not known how to find gauge theories with exceptional groups. These observations lead one to suspect that the M theory approach is more fundamental as it seems to underlie those two string theories. Is it possible then to derive the other three string theories directly from M theory? The answer is maybe for type IIB string theory where one must proceed via generalized geometry or F-theory. For the two $SO(32)$ theories, the answer is not yet known.

What we have described here is a how one should think of an electron, or quarks, or indeed any other elementary particle in string theory. In quantum field theory, these are all quantized field excitations. However, in string theory, everything is much more geometrical. There appear to many equivalent ways of looking at the same particle indicating a huge hidden symmetry in string theory that has yet to be completely explored. This is a huge challenge as the mathematics of string theory is hard. Nonetheless, humanity has in the past risen to such challenges and reached our current understanding of the nature of the Universe. Even 50 years ago, it would have been impossible to predict how much progress has been made. We can only hope that our civilization can continue to encourage and support progress in this field whose aim to understand what we are ultimately made of.

Our intention here was to write an essay about how to describe the quark, and other elementary particles, in M theory and why we are forced to follow such a route. We did not include any references for the simple reason that to single out any one reference would be unfair to the literally many hundreds of people who have contributed to our current knowledge. We therefore apologize to those who are so affected and beg their indulgence on this matter.

Acknowledgments

G. L. Kane is supported in part by the Department of Energy grant DE-SC0007859. M. J. Perry would like to thank the Michigan Center for Theoretical Physics for its hospitality and is supported in part by STFC rolling grant STJ000434/1 and by Trinity College, Cambridge.

Insights and Puzzles in Particle Physics

H. Leutwyler

Albert Einstein Center for Fundamental Physics,
Institute for Theoretical Physics, University of Bern,
Sidlerstr. 5, CH-3012 Bern, Switzerland

I briefly review the conceptual developments that led to the Standard Model and discuss some of its remarkable qualitative features. On the way, I draw attention to several puzzling aspects that are beyond the reach of our present understanding of the basic laws of physics.

1. Prehistory

Bohr's model of the hydrogen atom (1913) gave birth to quantum theory and eventually led to a very thorough understanding of the structure of atoms, molecules, solids, ... In that framework, the electrons and the nuclei represent the constituents of matter. Their properties are controlled by mass and charge — size and structure of the nuclei, magnetic moments, etc. manifest themselves only in fine details of the picture.

The first hint at the existence of particles other than electrons and nuclei occurred in β-decay, where the distribution of the decay products was puzzling because it violated energy conservation. Pauli solved the puzzle (1930): the observed spectrum can be explained if a yet unknown particle is emitted together with the electron. It must be neutral and escape detection. The experimental proof of existence for this particle became possible only much later. As Pauli wrote in his answer to the announcement of the discovery (Reines and Cowan, 1956): *Everything comes to him who knows how to wait.*

In today's terminology, Pauli predicted the electron neutrino, ν_e. In fact, we now know that both the electron and the neutrino have relatives, leptons, which come in three families:

$$\{e(1897),\ \nu_e(1956)\}\quad \{\mu(1936),\ \nu_\mu(1962)\}\quad \{\tau(1975),\ \nu_\tau(2000)\}\,.$$

When the μ was discovered, Rabi asked: *Who ordered that?* The existence of yet another charged lepton, the τ, did not shed any light on this puzzle: why are there three families of leptons?

The discovery of the neutron (Chadwick, 1932) simplified the picture considerably, as it reduced the number of constituents from over 90 to only 3: electron, proton, neutron. At the same time, it gave rise to a new puzzle: what forces the protons and neutrons to form nuclei?

Yukawa[1] and Stückelberg[2] realized that the Coulomb force $V \sim \frac{1}{r}$ is of long range because it is due to the exchange of massless particles, photons. They noticed that the exchange of a particle of mass m would instead give rise to a potential of finite range, $V \sim \frac{1}{r}e^{-r/r_0}$ and that the range is determined by the mass of the exchanged particle, $r_0 = \hbar/mc$. While Stückelberg considered massive particles of spin 1, Yukawa investigated the exchange of massive particles with spin 0. From the fact that the range of the nuclear force is of the order of a few fermi, he predicted the existence of a spinless particle, which strongly interacts with protons and neutrons and has a mass of the order of 100 MeV/c^2. As this is intermediate between the masses of electron and proton or neutron, he coined the term *meson* for this particle.

More than ten years later, the object was indeed found (Powell, 1947), at a mass of about 140 MeV/c^2. It is now referred to as the π-meson or *pion*. Around the same time, many other strongly interacting particles started showing up: K-mesons, hyperons, excited states of the nucleon, ...

Gradually, the understanding of the nuclear forces developed into a very successful framework based on nonrelativistic quantum mechanics, where the interaction among the nucleons is described by means of a refined version of the Yukawa potential. The structure of the nuclei, nuclear reactions, the processes responsible for the energy production in the sun, α-decay, etc. can all be understood on this basis. These phenomena concern interactions among nucleons with small relative velocities. Experimentally, it had become possible to explore relativistic collisions. A description in terms of nonrelativistic potentials cannot cover these.

2. Situation at the Beginning of the 1960s

General principles like Lorentz invariance, causality and unitarity had given deep insights: analyticity, dispersion relations, CPT theorem, relation between spin and statistics, for instance. Motivated by the successful predictions of Quantum Electrodynamics (QED), many attempts at formulating a theory of the strong interaction based on elementary fields for baryons and mesons were undertaken, but absolutely nothing worked even halfway. There was considerable progress in renormalization theory, but faith in quantum field theory was in decline, even concerning QED. I illustrate the situation at the beginning of the 1960s with the following quotations from Landau's assessment:[3]

- *We are driven to the conclusion that the Hamiltonian method for strong interaction is dead and must be buried, although of course with deserved honor.*
- *By now the nullification of the theory is tacitly accepted even by theoretical physicists who profess to dispute it. This is evident from the almost complete disappearance of papers on meson theory and particularly from Dyson's assertion that the correct theory will not be found in the next hundred years.*

The basis of this pessimistic conclusion is clearly spelled out:

- *... the effective interaction always diminishes with decreasing energy, so that the physical interaction at finite energies is always less than the interaction at energies of the order of the cutoff limit which is given by the bare coupling constant appearing in the Hamiltonian.*

In other words, all of the models of quantum field theory explored by that time had a positive β-function at weak coupling. Lagrangians involve products of the fields and their derivatives at the same space–time point. If the interaction grows beyond bounds when the distance shrinks, the Lagrangian is a questionable notion (for the asymptotically free theories to be discussed below, the β-function is negative at weak coupling, so that the problem does not arise, but these were discovered only later).

Many people doubted that the strong interaction could at all be described by means of a local quantum field theory. As a way out, it was suggested to give up quantum field theory and only rely on S-matrix theory — heated debates about this suggestion took place.[4] In short, fifty years ago, a theory of the strong interaction was not in sight. What was available was a collection of beliefs, prejudices and assumptions which where partly contradicting one another. As we now know, quite a few of these were wrong. The remaining ones are still with us ....

3. Quarks

The Bohr Model (1913) played a key role in unraveling the structure of the atoms. The discovery of the Quark Model (1964) represents the analogous step in the developments which led to a solution of the puzzle posed by the strong interaction. Indeed, our understanding of the laws of nature made remarkable progress in the eight years between that discovery and the formulation of Quantum Chromodynamics (1972), the keystone of the Standard Model.

Gell-Mann[5] and Ne'eman[6] independently pointed out that the many baryons and mesons observed by the beginning of the 1960s can be grouped into multiplets which form representations of an approximate symmetry. The proposal amounts to an extension of isospin symmetry, which is characterized by the group SU(2), to the larger group SU(3). As indicated by the name *Eightfold Way*, the Lie algebra of SU(3) contains eight independent elements, which play a role analogous to the three components of isospin.

On this basis, Gell-Mann predicted the occurrence of a baryon Ω^-, needed to complete the decuplet representation of the Eightfold Way. In contrast to the other members of the multiplet, which represent rapidly decaying resonances, this particle is long-lived and thus leaves a trace in bubble chamber pictures. In 1964, the Ω^- was indeed found at the predicted mass, in a Brookhaven experiment.

Zweig and Gell-Mann[7,8] then independently discovered that the multiplet pattern can qualitatively be understood if the strongly interacting particles are bound states formed with constituents of spin $\frac{1}{2}$, which transform according to the fundamental representation of SU(3): u, d, s. Zweig thought of them as real particles and called them 'aces'. In view of the absence of any experimental evidence for such constituents, Gell-Mann was more reluctant.[a] He introduced the name 'quark', borrowed from James Joyce[b] and suggested to treat the quarks like the veal in one of the recipes practiced in the Royal french cusine (the pheasant was baked between two slices of veal, which were then left for the less royal members of the court).

The Quark Model was difficult to reconcile with the spin-statistics theorem which implies that particles of spin $\frac{1}{2}$ must obey Fermi statistics. Greenberg proposed that the quarks obey neither Fermi-statistics nor Bose-statistics, but represent "para-fermions of order 3."[9] The proposal amounts to the introduction of a new internal quantum number. Indeed, in 1965, Bogoliubov, Struminsky and Tavkhelidze,[10] Han and Nambu[11] and Miyamoto[12] independently pointed out that some of the problems encountered in the quark model disappear if the u, d and s quarks occur in three different states. Gell-Mann coined the term "color" for the new quantum number.

[a] *Such particles [quarks] presumably are not real but we may use them in our field theory anyway ....*[13]

[b] According to a story told to me by Lochlainn O'Raifeartaigh, James Joyce once visited an agricultural exhibit in Germany. There he saw the advertisement "Drei Mark für Musterquark" (*Mark* was the currency used in Germany at the time, *Quark* is the German word for cottage cheese, *Muster* stands for 'exemplary,' 'model' or 'sample'). Joyce is said to have been fond of playing around with words and may have come up with the famous passage in Finnegans Wake in this way: *Three quarks for Muster Mark! | Sure he has not got much of a bark | And sure any he has it's all beside the mark.*

Today, altogether six quark *flavors* are needed to account for all of the observed mesons and baryons:

$$\{u(1964),\, d(1964)\} \quad \{c(1974),\, s(1964)\} \quad \{t(1995),\, b(1977)\}\,.$$

They also come in three families, but in contrast to the leptons, each one of the quarks comes in three versions, distinguished by the color quantum number.

In 1968 Bjorken pointed out that if the nucleons contain point-like constituents, then the *ep* cross-section should obey scaling laws in the deep inelastic region.[14] Indeed, the scattering experiments carried out by the MIT-SLAC collaboration in 1968/69 did show experimental evidence for such constituents.[15] Feynman called these *partons*, leaving it open whether they were the quarks or something else.

The operator product expansion turned out to be a very useful tool for the short distance analysis of the theory — the title of the paper where it was introduced,[16] "Non-Lagrangian models of current algebra," reflects the general skepticism towards Lagrangian quantum field theory discussed in Subsec. 2.

4. Gauge Fields

According to the Standard Model, all interactions except gravity are mediated by the same type of fields: gauge fields. The electromagnetic (e.m.) field is the prototype. The final form of the laws obeyed by this field was formulated by Maxwell, around 1860 — his formulation survived relativity and quantum theory, unharmed. While for the electrons, the particle aspect showed up first, the wave aspect of the e.m. field was thoroughly explored before the corresponding quanta, the photons, were discovered.

Fock pointed out that the Schrödinger equation for electrons in an e.m. field is invariant under a group of local transformations.[17] Weyl termed these *gauge transformations*. In fact, gauge invariance and renormalizability fully determine the form of the e.m. interaction. The core of Quantum Electrodynamics — photons and electrons — illustrates this statement. Gauge invariance and renormalizability allow only two free parameters in the Lagrangian of this system: e, m_e. Moreover, only one of these is dimensionless: $e^2/4\pi = 1/137.035\,999\,074\,(44)$. This shows that gauge invariance is the crucial property of the e.m. interaction: together with renormalizability, it fully determines the properties of the e.m. interaction, except for this number, which so far still remains unexplained.

The symmetry group that characterizes the e.m. field is the group U(1), but gauge invariance can be generalized to larger groups, such as SU(2) or SU(3).[18,19] Gauge invariance then requires the occurrence of more than one gauge field: three in the case of SU(2), eight in the case of SU(3), while a single gauge field is needed for U(1). Pauli had encountered this generalization of the concept of a gauge field earlier, when extending the Kaluza–Klein scenario to Riemann spaces of more than five dimensions. He did not consider this worth publishing, however, because he was convinced that the quanta of a gauge field are necessarily massless like the photon: gauge invariance protects these particles from picking up mass. In particular, inserting a mass term in the Lagrangian is not allowed, because such a term violates gauge invariance. Pauli concluded that the forces mediated by gauge fields are of long range and can therefore not possibly describe the strong or weak interactions — these are of short range.[20]

Ten years later, Englert and Brout,[21] Higgs,[22] and Guralnik, Hagen and Kibble[23] showed that Pauli's objection is not valid in general: in the presence of scalar fields, gauge fields can pick up mass, so that forces mediated by gauge fields can be of short range. The work of Glashow,[24] Weinberg[25] and Salam[26] then demonstrated that non-Abelian gauge fields are relevant for physics: the framework discovered by Englert *et al.* does lead to a satisfactory theory of the weak interaction.

5. Quantum Chromodynamics

One of the possibilities considered for the interaction that binds the quarks together was an Abelian gauge field analogous to the e.m. field, but this gave rise to problems, because the field would then interfere with the other degrees of freedom. Fritzsch and Gell-Mann pointed out that if the gluons carry color, then the empirical observation that quarks appear to be confined might also apply to them: the spectrum of the theory might exclusively contain color neutral states.

Gell-Mann's talk at the High Energy Physics Conference in 1972 (Fermilab), had the title "Current algebra: Quarks and what else?" In particular, he discussed the proposal to describe the gluons in terms of a *non-Abelian gauge field* coupled to color, relying on work done with Fritzsch.[27] As it was known already that the e.m. and weak interactions are mediated by gauge fields, the idea that color might be a local symmetry as well does not appear as far fetched. In the proceedings, Fritzsch and Gell-Mann mention unpublished work in this direction by Wess.

The main problem at the time was that for a gauge field theory to describe the hadrons and their interaction, it had to be fundamentally different from the quantum field theories encountered in nature so far. All of these, including the electroweak theory, have the spectrum indicated by the degrees of freedom occurring in the Lagrangian: photons, leptons, intermediate bosons, ... The Lagrangian of the strong interaction can be the one of a gauge field theory only if the spectrum of physical states of a quantum field theory can be qualitatively different from the spectrum of fields needed to formulate it: gluons and quarks in the Lagrangian, hadrons in the spectrum. In 1973, when the arguments in favor of QCD as a theory of the strong interaction were critically examined,[28] the idea that the observed spectrum of hadrons can fully be understood on the basis of a theory built with quarks and gluons still looked rather questionable and was accordingly formulated in cautious terms. That this is not mere wishful thinking became clear only when the significance of the fact that the β-function of a non-Abelian gauge field theory is negative at weak coupling was recognized. Interactions with this property are called *asymptotically free*. The following few comments refer to the history of this concept.[c] Further historical material concerning developments relevant for QCD is listed in Ref. 29.

Already in 1965, Vanyashin and Terentyev[30] found that the renormalization of the electric charge of a vector field is of opposite sign to the one of the electron (the numerical value of the coefficient was not correct). In the language of SU(2) gauge field theory, their result implies that the β-function is negative at one loop.

The first correct calculation of the β-function of a non-Abelian gauge field theory was carried out by Khriplovich, for the case of SU(2), relevant for the electroweak interaction.[31] He found that β is negative and concluded that the interaction becomes weak at short distance.

In his PhD thesis, 't Hooft performed the calculation of the β-function for an arbitrary gauge group, including the interaction with fermions and Higgs scalars.[32] He proved that the theory is renormalizable and confirmed that, unless there are too many fermions or scalars, the β-function is negative at weak coupling.

This demonstrates that there are exceptions to Landau's rule, according to which *the effective interaction always diminishes with decreasing energy*: non-Abelian gauge theories show the opposite behavior, asymptotic freedom.

---

[c]I thank Mikhail Vysotsky, Zurab Silagadze, Martin Lüscher, Jürg Gasser and Chris Korthals-Altes for information about this story.

Symanzik pointed out that for theories with a negative β-function at weak coupling, the behavior of the Green functions at large momenta is controlled by perturbation theory and hence computable.[33] The dimensions of the field operators are then the same as for free fields; the interaction only generates corrections that disappear at high energies, in inverse proportion to the logarithm of the momentum. He presented his results at a workshop in Marseille in 1972[34] and in the discussion that followed his talk, 't Hooft pointed out that non-Abelian gauge theories can have a negative β-function at weak coupling.

Parisi discussed the consequences of a negative β-function for the structure functions of deep inelastic scattering and suggested that this might explain Bjorken scaling.[35] Gross and Wilczek[36] and Politzer[37] then showed that — if the strong interaction is mediated by a non-Abelian gauge field — the asymptotic behavior of the structure functions is indeed computable. They predicted specific logarithmic modifications of the scaling laws. In the meantime, there is strong experimental evidence for these.

6. Standard Model

In the Standard Model, the interactions among the constituents of matter are generated by three distinct gauge fields:

Interaction		Group	Dim.	Particles	Source	Coupling
Electromagnetic	QED	U(1)	1	photon	charge	e
Weak	QFD	SU(2)	3	W^+W^-Z	flavor	$g_{\rm w}$
Strong	QCD	SU(3)	8	gluons	color	$g_{\rm s}$

The symmetry which underlies the weak interaction manifests itself in the fact that the families come in doublets: $\{e, \nu_e\}$, $\{u, d\}, \ldots$. The symmetry group SU(2), which characterizes the corresponding gauge field theory, Quantum Flavordynamics (QFD), takes the electron into a mixture of an electron and a neutrino, much like an isospin rotation mixes the members of the isospin doublet formed with the proton and the neutron. More precisely, the gauge transformations of QFD only affect the left-handed components of the doublets and leave the right-handed components alone. We do not know why that is so, nor do we understand why there is no gauge symmetry connecting the different families.

The symmetry group SU(3) acts on the color of the quarks, mixing the three color states. Since the leptons are left alone by the gauge group that underlies QCD, they do not participate in the strong interaction.

The statement that all of the interactions except gravity are mediated by the same type of field sounds miraculous. Since a long time, it is known that the electromagnetic, weak and strong interactions have qualitatively very different properties. How can that be if all of them are generated by the same type of field? For this puzzle, the Standard Model does offer an explanation — the following sections address this question.

7. Why is QED Different from QCD?

I first discuss the origin of the difference between the electromagnetic and strong interactions. It originates in the fact that the photons are electrically neutral while the gluons are colored objects, transforming according to the octet representation of the color group. The mathematical reason for the difference is that the gauge group of Quantum Electrodynamics, U(1), is an Abelian group while the gauge group of Quantum Chromodynamics, SU(3), is non-Abelian: for x_1, $x_2 \in$ U(1) the product $x_1 \cdot x_2$ is identical with $x_2 \cdot x_1$, but for x_1, $x_2 \in$ SU(3), the element $x_1 \cdot x_2$ in general differs from $x_2 \cdot x_1$.

The two plots below (Fig. 1) compare the surrounding of a lepton with the one of a quark. In either case, the interaction polarizes the vacuum. A positron generates a cloud of photons in its vicinity as well as a cloud of electrons and positrons. The electrons dominate, so that the charge density, which is shown in the plot on the left, receives a contribution that is of opposite sign to the bare charge inserted in the vacuum. Accordingly, the total charge of the positron is smaller than the bare charge, $e < e|_{\mathsf{bare}}$: the vacuum shields the charge.

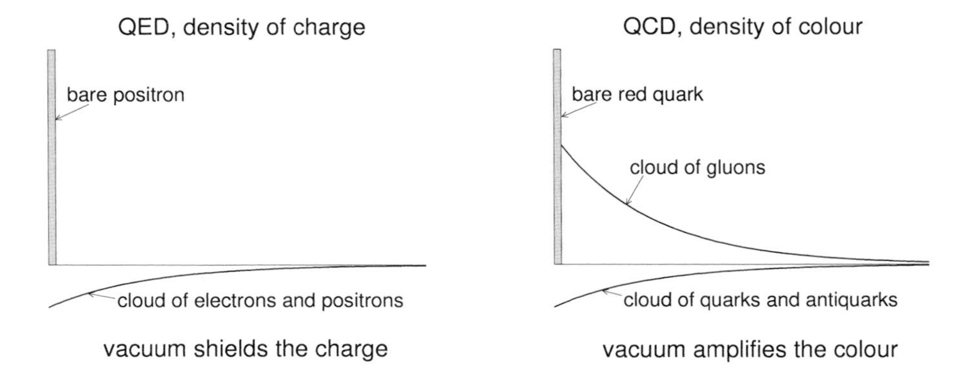

Fig. 1.

Likewise, a bare quark is surrounded by a cloud of gluons, quarks and antiquarks. The difference between QED and QCD manifests itself in the properties of these clouds: the photon cloud (the Coulomb field surrounding the bare positron) does not contribute to the charge density, because the photons are electrically neutral, but the gluons do show up in the density of color, because they are colored. The corresponding contribution to the color density turns out to be of the same sign as the one of the bare quark at the center. The gluon cloud thus amplifies the color rather than shielding it. Concerning the cloud of quarks and antiquarks, the situation is the same as in the plot on the left: their contribution to the color density is of opposite sign to the one from the bare quark. Unless there are too many quark flavors, the contribution from the gluons is more important than the one from the quarks and antiquarks: the vacuum amplifies the color. As it is the case with QED, vacuum polarization implies that the effective strength of the interaction depends on the scale, but while e shrinks with increasing size of the region considered, g_s grows: $g_s > g_s|_{\text{bare}}$.

In other words, the qualitative difference between QED and QCD arises because the two interactions polarize the vacuum differently.

8. Comparison with Gravity

A qualitatively very similar effect also occurs in gravity. The source of the gravitational field is the energy. The gravitational field itself carries energy, comparable to the situation in Chromodynamics, where the color is the source of the field and the gluons act as their own source, because they carry color. In application to the motion of a planet like Mercury, the planet is attracted only by the energy contained within its orbit — the forces generated by the gravitational field outside cancel out. This produces a tiny effect in the perihelion shift: if the orbit is calculated under the assumption that the attraction is produced by the total energy of the sun, the perihelion shift comes out larger than predicted by general relativity: 50" instead of 43" per century.

Figure 2 compares the orbit of Mercury with the charged pion, where two quarks are orbiting one another and where the effect is much stronger. Unfortunately, a quantitative comparison cannot be made, because gravity can be compared with Chromodynamics only at the classical level — a quantum field theory of gravity is not available and a description of the structure of the pion within classical field theory is not meaningful.

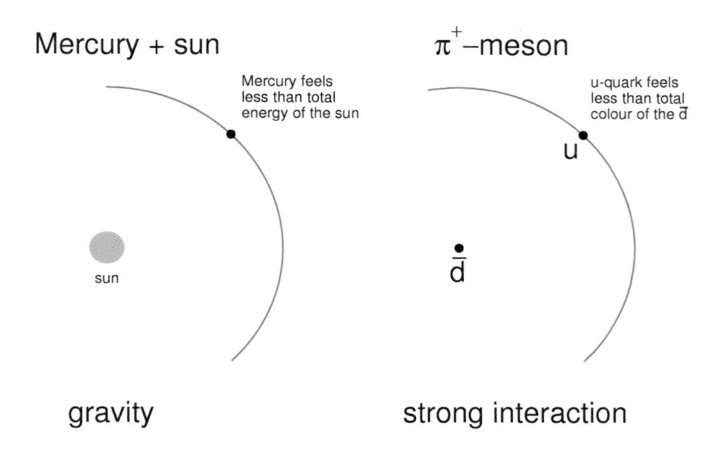

Fig. 2.

9. Physics of Vacuum Polarization

The fact that the vacuum reduces the electric field of a charged source but amplifies the gluonic field of a colored source has dramatic consequences: although the Lagrangians of QED and QCD are very similar, the properties of the electromagnetic and strong interactions are totally different.

The most important consequence of vacuum polarization is that QCD confines color while QED does not confine charge. Qualitatively, this is easy to understand: the energy density of the e.m. field that surrounds a positron falls off with the distance — the field energy contained in the region outside a sphere of finite radius is finite. Accordingly, only a finite amount of energy is needed to isolate a positron from the rest of the world: charged particles can live alone, the electric charge is not confined. For a quark, the situation is different, because the gluonic field that surrounds it does not fall off with the distance. In order to isolate a quark from the rest of the world, an infinite amount of energy would be needed. Only colorless states, hadrons, can live alone: mesons, baryons, nuclei.

As Zweig pointed out already in 1964, the forces between the colorless objects can be compared with the van der Waals forces between atoms.[7] At long distance and disregarding the e.m. interaction, the force between two hadrons is dominated by the exchange of the lightest strongly interacting particle, the π-meson: the Yukawa formula is indeed valid in QCD with $r_0 = \hbar/m_\pi c$, not for the force between two quarks, but for the force between two hadrons. It is of short range because the pion is not massless. The van der Waals force between two atoms, on the other hand, is of long range as it only drops off with a power of the distance: atoms can exchange photons and these are massless.

An analytic proof that QCD does indeed confine color is not available, but the results obtained by means of numerical simulations of QCD on a lattice represent overwhelming evidence that this is the case. The progress made with such simulations in recent years has made it possible to calculate the masses of the lowest lying bound states of QCD from first principles, in terms of the parameters occurring in the Lagrangian. The fact that the result agrees with the observed mass spectrum provides a very thorough test of the hypothesis that the quarks are bound together with a gauge field.

10. Why is QFD different from QCD?

The above discussion shows that the difference between QED and QCD arises from vacuum polarization, but this cannot explain why the weak and strong interactions are so different. The vacuum amplifies flavor for the same reason as it amplifies color: both of these theories are asymptotically free. Why, then, is color confined, flavor not?

The difference arises because, in addition to the constituents and the gauge fields that mediate the interaction between them, the Standard Model contains a third category of fields: *scalar fields*, referred to as *Higgs fields*. These affect the behavior of the weak interaction at long distance, but leave the strong and electromagnetic interactions alone. As shown by Glashow, Salam and Weinberg, an SU(2) doublet of Higgs fields is needed to describe the properties of the weak interaction: four real scalar fields are required.

Lorentz invariance allows scalar fields to pick up a vacuum expectation value. In particle terminology: spinless particles can form a *condensate*. For the gauge fields which mediate the weak interaction, the condensate feels like a medium which affects the W and Z bosons that fly through it.

If the frequency of a W-wave is less than a certain critical value ω_W, which is determined by the strength of the interaction between the W and the Higgs fields, it penetrates only a finite distance into the medium. For waves of low frequency, the penetration depth is given by $r_{\mathrm{w}} = c/\omega_W$. If the frequency is higher than ω_W, then the wave can propagate, but the group velocity is less than the velocity of light — the condensate impedes the motion. In particle language: the W-particles pick up mass. The size of the mass is determined by the critical frequency, $m_{\mathrm{w}} = \hbar\omega_W/c^2$.

In the presence of the condensate, the interaction mediated by the W-bosons thus involves the exchange of massive particles. Accordingly, the $\frac{1}{r}$-

potential characteristic of the exchange of massless particles is modified:

$$\frac{g_{\mathrm{W}}^2}{4\pi r} \Rightarrow \frac{g_{\mathrm{W}}^2}{4\pi r} \cdot e^{-\frac{r}{r_{\mathrm{W}}}} , \quad r_{\mathrm{W}} = \frac{\hbar}{m_{\mathrm{W}} c} .$$

The condensate thus gives the weak interaction a finite range, determined by the penetration depth. At low energies, only the mean value of the potential counts, which is given by

$$\int d^3 r \, \frac{g_{\mathrm{W}}^2}{4\pi r} \cdot e^{-\frac{r}{r_{\mathrm{W}}}} = g_{\mathrm{W}}^2 r_{\mathrm{W}}^2 .$$

In the Fermi theory of 1934,[38] the weak interaction is described by the contact potential $V_F = 4\sqrt{2}\, G_F \delta^3(\vec{x})$, for which the mean value is given by $4\sqrt{2} G_F$. At low energies, the exchange of W-bosons thus agrees with the Fermi theory, provided $g_{\mathrm{W}}^2 r_{\mathrm{W}}^2 = 4\sqrt{2} G_F$.

Experimentally, the gauge particles which mediate the weak interaction were discovered at the CERN SPS (UA1 and UA2) in 1983. The mass of the W turns out to be larger than the mass of an iron nucleus: $m_{\mathrm{W}} = 85.673 \pm 0.016\ m_{\mathrm{proton}}$. The Z, which is responsible for the weak interaction via neutral currents, is even heavier: $m_{\mathrm{Z}} = 97.187 \pm 0.002\ m_{\mathrm{proton}}$. This explains why the range of the weak interaction is very short, of the order of $2 \cdot 10^{-18}$ m, and why, at low energies, the strength of this interaction is so weak.

Massless gauge particles have only two polarization states, while massive particles of spin 1 have three independent states. The Higgs fields provide the missing states: three of the four real Higgs scalars are eaten up by the gauge fields of the weak interaction. Only the fourth one survives. Indeed a Higgs candidate was found in 2012, by the ATLAS and CMS teams working at the CERN LHC. As far as its properties revealed themselves by now, they are in accord with the theoretical expectations.

11. Transparency of the Vacuum

The above discussion of the qualitative differences between the electromagnetic, weak and strong interactions concerns the behavior at long distance. At short distance (10^{-19} m $\leftrightarrow$ 2 TeV), all of the forces occurring in the Standard Model obey the inverse square law. The interaction energy is of the form

$$V = \text{constant} \times \frac{\hbar c}{r} ,$$

where the constant is a pure number. As the gauge symmetry consists of three distinct factors, $G = \mathrm{U}(1) \times \mathrm{SU}(2) \times \mathrm{SU}(3)$, the strength of the interaction at short distances is characterized by three distinct numbers:

e.m.	weak	strong
$\frac{e^2}{4\pi}$	$\frac{g_{\mathrm{W}}^2}{4\pi}$	$\frac{g_{\mathrm{s}}^2}{4\pi}$

Vacuum polarization implies that each of the three constants depends on the scale. If the size of the region considered grows, then the e.m. coupling e diminishes (the vacuum shields the charge), while the coupling constant g_{W} of the weak interaction as well as the coupling constant g_{s} of the strong interaction grows (the vacuum amplifies flavor as well as color).

At distances of the order of the penetration depth of the weak interaction, however, the presence of a condensate makes itself felt. Since the particles in the condensate are electrically neutral, the photons do not notice them — for photons, the vacuum is a transparent medium and the mediation of the electromagnetic interaction is not impeded by it. The condensed particles do not have color, either, so that the gluons do not notice their presence — the vacuum is transparent also for gluons. The gauge field of the weak interaction, however, does react with the medium, because the condensed Higgs particles do have flavor: while the Higgs fields are invariant under the gauge groups U(1) and SU(3), they transform according to the doublet representation of SU(2). As discussed above, W- and Z-waves of low frequency cannot propagate through this medium: for such waves the vacuum is opaque. The amplification of the weak interaction due to the polarization of the vacuum only occurs at distances that are small compared to the penetration depth. At larger distances, the condensate takes over and shielding eventually wins — although the weak interaction is asymptotically free, flavor is not confined.

This also implies that, at low energies, the weak interaction and the Higgs fields freeze out: the Standard Model reduces to QED+QCD. In cold matter, only the degrees of freedom of the photon, the gluons, the electron and the u- and d-quarks manifest themselves directly. The remaining degrees of freedom only show up indirectly, through small, calculable corrections that are generated by the quantum fluctuations of the corresponding fields.

12. Masses of the Leptons and Quarks

The leptons and quarks also interact with the condensate. Gauge invariance does not allow the presence of corresponding mass terms in the Lagrangian, but they pick up mass though the same mechanism that equips the W- and Z-bosons with a mass. The size of the lepton and quark masses is determined by the strength of their interaction with the Higgs fields. Unfortunately, the symmetries of the Standard Model do not determine the strength of this interaction.

Indeed, the pattern of lepton and quark masses is bizarre: the observed masses range from 10^{-2} eV$/c^2$ to 10^{+11} eV$/c^2$. It so happens that the electron only interacts very weakly with the condensate and thus picks up only little mass. For this reason, the size of the atoms, which is determined by the Bohr radius, $a_{\rm Bohr} = 4\pi\hbar/e^2 m_e c$, is much larger than the size of the proton, which is of the order of the range of the strong interaction, $\hbar/m_\pi c$ — but why is the interaction of the electrons with the Higgs particles so weak?

13. Beyond the Standard Model

The Standard Model leaves many questions unanswered. In particular, it neglects the gravitational interaction. By now, quantum theory and gravity peacefully coexist for almost a century, but a theory that encompasses both of them and is consistent with what is known still remains to be found. This also sets an upper bound on the range of validity of the Standard Model: at distances of the order of the Planck length, $\ell_{\rm Planck} = \sqrt{G\hbar/c^3} = 1.6 \cdot 10^{-35}$ m, the quantum fluctuations of the gravitational field cannot be ignored.

Neither the electromagnetic interaction nor the interaction among the Higgs fields is asymptotically free. Accordingly, the Standard Model is inherently incomplete, even apart from the fact that gravity can be accounted for only at the classical level. A cutoff is needed to give meaning to the part of the Lagrangian that accounts for those interactions. The Standard Model cannot be the full story, but represents an effective theory, verified up to energies of the of order 1 TeV. It may be valid to significantly higher energies, but must fail before the Planck energy $E_{\rm Planck} \sim 1.2 \cdot 10^{28}$ eV is reached. Quite a few levels of structure were uncovered above the present resolution. It does not look very plausible that there are no further layers all the way to $\ell_{\rm Planck}$.

Astronomical observations show that the universe contains Dark Matter as well as Dark Energy. The latter may be accounted for with a cosmological constant, but we do not understand why this constant is so small. Anything that carries energy generates gravity. Why does gravity not take notice of the Higgs condensate? The Standard Model does not have room for Dark Matter, either. Supersymmetric extensions of the Standard Model do contain candidates for Dark Matter, but so far, not a single member of the plethora of superpartners required by this symmetry showed up.

We do not understand why the baryons dominate the visible matter in our vicinity. In fact, this is difficult to understand if the proton does not decay and until now, there is no upper bound on the proton lifetime. How could the observed excess of quarks over antiquarks have arisen if processes which violate baryon number do not occur? In particular, CP violation is necessary for baryogenesis. The phenomenon is observed and can be accounted for in the Standard Model, but we do not understand it, either.

Why are there so many lepton and quark flavors? What is the origin of the bizarre mass pattern of the leptons and quarks? The Standard Model becomes much more appealing if it is sent to the hairdresser, asking him or her to chop off all fields except the quarks and gluons. In the absence of scalar fields, a Higgs condensate cannot occur, so that the quarks are then massless: what survives the thorough cosmetic treatment is *QCD with massless quarks*.

This is how theories should be: massless QCD does not contain a single dimensionless parameter to be adjusted to observation. In principle, the values of all quantities of physical interest are predicted without the need to tune parameters (the numerical value of the mass of the proton in kilogram units cannot be calculated, because that number depends on what is meant by a kilogram, but the mass spectrum, the width of the resonances, the cross sections, the form factors, ... can be calculated in a parameter free manner from the mass of the proton, at least in principle). This theory does explain the occurrence of mesons and baryons and describes their properties, albeit only approximately — for an accurate representation, the quark masses cannot be ignored. Compared to this beauty of a theory, the Standard Model leaves much to be desired ....

References

1. H. Yukawa, *Proc. Phys. Math. Soc. Jpn.* **17**, 48 (1935).
2. E. C. G. Stueckelberg, *Helv. Phys. Acta* **11** (1938) 299.
3. L. D. Landau, in *Theoretical Physics in the Twentieth Century: A Memorial Volume to Wolfgang Pauli*, eds. M. Fierz and V. Weisskopf (Interscience Publishers, 1960), p. 245.
4. H. Pietschmann, *Eur. Phys. J. H* **36**, 75 (2011), arXiv:1101.2748.
5. M. Gell-Mann, *Phys. Rev.* **125**, 1067 (1962).

6. Y. Ne'eman, *Nucl. Phys.* **26**, 222 (1961).
7. G. Zweig, *An SU(3) model for strong interaction symmetry and its breaking*, CERN reports 8419/TH.401 and 8419/TH.412 (1964).
8. M. Gell-Mann, *Phys. Lett.* **8**, 214 (1964).
9. O. W. Greenberg, *Phys. Rev. Lett.* **13** (1964) 598.
10. N. N. Bogoliubov, B. V. Struminski and A. N. Tavkhelidze, Preprint JINR D-1968 (1964).
11. M. Han and Y. Nambu, *Phys. Rev. B* **139**, 1006 (1965).
12. Y. Miyamoto, *Prog. Theor. Phys. Suppl.* **E65**, 187 (1965).
13. M. Gell-Mann, *Physics* **I**, 63 (1964).
14. J. D. Bjorken, *Phys. Rev.* **179** (1969) 1547.
15. For an account of the experimental developments, see the *Nobel lectures in physics 1990* by J. I. Friedman, H. W. Kendall and R. E. Taylor, SLAC-REPRINT-1991-019.
16. K. G. Wilson, *Phys. Rev.* **179**, 1499 (1969).
17. V. A. Fock, *Z. Phys.* **39**, 226 (1926) [*Surveys H. E. Phys.* **5**, 245 (1986)]. For a discussion of the significance of this paper, see L. B. Okun, *Phys. Usp.* **53**, 835 (2010) [*Usp. Fiz. Nauk* **180**, 871 (2010)].
18. C.-N. Yang and R. L. Mills, *Phys. Rev.* **96**, 191 (1954).
19. R. Shaw, The problem of particle types and other contributions to the theory of elementary particles, Cambridge PhD thesis (1955) unpublished.
20. N. Straumann, *Space Sci. Rev.* **148**, 25 (2009), arXiv:0810.2213.
21. F. Englert and R. Brout, *Phys. Rev. Lett.* **13**, 321 (1964).
22. P. W. Higgs, *Phys. Lett.* **12**, 132 (1964).
23. G. S. Guralnik, C. R. Hagen and T. W. B. Kibble, *Phys. Rev. Lett.* **13**, 585 (1964).
24. S. L. Glashow, *Nucl. Phys.* **22**, 579 (1961).
25. S. Weinberg, *Phys. Rev. Lett.* **19**, 1264 (1967).
26. A. Salam, *Conf. Proc. C* **680519**, 367 (1968).
27. H. Fritzsch and M. Gell-Mann, *eConf C* **720906V2**, 135 (1972), hep-ph/0208010.
28. H. Fritzsch, M. Gell-Mann and H. Leutwyler, *Phys. Lett. B* **47**, 365 (1973).
29. G. 't Hooft, *Proc. Int. School of Subnuclear Physics* **36**, ed. A. Zichichi (World Scientific, 1998), arXiv:hep-th/9812203;
   H. Fritzsch, *The history of QCD*, CERN Courier, Sept. 27, 2012, `http://cerncourier.com/cws/article/cern/50796`;
   D. Gross and F. Wilczek, *A watershed: the emergence of QCD*, CERN Courier, Jan. 28, 2013, `http://cerncourier.com/cws/article/cern/52034`;
   G. Ecker, *The colorful world of quarks and gluons I: The shaping of Quantum Chromodynamics*, QCHS10, Munich, Germany, *PoS* Confinement, 344 (2012);
   H. Leutwyler, On the history of the strong interaction, in *Proc. Int. School of Subnuclear Physics*, Erice, Italy, 2012, ed. A.Zichichi, *Mod. Phys. Lett. A* **29**, 1430023 (2014), arXiv:1211.6777.
30. V. S. Vanyashin and M. T. Terentyev, *Sov. Phys. JETP* **21**, 375 (1965).
31. I. B. Khriplovich, *Yad. Fiz.* **10**, 409 (1969) [*Sov. J. Nucl. Phys.* **10**, 235 (1970)].
32. G. 't Hooft, *Nucl. Phys. B* **33**, 173 (1971); *ibid.* **35**, 167 (1971).

33. K. Symanzik, *Lett. Nuovo Cim.* **6**, 77 (1973).
34. K. Symanzik, On theories with massless particles, in *Proc. Colloquium on Renormalization of Yang–Mills Fields and Applications to Particle Physics*, Vol. 72, ed. C. P. Korthals-Altes (Centre Physique Theorique au CNRS, Marseille, France, 1972), p. 470.
35. G. Parisi, *Lett. Nuovo Cim.* **7**, 84 (1973).
36. D. J. Gross and F. Wilczek, *Phys. Rev. Lett.* **30**, 1343 (1973).
37. H. D. Politzer, *Phys. Rev. Lett.* **30**, 1346 (1973).
38. E. Fermi, *Z. Phys.* **88**, 161 (1934).

Quarks and QCD

Harald Fritzsch

Physics Department, Ludwig-Maximilians-University,
D-80333 Munich, Germany

In the decade after 1950 many new particles were discovered, in particular the four Δ resonances, the six hyperons and the four K mesons. The Δ resonances were observed in pion–nucleon collisions at the Radiation Laboratory in Berkeley. The hyperons and K mesons were discovered in cosmic-ray experiments. They were called "strange particles."

In order to understand the peculiar properties of the new hadrons, Murray Gell-Mann introduced in 1953 the new quantum number "strangeness." The nucleon has no strangeness. The Λ hyperon and the three Σ hyperons were assigned the strangeness (-1). The two Ξ hyperons have strangeness (-2), the K mesons strangeness $(+1)$.

Gell-Mann assumed that the strangeness quantum number was conserved by the strong and electromagnetic interactions, but violated by the weak interactions. In a hadronic collision particles with strangeness could be produced in pairs by the strong interaction, if the strangeness of the pair is zero, e.g. the production of a Λ hyperon and a K meson. But the decays of the strange particles into particles without strangeness, for example the decay of the Λ hyperon into a proton and a pion, could only proceed via the weak interaction.

In 1960 Gell-Mann proposed a new symmetry to describe the baryons and mesons, based on the unitary group SU(3), the group of unitary 3×3 matrices with determinant 1.[1] The SU(3)-symmetry is an extension of the isospin symmetry, which was introduced in 1932 by Werner Heisenberg.

The observed hadrons are members of specific representations of SU(3). The baryons and mesons were described by octet representations. The baryon octet contains the two nucleons, the Λ hyperon, the three Σ hyperons and the two Ξ hyperons. The members of the meson octet are the three pions, the eta meson, the four K mesons. The baryon resonances (spin 3/2) are members of a decuplet representation.

Only nine baryon resonances were known in 1960, the four Δ resonances, the three Σ resonances and the two Ξ resonances. Gell-Mann predicted that a negatively charged particle with strangeness (-3) should exist, which he called the Ω^- particle. This particle would be the tenth particle in the decuplet. Due to its strangeness it would have a long life time, since it could only decay by the weak interaction.

The Ω^- with a mass of 1672.5 MeV was found in 1964 by Nicholas Samios and his group at the Brookhaven National Laboratory. The mass had been predicted by Gell-Mann, using a simple model for the breaking of the SU(3)-symmetry — it is broken by a term, transforming under SU(3) as an octet.

In the SU(3) scheme the observed hadrons were octets and decuplets. There are no hadrons, described by the triplet or sextet representation. In 1964 Gell-Mann introduced the triplets as constituents of the baryons and mesons, the "quarks." This name appears in the novel of James Joyce "*Finnegans Wake*" on page 383 in the sentence "Three quarks for Muster Mark."

The triplet model was also considered by the Caltech graduate student George Zweig, who was visiting CERN. Gell-Mann published his idea in a short letter in Physics Letters,[2] Zweig published his work only as a CERN preprint.[3]

There are three quarks, the "up," "down" and "strange" quarks. The baryons are bound states of three quarks. For example the proton consists of two up-quarks and one down-quark: $p \sim (uud)$. The Λ hyperon has one strange quark: $\Lambda \sim (uds)$. The Ω^- is a bound state of three strange quarks: $\Omega^- \sim (sss)$.

The quarks have nonintegral electric charges:

$$u \sim \left(\frac{2}{3}e\right), \quad d \sim \left(-\frac{1}{3}e\right), \quad s \sim \left(-\frac{1}{3}e\right).$$

The strangeness of a particle denotes the number of strange quarks in the particle, but with a negative sign. Thus the Λ hyperon with the quark structure (uds) has strangeness (-1). It would have been better to use the other sign for the strangeness, but this became clear only after the introduction of the quark model in 1964, 11 years after the introduction of strangeness.

In 1968 the first experiments in the deep inelastic scattering of electrons and nuclei were carried out at the Stanford Linear Accelerator Center (SLAC). The cross-section depends on two variables, the mass of the virtual photon and the energy transfer from the electron to the nucleus.

In 1967 the SLAC theoretician James Bjorken predicted, that the cross-section for deep inelastic electron–proton scattering should show a scaling behavior — it should not depend on two variables, but only on the ratio of two variables. Thus the cross-section for deep inelastic scattering would be much larger than assumed at this time. One year later the scaling behavior of the cross-section was observed in the SLAC experiments.

Richard Feynman suggested that the scaling behavior implies that the electrons were deflected by point-like constituents in the proton, which he called "partons." These partons turned out to be the quarks. Thus in 1968 the quarks were observed inside the nucleons as point-like constituents.

In these experiments the nucleon matrix-element of the commutator of two electromagnetic currents is measured at nearly light-like distances. In 1971 Gell-Mann and I assumed that this commutator can be abstracted from the free-quark model, and we formulated the light-cone algebra of the currents.[4] Using this algebra, we could understand the scaling behavior. We obtained the same results as Richard Feynman in his parton model, if the partons are identified with the quarks. Later it turned out that the results of the light-cone current algebra are nearly correct in the theory of QCD. It follows from the asymptotic freedom of the theory.

The isospin symmetry is broken by the electromagnetic interaction. The symmetry breaking is about 1%, as expected. However the breaking of SU(3) is much stronger, about 20%, and there is no specific interaction, which breaks the symmetry. The symmetry breaking remained a puzzle until the introduction of the quarks.

In the quark model the symmetry breaking can be described by a mass term for the quarks. The mass of the strange quark is larger than the masses of the up or down quarks. The mass term is the sum of a term, which is a singlet under SU(3), and an octet term. One can calculate the mass differences in the octet or decuplet, in perfect agreement with the observed masses of the hadrons.

If the electromagnetic interaction is turned off, one expected originally that the isospin symmetry is perfect. However there is a problem with the mass of the neutron. If the mass difference between the neutron and the proton would be of electromagnetic origin, the proton mass should be larger than the neutron mass, due to the electromagnetic self-energy. But this is not the case. In the quark model the masses of the nucleons can be understood, if the mass of the d-quark is larger than the mass of the u-quark. Thus in the absence of the electromagnetic interaction there is still a violation of the isospin symmetry due to the quark mass term.

The Ω^- is a bound state of three strange quarks. Since this is the ground state, the space wave function should be symmetric. The three spins of the quarks are aligned to give the spin of the Ω^-. Thus the wave function of the Ω^- does not change if two quarks are interchanged. But the wave function must be antisymmetric due to the Pauli principle. This was a great problem for the quark model.

Another problem was related to the decay of the neutral pion. If the pion is considered as a bound state of a nucleon and an antinucleon, one can calculate the decay rate for the electromagnetic decay of the neutral pion into two photons, in perfect agreement with the observed rate. But in the quark model the decay rate is a factor nine smaller.

William Bardeen, Murray Gell-Mann and I introduced in 1971 a new quantum number for the quarks, which we called "color."[5] Each quark has three different colors: red, green or blue. The symmetry of the three colors is described by the group SU(3). The quarks are color triplets, two quarks can either be in a color sextet or in a color antitriplet. Three quarks can form a color singlet.

The simplest color singlets were the bound states of a quark and an antiquark (meson) or three quarks (baryon). We assumed that all hadrons are color singlets, and all configurations with color, e.g. the color triplet quarks, are permanently confined inside the hadrons. At this time we did not understand the reason for the confinement of color.

The three quarks in a baryon wave function form a color singlet, thus they are antisymmetric in the color index. The problem with the Pauli statistics disappears:

$$(qqq) \Rightarrow (q_r q_g q_b - q_g q_r q_b + q_b q_r q_g - q_r q_b q_g + q_g q_b q_r - q_b q_g q_r)\,.$$

The decay amplitude for the decay of the neutral pion into two photons is given by a triangle diagram, in which a quark–antiquark pair is created virtually and subsequently annihilates into two photons. We found that after the introduction of color the decay amplitude increases by a factor 3 — each color contributes to the amplitude with the same strength. The result agrees with the experimental value.

The cross-section for electron–positron annihilation into hadrons at high energies depends on the squares of the electric charges of the quarks and on the number of colors. For three colors one has:

$$\frac{\sigma(e^+ + e^- \Rightarrow \text{hadrons})}{\sigma(e^+ + e^- \Rightarrow \mu^+ + \mu^-)} \to 3\left[\left(\frac{2}{3}\right)^2 + \left(-\frac{1}{3}\right)^2 + \left(-\frac{1}{3}\right)^2\right] = 2\,.$$

Without colors this ratio would be 2/3. In 1971 the experimental data, however, were in agreement with a ratio of 2.

In 1972 Gell-Mann and I introduced a gauge theory for the description of the strong interactions.[6] The color degree of freedom was gauged, as the electric charge in QED. The gauge bosons were massless gluons, which transformed as an octet of the color group. Later we called this theory "Quantum Chromodynamics."

In 1972 Gell-Mann discussed this new theory of QCD at the Rochester conference at the Fermi National Laboratory. One year later Murray Gell-Mann, Heinrich Leutwyler and I described the advantages of QCD in a letter, published in *Physic Letters.*[7]

The gluons interact with the quarks, but also with themselves. This self-interaction leads to the interesting property of asymptotic freedom. The gauge coupling constant of QCD decreases logarithmically, if the energy is increased.

In QCD the scaling property of the cross-sections, observed in deep inelastic scattering, is not an exact property, but it is violated by small logarithmic terms. The scaling violations were observed and in good agreement with the theoretical predictions. They are determined by a free scale parameter Λ. The current experimental value is

$$\Lambda = 213^{+38}_{-35}\ \text{MeV}\,.$$

Experiments at SLAC, at DESY, at the Large Electron–Positron Collider in CERN and at the Tevatron in the Fermi National Laboratory have measured the decrease of the QCD coupling-constant. It agrees well with the prediction of QCD. With LEP it was also possible to determine the QCD coupling-constant at the mass of the Z boson rather precisely:

$$\alpha_s(M_Z) = 0.1184 \pm 0.0007\,.$$

It is useful to consider the theory of QCD with just one heavy quark Q. The ground-state meson in this hypothetical case would be a quark–antiquark bound state. The effective potential between the quark and its antiquark at small distances would be a Coulomb potential proportional to $1/r$, where r is the distance between the quark and the antiquark. However at large distances the self-interaction of the gluons becomes important. The gluonic field lines at large distances do not spread out as in electrodynamics — they attract each other. Thus the quark and the antiquark at large distances are connected by a string of gluonic field lines. The force between the quark and the antiquark is constant and the quarks are confined.

For light quarks the situation is different. If the quark and the antiquark inside a pion are moved away from each other, there is no string of gluonic field lines between the quark and the antiquark. Instead light quark–antiquark pairs are created. Probably this implies that the quarks are confined, but a rigorous proof for the confinement does not exist thus far.

In electron–positron annihilation the virtual photon creates a quark and an antiquark, which move away from each other with high speed. Pairs of light quarks and antiquarks are created, which form mostly pions, moving roughly in the same direction. The quark and the antiquark "fragment" to produce two jets of particles. The sum of the energies and momenta of the particles in each jet should be equal to the energy of the original quark, which is equal to the energy of each colliding lepton.

These quark jets were observed for the first time in 1978 at DESY. They had been predicted in 1975 by Richard Feynman (see e.g. Ref. 8). If a quark pair is produced in electron–positron annihilation, then QCD predicts that sometimes a high-energy gluon should be emitted from one of the quarks. The gluon would also fragment and produce a jet. Sometimes three jets should be produced. Such events were observed at DESY in 1979 and studied in detail later with the LEP-ring at CERN.

The basic quanta of QCD are the quarks and the gluons. Two color-octet gluons can form a color singlet. Such a state would be a neutral glue meson. In QCD one expects that the ground state of the glue mesons has a mass of about 1.4 GeV. In QCD with only heavy quarks this state would be stable, but in the real world it would mix with neutral quark–antiquark mesons and would decay quickly into pions. Thus far the glue mesons have not been identified clearly in experiments.

The simplest color-singlet hadrons in QCD are the baryons — consisting of three quarks — and the mesons, made of a quark and an antiquark. However there are other ways to form a color singlet. Two quarks can be in an antitriplet — they can form a color singlet together with two antiquarks. The result would be a meson consisting of two quarks and two antiquarks. Such a meson is called a tetraquark.

Three quarks can be in a color octet, as well as a quark and an antiquark. They can form a color-singlet hadron, consisting of four quarks and an antiquark. Such a baryon is called a pentaquark. Tetraquark mesons and pentaquark baryons have not been clearly observed in the experiments.

If QCD is right, the hadronic world depends in the absence of the quark masses only on the scale parameter, which determines the properties of the hadrons, e.g. the masses or the magnetic moments. The proton mass is a numerical constant, multiplied by the scale parameter Λ:

$$M = \text{const} \bullet \Lambda .$$

This constant can be calculated in QCD, e.g. by using lattice methods. Using the observed value of the scale parameter, one finds for the proton mass about 860 MeV.

The proton mass also depends on the quark masses. These contributions can be calculated, using the chiral symmetry. About 21 MeV of the proton mass are due to the mass of the u-quark. The mass of the d-quark contributes about 19 MeV. The mass of the strange quark is also relevant, since inside the proton there are many pairs of strange quarks and antiquarks — its contribution is about 36 MeV. The electromagnetic self-energy of the proton is about 2 MeV. The sum of these terms gives the correct mass of the proton.

The three quark flavors were introduced to describe the symmetry given by the flavor group SU(3). Now we know that in reality there are six quarks: the three light quarks u, d, s and the three heavy quarks c (charm), b (bottom) and t (top). These six quarks form three doublets of the electroweak symmetry group SU(2):

$$\begin{pmatrix} u \\ c \end{pmatrix} \Leftrightarrow \begin{pmatrix} c \\ s \end{pmatrix} \Leftrightarrow \begin{pmatrix} t \\ b \end{pmatrix} .$$

The masses of the quarks are arbitrary parameters in QCD, just as the lepton masses are arbitrary parameters in QED. Since the quarks do not exist as free particles, their masses cannot be measured directly. They can be estimated, using the chiral symmetry and the observed hadron masses. In QCD they depend on the energy scale under consideration. Typical values of the quark masses at the energy of 2 GeV are:

$$\begin{aligned} &u : 4\ \text{MeV}\,, \quad && c : 1.2\ \text{GeV}\,, \quad && t : 173\ \text{GeV}\,, \\ &d : 6\ \text{MeV}\,, \quad && s : 0.1\ \text{GeV}\,, \quad && b : 4.4\ \text{GeV}\,. \end{aligned}$$

The mass of the t-quark is large, similar to the mass of a gold atom. Due to this large mass the t-quark decays by the weak interaction with a lifetime that is less than the time needed to form a meson or baryon. Thus t-quarks do not form hadrons.

The theory of QCD is the correct field theory of the strong interactions and of the nuclear forces. Both hadrons and atomic nuclei are bound states of quarks, antiquarks and gluons. It is remarkable that a simple gauge theory can describe the complicated phenomena of the strong interactions in particle and nuclear physics.

References

1. M. Gell-Mann, *Phys. Rev.* **125**, 1067 (1962).
2. M. Gell-Mann, *Phys. Lett.* **8**, 214 (1964).
3. G. Zweig, CERN preprint 8182/TH401 (1964), unpublished.
4. H. Fritzsch and M. Gell-Mann, *Proc. Int. Conf. on Duality and Symmetry*, Tel Aviv, 1971 (Weizmann Science Press, 1971).
5. W. Bardeen, H. Fritzsch and M. Gell-Mann, *Scale and Conformal Symmetry in Hadron Physics* (John Wiley and Sons, 1973).
6. H. Fritzsch and M. Gell-Mann, *Proc. Int. Conf. on High Energy Physics*, Chicago, 1972.
7. H. Fritzsch, M. Gell-Mann and H. Leutwyler, *Phys. Lett. B* **47**, 365 (1973).
8. R. D. Field and R. P. Feynman, *Nucl. Phys. B* **136**, 1 (1978).

The Discovery of the Gluon

John Ellis

Theoretical Particle Physics and Cosmology Group,
Department of Physics, King's College London, London WC2R 2LS, UK
Theory Division, CERN, CH-1211 Geneva 23, Switzerland
John.Ellis@cern.ch

Soon after the postulation of quarks, it was suggested that they interact via gluons, but direct experimental evidence was lacking for over a decade. In 1976, Mary Gaillard, Graham Ross and the author suggested searching for the gluon via 3-jet events due to gluon bremsstrahlung in e^+e^- collisions. Following our suggestion, the gluon was discovered at DESY in 1979 by TASSO and the other experiments at the PETRA collider.

1. Quarks are Not Enough

When Murray Gell-Mann postulated quarks[1] and George Zweig postulated aces[2] in 1964 as the fundamental constituents of strongly-interacting particles, several questions immediately came to mind. What are the super-strong forces that bind them inside baryons and mesons? Are these forces carried by other particles, perhaps analogous the photons that bind electrons to nuclei inside atoms? But if so, why do we not see individual quarks? What is the important intrinsic difference between the underlying theory of the strong interactions and quantum electrodynamics (QED), the unified quantum theory of electricity and magnetism?

It was natural to suppose that the forces between quarks were mediated by photon-like bosons, which came to be known as gluons. If so, did they have spin one, like the photon? If so, did they couple to some charge analogous to electric charge? It was suggested that quarks carried a new quantum number called colour,[3] which could explain certain puzzles such as the apparent symmetry of the wave functions of the lightest baryons, the rate for $\pi^0 \to 2\gamma$ decay and (later) the rate for $e^+e^- \to$ hadrons. However, the postulation of colour raised additional questions. For example, did the quarks of different colours all have the same electric charge, and could individual coloured particles ever be observed, or would they remain confined inside hadrons?

Many of these issues were still being hotly debated when I started research in 1968. My supervisor was Bruno Renner, who had recently co-authored with Gell-Mann and Bob Oakes an influential paper[4] on the pattern of breaking of the approximate chiral SU(3) × SU(3), flavour SU(3) and chiral SU(2) × SU(2) symmetries of hadrons. I learnt from him that the gluons were presumably vector particles like photons, because otherwise it would be difficult to understand the successes of approximate chiral symmetry, but this posed another dilemma. On the one hand, if they were Abelian vector gluons as in QED, how could they confine quarks? On the other hand, non-Abelian gauge theories were known, but nobody knew how to calculate reliably their dynamics.

It was in 1968 that the results from deep-inelastic electron–proton scattering experiments at SLAC first became known.[5] Their (near-)scaling behaviour was interpreted by James Bjorken[6] and Richard Feynman[7] in terms of point-like constituents within the proton, called partons. It was natural to suppose that the partons probed by electrons though virtual photon exchange might be quarks, and this expectation was soon supported by measurements of the ratio of longitudinal and transverse scattering.[8] However, this insight immediately raised other questions. Might the proton contain other partons, such as gluons, that were not probed directly by photons? And what was the origin of the bizarre dynamics that enabled quarks to resemble quasi-free point-like particles when probed at short distances in the SLAC experiments, yet confined them within hadrons?

Chris Llewellyn-Smith pointed out that one could measure the total fraction of the proton momentum carried by quark partons,[9] and the experiments showed that this fraction was about half the total. Either the parton model was crazy, or the remaining half of the proton momentum had to be carried by partons without electric charges, such as neutral gluons. This was the first circumstantial evidence for the existence of gluons.

2. The Theory of the Strong Interactions

In 1971, Gerardus 't Hooft gave convincing arguments that unbroken (massless) non-Abelian gauge theories were renormalizable.[10] In retrospect, this should have triggered extensive theoretical studies of a non-Abelian gluon theory based on the SU(3) colour group,[11] but attention was probably distracted initially by 't Hooft's extension (together with Martinus Veltman) of his proof to spontaneously-broken (massive) non-Abelian gauge theories,[12] which opened the way to renormalizable theories of the weak interactions.[13]

Giorgio Parisi realized that the key to constructing a field theory of the strong interactions would be asymptotic freedom,[14] the property of renormalization driving the coupling of a field theory to smaller (larger) values at shorter (larger) distances, but the theories studied at that point did not have this property. Kurt Symanzik gave a talk at a conference in Marseille in the summer of 1972 where he pointed out that ϕ^4 theory would be asymptotically free if its coupling had an unphysical negative value.[15] After his talk, 't Hooft remarked that the coupling of a non-Abelian gauge theory would be driven towards zero, but apparently neither he, Symanzik nor anybody else who heard the remark made the connection between gauge theory and the strong interactions.

This connection was made in 1973 by David Politzer[16] and by David Gross and Frank Wilczek,[17] who not only demonstrated the asymptotic freedom of non-Abelian gauge theories but also proposed that the strong interactions be described by non-Abelian gluons whose interactions were specified by an unbroken SU(3) colour group. This proposal was supported strongly by Harald Fritzsch, Gell-Mann and Heinrich Leutwyler in a paper published later in 1973,[18] and came to be known as Quantum Chromodynamics (QCD).

It was possible to calculate in QCD logarithmic deviations from scaling in deep-inelastic scattering.[19] It was argued in 1974 that early results from deep-inelastic muon scattering at Fermilab were qualitatively consistent with these calculations,[20] since they suggested the characteristic features of falling structure functions at large quark–parton momentum fractions and rising structure functions at low momentum fractions.

However, the picture was clouded by the unexpected behaviour of the total cross-section for $e^+e^- \to$ hadrons, which was much larger at centre-of-mass energies above 4 GeV than was calculated in QCD with the three known flavours of quark.[21] The answer was provided in late 1974 by the discovery of the charm quark,[22] which boosted the $e^+e^- \to$ hadrons cross-section, as observed.

3. Where are the Gluons?

In the mid-1970s QCD was widely regarded as the only candidate theory of the strong interactions. It was the only realistic asymptotically-free field theory, capable of accommodating the (approximate) scaling seen in deep-inelastic scattering. QCD had qualitative success in fitting the emerging pattern of scaling violations through quark–parton energy loss via gluon bremsstrahlung and gluon splitting into quark–antiquark pairs. Moreover, QCD could explain semi-quantitatively the emerging spectrum of charmonia and had successes in calculating their decays in terms of quark–antiquark annihilations into two or three gluons.[23] No theorist seriously doubted the existence of the gluon, but direct proof of its existence, a "smoking gluon," remained elusive.

In parallel, jet physics was an emerging topic. Statistical evidence was found for two-jet events in low-energy electron–positron annihilation into hadrons at SPEAR at SLAC, which could be interpreted as $e^+e^- \to \bar{q}q$.[24] It was expected that partons should manifest themselves in high-energy hadron–hadron collisions via the production of pairs of large transverse-momentum jets. However, these were proving elusive at the Intersecting Storage Rings (ISR), CERN's pioneering proton–proton collider. It was known that the transverse-momentum spectrum of individual hadron production had a tail above the exponential fall-off seen in earlier experiments, but the shape of the spectrum did not agree with the power-law fall-off predicted on the basis of the hard scattering of quarks and gluons, and no jets were visible. Thus, there were no signs of quarks at the ISR, let alone gluons, and rival theories such as the constituent-interchange model,[25] in which the scattering objects were thought to be mesons, were being advocated.

4. The Three-Jet Idea

This was the context in 1976 when I was walking over the bridge from the CERN cafeteria back to my office one day. Turning the corner by the library, the thought occurred that the simplest experimental situation to search directly for the gluon would be through production via bremsstrahlung in electron–positron annihilation: $e^+e^- \to \bar{q}qg$. What could be simpler? Together with Graham Ross and Mary Gaillard, we calculated the gluon bremsstrahlung process in QCD,[26] see Fig. 1, arguing that under the appropriate kinematic conditions at high energies, asymptotic freedom would guarantee that this would be the leading correction to the process $e^+e^- \to \bar{q}q$ that gave two-jet final states. We showed that gluon bremsstrahlung would first manifest itself via broadening of one of the jets in a planar configuration, and that the appearance of three-jet events would provide the long-sought "smoking gluon," as seen in part (d) of Fig. 1. We also contrasted the predictions of QCD with a "straw-man" theory based on scalar gluons.

Two higher-energy collider projects were in preparation at the time, PETRA at DESY and PEP at SLAC, and we estimated that they should have sufficient energy to observe clear-cut three-jet events. I was already in contact with experimentalists at DESY, particularly my friend the late Bjørn Wiik, a leading member of the TASSO collaboration who was infected by

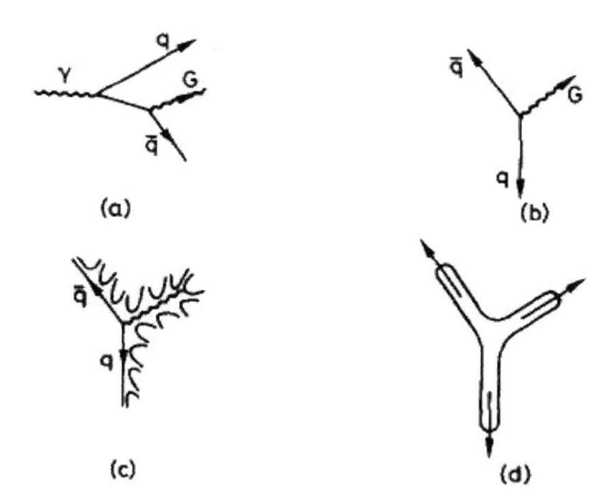

Fig. 1. (a) A Feynman diagram for gluon bremsstrahlung. (b) Wide-angle gluon bremsstrahlung visualized in momentum space in the centre of mass. (c) Hadronization of wide-angle gluon bremsstrahlung. (d) The resulting 3-jet final state. Figure from Ref. 25.

our enthusiasm for the three-jet idea. Soon after Mary, Graham and I had published our paper, I made a trip to DESY to give a seminar about it. The reception from the DESY theorists of that time was one of scepticism, bordering on hostility, and I faced fierce questioning why the short-distance structure of QCD should survive the hadronization process. My reply was that hadronization was expected to be a soft process involving small exchanges of momenta, as seen in part (c) of Fig. 1, and that the appearance of two-jet events at SPEAR supported the idea. At the suggestion of Bjørn Wiik, I also went to the office of Günter Wolf, another leader of the TASSO collaboration, to discuss with him the three-jet idea. He listened much more politely than the theorists.

The second paper on three-jet events was published in mid-1977 by Tom Degrand, Jack Ng and Henry Tye,[27] who contrasted the QCD prediction with those of the constituent-interchange model and a quark-confining string model. Then, later in 1977, George Sterman and Steve Weinberg published an influential paper[28] showing how jet cross-sections could be defined rigorously in QCD with a careful treatment of infrared and collinear singularities. In our 1976 paper, we had contented ourselves with showing that these were unimportant in the three-jet kinematic region of interest to us. The paper by Sterman and Weinberg opened the way to a systematic study of variables describing jet broadening and multi-jet events, which generated an avalanche of subsequent theoretical papers.[29] In particular, Alvaro De Rújula, Emmanuel Floratos, Mary Gaillard and I wrote a paper in early 1978[30] showing how "antenna patterns" of gluon radiation could be calculated in QCD and used to extract statistical evidence for gluon radiation, even if individual three-jet events could not be distinguished.

Meanwhile, the PETRA collider was being readied for high-energy data-taking with its four detectors, TASSO, JADE, PLUTO and Mark J. Bjørn Wiik, one of the leaders of the TASSO collaboration, and I were in regular contact. He came frequently to CERN around that time for various meetings, and I was working with him on electron–proton colliders. He told me that Sau Lan Wu had joined the TASSO experiment, and asked my opinion (which was enthusiastic) on the suggestion that she prepare a three-jet analysis for the collaboration. In early 1979, she and Haimo Zobernig wrote a paper[31] describing an algorithm for distinguishing three-jet events.

Another opportunity for detecting gluon effects had appeared in 1977, with the discovery of the Υ bottomonium states.[32] According to perturbative QCD, the dominant decay of the vector $\Upsilon(1S)$ state should be into three gluons, so the final state should look different from the dominant two-jet final states seen in the $e^+e^- \to \bar{q}q$ continuum.[33] A difference was indeed seen,[34] providing important circumstantial evidence for the gluon.

5. Proof at Last

During the second half of 1978 and the first half of 1979, the machine team at DESY was increasing systematically the collision energy of PETRA, getting into the energy range where we expected three-jet events due to gluon bremsstrahlung to be detectable. The first three-jet news came in June 1979 at the time of a neutrino conference in Bergen, Norway. The weekend before that meeting I was staying with Bjørn Wiik at his father's house beside a fjord outside Bergen, when Sau Lan Wu arrived over the hills bearing printouts of the first three-jet event. Bjørn included the event in his talk at the conference[35] and I also mentioned it in mine.[36] I remember Don Perkins asking whether one event was enough to prove the existence of the gluon: my tongue-in-cheek response was that it was difficult to believe in eight gluons on the strength of a single event! More scientifically, a single event might always be a statistical fluctuation, and more events would be needed to cement the interpretation.

The next outing for three-jet events was at the European Physical Society conference in Geneva in July 1979. Three members of the TASSO collaboration, Roger Cashmore, Paul Söding and Günter Wolf, spoke at the meeting and presented several clear three-jet events.[37] The hunt for gluons was looking good!

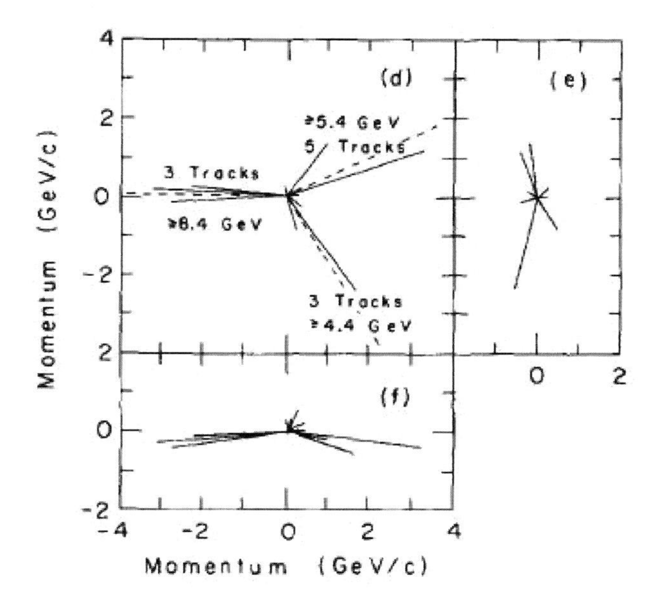

Fig. 2. An event detected by the TASSO collaboration,[35,37] exhibiting the planar 3-jet structure predicted in Ref. 26. Figure from Ref. 36.

The public announcement of the gluon discovery came at the Lepton/Photon Symposium held at Fermilab in August 1979. All four PETRA experiments showed evidence: JADE and PLUTO followed TASSO in presenting evidence for jet broadening and three-jet events as suggested in our 1976 paper, while the Mark J collaboration led by Sam Ting presented an analysis of antenna patterns along the lines of our 1978 paper. There was a press conference at which one of the three-jet events was presented, and a journalist asked which jet was the gluon. He was told that the smart money was on the jet on the left (or was it the right?). Refereed publications by TASSO[37] and the other PETRA collaborations[38] soon appeared, and the gluon finally joined the Pantheon of established particles as the second gauge boson to be discovered, joining the photon.

6. After the Discovery

An important question remained: was the gluon a vector particle, as predicted by QCD and everybody expected, or was it a scalar boson as nobody expected? Back in 1978, Inga Karliner and I had written a paper[39] that proposed a method for distinguishing the two possibilities by looking at a distribution in the angles between the three jets, based on our intuition about the nature of gluon bremsstrahlung. This method was used in 1980 by the

TASSO collaboration to prove that the gluon was indeed a vector particle,[40] a result that was confirmed by the other experiments at PETRA in various ways.

Studies of gluon jets have developed into a precision technique for testing QCD. One-loop corrections to three-jet cross-sections were calculated by Keith Ellis, Douglas Ross and Tony Terrano in 1980.[41] They have subsequently been used, particularly by the LEP collaborations working at the $e^+e^- \to Z^0$ peak, to measure the strong coupling. The LEP collaborations also used four-jet events to verify the QCD predictions for the three-gluon coupling, a crucial consequence of the non-Abelian nature of QCD. More recently, two-loop corrections to three-jet production in e^+e^- annihilation have been calculated,[42] enabling jet production to become a high-precision tool for testing QCD predictions and measuring accurately the QCD coupling and its asymptotically-free decrease with energy.[43]

Large transverse-momentum jets in hadron–hadron collisions emerged clearly at the CERN proton–antiproton collider in 1982.[44] Detailed studies of gluon collisions and jets then went on to become a staple of tests of the Standard Model at subsequent colliders. Most recently, the dominant mode of production of the Higgs boson at the LHC is via gluon–gluon collisions.[45]

The discovery of the gluon is a textbook example of how theorists and experimentalists, working together, can advance knowledge. The discovery of the Higgs boson at the LHC has been another example for the textbooks, with the gluon progressing from a discovery to a tool for making further discoveries.

Acknowledgments

The author thanks Chris Llewellyn-Smith for his comments on the text. This work was supported in part by the London Centre for Terauniverse Studies (LCTS), using funding from the European Research Council via the Advanced Investigator Grant 267352 and from the UK STFC via the research grant ST/J002798/1.

References

1. M. Gell-Mann, *Phys. Lett.* **8**, 214 (1964).
2. G. Zweig, An SU(3) model for strong interaction symmetry and its breaking, Versions 1 and 2, CERN-TH-401, CERN-TH-410 (1964) [See also A. Peterman, *Nucl. Phys.* **63**, 349 (1965), which was submitted on Dec. 30, 1963].
3. O. W. Greenberg, *Phys. Rev. Lett.* **13**, 598 (1964); M. Y. Han and Y. Nambu, *Phys. Rev.* **139**, B1006 (1965).

4. M. Gell-Mann, R. J. Oakes and B. Renner, *Phys. Rev.* **175**, 2195 (1968).
5. E. D. Bloom *et al.*, *Phys. Rev. Lett.* **23**, 930 (1969); J. D. Bjorken, *Phys. Rev.* **148**, 1467 (1966) [Scaling in deep-inelastic scattering was foreseen in J. D. Bjorken].
6. J. D. Bjorken and E. A. Paschos, *Phys. Rev.* **185**, 1975 (1969); J. D. Bjorken, Theoretical ideas on inelastic electron and muon scattering, in *Proc. 1967 Int. Symp. on Electron and Photon Interactions at High Energies*, I.U.P.A.P. & U.S. A.E.C., p. 109; J. D. Bjorken, *Conf. Proc.* **C670717**, 55 (1967).
7. R. P. Feynman, *Phys. Rev. Lett.* **23**, 1415 (1969).
8. C. G. Callan, Jr. and D. J. Gross, *Phys. Rev. Lett.* **22**, 156 (1969).
9. C. H. Llewellyn Smith, *Phys. Rev. D* **4**, 2392 (1971).
10. G. 't Hooft, *Nucl. Phys. B* **33**, 173 (1971).
11. W. A. Bardeen, H. Fritzsch and M. Gell-Mann, Light cone current algebra, π^0 decay, and e^+e^- annihilation, in *Proc. Topical Meeting on the Outlook for Broken Conformal Symmetry in Elementary Particle Physics*, Frascati, Italy, May 4–5 1972, ed. R. Gatto (Wiley, 1973), arXiv:hep-ph/0211388.
12. G. 't Hooft, *Nucl. Phys. B* **35**, 167 (1971); G. 't Hooft and M. J. G. Veltman, *Nucl. Phys. B* **44**, 189 (1972).
13. S. Weinberg, *Phys. Rev. Lett.* **19**, 1264 (1967); A. Salam, Elementary particle theory, in *Proc. Nobel Symposium*, Lerum, Sweden, 1968, ed. N. Svartholm (Stockholm, 1968), pp. 367–377.
14. G. Parisi, *Lett. Nuovo Cimento* **7**, 84 (1973).
15. K. Symanzik, *Lett. Nuovo Cimento* **6**, 77 (1973).
16. H. D. Politzer, *Phys. Rev. Lett.* **30**, 1346 (1973).
17. D. J. Gross and F. Wilczek, *Phys. Rev. Lett.* **30**, 1343 (1973).
18. H. Fritzsch, M. Gell-Mann and H. Leutwyler, *Phys. Lett. B* **47**, 365 (1973).
19. H. Georgi and H. D. Politzer, *Phys. Rev. D* **9**, 416 (1974); D. J. Gross and F. Wilczek, *Phys. Rev. D* **8**, 3633 (1973); D. J. Gross and F. Wilczek, *Phys. Rev. D* **9**, 980 (1974).
20. D. Gross, in *Proc. 17th International Conference on High-Energy Physics*, London, England, July 1–July 10, 1974, ed. J. R. Smith (Rutherford Laboratory, Chilton, England, 1974), p. III.65; J. Iliopoulos, *ibid.*, p. III.89.
21. G. Tarnopolsky, J. Eshelman, M. E. Law, J. Leong, H. Newman, R. Little, K. Strauch and R. Wilson, *Phys. Rev. Lett.* **32**, 432 (1974).
22. E598 Collab. (J.-J. Aubert *et al.*), *Phys. Rev. Lett.* **33**, 1404 (1974); SLAC-SP-017 Collab. (J.-E. Augustin *et al.*), *Phys. Rev. Lett.* **33**, 1406 (1974).
23. T. Appelquist and H. D. Politzer, *Phys. Rev. Lett.* **34**, 43 (1975).
24. G. Hanson *et al.*, *Phys. Rev. Lett.* **35**, 1609 (1975).
25. J. F. Gunion, S. J. Brodsky and R. Blankenbecler, *Phys. Rev. D* **6**, 2652 (1972); P. V. Landshoff and J. C. Polkinghorne, *Phys. Rev. D* **8**, 927 (1973).
26. J. R. Ellis, M. K. Gaillard and G. G. Ross, *Nucl. Phys. B* **111**, 253 (1976) [Erratum: *ibid.* **130**, 516 (1977)].
27. T. A. DeGrand, Y. J. Ng and S. H. H. Tye, *Phys. Rev. D* **16**, 3251 (1977).
28. G. F. Sterman and S. Weinberg, *Phys. Rev. Lett.* **39**, 1436 (1977).

29. H. Georgi and M. Machacek, *Phys. Rev. Lett.* **39**, 1237 (1977); E. Farhi, *Phys. Rev. Lett.* **39**, 1587 (1977).
30. A. De Rújula, J. R. Ellis, E. G. Floratos and M. K. Gaillard, *Nucl. Phys. B* **138**, 387 (1978).
31. S. L. Wu and G. Zobernig, *Z. Phys. C* **2**, 107 (1979).
32. S. W. Herb *et al.*, *Phys. Rev. Lett.* **39**, 252 (1977).
33. K. Koller and T. F. Walsh, *Phys. Lett. B* **72**, 227 (1977) [Erratum: *ibid.* **73B**, 504 (1978)].
34. PLUTO Collab. (C. Berger *et al.*), *Phys. Lett. B* **82**, 449 (1979).
35. B. Wiik, *Proc. 7th Int. Conf. on Neutrinos, Weak Interactions and Cosmology — Neutrino 79*, Bergen, Norway, June 18–22 1979, eds. A. Haatuft and C. Jarlskog (Bergen University, Bergen, Norway, 1979), p. 1.113.
36. J. Ellis, *ibid.*, p. 1.451.
37. TASSO Collab. (R. Brandelik *et al.*), *Phys. Lett. B* **86**, 243 (1979).
38. D. P. Barber *et al.*, *Phys. Rev. Lett.* **43**, 830 (1979); PLUTO Collab. (C. Berger *et al.*), *Phys. Lett. B* **86**, 418 (1979); JADE Collab. (W. Bartel *et al.*), *Phys. Lett. B* **91**, 142 (1980).
39. J. R. Ellis and I. Karliner, *Nucl. Phys. B* **148**, 141 (1979).
40. TASSO Collab. (R. Brandelik *et al.*), *Phys. Lett. B* **97**, 453 (1980).
41. R. K. Ellis, D. A. Ross and A. E. Terrano, *Nucl. Phys. B* **178**, 421 (1981).
42. L. W. Garland, T. Gehrmann, E. W. N. Glover, A. Koukoutsakis and E. Remiddi, *Nucl. Phys. B* **627**, 107 (2002), arxiv:hep-ph/0112081.
43. Particle Data Group (K. A. Olive *et al.*), *Chin. Phys. C* **38**, 090001 (2014).
44. UA2 Collab. (M. Banner *et al.*), *Phys. Lett. B* **118**, 203 (1982).
45. ATLAS Collab. (G. Aad *et al.*), *Phys. Lett. B* **716**, 1 (2012), arXiv:1207.7214 [hep-ex]; CMS Collab. (S. Chatrchyan *et al.*), *Phys. Lett. B* **716**, 30 (2012), arXiv:1207.7235 [hep-ex].

Discovery of the Gluon

Sau Lan Wu

University of Wisconsin-Madison, Madison, WI 53706, USA

This article "Discovery of the Gluon" is dedicated to Professor Chen Ning Yang. The Gluon is the first Yang–Mills non-Abelian gauge particle discovered experimentally. The Yang–Mills non-Abelian gauge field theory was proposed by Yang and Mills in 1954, sixty years ago.

The experimental discovery of the first Yang–Mills non-Abelian gauge particle — the gluon — in the spring of 1979 is summarized, together with some of the subsequent developments, including the role of the gluon in the recent discovery of the Higgs particle.

In 1977, I became a faculty member as an assistant professor of physics at the University of Wisconsin at Madison. With support from the US Department of Energy, I formed a group of three post-docs and one graduate student. The first major decisions that I had to make were: what important problem of particle physics should I work on, and which experiment should I participate in? At that time, I was interested in positron–electron colliding experiments for various reasons, one of the most important ones being that the events are quite clean. The best choices were therefore the PEP at SLAC, Stanford University in California and the PETRA in Hamburg, Germany.

After discussions with several members of the experiments at these two accelerators, I went to talk to Björn Wiik of the TASSO Collaboration at PETRA. Getting to know Wiik constitutes a major lucky break at the beginning of my career in physics as a faculty member. Wiik introduced me to his colleagues Paul Söding and Günter Wolf, and brought me into the TASSO Collaboration.

After becoming a member of the TASSO Collaboration, my group and I concentrated on the construction of the drift chamber and the Čerenkov counters, two of the most interesting components, together with the F35 group of the TASSO Collaboration.

1. Particle Physics in 1977

Even before becoming a member of the TASSO Collaboration, I spent a great deal of time thinking about what physics to work on. For my thinking, it is useful to review some of the historical aspects of physics.

One of the most far-reaching conceptional advances in physics during the nineteenth century is that there is no action-at-a-distance. The best example is that the interaction between two charged particles, such as two electrons, is not correctly described in general by the Coulomb potential; instead, these two electrons interact through the intermediary of an electromagnetic field. This is the essence of the Maxwell equations. Thus the Coulomb potential by itself gives a correct description only for the static limit, i.e., when the two electrons are not moving with respect to each other.

Quantum mechanics, which was developed at the beginning of the twentieth century, taught us that particles are waves and waves are particles. In one of the epoch-making 1905 papers of Einstein,[1] he predicted that electromagnetic waves are also particles, called the photons. This startling prediction was verified experimentally by Compton[2] through photon–electron elastic scattering, a process that has since borne his name. Therefore charged particles such as electrons interact with each other through the exchange of photons.

The photon is the first known particle that is the carrier of force, in this case the electromagnetic force. Such particles are called gauge particles.

The next conceptual advance was due to the genius of Yang and Mills,[3,4] where for the first time the idea of the electromagnetic field was generalized. These papers were published almost exactly sixty years ago. While the underlying group for the electromagnetic theory is U(1), Yang and Mills considered the more complicated case of isotopic spin described by the group SU(2). Once this case of SU(2) is understood, further generalizations to other groups are immediate. While U(1) is an Abelian group, SU(2) is a non-Abelian group. From the point of physics, the difference is:

(1) For the case of the Abelian group, the electromagnetic field itself is neutral, and therefore is not the source of further electromagnetic fields; but
(2) For the case of any non-Abelian group, the corresponding "electromagnetic field", called the Yang–Mills field, is itself a source of further Yang–Mills field.

Another way of saying the same is that, while the Maxwell equations are linear in the electromagnetic field, the corresponding Yang–Mills equations are non-linear.

At that time, I was fascinated by this Yang–Mills generalization of the electromagnetic field. This was to be contrasted with the experimental situation at that time: while the photons were everywhere in the detector, no Yang–Mills particle had ever been observed in any experiment.

I therefore formulated the following problem for myself: how could I discover experimentally the first Yang–Mills gauge particle using the TASSO detector?

2. PETRA and TASSO

PETRA (Positron–Electron Tandem Ring Accelerator) was a 2.3 kilometer electron and positron storage ring located in the German National Laboratory called Deutsches Elektronen-Synchrotron (DESY) in a west suburb of Hamburg, Germany. The name, Deutsches Elektronen-Synchrotron, is derived from the first accelerator in the Laboratory, a 7-GeV alternating gradient synchrotron. This laboratory was established in 1959 under the direction of Willibald Jentschke, and has played a crucial role in the re-emergence of Germany as one of the leading countries in physics.

The proposal[5] for the project for the construction of PETRA was submitted to the German government in November 1974, and was approved one year later. The hero of the construction of PETRA was Gustav Voss; under his leadership, the electron beam was first stored on July 15, 1978, more than nine months earlier than originally scheduled — a feat that was unheard of. There are many impressive stories how Voss accomplished this feat.

PETRA started its operation as a collider in November 1978; the center-of-mass energy was initially 13 GeV, but increased to 27 GeV in the spring of 1979. PETRA, together with its five detectors in four experimental halls are shown in Fig. 1. Note that, in this figure, "DESY" refers to the original 7-GeV synchrotron, not the German laboratory.

As seen from this Fig. 1, TASSO (Two-Arm Spectrometer SOlenoid) was one of the detectors; together with two other detectors, it was moved into place in October 1978. The first hadronic event at PETRA was observed in November; since then until its eventual shutdown, PETRA had been operating regularly, reliably and well.

A feature of the TASSO detector is the two-arm spectrometer, which leads to the name TASSO. The end view of this detector, i.e., the view along the beam pipe of the completed detector, is shown in Fig. 2. When TASSO was first moved into the PETRA beams in 1978, not all of the detector components shown in Fig. 2 were in working order. In particular, the central detector, especially the drift chamber, was functioning properly.

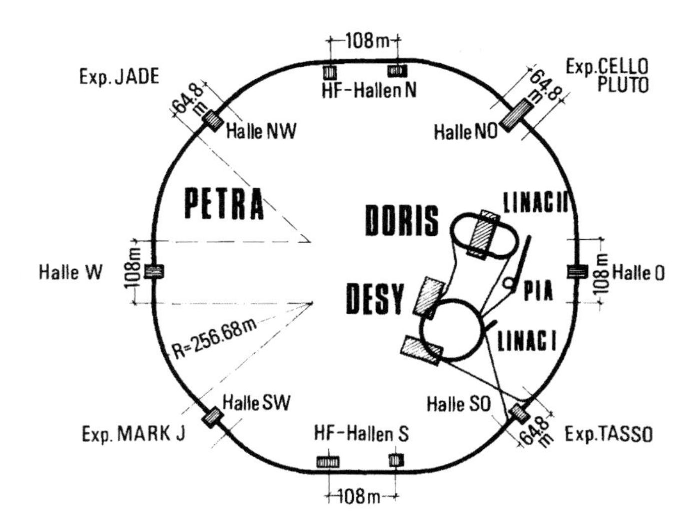

Fig. 1. PETRA (Positron-Electron Tandem Ring Accelerator).

So far as softwares are concerned, it should be emphasized that computers at that time were very slow by modern standard. As to be discussed later, this slowness of the available computers must be taken into account and played a deciding role in figuring out the practical methods of data analysis, including especially the ideas of how to discover the first Yang–Mills gauge particle.

Figure 3 shows Wiik and the author relaxing after installing some of the drift-chamber cables. This large drift chamber, which was designed and constructed under the direction of Wiik and Ulrich Kötz, had a sensitive length of 2.23 m with inner and outer diameters of 0.73 and 2.44 m. There are in total 15 layers, nine with the sense wires parallel to the axis of the chamber and six with the sense wires oriented at an angle of approximately $\pm 4°$. These six layers make it possible to measure not only the transverse momenta of the produced charged particles but also their longitudinal momenta.

3. Gluon Bremsstrahlung

The gluon[a] is the Yang–Mills non-Abelian gauge particle for strong interactions, i.e., the strong interactions between quarks[7,8] are mediated by the gluon, in much the same way as the electromagnetic interactions between electrons are mediated by the photon. For this reason, I concentrated on the

[a]The word "gluon" was originally introduced by Gell-Mann[6] to designate a hypothetical neutral vector field coupled strongly to the baryon current, without reference to color. Nowadays, this word is used exclusively to mean the Yang–Mills non-Abelian gauge particle for strong interactions.

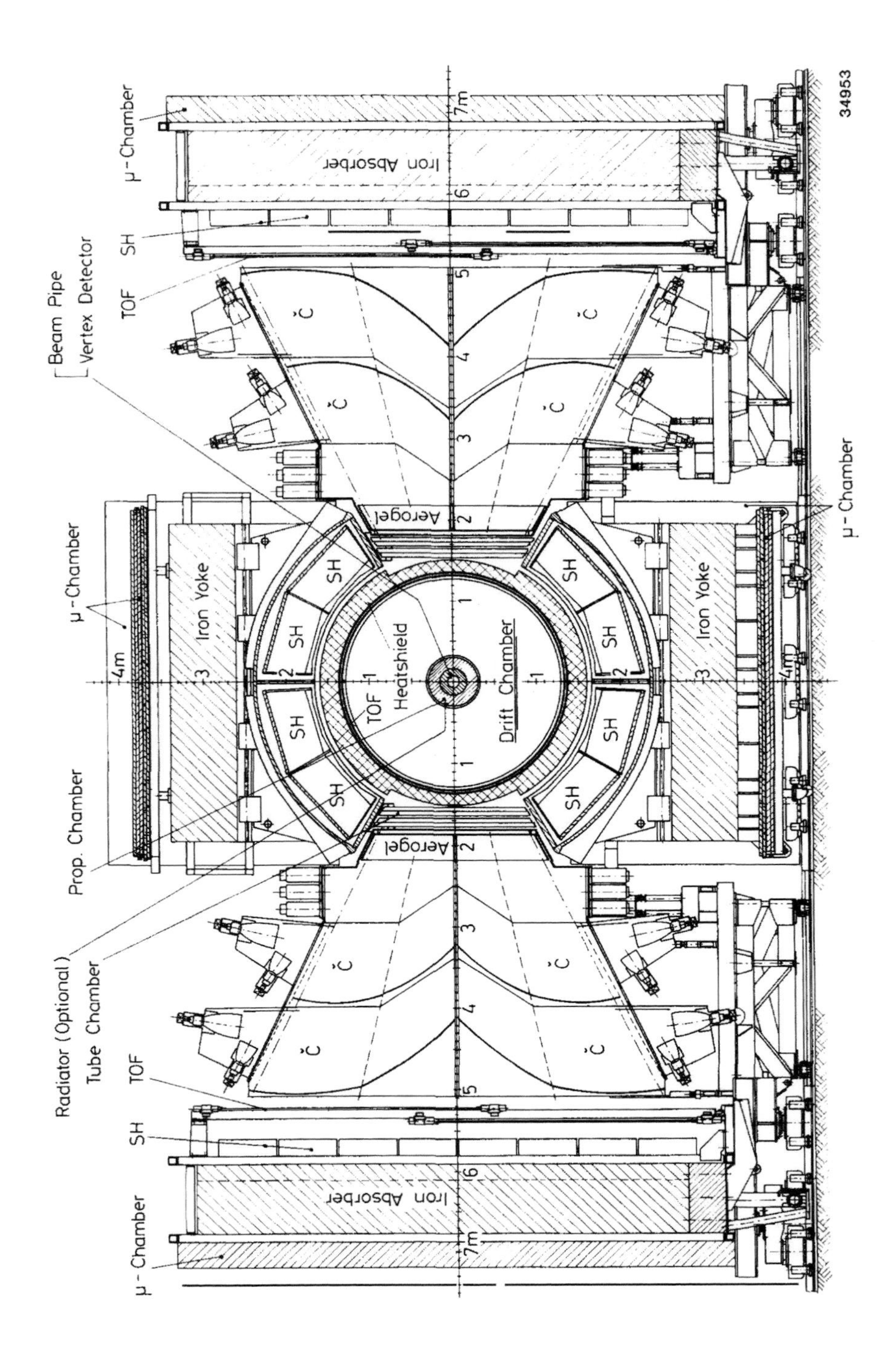

Fig. 2. End view of the TASSO detector.

Fig. 3. Björn Wiik and the author relaxing after tedious work on drift-chamber cabling. The cables in the photograph carried the timing information from the preamplifier boxes of the large central drift chamber of the TASSO detector at PETRA. They were to be connected to the read-out electronics. Dr. Ulrich Kötz of DESY/TASSO took this photograph in 1978.

gluon as the first Yang–Mills gauge particle, and the second gauge particle after the photon, to be discovered using the TASSO detector.

Indirect indication of gluons had been first given by deep inelastic electron scattering and neutrino scattering. The results of the SLAC-MIT deep inelastic scattering experiment[9] on the Callan–Gross sum rule were inconsistent with parton models that involved only quarks. The neutrino data from Gargamelle[10] showed that 50% of the nucleon momentum is carried by isoscalar partons or gluons. Further indirect evidence for gluons was provided by observation of scale breaking in deep inelastic scattering.[11] The very extensive neutrino scattering data from the BEBC and CDHS Collaborations[12] at CERN made it feasible to determine the distribution functions of the quark and gluon by comparison what was expected from QCD, and it was found that the gluon distribution function is sizeable. This information about the gluon is interesting but indirect, similar to that for the Z through the $\mu^+\mu^-$ asymmetry in electron–positron annihilation.[13]

At the time of PETRA turn-on in 1978–79, several groups were interested in looking for jet broadening following the suggestion of Ellis, Gaillard, and Ross.[14] The idea was to look for two-jet events where one of the jets becomes broadened due to the effect of the gluon. However, I considered this to provide just one more indirect indication of the gluon.

The discovery of the gluon requires direct observation.

Since the gluon is gauge particle for strong interactions, the most direct way to produce a gluon is by the gluon bremsstrahlung process from a quark-antiquark pair:

$$e^+e^- \to q\bar{q}g\,. \tag{1}$$

At high energy, the limited transverse-momentum distribution of the hadrons with respect to the quark direction, characteristic of strong interactions, results in back-to-back jets of hadrons[15] from the process $e^+e^- \to q\bar{q}$. In 1975, MARK I at SPEAR (SLAC) was the first to observe a two-jet structure[16] in e^+e^- annihilation as well as the spin $\frac{1}{2}$ nature of the produced quarks.[17]

How do the quark and the antiquark in the final state of process $e^+e^- \to q\bar{q}$ become jets? The heuristic picture is as follows. As the quark and the antiquark fly apart from each other, additional quark-antiquark pairs are excited; the quarks and antiquarks from these pairs then combine with the original antiquark and the original quark to form mesons (and some baryons). These mesons and baryons form the two jets. This process of jet formation is shown schematically in Fig. 4.

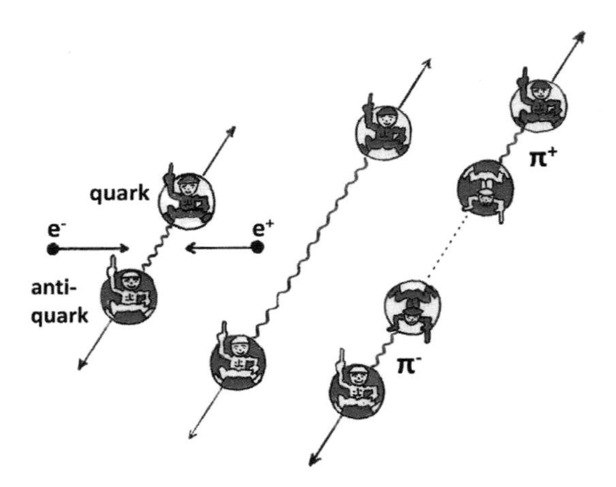

Fig. 4. Schematic picture of jet formation in $e^+e^- \to q\bar{q}$.

With this heuristic understanding of the experimental results from MARK I, I had to make my best guess as to how the gluon bremsstrahlung process (1) would look like in the PETRA detectors, including TASSO. Since the gluon is the Yang–Mills non-Abelain gauge particle for strong interactions, it is itself a source for gluon fields; see (2) of Sec. 1. It therefore seems reasonable to believe that the gluon in the final state of the process (1) would be seen in the detector as a jet, just like the quark and the antiquark.

Therefore the gluon bremsstrahlung process leads to three-jet events.

4. Three-Jet Analysis

Using the SPEAR information on the quark jets from the process, $e^+e^- \to q\bar{q}$, I convinced myself that three-jet events, if they were produced, could be detected once the PETRA energy went above three times the SPEAR energy i.e., $3 \times 7.4 \sim 22$ GeV. The arguments were as follows:

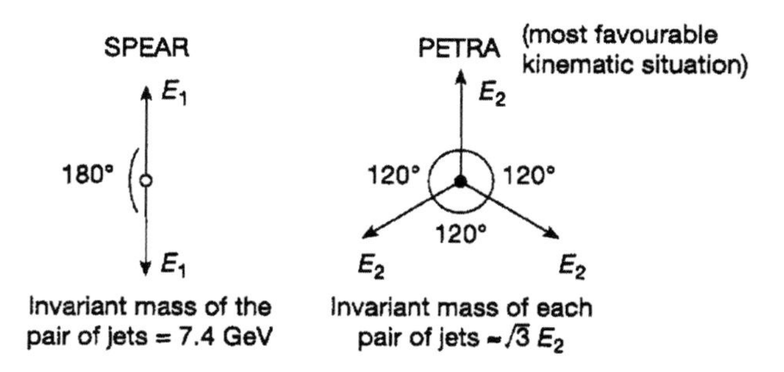

Fig. 5. Two-jet and three-jet configurations at SPEAR and PETRA respectively.

Figure 5 shows a comparison of the two-jet configuration at SPEAR with the most favorable kinematic situation of the three-jet configuration at PETRA. If the two invariant masses are taken to be the same, i.e., $\sqrt{3}E_2 \approx 7.4$ GeV, then the total energy of the three jets is $3E_2 \approx 13$ GeV, which must be further increased because each jet has to be narrower than the SPEAR jets. This additional factor is estimated to be $180°/120° = 1.5$, leading to about 20 GeV. Phase space considerations further increase this energy to about 22 GeV.

This estimate of 22 GeV was very encouraging because PETRA was expected to exceed it soon; indeed, it provided the main impetus for me to continue the project to discover the first Yang–Mills non-Abelian gauge particle — the gluon.

At the same time, I had to address the problem: how could I find three-jet events at PETRA? I made a number of false starts until I realized the power of the following simple observation. By energy-momentum conservation, the two jets in $e^+e^- \to q\bar{q}$ must be back-to-back. Similarly, the three jets in $e^+e^- \to q\bar{q}g$ must be coplanar. Therefore, the search for the three jets can be carried out in the two-dimensional event plane, the plane formed by the momenta of $q, \bar{q}$ and g. Figure 6 shows a few pages of my notes written in June 1978.

The concept of the event plane for three-jet event deserves some discussion. As an analogous but simpler situation, consider an $\omega(782)$ at rest. Its dominant decay mode is

$$\omega \to \pi^+\pi^-\pi^0$$

followed by

$$\pi^0 \to \gamma\gamma.$$

The momenta of the π^+ and the π^- in this decay are well defined, and the momentum of π^0 is the sum of those of the two gammas. Since these momenta of π^+, π^-, π^0 must add together to zero, they in general define a plane — the event plane.

Let this consideration be applied to the gluon bremsstrahlung process $e^+e^- \to q\bar{q}g$, where the quark q, the antiquark $\bar{q}$, and the gluon g each metamorphoses into a jet. Thus the momenta of q, $\bar{q}$, and g are each given by the sum of the momenta in the respective jet. However, there is no well-defined way of assigning which of the mesons (and baryons) to which jet. Therefore, in the case of three-jet events (1), the event plane is not completely well defined. In spite of this slight ambiguity, it serves well for the purpose of three-jet analysis.

How did the event plane help me to discover the gluon? Consider first a two-jet event from $e^+e^- \to q\bar{q}$. In this case, the two jets are back-to-back in

S.L. Wu
June 1978

Analysis of Jets

1. Two opposite jets
2. Two non-opposite jets
3. Planar case
4. Three jets

P9
4-9

the analysis of 3 jet events

A With suitable χ^2 cuts, pick out the events from the region marked "Planar Distribution". For these events, T_3 is relatively small. Let $\hat{n}_3$ be eigenvector corresponding to T_3, and we project out this $\hat{n}_3$ component.

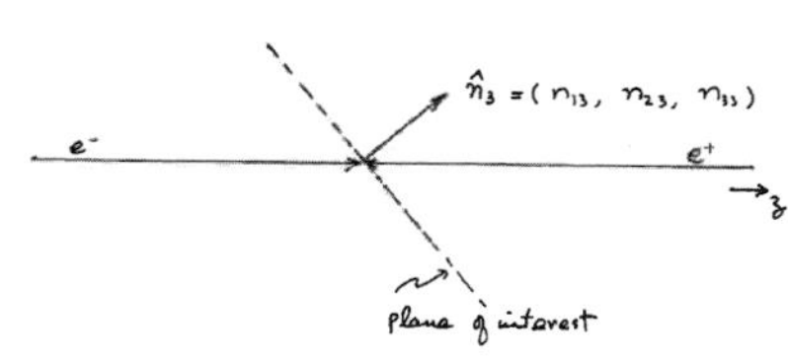

We choose n_{33} to be positive.
Let the X-axis be determined by projecting the x-axis to the $\hat{n}_1 - \hat{n}_2$ plane

$$\hat{X} = \frac{\hat{x} - \hat{n}_3(\hat{x}\cdot\hat{n}_3)}{|\hat{x} - \hat{n}_3(\hat{x}\cdot\hat{n}_3)|}$$

and the Y-axis by orthogonality

$$\hat{Y} = \hat{n}_3 \times \hat{X} = \frac{\hat{n}_3 \times \hat{x}}{|\hat{x} - \hat{n}_3(\hat{x}\cdot\hat{n}_3)|}$$

P10
4-10

Instead of $\vec{P}_i$, we now have a list

$$\{P_{iX}, P_{iY}\} \quad i = 1, 2, \cdots N$$

Rearrange the i's such that θ_i is in asending order: (w.r.t. X-axis)

$$0 \le \theta_1 \le \theta_2 \le \theta_3 \le \cdots \le \theta_N \le 2\pi$$

where $P_{iX} + iP_{iY} = \sqrt{P_{iX}^2 + P_{iY}^2}\, e^{i\theta_i}$

B. A possible procedure for numerically treat these projected momenta is as follows

Choose three integers N_1, N_2, N_3 such that

a) $1 \le N_1 < N_2 < N_3 \le N$

N = no of measured tracks

b) $\theta_{N_2-1} - \theta_{N_1} \le \pi$ (θ_{N_2-1}: last track of group 1; θ_{N_1}: first track of group 1)

c) $\theta_{N_3-1} - \theta_{N_2} \le \pi$ (θ_{N_3-1}: last track of group 2; θ_{N_2}: first track of group 2)

d) $2\pi + \theta_{N_1-1} - \theta_{N_3} \le \pi$ (θ_{N_1-1}: last track of group 3; θ_{N_3}: first track of group 3)

These conditions are imposed so that each jet goes more or less in a definite direction not in both direction

P11
4-11

In this way, the N momenta have been split into three sets

$$\mathcal{A}_1 = \{N_1, N_1+1, \cdots, N_2-1\}$$
$$\mathcal{A}_2 = \{N_2, N_2+1, \cdots, N_3-1\}$$
$$\mathcal{A}_3 = \{N_3, N_3+1, \cdots, N_1-1\}$$

For each set we essentially go through the procedure of Hansen. More precisely, define

$$M_{jk}^{(\ell)} = \sum_{i \in \mathcal{A}_\ell} P_{ij} P_{ik} \qquad j, k = X, Y \qquad \ell = 1, 2, 3 \text{ groups.}$$

Find the larger eigenvalue of the 2×2 matrix

$$M^{(\ell)} = \begin{bmatrix} M_{XX}^{(\ell)} & M_{YX}^{(\ell)} \\ M_{XY}^{(\ell)} & M_{YY}^{(\ell)} \end{bmatrix}$$

Call it $\lambda^{(\ell)}$. More precisely

$$\lambda^{(\ell)2} - (M_{XX}^{(\ell)} + M_{YY}^{(\ell)})\lambda^{(\ell)} - (M_{XX}^{(\ell)} M_{YY}^{(\ell)} - M_{XY}^{(\ell)2}) = 0$$

$$\lambda^{(\ell)} = \frac{1}{2}\left\{ M_{XX}^{(\ell)} + M_{YY}^{(\ell)} \oplus \sqrt{(M_{XX}^{(\ell)} + M_{YY}^{(\ell)})^2 - 4(M_{XX}^{(\ell)} M_{YY}^{(\ell)} - M_{XY}^{(\ell)2})} \right\}$$

(choose the larger value)

$$= \frac{1}{2}\left\{ M_{XX}^{(\ell)} + M_{YY}^{(\ell)} \pm \sqrt{(M_{XX}^{(\ell)} - M_{YY}^{(\ell)})^2 + 4M_{XY}^{(\ell)2}} \right\}$$

Fig. 6. Four of the pages from my notes of June 1978 on jet analysis.

the detector. While the event, that may consist of charged tracks and signals from the ECAL and the HCAL due to neutral particles, is three-dimensional, only two-dimensional views were available in the late seventies. Because of the slowness of the computers at that time, it was not yet possible at that time to rotate the events for viewing on the computer screen.

For a two-jet event, viewing in any direction shows the back-to-back jets, the exception being views from near the direction of the two jets. Therefore, if two orthogonal views are given, at least one of them shows the two-jets structure clearly.

This is not the case for three-jet events from the gluon bremsstrahlung process $e^+e^- \to q\bar{q}g$. Contrary to the two-jet case, a three-jet event can be seen clearly as such only when viewed from certain limited directions. The best view is obtained from a direction perpendicular to the event plane. This is the underlying reason for the importance of the event plane.

Equipped with my estimate of 22 GeV center-of-mass energy as described above and the idea of viewing the events in their event planes, I was in a good position to look for three-jet events. The development of the method of data analysis for such events then proceeded rapidly. The main remaining job was to find a way of identifying the three jets; this is essential in itself, including in particular in order to compare the properties of the jets in these three-jet events with those in the two-jet events.

Projected into this two-dimensional event plane, the momenta of the particles in the detector can be naturally ordered cyclically. Thus, if the polar coordinates of N vectors $\vec{q}_j$ are (q_j, θ_j) then these $\vec{q}_j$ can be relabelled such that

$$0 \leqq \theta_1 \leqq \theta_2 \leqq \theta_3 \leqq \cdots \leqq \theta_N < 2\pi\,. \tag{2}$$

With this cyclic ordering, the $N\vec{q}_j$'s can be split up into three sets of contiguous vectors, and these three sets are to be identified as the three jets. There are of course a number of ways of carrying out this splitting, and, with suitable restrictions, the one with smallest average transverse momentum is chosen as the best approximation to the correct way of identifying the three jets.

This cyclic ordering (2) reduces greatly the number of combinations that need to be studied to identify the three jets. Without any ordering, the number of ways to partition the N observed tracks into three sets, each consisting of at least one track, is

$$(3^{N-1} - 2^N + 1)/2\,.$$

For example, this is 2.4×10^6 for $N = 15$, 5.8×10^8 for $N = 20$, 1.4×10^{11} for $N = 25$, and 3.4×10^{13} for $N = 30$. With this ordering, it is sufficient to consider three sets of contiguous vectors, and the number of such partitions is

$$\binom{N}{3} = N(N-1)(N-2)/6.$$

This is 455 for $N = 15$, 1140 for $N = 20$, 2300 for $N = 25$ and 4060 for $N = 30$; or reductions by factors of 5×10^3, 5×10^5, 6×10^7 and 8×10^9 respectively. Such is the power of using the event plane.

It should be emphasized that this large reduction should not be described merely as an improvement in efficiency, rather it shortened enormously long computer runs so that they became manageable.

This analysis of the PETRA events for three jets then proceeds as follows.[18] First, using the momentum tensor (This tensor is essentially the same as the one used previously for two-jet analysis.)

$$M_{\alpha\beta} = \sum_j p_{j\alpha} p_{j\beta}\,, \tag{3}$$

the event plane is determined as the plane with the smallest transverse momentum, i.e., the plane perpendicular to the eigenvector $\hat{n}_3$ that corresponds to the smallest eigenvalue λ_3 of M. Then all the measured momenta of the produced particles are projected into this event plane. Using (2) above, rearrange these projected momenta $\vec{q}_j$ into a cyclic order. For $N \geqslant 3$, split into contiguous sets by choosing three numbers N_1, N_2 and N_3 that satisfy

$$1 \leqq N_1 < N_2 < N_3 \leqq N\,. \tag{4}$$

The three sets mentioned above then consist respectively of

$$\begin{gathered}\{N_1, N_1+1, \ldots, N_2-1\}\\ \{N_2, N_2+1, \ldots, N_3-1\}\\ \{N_3, N_3+1, \ldots, N, 1, 2, \ldots, N_1-1\}\,.\end{gathered} \tag{5}$$

Each of these three sets is required to span an angle of less than π, i.e.,

$$\begin{aligned}\theta_{N_2-1} - \theta_{N_1} &< \pi\,,\\ \theta_{N_3-1} - \theta_{N_2} &< \pi\,,\end{aligned}$$

and

$$\theta_{N_1-1} + 2\pi - \theta_{N_3} < \pi\,. \tag{6}$$

For each of these sets $S^{(\tau)}$, $\tau = 1, 2, 3$, define a two-dimensional analog of the momentum tensor (3); let $\Lambda^{(\tau)}$ be the larger eigenvalue and $\hat{m}^{(\tau)}$ the corresponding normalized eigenvector.

Since each jet can contain only particles in one direction, not simultaneously in both directions, the requirement is imposed that the signs of $\hat{m}^{(\tau)}$ can be chosen so that

$$\vec{q}_j \cdot \hat{m}^{(\tau)} > 0 \tag{7}$$

for each j in the corresponding set listed in (5). These conditions (7) actually consist of the following six:

$$\begin{aligned}
\vec{q}_{N_1} \cdot \hat{m}^{(1)} &> 0\,, \\
\vec{q}_{N_2-1} \cdot \hat{m}^{(1)} &> 0\,, \\
\vec{q}_{N_2} \cdot \hat{m}^{(2)} &> 0\,, \\
\vec{q}_{N_3-1} \cdot \hat{m}^{(2)} &> 0\,, \\
\vec{q}_{N_3} \cdot \hat{m}^{(3)} &> 0\,,
\end{aligned}$$

and

$$\vec{q}_{N_1-1} \cdot \hat{m}^{(3)} > 0\,. \tag{8}$$

These conditions (8) are stronger than the (6) above.

For each admissible way of splitting into three contiguous sets, calculate the sum of these three largest eigenvalues:

$$\Lambda(N_1, N_2, N_3) = \Lambda^{(1)} + \Lambda^{(2)} + \Lambda^{(3)}\,. \tag{9}$$

This $\Lambda(N_1, N_2, N_3)$ is maximized over N_1, N_2, N_3 that satisfy (4) and (8). This maximizing partition gives the three jets and the corresponding $\hat{m}^{(1)}$, $\hat{m}^{(2)}$, $\hat{m}^{(3)}$ yield the directions of the jet axes.

In short, the event plane is used to put the projections of the measured momenta of the produced particles into cyclic order. For each way of splitting into three contiguous sets, the sum of the larger eigenvalues corresponding to the two-dimensional momentum tensors for the three sets is evaluated. The particular splitting with the largest value of this sum corresponds to the smallest average momentum transverse to the three axes, and is therefore chosen as the way to identify the three jets. It is then straightforward to study the various properties of the jets, such as the average transverse momentum of each jet in three-jet events, and compare them with the corresponding properties in two-jet events.

This procedure has a number of desirable features. First, all three jet axes are determined, and they are in the same plane. This is the feature that plays a central role in the determination of the spin of the gluon, see Sec. 6.

Secondly, particle identification is not needed, since there is no Lorentz transformation.

Thirdly, the computer time is moderate even when all the measured momenta are used.

Finally, it is not necessary to have the momenta of all the produced particles; it is only necessary to have at least one momentum from each of the three jets. Thus, for example, my procedure works well even when no neutral particles are included.

This last advantage is important, and it is the reason why this procedure is a good match to the TASSO detector at the time of the PETRA turn-on. As explained in Sec. 2, when the TASSO detector was first moved into the PETRA beams in 1978, some of the detector components were not yet in working order.

This was true not only for TASSO but also for all the other detectors at PETRA at that time. Fortunately for my three-jet analysis described here, what was needed was the TASSO central detector, which was already functioning properly.

I had at that time Georg Zobernig as my post-doc; he was excellent in working with computers. My procedure of identifying the three-jet events in order to discover the gluon, programmed by Zobernig on an IBM 370/168 computer, was ready before the turn-on of PETRA in September of 1978.

Shortly thereafter, I showed the procedure and the program to Wiik; he was very excited about them and happy that I had found and solved such an important problem. I presented my analysis method in a TASSO meeting and later had it published with Zobernig.[18]

5. Discovery of the Gluon

When we had obtained data for center-of-mass energies of 13 GeV and 17 GeV, Zobernig and I looked for three-jet events. It was not until just before the Neutrino 79 (International Conference on Neutrino, Weak Interactions and Cosmology at Bergen, Norway) in the late spring of 1979 that we started to obtain data at the higher center-of-mass energy of 27.4 GeV. We found one clear three-jet event from a total of 40 hadronic events at this center-of-mass energy. This first three-jet event of PETRA, as seen in the event plane, is shown in Fig. 7(a). When this event was found, Wiik had already left Hamburg to go to the Bergen Conference. Therefore, during the

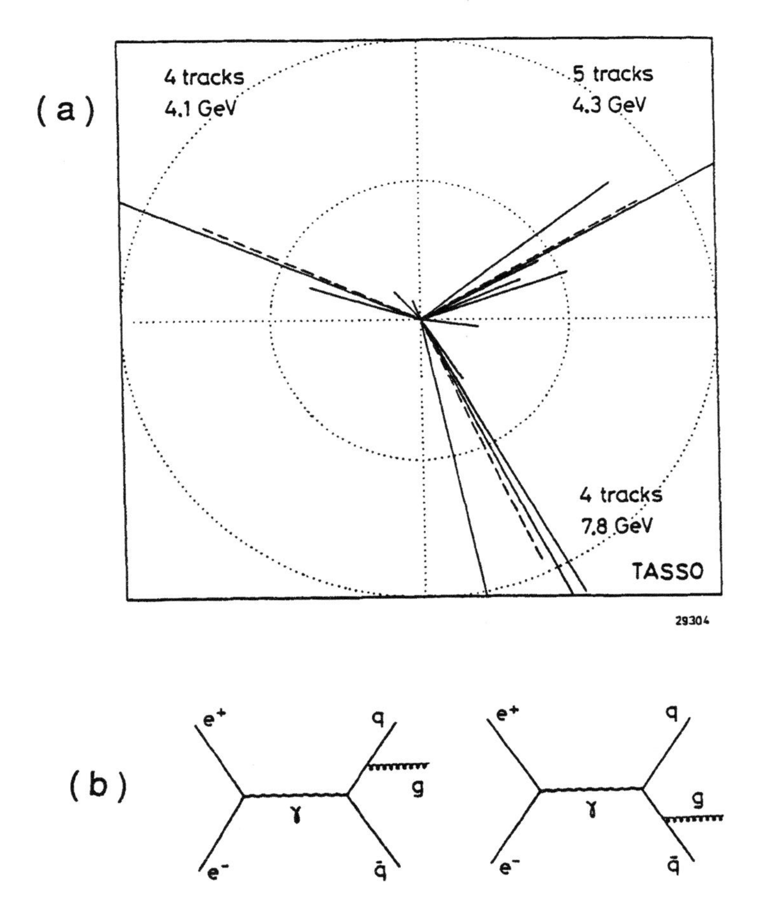

Fig. 7. (a) The first three-jet event from electron–positron annihilation, as viewed in the event plane. It has three well separated jets.[19,20] (b) Feynman diagrams for the gluon bremsstrahlung process $e^+e^- \to q\bar{q}g$.

weekend before the conference, I took the display produced by my procedure for this event to Norway to meet Wiik at his house near Bergen. It turned out that John Ellis was visiting Wiik also; after seeing my event, Ellis described this event as "gold-plated". During this weekend, I also telephoned Günter Wolf, the TASSO spokesman, at his home in Hamburg and told him of the finding. Wiik showed the event in his plenary talk "First Results from PETRA", acknowledging that it was my work with Zobernig by putting our names on his transparency of the three-jet event, and referred to me for questions. Donald Perkins took this offer and challenged me by wanting to see all forty TASSO events. I showed him all forty events, and, after we had spent some time together studying the events, he was convinced.

The following is quoted from the write-up of Wiik's talk:[19]

"If hard gluon bremsstrahlung is causing the large $p_\perp$ values in the plane then a small fraction of the events should display a three jet structure. The events were analyzed for a three jet structure using a method proposed by Wu and Zobernig[27)] ... A candidate for a 3 jet event, observed by the TASSO group at 27.4 GeV, is shown in Fig. 21 viewed along the $\hat{n}_3$ direction. Note that the event has a three clear well separated jet and is just not a widening of a jet."

As soon as I returned from Bergen, I wrote a TASSO note with Zobernig on the observation of this three-jet event.[20] The first page of this TASSO note is shown in Fig. 8. Both in Wiik's talk[19] and in the TASSO note,[20] this three-jet event was already considered to be due to the hard gluon bremsstrahlung process (1), described by the Feynman diagrams of Fig. 7(b). As seen from Fig. 7(a), this first three-jet event had three clear, well separated jets, and was considered to be more convincing than a good deal of statistical analysis. Indeed, before the question of statistical fluctuation could be seriously raised, events from $E_{\text{cm}} = 27.4$ GeV rolled in and we found a number of other three-jet events.

Less than two weeks after the Bergen Conference, four of the TASSO three-jet events, as given in Fig. 9(a), were shown by Paul Söding of DESY/TASSO at the European Physical Society (EPS) Conference in Geneva.[21] Figure 9(b) gives several plots of the various transverse momentum distributions. The first one is the distribution of

$$\langle p_\perp^2 \rangle_{\text{out}} = \frac{1}{N} \sum_j (\bar{p}_j \cdot \hat{n}_3)^2 \tag{10}$$

(= square of momentum component normal to the event plane averaged over the charged particles in one event), while the second one is that of

$$\langle p_\perp^2 \rangle_{\text{in}} = \frac{1}{N} \sum_j (\bar{p}_j \cdot \hat{n}_2)^2 \tag{11}$$

(= square of momentum component in the event plane and perpendicular to the sphericity axis averaged the same way). A comparison of these two plots shows that the major difference is the absence of a tail for $\langle p_\perp^2 \rangle_{\text{out}}$ and the presence of one for $\langle p_\perp^2 \rangle_{\text{in}}$. Since three-jet events tend to have a small $\langle p_\perp^2 \rangle_{\text{out}}$ but a much larger $\langle p_\perp^2 \rangle_{\text{in}}$, this distribution $\langle p_\perp^2 \rangle_{\text{in}}$ shows a continuous transition from two-jet events to three-jet events.

Also shown in Fig. 9(b) is $\langle p_\perp^2 \rangle_{\text{in, 3 jet axes}}$, which is defined the same way as (11) but, for each jet, the jet axis found by my method is used.

TASSO Note No. 84
26.6.1979

From Sau Lan Wu and Haimo Zobernig

On: A three-jet candidate (run 447 event 13177)

We have made a three jet analysis to all the hadronic candidates (43 events for $\Sigma_i|P_i| \geq 9$ GeV) of the May 1979 data at $E_{cm} = 27.4$ GeV using our method described in DESY 79/23 (A method of three jet analysis in e^+e^- annihilation).

Fig. 1 gives the triangular plot of the normalized eigenvalues Q_1, Q_2 and Q_3 ($Q_1 \leq Q_2 \leq Q_3$) of the momentum matrix

$$M_{\alpha\beta} = \sum_j (P_{j\alpha} \quad P_{j\beta})$$

(See equation (1) and Fig. 1 of DESY 79/23). We find two three jet candidates

run 447 event 13177

run 439 event 12845

We then display each event on the 3 planes

plane 1: normal to $\hat{n}_1$, the normalized eigenvector corresponding to Q_1. $\sum_i |P_{i\perp}|^2$

with respect to this plane is minimized.

plane 2: normal to $\hat{n}_2$, the normalized eigenvector ~~to $\hat{n}_2$~~ corresponding to Q_2

plane 3: normal to $\hat{n}_3$, the normalized eigenvector corresponding to Q_3.

Fig. 2 displays the three jet candidate (run 447 event 13177) on planes 1, 2, 3.

Fig. 3 displays plane 1 of this event in a blow up scale.

The axis for each of the three jets are found. Given the axes and $\sum_i |P_i|$ of each jet, the total energy of each jet is determined assuming the mass of each quark (or gluon) is zero.

Fig. 4 displays the event run 439 event 12845. This event looks like two charged jets and one neutral jet.

Fig. 8. The first page of TASSO Note No. 84, June 26, 1979, by Wu and Zobernig.[20]

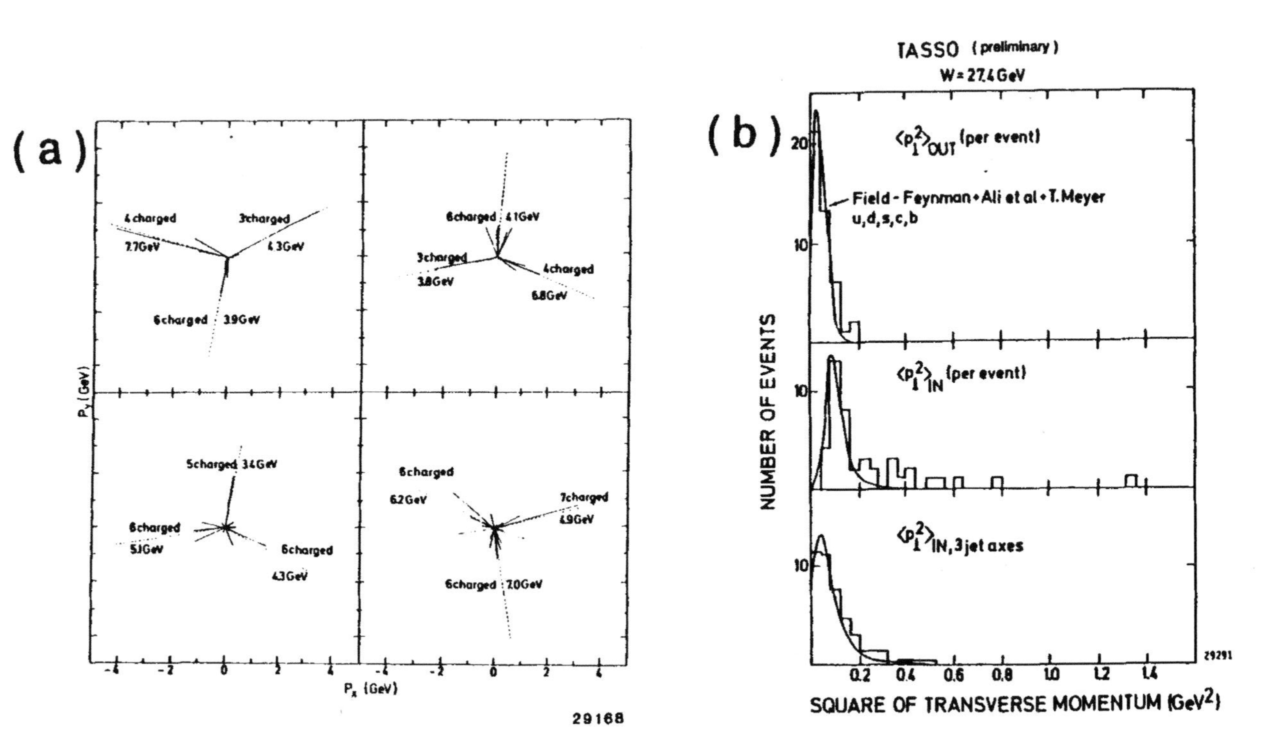

Fig. 9. (a) Four TASSO three-jet events as seen in the event plane. (b) Distribution of the average squared transverse momentum component out of the event plane (top), and in the event plane (center), for the early TASSO events at the center-of-mass energy of 27.4 GeV (averaging over charged hadrons only). The curves are for $q\bar{q}$ jets without gluon bremsstrahlung. Comparison of these distributions gives evidence that broadening (compared to $q\bar{q}$ jets) occurs in one plane. The bottom figure shows $\langle p_\perp^2 \rangle$ per jet when 3 jet axes are fitted, again compared with the 2-jet model.[21]

The absence of a tail and the similarity to the first distribution means that the jets in three-jet events are similar to those in two-jet events, justifying the use of the same word "jet" in both cases. At this time of the EPS Conference in Geneva, no other experiment at PETRA mentioned anything about three-jet events.

On July 31, 1979, at the presentations by each of the PETRA experiments at the open session of the DESY Physics Research Committee, again only TASSO (represented by Peter Schmüser of the University of Hamburg) gave evidence of three-jet events.

With these three-jet events, the question is: what are the three jets? Since quarks are fermions, and two fermions (electron and positron) cannot become three fermions, it immediately follows that these three jets cannot all be quarks and antiquarks. In other words, *a new particle has been discovered.*

Secondly, since this new particle, similar to the quarks, also hadronizes into a jet, it cannot be a color singlet. Color singlets, such as the pion, the kaon, and the proton, either leave a track (if charged), or give an energy deposition in a calorimeter, or decay into well-defined final states, but do not metamorphose into jets. Therefore, the abundance of three-jet events means that the carrier of strong forces, unlike the photon, is not an Abelian gauge particle (which must be colorless).

For these reasons, it was readily accepted by most of the high-energy physicists that the three-jet events are due to $e^+e^- \to q\bar{q}g$.

6. Confirmations and the Spin of the Gluon

As more data were obtained, by the end of August other experiments at PETRA began to have their own three-jet analysis ready. It is quite natural that a discovery is corroborated by higher statistics and by other experiments at a later date. At the Lepton–Photon Conference at FNAL in late August of 1979, all four experiments at PETRA gave more extensive data, confirming the earlier observation of TASSO. Since these experiments were run simultaneously, the amounts of data were similar and there is not much difference in their statistical significance. Since this was the highlight of the conference, Leon Lederman, Director of FNAL, called for a press conference on the discovery of the gluon. In a period of three months, between August 29 and December 7, these more extensive data were submitted for publication by TASSO,[22] MARK J,[23] PLUTO[24] and JADE,[25] in this order. The gluon remains one of the most interesting discoveries from PETRA.[26]

The early papers related to the PETRA three-jet events are the following.

(1) Sau Lan Wu and Georg Zobernig, *Z. Phys. C — Particles and Fields* **2**, 107 (1979).

(2) B. H. Wiik, *Proceedings of International Conference on Neutrinos, Weak Interactions and Cosmology*, Bergen, Norway, June 18–22, 1979, p. 113.

(3) Sau Lan Wu and Haimo Zobernig, *TASSO Note 84*, June 26, 1979.

(4) P. Söding, *Proceedings of European Physical Society International Conference on High Energy Physics*, Geneva, Switzerland, 27 June–4 July, 1979, p. 271.

(5) TASSO Collaboration (R. Brandelik *et al.*), *Phys. Lett. B* **86**, 243 (1979) [received on August 29, 1979].

(6) MARK J Collaboration (D. P. Barber *et al.*), *Phys. Rev. Lett.* **43**, 830 (1979) [received on August 31, 1979].

(7) PLUTO Collaboration (Ch. Bergen *et al.*), *Phys. Lett. B* **86**, 418 (1979) [received on September 13, 1979].

(8) JADE Collaboration (W. Bartel *et al.*), *Phys. Lett. B* **91**, 142 (1980) [received on December 7, 1979].

(9) JADE Collaboration, MARK J Collaboration, PLUTO Collaboration and TASSO Collaboration, *Proceedings of 1979 International Symposium on Lepton and Photon Interactions at High Energies*, Fermi National Accelerator Laboratory (FNAL), Batavia, Illinois, August 23–29, 1979.

In the above list, the first one provides the method of analysis used in the TASSO papers, and the others are all experimental.

Since the gluon is a Yang–Mills non-Abelian gauge particle, it must have spin 1. It is nevertheless nice to have a measurement of its spin, especially since it is the first such particle ever seen experimentally. As described in Sec. 3, my procedure gives all three jet axes, and they are in the same plane. This ability to isolate individual jets in high-energy three-jet events makes it possible to measure the correlation between the directions of the three jets. Since the quark spin[16,17] is known to be $\frac{1}{2}$ and this correlation depends on the gluon spin, it can be used to give the desired determination of the spin of the gluon.

The amount of data collected at PETRA up to the time of the FNAL Conference was not quite enough for this spin determination. As soon as there were enough data, in September 1980, the TASSO Collaboration determined the spin of the gluon.[27] In this determination, the Ellis–Karliner

Fig. 10. The four Prize Recipients at the ceremony of the 1995 European Physical Society High Energy and Particle Physics Prize in Brussels, Belgium. Front row: Günter Wolf and Sau Lan Wu; second row: Björn Wiik and Paul Söding.

angle[28] was used with my three-jet analysis. Not surprisingly, the spin of the gluon was found to be indeed 1. One month later, the PLUTO Collaboration reached the same conclusion.[29] This result was confirmed subsequently by the other PETRA Collaborations.

Because of this discovery of the gluon by the TASSO Collaboration, Söding, Wiik, Wolf, and I were awarded the 1995 European Physical Society High Energy and Particle Physics Prize (Figs. 10 and 11). Because of my leading role in this discovery, I was chosen to give the acceptance speech at the EPS award ceremony.[30]

7. Recent Developments

The gluon was discovered thirty five years ago. During these thirty five years, it has become more and more important, one prominent example being its role in the recent discovery of the Higgs particle[31] by the ATLAS Collaboration[32] and the CMS Collaboration[33] using the Large Hadron Collider (LHC) at CERN. This observation of the Higgs particle in 2012 has led to the award of the 2013 Nobel Prize in physics to Englert and Higgs. Brout had unfortunately died a year and half earlier.

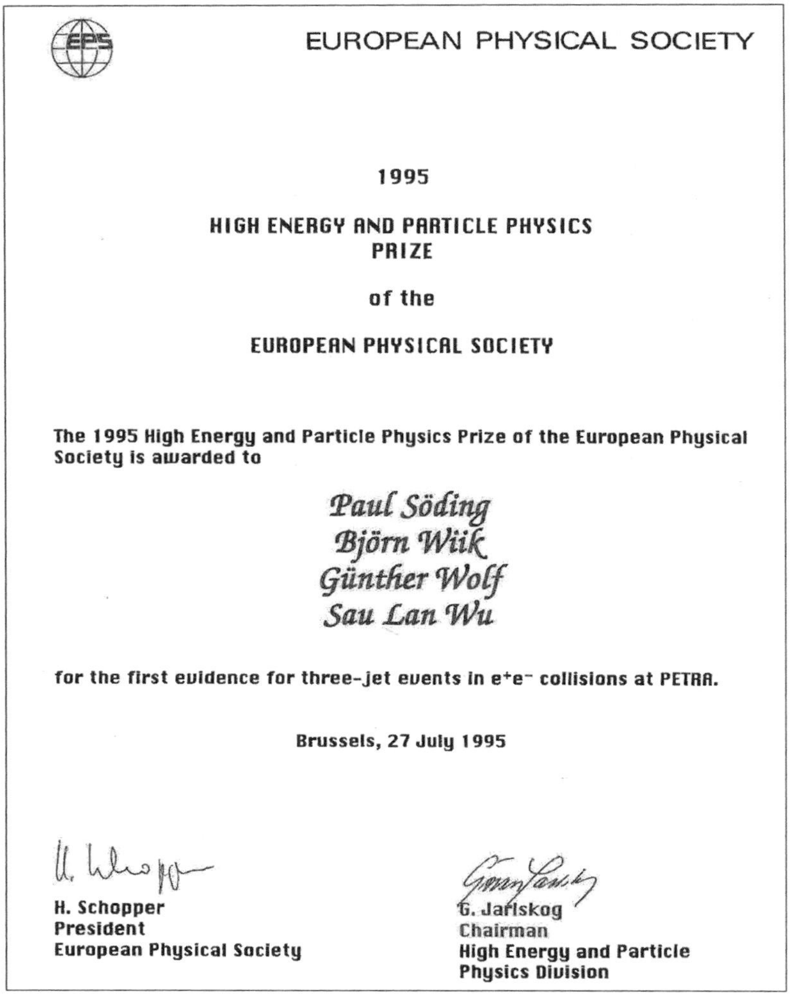

EUROPEAN PHYSICAL SOCIETY

1995

HIGH ENERGY AND PARTICLE PHYSICS
PRIZE

of the

EUROPEAN PHYSICAL SOCIETY

The 1995 High Energy and Particle Physics Prize of the European Physical Society is awarded to

Paul Söding
Björn Wiik
Günther Wolf
Sau Lan Wu

for the first evidence for three-jet events in e^+e^- collisions at PETRA.

Brussels, 27 July 1995

H. Schopper
President
European Physical Society

G. Jarlskog
Chairman
High Energy and Particle Physics Division

Fig. 11. European Physical Society High Energy and Particle Physics Prize, 1995.

Since the gluon is the Yang–Mills gauge particle for strong interactions, to a good approximation a proton consists of a number of gluons in addition to two u quarks and a d quark. Since the coupling of the Higgs particle to any elementary particle is proportional to its mass, there is little coupling between the Higgs particle and these constituents of the proton. Instead, some heavy particle needs to be produced in a proton-proton collision at LHC, and is then used to couple to the Higgs particle. Among all the known elementary particles, the top quark t, with a mass of 173 GeV/c^2, is the heaviest.[34]

The top quark is produced predominantly together with an anti-top quark or an anti-bottom quark. Since the top quark has a charge of $+2/3$ and is a color triplet, such pairs can be produced by

(a) a photon: $\gamma \to t\bar{t}$;
(b) a Z: $Z \to t\bar{t}$;
(c) a W: $W^+ \to t\bar{b}$; or
(d) a g: $g \to t\bar{t}$.

As discussed in the preceding paragraph, there is no photon, or Z, or W as a constituent of the proton. Since, on the other hand, there are gluons in the proton, (d) is by far the most important production process for the top quark.

Because of color conservation — the gluon has color but not the Higgs particle — the top and anti-top pair produced by a gluon cannot annihilate into a Higgs particle. In order for this annihilation into a Higgs particle to occur, it is necessary for the top or the anti-top quark to interact with a second gluon to change its color content. It is therefore necessary to involve two gluons, one each from the protons of the two opposing beams of LHC, and we are led to the diagram of Fig. 12 for Higgs production.

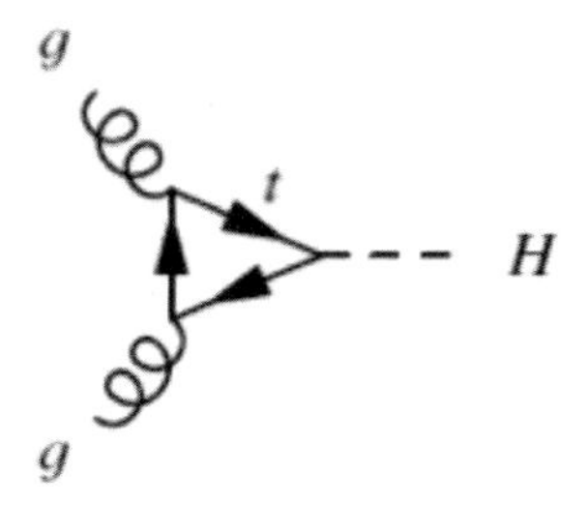

Fig. 12. Feynman diagram for the Higgs (H) production by gluon–gluon fusion (also called gluon fusion).

This production process is called "gluon–gluon fusion" (also called "gluon fusion"). As expected from the large mass of the top quark, this gluon–gluon fusion is by far the most important Higgs production process, and shows the central role played by the gluon in the discovery of the Higgs particle in 2012.

It is desirable to make this statement more quantitative. It is shown in Fig. 13 the various Higgs production cross sections from calculation as functions of the Higgs mass.[35] The top curve is for gluon–gluon fusion, and next one is for vector boson fusion (VBF); note that the vertical scale is logarithmic.

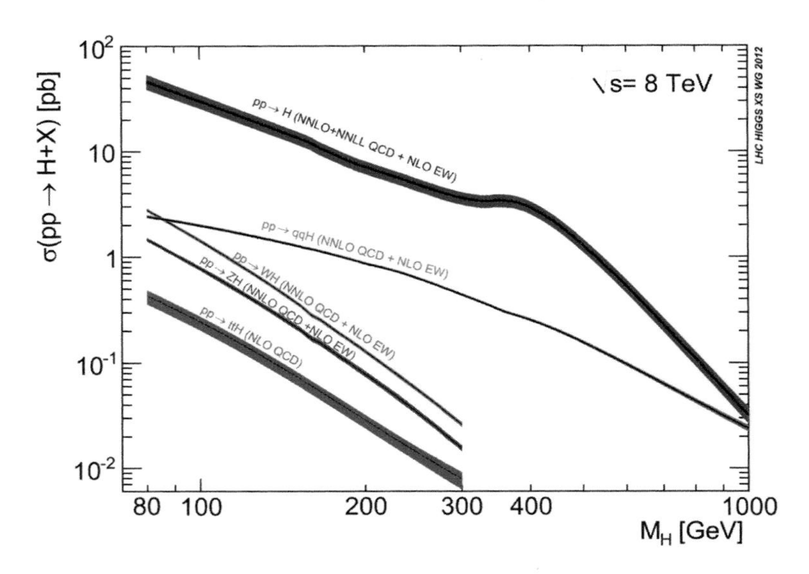

Fig. 13. Higgs production cross sections from gluon–gluon fusion (top curve), vector boson fusion, and three associated production processes at the LHC center-of-mass energy of 8 TeV.[35] It is seen that gluon–gluon fusion is the most important production process.

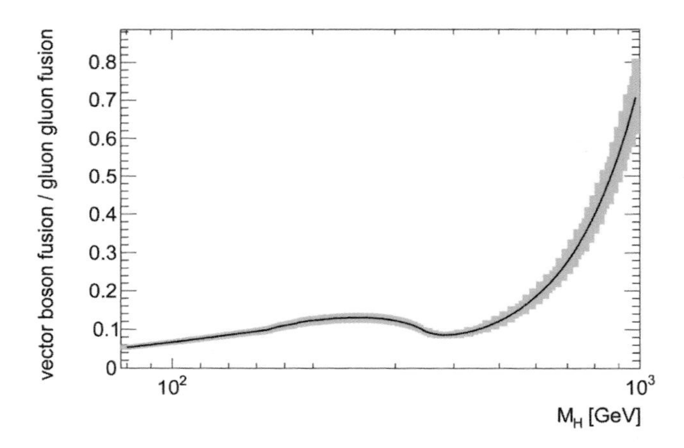

Fig. 14. Ratio of VBF cross section divided by the gluon–gluon fusion cross section.

In order to show even more clearly the importance of the gluon in the production of the Higgs particle, Fig. 14 shows the ratio of the second most important production process VBF to gluon–gluon fusion. It is seen that, for the relatively low masses of the Higgs particle, VBF cross section is less than 10% of that of gluon–gluon fusion. It is known that the Higgs particle[32,33] indeed has a mass in this range. Thus, through gluon–gluon fusion, the gluon contributes about 90% of Higgs production at the Large Hadron Collider. A more dramatic way of saying the same thing is that, if there were no gluon, the Higgs particle could not have been discovered for years!

8. Summary

In summary, since two fermions cannot turn into three fermions, the experimental observation of three-jet events in e^+e^- annihilation, first accomplished by the TASSO Collaboration in the spring of 1979 and confirmed by the other Collaborations at PETRA two months later in the summer, implies the discovery of a new particle. Similar to the quarks, this new particle hadronizes into a jet, and therefore cannot be a color singlet. These three-jet events are most naturally explained by hard non-collinear bremsstrahlung $e^+e^- \to q\bar{q}g$. One year later, the spin of the gluon was determined experimentally to be indeed 1, as it should be for a gauge particle.

Thus the 1979 discovery of the second gauge particle, the gluon, occurred more than half a century after that by Compton[2] of the first, the photon. This second gauge particle is also the first Yang–Mills non-Abelian gauge particle,[3,4] i.e., a gauge particle with self-interactions. Four years later, in 1983, the second and the third non-Abelian gauge particles, the W and the Z, were discovered at the CERN proton–antiproton collider by the UA1 Collaboration of Carlo Rubbia, Simon van der Meer *et al.*[36] and by the UA2 Collaboration of Pierre Darriulat *et al.*[37]

Since its first observation thirty five years ago, the importance of the gluon in particle physics has grown significantly. An especially noticeable example is its essential role in the discovery of the Higgs particle in 2012 by the ATLAS Collaboration[32] and the CMS Collaboration[33] at the Large Hadron Collider at CERN.

Acknowledgments

I am most grateful to the support, throughout many years, of the United States Department of Energy (Grant No. DE-FG02-95ER40896) and the University of Wisconsin through the Wisconsin Alumni Research Foundation and the Vilas Foundation.

I would very much like to thank Professor K. K. Phua and Professor H. Fritzsch for the invitation to contribute to this book "50 Years of Quarks".

References

1. Albert Einstein, "Über einen die Erzeugung und Verwandlung des Lichtes betreffenden heuristischen Gesichtspunkt" (On a Heuristic Point of View Concerning the Production and Transformation of Light) *Annalen der Physik* **17**, 132 (1905).
2. Arthur Holly Compton, "The Total Reflexion of X-Rays", *Philosophical Magazine* **45**, 1121 (1923).
3. Chen Ning Yang and Robert L. Mills, "Isotopic Spin Conservation and a Generalized Gauge Invariance, *Phys. Rev.* **95**, 631 (1954).
4. Chen Ning Yang and Robert L. Mills, "Conservation of Isotopic Spin and Isotopic Gauge Invariance", *Phys. Rev.* **96**, 191 (1954).
5. Deutsches Elektronen-Synchrotron, PETRA — a proposal for extending the storage-ring facilities at DESY to higher energies, DESY, Hamburg, November 1974; Deutsches Elektronen-Synchrotron, PETRA — updated version of the PETRA proposal, DESY, Hamburg, February 1976.
6. Murray Gell-Mann, "Symmetries of Baryons and Mesons", *Phys. Rev.* **125**, 1067 (1962).
7. Murray Gell-Mann, "A Schematic Model of Baryons and Mesons", *Phys. Lett.* **8**, 214 (1964).
8. George Zweig, "An SU3 Model for Strong Interaction Symmetry and its Breaking", CERN Preprints TH401 (January 17, 1964) and TH412 (February 21, 1964).
9. SLAC-MIT Collaboration (Elliott D. Bloom, *et al.*), *Phys. Rev. Lett.* **23**, 930 (1969); SLAC-MIT Collaboration (Martin Breidenbach, *et al.*), *Phys. Rev. Lett.* **23**, 935 (1969); SLAC-MIT Collaboration, data presented by H. Kendall, *Symp. Electron and Photon Interactions*, 1971, Cornell Univ., Ithaca, p. 248; J. Friedman and H. Kendall, *Ann, Rev. Nucl. Sci.* **22**, 203 (1972); SLAC-MIT Collaboration (G. Miller *et al.*), *Phys. Rev. D* **5**, 528 (1972).
10. Gargamelle Collaboration (T. Eichten *et al.*), *Phys. Lett. B* **46**, 274 (1973).
11. Y. Watanabe *et al.*, *Phys. Rev. Lett.* **35**, 898 (1975); C. Chang *et al.*, *Phys. Rev. Lett.* **35**, 901 (1975); W. B. Atwood *et al.*, *Phys. Lett. B* **64**, 479 (1976).
12. P. C. Bosetti, *et al.*, *Nucl. Phys. B* **142**, 1 (1978); J. G. H. de Groot, *et al.*, *Phys. Lett. B* **82**, 456 (1979); J. G. H. de Groot, *et al.*, *Z. Phys. C: Particles and Fields* **1**, 143 (1979).
13. JADE Collaboration (W. Bartel, *et al.*), *Phys. Lett. B* **108**, 140 (1982); TASSO Collaboration (R. Brandelik, *et al.*), *Phys. Lett. B* **110**, 173 (1982); MARK J Collaboration (B. Adeva, *et al.*), *Phys. Rev. Lett.* **48**, 1701 (1982); CELLO Collaboration (H. J. Behrend, *et al.*), *Z. Phys. C: Particles and Fields* **14**, 283 (1982).
14. John Ellis, Mary K. Gaillard, and Graham Ross, *Nucl. Phys. B* **111**, 253 (1976); [Erratum *ibid.* **130**, 516 (1977)]. Many experimentalists, in preparing for the data analysis at the beginning of the PETRA run, were influenced by the following sentences in this paper: "The first observable effect should be a tendency for the two-jet cigars to be unexpectedly oblate, with a high large p_T cross section. Eventually, events with large p_T would have a three-jet structure,

without local compensation of p_T."

15. S. D. Drell, D. J. Levy, and T. M. Yan, *Phys. Rev.* **187**, 2159 (1969), and *Phys. Rev. D* **1**, 1617 (1970); N. Cabibbo, G. Parisi, and M. Testa, *Nuovo Cimento* **4**, 35 (1970); J. D. Bjorken and S. J. Brodsky, *Phys. Rev. D* **1**, 1416 (1970); R. P. Feynman, *Photon-Hadron Interactions* (Benjamin, Reading, Mass., 1972), p. 166.
16. G. Hanson, *et al.*, *Phys. Rev. Lett.* **35**, 1609 (1975).
17. R. F. Schwitters, *et al.*, *Phys. Rev. Lett.* **35**, 1320 (1975).
18. Sau Lan Wu and Georg Zobernig, "A Method of Three-jet Analysis in e^+e^- Annihilation", *Z. Phys. C: Part. Fields* **2**, 107 (1979).
19. Björn H. Wiik, First Results from PETRA, in *Proceedings of Neutrino 79, International Conference on Neutrinos, Weak Interactions and Cosmology,* Volume **1**, Bergen June 18–22 1979, 113–154. [see pp. 127–128 for the quotation].
20. Sau Lan Wu and Haimo Zobernig, "A three-jet candidate (run 447, event 13177)", TASSO Note No. 84 (June 26, 1979).
21. Paul Söding, "Jet Analysis", *Proceedings of European Physical Society International Conference on High Energy Physics*, Geneva, Switzerland, 27 June–4 July, 1979, pp. 271–281.
22. TASSO Collaboration (R. Brandelik *et al.*), "Evidence for Planar Events in e^+e^- Annihilation at High Energies", *Phys. Lett. B* **86**, 243 (1979), received on August 29, 1979.
23. MARK J Collaboration (D.P. Barber *et al.*), "Discovery of Three-Jet Events and a Test of Quantum Chromodynamics at PETRA", *Phys. Rev. Lett.* **43**, 830 (1979), received on August 31, 1979.
24. PLUTO Collaboration (Ch. Berger *et al.*), "Evidence for Gluon Bremsstrahlung in e^+e^- Annihilation at High Energies", *Phys. Lett. B* **86**, 418 (1979), received on September 13, 1979.
25. JADE Collaboration (W. Bartel *et al.*), "Observation of Planar Three-Jet Events in e^+e^- Annihilation and Evidence for Gluon Bremsstrahlung", *Phys. Lett. B* **91**, 142 (1980), received on December 7, 1979.
26. Sau Lan Wu, "e^+e^- Physics at PETRA – the First Five Years", *Phys. Rep.* **107**, 59–324 (1984).
27. TASSO Collaboration (R. Brandelik, *et al.*), *Phys. Lett. B* **97**, 453 (1980).
28. J. Ellis and I. Karliner, *Nucl. Phys. B* **148**, 141 (1979).
29. PLUTO Collaboration (Ch. Berger, *et al.*), *Phys. Lett. B* **97**, 459 (1980).
30. P. Söding, B. Wiik, G. Wolf and S. L. Wu (presented by S. L. Wu) for the 1995 EPS High Energy and Particle Physics Prize, in *Proceedings of the International Europhysics Conference on High Energy Physics*, Brussels, Belgium 27 Jul.–2 Aug. 1995, pp. 3–10.
31. F. Englert and R. Brout, *Phys. Rev. Lett.* **13**, 321 (1964); P. W. Higgs, *Phys. Lett.* **12**, 132 (1964); P. W. Higgs, *Phys. Rev. Lett.* **13**, 508 (1964); G. S. Guralnik, C. R. Hagen and T. W. B. Kibble,*Phys. Rev. Lett.* **13**, 585 (1964); P. W. Higgs, *Phys. Rev.* **145**, 1156 (1966); T. W. B. Kibble, *Phys. Rev. Lett.* **155**, 1554 (1967).

32. ATLAS Collaboration (G. Aad *et al.*), "Observation of a new particle in the search for the Standard Model Higgs boson with the ATLAS detector at the LHC", *Phys. Lett. B* **716**, 1 (2012).
33. CMS Collaboration (S. Chatrchyan *et al.*), "Observation of a new boson at a mass of 125 GeV with the CMS experiment at the LHC", *Phys. Lett. B* **716**, 30 (2012).
34. CDF Collaboration (F. Abe *et al.*), "Observation of Top Quark Production in p anti-p Collisions with the Collider Detector at Fermilab", *Phys. Rev. Lett.* **74**, 2626 (1995); D0 Collaboration (S. Abachi *et al.*), "Observation of the Top Quark", *Phys. Rev. Lett.* **74**, 2632 (1995)
35. S. Heinemeyer, C. Mariotti, G. Passarino and R. Tanaka (eds.), CERN Yellow Report — Handbook of LHC Higgs cross sections: Books 1, 2 and 3. Report of the LHC Higgs Cross Section Working Group, arXiv:1101.0593v3 [hep-ph] 20 May 2011, arXiv:1201.3084v1 [hep-ph] 15 Jan. 2012, arXiv:1307.1347v1 [hep-ph] 4 Jul. 2013. See also The Higgs Cross Section Working Group web page: `https://twiki.cern.ch/twiki/bin/view/LHCPhysics/CrossSections`, 2013.
36. UA1 Collaboration (G. Arnison *et al.*), "Experimental Observation of Isolated Large Transverse Energy Electrons with Associated Missing Energy at $s =$ 540 GeV", *Phys. Lett. B* **122**, 103 (1983); "Experimental Observation of Lepton Pairs of Invariant Mass around 95 GeV/c^2 at the CERN SPS Collider", *Phys. Lett. B* **126**, 398 (1983).
37. UA2 Collaboration (M. Banner *et al.*), "Observation of Single Isolated Electrons of High Transverse Momentum in Events with Missing Transverse Energy at the CERN SPS Collider", *Phys. Lett. B* **122**, 476 (1983); UA2 Collaboration (P. Bagnaia *et al.*), "Evidence for $Z^\circ \to e^+e^-$ at the CERN SPS Collider", *Phys. Lett. B* **129**, 130 (1983).

The Parton Model and Its Applications

Tung-Mow Yan

Laboratory for Elementary Particle Physics,
Cornell University, Ithaca, NY 14853, USA

Sidney D. Drell

Stanford Linear Accelerator Center,
Stanford University, Stanford, CA 94305, USA

This is a review of the program we started in 1968 to understand and generalize Bjorken scaling and Feynman's parton model in a canonical quantum field theory. It is shown that the parton model proposed for deep inelastic electron scatterings can be derived if a transverse momentum cutoff is imposed on all particles in the theory so that the impulse approximation holds. The deep inelastic electron–positron annihilation into a nucleon plus anything else is related by the crossing symmetry of quantum field theory to the deep inelastic electron–nucleon scattering. We have investigated the implication of crossing symmetry and found that the structure functions satisfy a scaling behavior analogous to the Bjorken limit for deep inelastic electron scattering. We then find that massive lepton pair production in collisions of two high energy hadrons can be treated by the parton model with an interesting scaling behavior for the differential cross-sections. This turns out to be the first example of a class of hard processes involving two initial hadrons.

1. Introduction

In the 1950s and 1960s many new particles were found experimentally. M. Gell-Mann[1] and Y. Ne'eman[2] showed that these particles can be fit into an octet (mesons and baryons) or a decuplet (hyperons) representation of a symmetry group called SU(3). A peculiar feature is that the simplest representation of the group, a triplet, was not realized in Nature. Furthermore, it is hard to imagine that these particles were all elementary. Gell-Mann[3] and Zweig[4] independently discovered that if one proposes the existence of three spin-1/2 fundamental particles which Gell-Mann called quarks, then a meson can be treated as a bound state of a quark and an antiquark, and a baryon can be treated as a bound state of three quarks. To accomplish this feat, however, the fundamental constituents quarks must possess very strange properties: their electric charge must be fractional (1/3 or 2/3 of an electronic charge), and they must violate the spin-statistics connection. Later, the new quantum number color[5] was proposed to resolve these difficulties. This was the birth of the quark model in 1964.

In the mean time, the popularity of canonical quantum field theories was in decline due to the absence of a viable field theory for the strong interactions. Instead, Gell-Mann[6] postulated that the commutation relations derived from spin-1/2 quark fields for the vector and axial vector currents are exact whether or not the quarks exist and whether the underlying symmetry $\mathrm{SU}(3)_L \times \mathrm{SU}(3)_R$ is exact or not. These current algebras, in combination with PCAC (partially conserved axial current)[7] and soft pion theorems provide a framework for extracting dynamical information on strong and weak interactions. Initial applications focused on low energy phenomena. The first application of current algebra to high energy processes was made by S. Adler[8] who derived sum rules for high energy neutrino and antineutrino scatterings. At the time the prospect for neutrino scatterings was quite remote. J. Bjorken[9] obtained from these sum rules an inequality for high energy electron–nucleon scatterings by an isospin rotation. The inequality showed that the cross-sections for electron–nucleon scatterings is of comparable size with that for a point-like target, and this could be tested by the ongoing SLAC-MIT experiment at SLAC.

In an attempt to understand Adler's sum rules and Bjorken's inequality, Bjorken proposed that the structure functions that describe the cross-sections for the inelastic electron scatterings satisfy a scaling property known as Bjorken scaling.[10] There are two Lorentz invariant kinematic variables for the inelastic electron scatterings. The structure functions depend on the two variables. Bjorken scaling means that in the large momentum transfers, these structure functions become a function of the ratio of the two variables. Bjorken scaling was quickly confirmed by the experiments at SLAC.[11] Feynman[12] interpreted the Bjorken scaling as the point-like nature of the nucleon's constituents when they were incoherently scattered by the incident electron. Feynman named the point-like constituents partons. This is the parton model. Feyman left open the possibility that the partons need not be the quarks. However, theorists quickly identified the partons with quarks (in the late 1960s and early 1970s QCD did not exist, and so gluons did not enter the picture). A nucleon consists of three "valence" quarks which carry the nucleon's quantum numbers and a "sea" of quark–antiquark pairs.

This identification led to many predictions for electron and neutrino (and antineutrino) scatterings from a nucleon.[13]

In the fall of 1968 soon after Feynman's parton model was proposed, we embarked on a comprehensive program[14–18] to understand and apply the parton model in a quantum field theory framework. First, we would like to know under what conditions the parton model could be derived from a quantum field theory. Second, was it possible to apply the parton model to other processes? Let us state briefly the conclusions of our investigations here. We showed that the parton model could be derived if the impulse approximation was valid: that during the scattering the constituents behave as if they were free. To accomplish this, we had to impose a transverse momentum cutoff for the particles that appeared in the quantum field theory.[14–16] Crossing symmetry in quantum field theory relates deep inelastic electron–nucleon scatterings and deep inelastic electron–positron annihilation into a nucleon plus anything else. We have found the parton model can be applied to the crossed channel reaction. The structure functions are found to have Bjorken scaling as in the scattering case. In search for other processes to apply the parton model, we found at least one: namely, the lepton pair production by proton–proton collision which was under study by Christenson *et al.*[19] at BNL. The conditions for applying the parton model were satisfied if the lepton pair was produced by the annihilation of a parton from one proton and an antiparton from the other proton and the lepton pair mass is sufficiently high. This is now known as Drell–Yan mechanism.[20]

Finally, we must specify a particular quantum field theory. At the time, there was no good candidate for a quantum field theory for the strong interactions. So our choice to a large extent is arbitrary. We were guided by simplicity. If we were to include particles with isospin, then we had to exclude vector mesons. Otherwise, we would have to deal with non-Abelian massive vector mesons, and no one knew how to do that. Thus, we settled on a quantum field theory involving a nucleon isodoublet and an isotriplet pions, so that there will be both a spin-1/2 charge carrier and a spin-0 charge carrier. They interact through a pseudoscalar coupling.

In the following we will apply our model quantum field theory to elaborate on the points we mentioned above. It should be pointed out that our emphasis will be on the motivations, principles, the most general results, and what modifications that have to be made after the discovery of QCD. We should also mention that we have also found a relation between the threshold behavior of the structure functions in the deep inelastic scatterings and the asymptotic behavior of the elastic electromagnetic form factor (Ref. 21;

G. B. West independently found the connection: Ref. 22), and have studied the semi-inclusive deep inelastic scatterings $e + P \to e' + h + X$ which gives rise to fragmentation functions.[23] But we will not discuss these last two topics any further.

2. Derivation of the Parton Model from a Canonical Quantum Field Theory

The differential cross-section for the inelastic electron scattering from a nucleon

$$e + P \to e' + \text{anything} \tag{1}$$

is described by the tensor:

$$\begin{aligned} W_{\mu\nu} &= 4\pi^2 \frac{E_p}{M} \sum_n \langle P|J_\mu(0)|n\rangle\langle n|J_\nu(0)|P\rangle(2\pi)^4\delta^4(q + P - P_n) \\ &= -\left(g_{\mu\nu} - \frac{q_\mu q_\nu}{q^2}\right)W_1(q^2,\nu) \\ &\quad + \frac{1}{M^2}\left(P_\mu - \frac{P\cdot q}{q^2}q_\mu\right)\left(P_\nu - \frac{P\cdot q}{q^2}q_\nu\right)W_2(q^2,\nu)\,, \end{aligned} \tag{2}$$

where $|P\rangle$ is a one nucleon state with four momentum P, J is the electromagnetic current, q is the four momentum of the virtual photon, $q^2 = -Q^2 < 0$ is the square of the virtual photon mass, and $M\nu = P\cdot q$ is the energy transfer to the photon in the laboratory system. An average over the proton spin is understood. The kinematics is depicted in Fig. 1.

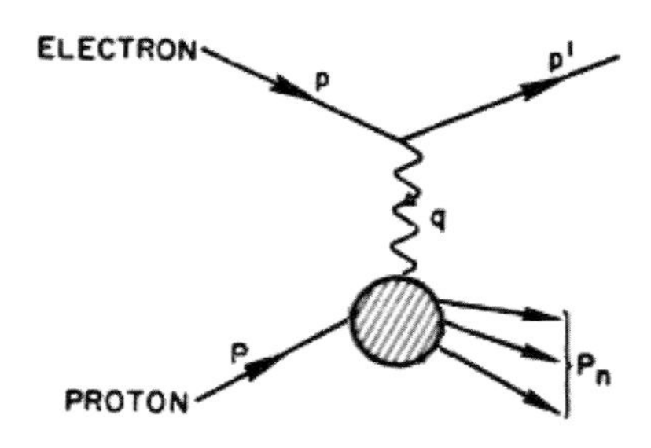

Fig. 1.

The differential cross-section in the rest frame of the proton is given by

$$\frac{d^2\sigma}{d\epsilon' d\cos\theta} = \frac{8\pi\alpha^2}{(Q^2)^2}(\epsilon')^2\left[W_2(q^2,\nu)\cos^2(\theta/2) + 2W_1(q^2,\nu)\sin^2(\theta/2)\right], \tag{3}$$

where ϵ and ϵ' are the initial and final energy of the electron and θ is its scattering angle.

We are interested in the Bjorken limit of large Q^2 and ν with the ratio $\xi = Q^2/2M\nu$ fixed. Let us work in the infinite-momentum center-of-mass frame of the electron and proton. Then

$$q^0 = \frac{2M\nu - Q^2}{4P}, \quad q_3 = \frac{-2M\nu - Q^2}{4P}, \quad |\mathbf{q}_\perp| = \sqrt{Q^2} + \mathcal{O}(1/P^2), \quad (4)$$

with the nucleon momentum $\mathbf{P}$ along the 3-axis. We go to the interaction picture with the familiar U-matrix transformation

$$J_\mu(x) = U^{-1}(t) j_\mu(x) U(t), \quad (5)$$

where $J(x)$ and $j(x)$ are the fully interacting and bare electromagnetic currents, respectively.

Equation (2) can be rewritten as

$$W_{\mu\nu} = 4\pi^2 \frac{E_p}{M} \sum_n \langle UP|j_\mu(0)U(0)|n\rangle\langle n|U^{-1}(0)j_\nu(0)|UP\rangle$$
$$\times (2\pi)^4 \delta^4(q + P - P_n), \quad (6)$$

where $|UP\rangle = U(0)|P\rangle$. In the old fashioned perturbation theory, the state $|UP\rangle$ is represented by all the particles that appear just before the electromagnetic vertex, and the state $U(0)|n\rangle$ is represented by all the particles that appear right after the electromagnetic vertex (see Fig. 2).

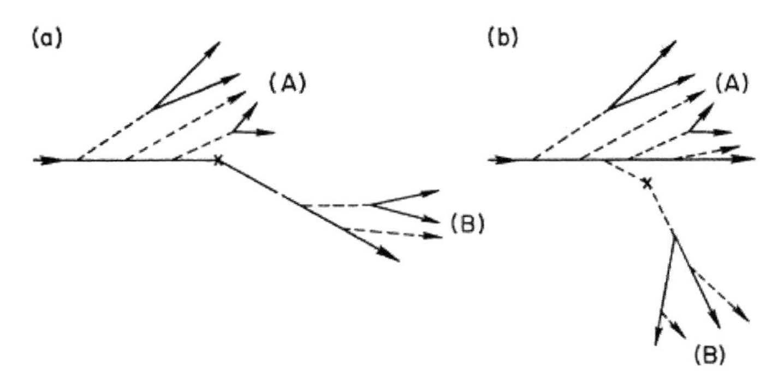

Fig. 2.

From Fig. 2, it is seen that there are two groups of particles after the scattering by the photon, group (A) moves along the direction of the initial nucleon momentum $\mathbf{P}$, and group (B) moves along the scattered momentum $\mathbf{p}+\mathbf{q}$, where $\mathbf{p}$ is the momentum of the parton to be scattered by the photon.

Each group will have limited transverse momentum relative to their large momentum $\mathbf{P}$ and $\mathbf{p}+\mathbf{q}$, respectively.

If we schematically denote the energy of the a particular component of $|UP\rangle$ by E_{UP}, and the energy of a particular component of $U(0)|n\rangle$ by E_{Un}, then with a transverse momentum cutoff introduced for each field particle in the theory, the energy differences $E_P - E_{UP}$ and $E_n - E_{Un}$ are

$$E_P - E_{UP} = \mathcal{O}\left(\frac{k_T^2 + M^2}{P}\right), \tag{7a}$$

$$E_n - E_{Un} = \mathcal{O}\left(\frac{k_T^2 + M^2}{P}\right), \tag{7b}$$

where k_T and M are typical transverse momentum and mass scales, and Eqs. (7a), (7b) are small compared with q^0, so we can make the substitution in the overall energy–momentum conserving Dirac delta function in Eq. (6),

$$q^0 + E_P - E_n = q^0 + E_{UP} - E_{Un} = q^0 + E_a - E_{a'}\,, \tag{8}$$

where a is a constituent in $|UP\rangle$ with momentum p_a which is scattered into a constituent a' with momentum $p_{a'}$. In other words, the overall energy conservation implies that the energy is conserved across the electromagnetic vertex. We should emphasize that we have accomplished this by imposing a transverse momentum cutoff in the underlying quantum field theory and the Bjorken limit gives Q^2, and $M\nu \gg k_T^2$. Equation (8) is a statement of impulse approximation: Because the scattering event occurs so suddenly that the constituents involved can be treated as free. We now make use of the translation operators, completeness of the states $|n\rangle$ and unitarity of the U matrix to obtain:

$$\begin{aligned}
&\lim_{P\to\infty;\, q^2,\, M\nu\to\infty,\, \omega \text{ fixed}} W_{\mu\nu} \\
&= 4\pi^2 \frac{E_p}{M} \sum_n \int (dx) e^{+iqx} \langle UP|j_\mu(x)U(0)|n\rangle \\
&\quad \times \langle n|U^{-1}(0)j_\nu(0)|UP\rangle \\
&= 4\pi^2 \frac{E_p}{M} \int (dx) e^{+iqx} \langle UP|j_\mu(x)U(0)U^{-1}(0)j_\nu(0)|UP\rangle \\
&= 4\pi^2 \frac{E_p}{M} \int (dx) e^{+iqx} \langle UP|j_\mu(x)j_\nu(0)|UP\rangle\,.
\end{aligned} \tag{9}$$

Energy–momentum conservation across the electromagnetic vertex gives

$$(p_a + q)^2 = (p_{a'})^2 = 0\,, \tag{10}$$

or

$$2p_a \cdot q + q^2 = 0\,.$$

If we denote by ξ the fraction of the longitudinal momentum carried by the initial parton by

$$p_a = \xi P\,, \tag{11}$$

then

$$\xi = \frac{1}{\omega} = \frac{Q^2}{2M\nu}\,. \tag{12}$$

Thus, the Bjorken scaling variable is identified as the fractional longitudinal momentum carried by the scattered parton. The structure functions W_1 and W_2 are related to the longitudinal momentum distribution functions of the initial proton. Working out the tensor structure of Eq. (9), we find that the structure functions W_1 and W_2 depend only on the Bjorken variable ω,

$$MW_1(q^2, \nu) = F_1(\omega)\,, \tag{13a}$$

$$W_2(q^2, \nu) = F_2(\omega)\,, \tag{13b}$$

and the relations between W_1 and W_2 depending on the spin of the current,

$$F_1(\omega) = \frac{1}{2}\omega F_2(\omega) \quad \text{(spin-1/2 current)}\,, \tag{14a}$$

$$F_1(\omega) = 0 \qquad \text{(spin-0 current)}\,. \tag{14b}$$

This completes the derivation of the parton model.

3. QCD and the Improved Parton Model

Our studies preceded the discovery of Quantum Chromodynamics (QCD) which is a non-Abelain gauge theory[24] with octet colored gluons and triplet colored quarks (at present, there are three generations of triplet colored quarks: u, d, c, s, t and b). It has the unique property of asymptotic freedom[25] that its coupling constant decreases logarithmically with momentum scale Q. Since 1973, QCD has been accepted as the correct theory of strong interactions. In this theory deep inelastic scatterings can be analyzed rigorously. The main results are:

(1) The moments of the structure functions W_1, and W_2 are no longer independent of Q as they would be if Bjorken scaling is exact. These moments decrease with Q logarithmically with certain powers which are called anomalous dimensions of twist 2 operators and are calculable in QCD.[a]
(2) It is possible to relate the more formal analysis of QCD to the more intuitive parton model. The relations are provided by a set of Altarelli–Parisi equations.[27] These equations describe as we increase Q how a gluon evolves into a gluon pair or a quark–antiquark pair, or a quark evolves into a quark plus a gluon. These evolutions are described by a set of quantities which are called splitting functions. Moments of these splitting functions turn out to be the anomalous dimensions which appear in the moments of the structure functions W_1 and W_2. Thus, the Altarelli–Parisi equations offer further insight into the parton model and the working of QCD.

4. Deep Inelastic Electron–Positron Annihilation

The process

$$e^+ + e^- \to P + \text{anything} \tag{15}$$

is related to the inelastic electron scattering (1) by the crossing symmetry of relativistic quantum field theory. The kinematics is shown in Fig. 3.

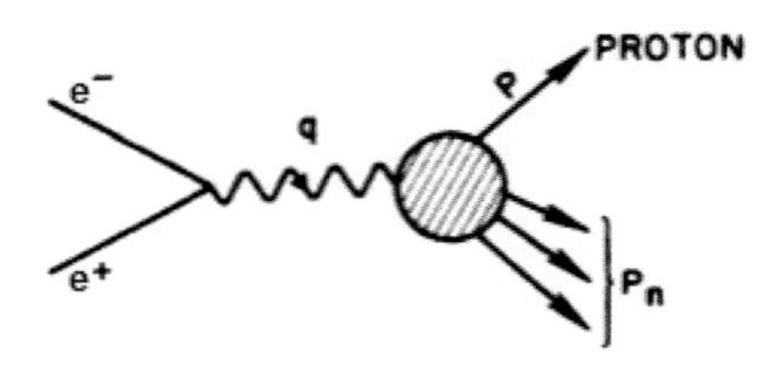

Fig. 3.

It is therefore interesting to ask if the parton model ideas can be applied to this process. We will indicate in this section that the answer indeed is yes. We will only sketch the main ideas. The details can be found in Refs. 15 and 17. The cross-section for the process (15) is summarized by the two structure functions defined by[15,17]

$$\bar{W}_{\mu\nu} = 4\pi^2 \frac{E_p}{M} \sum_n \langle 0|J_\mu(0)|Pn\rangle\langle nP|J_\nu(0)|0\rangle(2\pi)^4\delta^4(q - P - P_n)$$

[a]See the discussions in two excellent text books.[26]

$$= -\left(g_{\mu\nu} - \frac{q_\mu q_\nu}{q^2}\right)\bar{W}_1(q^2,\nu)$$
$$+ \frac{1}{M^2}\left(P_\mu - \frac{P\cdot q}{q^2}q_\mu\right)\left(P_\nu - \frac{P\cdot q}{q^2}q_\nu\right)\bar{W}_2(q^2,\nu)\,. \quad (16)$$

Then the differential cross-section is given by[15,17]

$$\frac{d^2\sigma}{dEd\cos\theta} = \frac{4\pi\alpha^2}{(q^2)^2}\frac{M^2\nu}{\sqrt{q^2}}\left(1-\frac{q^2}{\nu^2}\right)^{1/2}$$
$$\times\left[2\bar{W}_1(q^2,\nu) + \frac{2M\nu}{q^2}\left(1-\frac{q^2}{\nu^2}\right)\frac{\nu\bar{W}_2(q^2,\nu)}{2M}\sin^2\theta\right], \quad (17)$$

where E is the energy of the detected proton and θ is the angle of the proton momentum $\mathbf{P}$ with respect to the colliding e^+ and e^- beams in the center-of-mass system. The two Lorentz invariant kinematic variables are defined by

$$Q^2 = q^2 > 0\,, \quad (18a)$$
$$M\nu = P\cdot q\,. \quad (18b)$$

In the present case the ratio $0 < 2M\nu/Q^2 < 1$. If we follow a similar analysis given to deep inelastic scattering in Sec. 2, we will find that the two structure functions satisfy Bjorken scaling:

$$\lim_{\mathrm{Bj}} M\bar{W}_1(q^2,\nu) = -\bar{F}_1(\omega)\,,$$
$$\lim_{\mathrm{Bj}} \nu\bar{W}_2(q^2,\nu) = \bar{F}_2(\omega)\,, \quad (19)$$

where $\omega = 2M\nu/q^2$. Furthermore, we have

$$\frac{\bar{F}_1(\omega)}{\bar{F}_2(\omega)} = \frac{1}{2}\omega \quad \text{(spin-1/2 current)}\,, \quad (20a)$$
$$\bar{F}_1(\omega) = 0 \qquad \text{(spin-0 current)}\,. \quad (20b)$$

The above relations are similar to the results Eq. (14a) for deep inelastic scattering. In the deep inelastic scattering case, the Bjorken's scaling variable ξ is identified with the longitudinal momentum fraction carried by the scattered parton inside the nucleon. What is the meaning for the corresponding scaling variable $Q^2/2M\nu$ in the deep inelastic annihilation? Let us follow similar steps as in Eqs. (10)–(12). The annihilation process proceeds through the creation of a parton a and its antiparton a', followed by the decay of parton a into a group of particles which contains the detected nucleon with momentum P. Energy–momentum conservation at the electromagnetic vertex gives

$$(q - p_a)^2 = p_{a'}^2 , \tag{21}$$

or

$$Q^2 - 2p_a \cdot q = 0 . \tag{22}$$

If we denote by η the ratio of the momentum p_a of the parent parton a to the nucleon momentum P,

$$p_a = \eta P , \tag{23}$$

then Eq. (22) gives

$$\eta = Q^2/2M\nu > 1 . \tag{24}$$

The ratio is larger than unity as it should be since the nucleon can only carry a fraction of the momentum of its parent parton a. Finally, these scaling predictions will receive QCD's logarithmic corrections in Q,[28] just as in the case of deep inelastic scatterings.

5. Lepton Pair Production

The field on lepton pair production began with the experiment at BNL by Christenson *et al.*[19] They studied the reaction

$$p + U \to \mu^+\mu^- + X \tag{25}$$

for proton energies 22–29 GeV, and the muon pair mass 1–6.7 GeV. Two features of the data stand out: (1) the shoulder-like structure near the muon pair mass of 3 GeV, and (2) the rapid fall-off of the cross-section with the muon pair mass. We know now that the shoulder-like structure is due to the J/Ψ which was discovered in 1974 by a muon pair production experiment at BNL[29] and an e^+e^- colliding beam experiment at SLAC.[30]

We got interested in the process (25) for two reasons: (1) we were looking for applications of the parton model outside deep inelastic lepton scatterings, and (2) we wanted to understand if the rapid decrease of the cross-section with the muon pair mass could be reconciled with the point-like cross-sections observed in the deep inelastic electron scatterings.

The key idea in our approach was once again the impulse approximation. First, we picked an appropriate infinite momentum frame to exploit the time dilation. In this frame, if we were able to establish that the time duration of the external probe is much shorter than the lifetimes of the relevant intermediate states, i.e.

$$\tau_{\text{probe}} \ll \tau_{\text{int. states}} , \tag{26}$$

then the constituents could be treated as free. Thus, the cross-section in the impulse approximation is a product of the probability to find the particular parton configuration and the cross-section for the free partons. In the case of lepton pair production from two initial hadrons,

$$P_1 + P_2 \to l^+ l^- + X , \tag{27}$$

the pair production by the parton–antiparton annihilation satisfies the criteria of impulse approximation[20] (see Fig. 4).

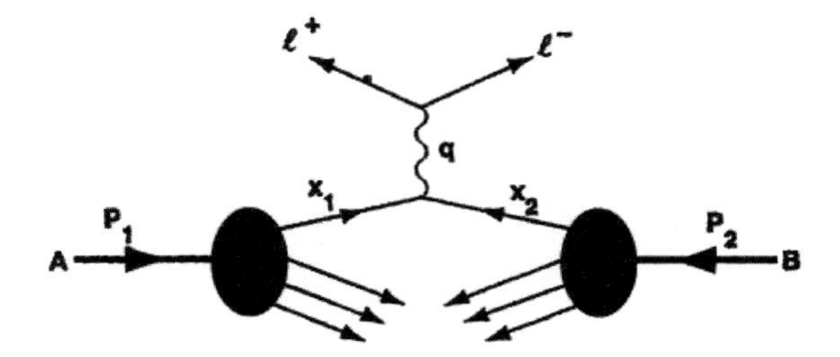

Fig. 4.

It is easily shown that the fractional longitudinal momenta of the annihilating partons satisfy

$$\tau = x_1 x_2 = \frac{Q^2}{s} , \tag{28}$$

where Q^2 and s are, respectively, the pair mass squared and the square of the center-of-mass energy of the initial energy of the initial hadrons. The rapidity of the pair is given by

$$y = \frac{1}{2} \ln \frac{x_1}{x_2} . \tag{29}$$

The predictions stated in our original paper[20] are:

(1) The magnitude and shape of the cross-section are determined by the parton and antiparton distributions measured in deep inelastic lepton scatterings:

$$\frac{d\sigma}{dQ^2 dy} = \frac{4\pi\alpha^2}{3Q^4} \frac{1}{N_c} \sum_p x_1 f p(x_1) x_2 f \bar{p}(x_2) , \tag{30}$$

where a color factor N_c is included in anticipating QCD;

(2) The cross-section $Q^4 d\sigma/dQ^2$ depends only on the scaling variable $\tau = Q^2/s$;
(3) If a photon, pion, kaon, or antiproton is used as the projectile, its structure functions can be measured by lepton pair production.[b] This is the only way we know of to study the parton structure of a particle unavailable as a target for lepton scatterings;
(4) The transverse momentum of the pair should be small ($\sim$ 300–500 MeV);
(5) In the rest frame of the lepton pair, the angular distribution is $1 + \cos^2\theta$ with respect to the hadron collision axis, typical of the spin-1/2 pair production from a transversely polarized virtual photon;
(6) The same model can be easily modified to account for W boson productions.

In this model, the rapid decrease of the cross-section with Q^2 as seen in (25) is related to the rapid fall-off of structure functions as $x \to 1$ in deep inelastic electron scatterings.

The lepton pair production considered here is the first example of a class of hard processes involving two initial hadrons. These processes are not dominated by short distances or light cone. So the standard analysis using operator product expansion is not applicable. But the parton model works. Soon after our work, Berman, Bjorken, and Kogut[32] applied similar ideas to large transverse momentum processes

$$h_1 + h_2 \to h(\text{large } P_T) + X \tag{31}$$

induced by deep inelastic electromagnetic interactions. At that time, it was believed that strong interactions severely suppressed large transverse momenta, therefore, electromagnetic interactions would quickly dominate the large transverse momentum processes. This was the precursor of the point-like gluon exchanges in QCD.

After the advent of QCD, the basic picture of lepton pair production has been confirmed theoretically and the details have been greatly improved.[c] It is no longer a model. That lepton pairs are produced by parton–antiparton annihilation is a consequence of QCD. In QCD, the partons are quarks, antiquarks, and gluons, and the number of color $N_c = 3$. The unique property of QCD being an asymptotically free gauge theory makes the parton model

[b]For a review of the status of pion structure functions measured by pion induced lepton pair production, see Ref. 31.

[c]There has been a vast amount of literature on QCD improved lepton pair production and related processes. Here we list two comprehensive reviews on theoretical and experimental status up to the mid 1990s, the reader can also consult the references cited therein: Refs. 33 and 34.

almost correct, namely for deep inelastic processes we have

$$\text{QCD} = \text{parton model} + \text{small corrections}\,. \tag{32}$$

In the modern language, the impulse approximation is replaced by the more precise concept of factorization which separate the long distance and short distance physics and the condition (26) now becomes

$$Q^2 \gg \Lambda^2_{\text{QCD}}\,. \tag{33}$$

The constituents are almost free leading to logarithmic corrections to the structure functions

$$f_i \implies f_i(x, \ln Q^2)\,. \tag{34}$$

Factorization for the lepton pair production works in QCD, but in a more complicated manner and it has taken the hard work of many people and many years to establish.[35,36] The main complication arises from the new feature of initial and final state interactions between the hadrons.[d] The result is fairly simple to state

$$\frac{d\sigma^{AB}}{dQ^2dy} = \sum_{a,b} \int_{x_A}^{1} d\xi_A \int_{x_B}^{1} d\xi_B\, f_{a/A}(\xi_A, Q^2) f_{b/B}(\xi_B, Q^2) H_{ab}\,, \tag{35}$$

where the sum over a and b are over parton species. The parton distribution functions are the same as those in deep inelastic lepton scatterings with the understanding that Q^2 is its absolute value. The function H_{ab} is the parton level hard scattering cross-section computable in perturbative QCD and is often written as

$$H_{ab} = \frac{d\hat{\sigma}}{dQ^2\,dy}\,. \tag{36}$$

Beside the logarithmic scaling violation, a large transverse momentum of the lepton pair can be produced by recoil of quarks or gluons. A simple dimensional analysis gives

$$\langle k_T^2 \rangle = a + \alpha_s(Q^2) s f(\tau, \alpha_s)\,. \tag{37}$$

The constant a is related to the primordial or intrinsic transverse momentum of the partons.

The full angular distributions in both θ and ϕ depend on input quark and gluon densities and are rather complicated.[38] For small k_T the θ dependence

[d] For a most recent review on the subject of factorization and related topics, see the rapporteur's talk by M. Neubert (Ref. 37, and references cited therein).

is close to $1+\cos^2\theta$ even when high order corrections are taken into account. For large k_T, the θ dependence is expected to be substantially modified.[e]

Many of the predictions have been tested and confirmed by many experiments at Fermilab and CERN and elsewhere.[39] We will not go into the details. We will only point out that the model is so successful that its data have become an integral component of the global fit together with the deep inelastic lepton scatterings in determining the parton distributions inside a nucleon.

6. The Process as a Tool for New Discoveries

It seems natural to broaden the definition of Drell–Yan process to mean a class of high energy hadron–hadron collisions in which there is a subhard process involving one constituent from each of the two incident hadrons. New physics always manifests itself in the production of new particle(s), and the ordinary particles do not carry the new quantum number of the new physics. To discover new physics in a hadron–hadron collider therefore requires annihilation of the ordinary particles to create these new particles. Thus, the Drell–Yan mechanism is an ideal tool for the new discoveries. Let us mention three important discoveries in the recent past which had employed this process to help:

(1) It was used to design the experiments at CERN that discovered the W and Z bosons.[40]
(2) The process was also crucial in the discovery of the top quark at Fermilab.[41]
(3) The discovery of the Higgs Boson at CERN in 2012[42] was perhaps the most dramatic example of the utility of the process. The Higgs Boson is the last particle that appears in the Standard model to have been found.

Since the first experiment at BNL and the naive model proposed to understand it, both experiments and theory have come a long way. It is interesting to note that our original crude fit[20] did not remotely resemble the data. We went ahead to publish our paper because of the model's simplicity and our belief that future experiments would be able to definitely confirm or demolish the model. It is gratifying to see that the successor of the naive model, the QCD improved version, has been confirmed by the experiments carried out in the last forty years. Lepton pair production process has been an important and active theoretical arena to understand various theoretical issues such as

[e] For a review and analysis of the situation, see Ref. 39.

infrared divergences, collinear divergences leading to the factorization theorem in QCD for hard processes involving two initial hadrons. The process has been so well understood theoretically that it has become a powerful tool for discovering new physics. We can expect to find new applications of this process in the future.

7. Conclusion

In this article, we have reviewed the development of our effort to understand Bjorken scaling and Feynman's parton model immediately after these ideas were proposed. We chose to study these topics in a relativistic canonical quantum field theory to take advantage of its crossing symmetry, unitarity, etc. An important input for our program is the impulse approximation. We discovered quickly that the impulse approximation is severely violated due to large logarithms that present at high energies. To restore the validity of the impulse approximation, we have to impose a transverse momentum cutoff for each of the particle in the theory. We then pick the simplest quantum field theory describing nucleons and pions in a pseudoscalar coupling. We started our program in 1968, long before QCD and asymptotic freedom were known, and though quarks had been proposed, they were not yet widely accepted as the fundamental constituents of matter. Within this framework, we derived the parton model for deep inelastic lepton scatterings, found that the parton model is applicable to crossed channel of the deep inelastic inclusive electron–positron annihilation into hadrons, and an additional application of the parton model to massive lepton pair production in hadron–hadron collisions. Some of the interesting results are presented in this article. In spite of an unrealistic quantum field theory and the transverse momentum cutoff imposed by hand, most of our results remain valid today. Of course, we could not have anticipated the subsequent discovery of QCD and its consequences. This remarkable fact is mainly due to our attempt to extract results from general properties of the underlying theory rather than from its specifics. After more than forty years, the subjects that we studied: deep inelastic scatterings, deep inelastic electron–positron annihilation, and lepton pair production, are still very active players in the arena in our quest for our understanding of the inner structure of the elementary particles. We were fortunate that we had the opportunity to play a small part in the endeavor.

Acknowledgments

We thank Professor Jen-Chieh Peng and Professor Matthias Neubert for useful communications.

References

1. M. Gell-Mann, California Institute of Technology Synchrotron Laboratory Report No. CTSL-20, (1961), unpublished.
2. Y. Ne'eman, *Nucl. Phys.* **26**, 222 (1961).
3. M. Gell-Mann, *Phys. Lett.* **8**, 214 (1964).
4. G. Zweig, CERN preprints, 401, 412 (1964), unpublished.
5. O. W. Greenberg, *Phys. Rev. Lett.* **13**, 598 (1964); M. Y. Han and Y. Nambu, *Phys. Rev.* **139**, B1006 (1965).
6. M. Gell-Mann, *Physics* **1**, 63 (1964).
7. Y. Nambu, *Phys. Rev. Lett.* **4**, 380 (1960); K. C. Chou, *Sov. Phys. JETP* **12**, 492 (1961); M. Gell-Mann and M. Levy, *Nuovo Cimento* **16**, 705 (1960).
8. S. L. Adler, *Phys. Rev.* **143**, 1144 (1966).
9. J. D. Bjorken, *Phys. Rev. Lett.* **16**, 408 (1966).
10. J. D. Bjorken, *Phys. Rev.* **179**, 1547 (1969).
11. W. K. H. Panofsky, *Proc. 14th Int. Conf. on High Energy Physics*, Vienna, eds. J. Prentki and J. Steinberger (CERN, Geneva, 1968).
12. R. P. Feynman, *Phys. Rev. Lett.* **23**, 1415 (1969); J. D. Bjorken and E. A. Paschos, *Phys. Rev.* **185**, 1975 (1969), these authors had worked on similar ideas independently of Feynman.
13. F. Close, *An Introduction to Quarks and Partons* (Academic Press, 1979).
14. S. D. Drell, D. J. Levy and T.-M. Yan, *Phys. Rev. Lett.* **22**, 744 (1969).
15. S. D. Drell, D. J. Levy and T.-M. Yan, *Phys. Rev.* **187**, 2159 (1969).
16. S. D. Drell, D. J. Levy and T.-M. Yan, *Phys. Rev. D* **1**, 1035 (1970).
17. S. D. Drell, D. J. Levy and T.-M. Yan, *Phys. Rev. D* **1**, 1617 (1970).
18. T.-M. Yan and S. D. Drell, *Phys. Rev. D* **1**, 2402 (1970).
19. J. H. Christenson, G. H. Hicks, L. Lederman, P. J. Limon, B. G. Pope and E. Zavatini, *Phys. Rev. Lett.* **25**, 1523 (1970); *Phys. Rev. D* **8**, 2016 (1973).
20. S. D. Drell and T.-M. Yan, *Phys. Rev. Lett.* **25**, 316 (1970); *Ann. Phys. (N.Y.)* **66**, 578 (1971).
21. S. D. Drell and T.-M. Yan, *Phys. Rev. Lett.* **24**, 181 (1970).
22. G. B. West, *Phys. Rev. Lett.* **24**, 1206 (1970).
23. S. D. Drell and T.-M. Yan, *Phys. Rev. Lett.* **24**, 855 (1970).
24. C. N. Yang and R. Mills, *Phys. Rev.* **96**, 191 (1954).
25. D. Gross and F. Wilczek, *Phys. Rev. Lett.* **30**, 1343 (1973); H. D. Politzer, *Phys. Rev. Lett.* **30**, 1346 (1973).
26. T.-P. Cheng and L.-F. Li, *Gauge Theory of Elementary Particle Physics* (Oxford University Press, New York, 1984); M. E. Peskin and D. V. Schroeder, *An Introduction to Quantum Field Theory* (Addison-Wesley, 1995).
27. G. Altarelli and G. Parisi, *Nucl. Phys. B* **126**, 298 (1977); V. N. Gribov and L. N. Lipatov, *Sov. J. Nucl. Phys.* **15**, 438, 675 (1972).

28. A. H. Mueller, *Phys. Rev. D* **18**, 3705 (1978).
29. J. J. Albert *et al.*, *Phys. Rev. Lett.* **33**, 1404 (1974).
30. J. E. Augustin *et al.*, *Phys. Rev. Lett.* **33**, 1406 (1974).
31. W. C. Chang and D. Dutta, *Int. J. Mod. Phys. E* **22**, 1330020 (2013), arXiv:1306.3971.
32. S. Berman, J. Bjorken and J. Kogut, *Phys. Rev. D* **4**, 3388 (1971).
33. G. F. Sterman *et al.*, *Rev. Mod. Phys.* **67**, 157 (1995).
34. R. K. Ellis, W. J. Stirling and B. R. Weber, *QCD and Collider Physics* (Cambridge University Press, 1996).
35. J. C. Collins, D. E. Soper and G. F. Sterman, *Adv. Ser. Direct. High Energy Phys.* **5**, 1 (1988), arXiv:hep-ph/0409313.
36. J. C. Collins, *Foundations of Perturbative QCD* (Cambridge University Press, Cambridge, 2011).
37. M. Neubert, talk at LHCP 2014, Columbia University, New York, 2–7 June, 2014, https://indico.cern.ch/event/279518/other-view?view=standard.
38. J. C. Collins and D. E. Soper, *Phys. Rev. D* **16**, 2219 (1977); C. S. Lam and W. K. Tung, *Phys. Rev. D* **18**, 2447 (1978); *ibid.* **21**, 2712 (1980); E. L. Berger and S. J. Brodsky, *Phys. Rev. Lett.* **42**, 940 (1979).
39. J. C. Peng and J. W. Qiu, *Prog. Part. Nucl. Phys.* **76**, 43 (2014), arXiv:1401.0934.
40. UA1 Collab. (G. Arnison *et al.*), *Phys. Lett. B* **122**, 103 (1983); UA2 Collab. (G. Banner *et al.*), *Phys. Lett. B* **122**, 476 (1983).
41. CDF Collab. (F. Abe *et al.*), *Phys. Rev. Lett.* **74**, 2626 (1995); D0 Collab. (S. Abachi *et al.*), *Phys. Rev. Lett.* **74**, 2632 (1995).
42. CMS Collab., *Phys. Lett. B* **716**, 30 (2012); ATLAS Collab., *Phys. Lett. B* **716**, 1 (2012).

From Old Symmetries to New Symmetries: Quarks, Leptons and $B - L$

Rabindra N. Mohapatra
Maryland Center for Fundamental Physics and Department of Physics, University of Maryland, College Park, Maryland 20742, USA
rmohapat@physics.umd.edu

The Baryon–Lepton difference ($B-L$) is increasingly emerging as a possible new symmetry of the weak interactions of quarks and leptons as a way to understand the small neutrino masses. There is the possibility that current and future searches at colliders and in low energy rare processes may provide evidence for this symmetry. This paper provides a brief overview of the early developments that led to $B-L$ as a possible symmetry beyond the standard model, and also discusses some recent developments.

1. Early History

Progress in physics comes in many ways. Sometimes theories follow experiments and sometimes it comes the other way. Classic examples of the first type are, e.g. Faraday's law and Oersted's discovery of connection between electricity and magnetism to name but two. There are also equally illustrious example of experiments following theory: Hertz's discovery of electromagnetic waves following the suggestion of Maxwell, and a more recent example of neutrino being discovered almost 25 years after the suggestion made by Pauli. The same pattern of close entanglement between theory and experiment, with one influencing the other, has continued in the 20th and onto the current century. Theoretical insights into physical phenomena have often followed from the application of novel mathematical techniques, e.g differential equations, algebras, group theory, to name a few. Again in this area too, mathematical developments have followed from physics and vice versa (consider for example the development of calculus by Newton as a way to describe motion).

In the second half of the 20th century, group theory has played an important role in the development of physics as new symmetries were discovered in a variety of physical systems. These provided the primary guiding light for fundamental areas of physics such as elementary particles and condensed matter physics and even the *a posteriori* understanding of some results in quantum mechanics, e.g. the role of $O(4)$ symmetry in the hydrogen atom. In the domain of elementary particle physics, the discovery of the quark model of hadrons and of the standard model (SM) of electroweak interactions are examples where symmetries played a triumphant role. This success strengthened the belief that there may be newer symmetries in nature that will be manifested as we move to uncover physics at ever smaller distances. Many attempts were made to combine space–time symmetries with internal symmetries, e.g. theories based on $SU(6)$ and in the 1970s the emergence of supersymmetry, which from 1980s became the dominant theme in both theory and experiment. Although experiments ultimately decide which symmetries live and which die, either way they leave a lasting impact on the field. In this article, I will focus on a new symmetry of particle physics, the $B-L$ symmetry, which is a global symmetry of the standard model and appears to be emerging as a local symmetry designed for understanding the physics of neutrino masses.

This article is organized as follows. In Sec. 2, I review the history of how symmetries started to enter particle and nuclear physics, how they slowly determined the subsequent developments in the field and how $B-L$ started to make its appearance from the apparent similarity between hadrons and leptons. Section 3 briefly discusses the SM of electroweak interactions and the clear indication of a new $B-L$ symmetry in nature. Section 4, discusses the suggestion that $B-L$ plays the role of a gauge symmetry, once the right-handed neutrinos are introduced into the SM, a realization that followed only after the left–right symmetric models of weak interactions were introduced making the existence of the right-handed neutrino automatic and its defining role in making $B-L$ gauge symmetry theoretically consistent. This preceded by several years the developments in understanding of its role in neutrino mass. Section 5 discusses the connection between small neutrino masses and the breaking of $B-L$ symmetry, followed in Sec. 6, by the prediction of a new baryon number violating process once the left–right symmetric models are embedded into a quark–lepton unified version of the model. Section 7, is devoted to a model-independent connection between proton decay, neutron oscillation and Majorana neutrinos, where $B-L$ breaking plays an important role. In the concluding section, we note briefly searches for the $B-L$ symmetry in various experiments and future prospects for success of such searches.

2. Old Symmetries and Quark–Lepton Similarity

The field of particle physics was born in the 1950s as more and more particles beyond the familiar neutron, proton and the π meson were discovered in cyclotrons and in cosmic rays. They included both neutral and charged K-mesons (discovered in 1947), the hyperon Λ (discovered in 1950) and $\Sigma^{\pm,0}$ and Ξ (both in 1953). The ρ meson was discovered in 1961 following theoretical suggestions (an early example of experiment following theory). This was followed shortly thereafter by ϕ, K^* and the ω vector mesons. As the number of new particles kept increasing, there clearly was need to understand their fundamental nature and systematize their study and possibly predict more new particles from such studies. That is precisely what happened in the 1960s when, following the isospin symmetry suggested by Heisenberg, Gell-Mann and Ne'eman proposed the $SU(3)$ symmetry of strong interactions as a way for classifying the new particles and studying their properties. While isospin was based on the internal symmetry $SU(2)$, which covered only the particles p, n, $\pi^{\pm,0}$, the goal of $SU(3)$ was more ambitious; it was supposed to explain many of the newly discovered mesons and baryons in terms of irreducible representations of an internal $SU(3)$ symmetry and in that process understand their masses and decay properties. The Gell-Mann–Okubo mass formula introduced to understand the masses was a phenomenal success, and predicted the Ω^- particle which was discovered in 1964 at the Brookhaven National Laboratory, providing thereby a striking confirmation of the relevance of the symmetry approach to particle physics. This symmetry approach ultimately led to the suggestion by Gell-Mann and Zweig[1,2] in 1964 that the fundamental building blocks of all hadrons (e.g. p, n, $\pi, \ldots$) are tinier particles called quarks.

As developments were taking place in hadron physics, a quiet revolution was taking shape in the domain of leptons, i.e. electrons, neutrinos (ν_e), etc. The electron neutrino, proposed by Pauli in 1930, was discovered in 1957 by Reines and Cowan. The muon, a close but heavier cousin of the electron was discovered in cosmic rays in 1947. It was realized already by Pauli and Fermi that in nuclear β decay an electron is accompanied by the antineutrino ($\bar{\nu}_e$). Also bombarding the nucleus by this produced neutrino state only produced positrons and not electrons suggesting that there was a difference between neutrino and its antiparticle $\bar{\nu}_e$ produced in nuclear β decay. Similar situations had been encountered before with the proton in that the hydrogen atom was stable but there was no apparent reason for it to be that way. This led Stuckelberg to propose in 1938 that there must be a new conserved quantum number, the baryon number (denoted by B). This meant that in any physical process, both the initial and the final states must have the same baryon number. This also keeps proton as a stable particle, in agreement with observations. Subsequent discovery of the other baryons, e.g. Λ, Σ, etc., which decay only transform to other baryons, e.g. protons and neutrons added richness to the concept of the baryon number. The fundamental origin of this quantum number has been one of the mysteries of theoretical physics for a very long time.

The fact that the electron in the hydrogen atom remains stable can also be understood in a similar manner by postulating another quantum number (called lepton number L). Of course, if only the electron carried a lepton number, it would not have been very interesting but, as noted above, the antineutrino, which is emitted simultaneously with the electron in β decay, also seemed to have this property that, in its subsequent scattering from nuclei produces only a positron and not an electron. This could be understood easily if the neutrino also carried the same lepton number as the electron, with the positron and the $\bar{\nu}_e$ carrying a negative lepton number. This way, nuclear β decay, where $n \to p + e^- + \bar{\nu}_e$, both baryon number and lepton numbers are conserved. Thus far, we have not assigned any particular value to B and L for the above particles, and, without any loss of generality, we can assign $B = 1$ to $p, n, \Lambda, \Sigma, \ldots$, and $L = +1$ to $\nu_e, e, \ldots$. According to the quark model, three quarks would form a baryon such as the proton implying that a quark would carry $B = \frac{1}{3}$. No process involving elementary particles has been observed to violate either baryon or lepton number conservation. There are nevertheless, circumstantial indications, that these laws must be broken. Most compelling of them is the fact the universe seems to have an excess of baryons (matter) over antibaryons (antimatter) and the theoretical recognition that nonperturbative effects in the standard model can lead to the violation of both baryon and lepton number.

As $SU(3)$ symmetry of hadrons was gaining a firm hold in the physics of baryons and mesons, a question was being raised regarding whether there were any similar symmetries in the domain of leptons. However unlike the multiplicity of baryons and mesons, until the beginning of 1960s, only three leptons were known to exist: e^-, μ^- and ν_e and their antiparticles and no more. Clearly, there was no need for a symmetry like $SU(3)$. A curious feature nonetheless was noted in 1959 by Gamba, Marshak and Okubo[3] that the three baryons (p, n, Λ) were arranged in electric charge (and very crudely in mass) roughly the same way as the leptons (ν_e, e^-, μ^-). This led them to suggest that there was a symmetry between baryons and leptons (modulo an overall shift in the charge values). They called this baryon–lepton symmetry. The shifted charge pattern could be understood by writing the following formula:

$$Q = I_3 + (T + B - L)/2\,, \tag{1}$$

where (p, n) and (ν_e, e) are assigned to some new isospin group, with I_3 being the third isospin generator; the μ^- and Λ in this approach are considered iso-singlets. The quantum number T is like strangeness with $T = -1$ for both Λ and μ^-.[a] Equation (1) is a unified formula for both hadrons and leptons. Once the quark model was introduced, quark lepton symmetry could be used instead of baryon lepton symmetry and one would write (u, d, s) instead of (p, n, Λ) and the same electric charge formula would apply. Quark–lepton symmetry then became a reflection of the symmetry

$$\begin{pmatrix} u \\ d \\ s \end{pmatrix} \leftrightarrow \begin{pmatrix} \nu_e \\ e \\ \mu \end{pmatrix}. \tag{2}$$

A beautiful aspect of quark–lepton symmetry is that it connects two kinds of elementary particles and suggests that there is a separate weak $SU(2)$ symmetry (to be contrasted with the familiar one operating on protons, neutrons, pions, etc.) that also operates on leptons. For the first time, a symmetry appeared in the lepton sector. A new symmetry $SU(2)_W$ was born from the old hadronic $SU(2)$ of strong interactions. Eventually, this new $SU(2)_W$ was identified as a local symmetry and became the corner stone of the standard model of Glashow, Weinberg and Salam.[5–7]

A major transformation came over this formulation of quark–lepton symmetry when in 1962, Lederman *et al.* discovered the muon neutrino ν_μ. It appeared distinct from the ν_e, and therefore the three quarks needed another partner for quark–lepton symmetry still was to work. This led Bjorken and Glashow[8] in 1964 to postulate that there must be a heavier up-like quark which now a days we know to be the charm quark. Discovery of charm quark at SLAC and Brookhaven maintained the quark–lepton symmetry in its intended form. Clearly, the formula for electric charge had to be rewritten. Is $B - L$ going to remain as part of the formula? The answer had to be postponed. Meanwhile, these and other developments were slowly ushering in a new era in particle physics, which not only determined the spectacular developments that dominated the field for the next fifty years, but are also likely to continue their impact in the future.

---

[a]For a review and history of these ideas, see Ref. 4.

3. $B - L$ Symmetry and the Standard Model

Glashow, Weinberg and Salam[5–7] recognized that this new $SU(2)_W$ (or $SU(2)_L$) symmetry is a local symmetry whose associated gauge bosons can mediate the weak interactions. They proposed a gauge theory based on the $SU(2)_L \times U(1)_Y$ group that at the lowest tree level had the right properties to describe the known $V - A$ form for the charged current weak interactions; the extra $U(1)_Y$ group was needed to unify electromagnetism with weak interactions. We now recognize this theory as the standard model of weak and electromagnetic interactions. This model has been confirmed by experiments, the latest being the discovery of the Higgs boson at the LHC. Below we give a brief overview of some of the symmetry aspects of the model. Under the weak $SU(2)_L \times U(1)_Y$ group, the fermions of one generation are assigned as follows:

$$\begin{gathered} Q_L = \begin{pmatrix} u_L \\ d_L \end{pmatrix} \equiv (1/2, 1/3)\,, \quad L = \begin{pmatrix} \nu_L \\ e_L \end{pmatrix} \equiv (1/2, -1)\,, \\ u_R \equiv (1, 4/3)\,, \quad d_R \equiv (1, -2/3)\,, \quad e_R \equiv (1, -2)\,, \end{gathered} \tag{3}$$

where u, d, ν, e are the up and down quarks and the neutrino and electron fields, respectively. The subscripts L, R stand for the left- and right-handed spin chiralities of the corresponding fermion fields. The numbers in the brackets denote the $SU(2)_L$ and Y quantum numbers. There are four gauge bosons $W^{\pm}_{\mu}$, W^3_{μ} and B_{μ} associated with the four generators of the gauge group. The interactions of these gauge fields with matter (the quarks and leptons) are determined by the symmetry of the theory and lead to the current–current form for weak interactions via exchange of the $W^{\pm}$ gauge boson. Before symmetry breaking, the gauge bosons and fermions are massless. The masslessness of the gauge bosons follows in a way similar to that of photon being massless in QED. Since fermion mass terms correspond to bilinears of the form $\bar{\psi}_L \psi_R$ that connect the left and right chirality states of the fermion, such terms are forbidden by gauge invariance since the left chiral states of fermions in SM are $SU(2)_L$ doublets whereas the right chirality ones are singlets.

To give them mass, we adopt the model for spontaneous breaking of gauge symmetry through the inclusion of scalar fields,[9–11] $\phi(1/2, 1)$ in the theory which transform as doublets of the gauge group (or weak isospin 1/2). This allows Yukawa couplings of the form $\bar{Q}\phi d_R$, $\bar{\psi}_L \phi e_R$ and $\bar{Q}\tilde{\phi} u_R$ where Q refers to doublets of quarks and ψ refers to doublets of leptons, and $\tilde{\phi} = i\tau_2 \phi^*$ ($\tau_{1,2,3}$ denote the three Pauli matrices). If the gauge symmetry is broken by assigning a ground state value of the field ϕ as $\langle\phi\rangle = \binom{0}{v}$, this gives mass not only to all the fermions in the theory but also to the gauge bosons $W^{\pm}$ and to $Z \equiv \cos\theta_W W_3 + \sin\theta_W B$ where $\theta_W = \tan^{-1}\frac{g'}{g}$ is the weak mixing angle. It should be noted that $\langle\phi\rangle$ leaves one gauge degree of freedom unbroken, i.e.

$$Q = I_3 + \frac{Y}{2}, \tag{4}$$

(where I_3 denotes the third weak isospin generator of the $SU(2)_L$ gauge group). since $Q\langle\phi\rangle = 0$, Q remains an unbroken symmetry and can be identified as the electric charge. Given the quantum numbers assigned to different particles, it reproduces the observed electric charges of all the particles of SM. There is however a very unsatisfactory aspect to this electric charge formula due to the presence of the "fluttering" Y term which can be assigned at random. As a result, origin of the electric charge remains mysterious in the SM.

A point worth noting is that to make SM consistent with low energy observations such as extreme suppression of flavor changing neutral current (FCNC) process $K_L^0 \to \mu^+\mu^-$, the charm quark had to be invoked. This is the celebrated Glashow–Illiopoulos–Maiani (GIM) mechanism[12] for suppressing the FCNC, which provided the true dynamical role of the charm quark beyond just, being the quark partner of the ν_μ as noted above.

A major expectation of the standard model was that it should be renormalizable so that as in quantum electrodynamics, it can make testable predictions. Chiral gauge theories however were known from late 1960s to have the Adler–Bell–Jackiw anomalies.[13–16] The existence of such anomalies implies that an apparently conserved current at the tree level is no more conserved once quantum effects are taken into account and would thereby undo the renormalizability of the theory. The model as constructed above turns out to be free of these gauge anomalies. Asking whether there are any other symmetries that are exact, brings to mind two natural candidates of great interest: the baryon (B) and lepton (L) numbers. Clearly, the tree level Lagrangian respects these symmetries separately; however at the quantum level (one loop), there are triangle diagrams that make both the B and L currents anomalous, i.e. for one generation of fermions, keeping only the $SU(2)_L$ contributions, we have

$$\partial^\mu J_{B,\mu} = \partial^\mu J_{L,\mu} = \frac{g^2}{32\pi^2} W^{\alpha\beta}\tilde{W}_{\alpha\beta}, \tag{5}$$

which means that they are separately not conserved but interestingly, the combination $B-L$ is conserved, i.e.

$$\partial^{\mu}(J_{B,\mu}-J_{L,\mu})=0\,. \tag{6}$$

The freedom from anomalies implies that no nonperturbative effect can break this symmetry. Thus $B-L$ symmetry is back in play as a symmetry for quark–lepton physics and from a completely different framework. In the standard model, $B-L$ is not free of cubic, i.e. anomalies in $\mathrm{Tr}(B-L)^3\neq 0$, when summed over all fermions in the SM. Rather it is only free of linear anomalies, i.e. $\mathrm{Tr}(U(1)_{B-L}[SU(2)_L]^2)=0$ and $\mathrm{Tr}(U(1)_{B-L}[U(1)_Y]^2)=0$. This means that $B-L$ is not a hidden local symmetry of standard model but rather just an exact global symmetry as noted.

One important consequence of $B-L$ being an exact global symmetry is that the nonperturbative effects in the SM known as sphaleron effects do break B and L separately but not $B-L$, as pointed out by 't Hooft.[17] They can be represented as a twelve fermion gauge invariant operator $QQQQQQQQQLLL$ which breaks both baryon and lepton number but conserves $B-L$. The strength of this interaction is very weak at zero temperature but is much stronger in the early universe and has important implications for cosmology.

4. $B-L$ as the $U(1)$ Gauge Generator of Weak Interactions and New Electric Charge Formula

A key prediction of the standard model (SM) is that neutrino masses vanish since, unlike other fermions, which have both left- and right-handed chiralities in the theory, there is no right-handed neutrino but just the left-handed $SU(2)_L$ partner of e_L. Because several experiments have confirmed since 1990s that neutrinos have mass, the simplest extension of standard model is to add to it one right-handed neutrino per generation to account for this fact. As soon as this is done, one not only has $\mathrm{Tr}(U(1)_{B-L}[SU(2)_L]^2)=0$ and $\mathrm{Tr}(U(1)_{B-L}[U(1)_Y]^2)=0$ but also $\mathrm{Tr}(B-L)^3=0$. This allows for the possibility of gauging the $U(1)_{B-L}$ quantum number, which gives $B-L$ a dynamical role.

Addition of a right-handed neutrino suggests that weak interaction theory is not only quark–lepton symmetric but can also be written in a way that it conserves parity. These are the left–right symmetric models (LRS) which were written down in 1974–75.[18–20] The gauge group of the LRS model is: $SU(2)_L\times SU(2)_R\times U(1)_{B-L}$ which includes discrete parity symmetry and fermion assignments given by:

$$\begin{aligned} Q_{L,R} &= \begin{pmatrix} u \\ d \end{pmatrix}_{L,R} (1/2, 0, 1/3) \quad \text{or} \quad (0, 1/2, 1/3)\,; \\ \psi_{L,R} &= \begin{pmatrix} \nu \\ e \end{pmatrix}_{L,R} \equiv (1/2, 0, -1) \quad \text{or} \quad (0, 1/2, -1)\,. \end{aligned} \tag{7}$$

That is because, under parity inversion left-handed fermions go to right-handed fermions, the above assignment is parity symmetric. The resulting weak interaction Lagrangian is given by:

$$\mathcal{L}_{wk} = i\frac{g}{2}\left(\bar{Q}_L \boldsymbol{\tau} \cdot \mathbf{W}^{\mu}_L \gamma_\mu Q_L + \bar{\psi}_L \boldsymbol{\tau} \cdot \mathbf{W}^{\mu}_L \gamma_\mu \psi_L\right) + L \leftrightarrow R\,. \tag{8}$$

Clearly under parity inversion, if we transform $W_L \to W_R$, the Lagrangian is parity conserving. However once symmetry breaking is turned on, W_R will acquire a higher mass and introduce parity violation into low energy weak interaction. The effective weak interaction Hamiltonian below the W boson mass can be written as:

$$\mathcal{H}_I = \frac{g^2}{2M^2_{W_L}}\left(\mathcal{J}^{+,\mu}_L \mathcal{J}^{-}_{\mu,L}\right) + \frac{g^2}{2M^2_{W_R}}\left(\mathcal{J}^{+,\mu}_R \mathcal{J}^{-}_{\mu,R}\right) + \text{h.c.} \tag{9}$$

Note that if $m_{W_R} \gg m_{W_L}$, the above weak interactions violate parity almost maximally since the right-handed current effects are suppressed by a factor $\frac{m^2_{W_L}}{m^2_{W_R}}$. This is a fundamentally different approach to observed parity violation than the one espoused in the SM.

In fact, it was pointed out independently by Marshak and me[21] and Davidson,[22] that the electric charge formula now becomes

$$Q \;=\; I_{3L} + I_{3R} + \frac{B-L}{2}\,. \tag{10}$$

This is a considerable improvement over the SM electric charge formula of Eq. (4) in the sense that all terms in the formula are determined through physical considerations of weak, left and right isospin, and baryon and lepton number, that reflect independent characteristics of the various elementary particles. No freely floating parameters are needed to fix electric charges as in the standard model. Electric charge is no more a free parameter but is connected to other physical quantum numbers in the theory. As a result, a number of interesting implications follow. I discuss them below.

5. Neutrino Mass and $B - L$ Symmetry

Discovery of neutrino oscillations has confirmed that neutrinos have mass requiring therefore an extension of the standard model. If we simply add a ν_R and construct the usual Dirac mass for the neutrino in the same way as for the other fermions in SM, the accompanying Yukawa coupling h_ν has to be order 10^{-12} to match observations. Perhaps this suggests that there is some new physics even beyond adding the ν_R to SM that will not require such small parameters. This is where the seesaw mechanism enters,[23–27] which demands that the ν_R's have a large Majorana mass.[28] Since the neutrinos have no electric charge, a Majorana mass for them is compatible with electric charge conservation, which seems to be an absolute symmetry of nature. Once this is done, the $\nu_L - \nu_R$ mass matrix is given by:

$$\mathcal{M}_{\nu,N} = \begin{pmatrix} 0 & m_D \\ m_D^T & M_R \end{pmatrix}, \tag{11}$$

where each of the entries are 3×3 matrices corresponding to three generations of fermions observed in nature. The Feynman diagram responsible for this is given in Fig. 1.

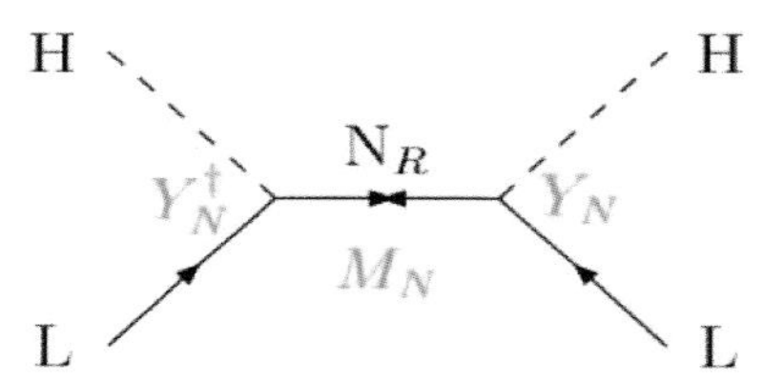

Fig. 1. Feynman diagram for the seesaw mechanism for understanding small neutrino masses. Y_N are the Yukawa couplings of the right-handed neutrino N_R. See Fig. 2 as a cartoon illustration of the seesaw idea.

Fig. 2. Cartoon illustration of the seesaw mechanism.

Diagonalization of this mass matrix leads to a mass formula for the light neutrinos of the form

$$\mathcal{M}_\nu = -m_D^T M_R^{-1} m_D \,. \tag{12}$$

As M_R corresponds to the right-handed neutrino mass, it is not restricted by physics of the SM and can be large whereas m_D is proportional to the scale of standard electroweak symmetry breaking and therefore of the same order as the quark–lepton masses. Thus by making M_R large, we can obtain a very tiny neutrino mass.

This raises two questions: (i) whether it is a bit *ad hoc* to add the right-handed neutrinos? and (ii) is there a physical origin of the seesaw scale or we just accept it as an arbitrary input into the theory? We see below that both these questions are answered in the left–right symmetric extension of the SM, introduced for the purpose of understanding origin of parity violation in Nature.

The left–right model provides a reason why the ν_R should exist and the seesaw scale is given by the scale of parity breaking. In terms of the electric charge formula, we see that since $\Delta Q = 0$ and above the SM scale, $\Delta I_{3L} = 0$, we get from Eq. (12)

$$\Delta I_{3R} \simeq -\Delta\left(\frac{B-L}{2}\right). \tag{13}$$

Because neutrino mass does not involve any hadrons, it has $\Delta B = 0$ and therefore, parity violation (i.e. $\Delta I_{3R} \neq 0$) implies that $\Delta L \neq 0$, i.e. the neutrino is a Majorana particle and its small mass is connected to largeness of the parity breaking scale (or the smallness of the strength of $V+A$ currents in weak interaction). A detailed implementation of seesaw and its connection to left–right symmetry breaking can be seen as follows: Suppose that the full gauge group is broken down to the SM group by a Higgs field belonging to $\Delta_R(1,3,2) \oplus \Delta_L(3,1,2)$ with VEV $\langle\Delta_R^0\rangle = v_R$, then the Yukawa coupling $f\psi_R\psi_R\Delta_R$ leads to a Majorana mass term for the right-handed neutrino of magnitude fv_R. The Dirac masses arise once the SM gauge group is broken as in GWS model leading to the seesaw formula. This immediately makes it clear how intimately the smallness of neutrino mass in this framework is connected to the parity breaking as well as $B-L$ scale. This makes left–right models a compelling platform for discussing neutrino masses.

Again, to connect with the main message of this article, neutrino mass enhances the case for $B-L$ being the next symmetry of nature and hence possibly the left–right symmetric nature of weak interaction at higher energies.

It must be emphasized that until direct experimental evidence for left–right symmetric theories such as the signal for a right-handed W_R and a heavy right-handed neutrino is found at the LHC or alternatively a Z' boson coupled to the $B-L$ current is discovered, the possibility remains open that $B-L$ is not a local but a global symmetry of nature whose breaking could still be at the heart of neutrino masses. In this case, however, there must be a massless Nambu–Goldstone boson present in nature. This particle, called the "majoron"[29] in literature, can manifest itself in neutrinoless double beta decay process and is being searched for in various experiments. Another signal could be new invisible decay modes of the 125 GeV Higgs boson.

6. Neutron–Antineutron Oscillation, Majorana Neutrino Connection

It is clear from Eq. (13), that parity violation ($\Delta I_{3R} \neq 0$) can also lead to baryon number violation since B is part of the electric charge formula. In fact if $\Delta I_{3R} = 1$ which is true if the Higgs field that breaks parity is an $SU(2)_R$ triplet with $B-L = 2$ as in the above derivation of seesaw formula for neutrinos, then in principle this theory could lead to $\Delta B = 2$ baryon number violating process. There are several such processes, e.g. $pp \to K^+K^+$, $\pi^+\pi^+$, as well as neutron–antineutron oscillation. The last process is quite interesting as it implies that neutrons traveling in free space can spontaneously convert to antineutrons and current bounds on the oscillation time for this process is $\tau_{n\bar{n}} \geq 0.8 \times 10^8$ sec. It is interesting that even though the oscillation time is about 2 years, all nuclei are stable due to a potential energy difference between neutron and antineutron in the nucleus. For a discussion of this and other issues related to neutron–antineutron oscillation, see the review Ref. 31. There is now a plan to redo the search for this process at a higher level of sensitivity.[32]

The question that has to be tackled is whether there exist a theory that combines neutrino mass via the seesaw mechanism which predicts an observable $\tau_{n\bar{n}}$ and yet keeps the proton is stable. One example of such a theory was presented in 1980 in Ref. 30. This model presents an embedding of the left–right seesaw model into a quark–lepton unified framework using the gauge group[33] $SU(2)_L \times SU(2)_R \times SU(4)_c$ with the symmetry breaking suggested in Refs. 18–20 rather than in Refs. 23–27. The unified quarks–lepton multiplet is given by $\Psi_L(2,1,4) \oplus \Psi_R(1,2,4)$ with Ψ given by

$$\Psi = \begin{pmatrix} u_1 & u_2 & u_3 & \nu_e \\ d_1 & d_2 & d_3 & e \end{pmatrix}, \tag{14}$$

where subscripts $(1,2,3)$ denote the color index. The 16 chiral fermions of the $SU(2)_L \times SU(2)_R \times SU(4)_c$ model fit into the 16-dimensional spinor representation of the $SO(10)$ group[34,35] which can be the final grand unification group for left–right symmetry as well as $B-L$ gauge symmetry. The symmetry breaking from the group $SU(2)_L \times SU(2)_R \times SU(4)_c$ down to the SM group is achieved by the Higgs fields $\Delta(1,3,10)$ which is the $SU(4)_c$ generalization of the seesaw generating Higgs field $\Delta_R(1,3,+2)$ discussed in the previous section. Without getting into too much group theory details, one can see that the Δ_R fields must have their quark partners present inside them (denote them by Δ_{qq}), which couple to two quarks. Combined with $\Delta_{\nu_R\nu_R}$ VEV breaking $B-L$ symmetry to give seesaw structure for neutrino masses, this leads to the six quark operator $u_Rd_Rd_Ru_Rd_Rd_R$ via the diagram in Fig. 3 below to lead to nonvanishing neutron oscillation amplitude. For multi-TeV scale seesaw, the strength of this operator is of order $G_{\Delta B=2} \sim \frac{\lambda f^3 v_R}{M^6_{\Delta_{qq}}} \sim 10^{-28}$ GeV^{-5} for $f \sim \lambda 10^{-2}$ and $v_R \sim M_\Delta \sim 10$ TeV. Once this operator is hadronically dressed, it gives $\tau_{n\bar{n}} \sim 10^{8-10}$ sec, which is in the observable range with currently available neutron sources around the world.

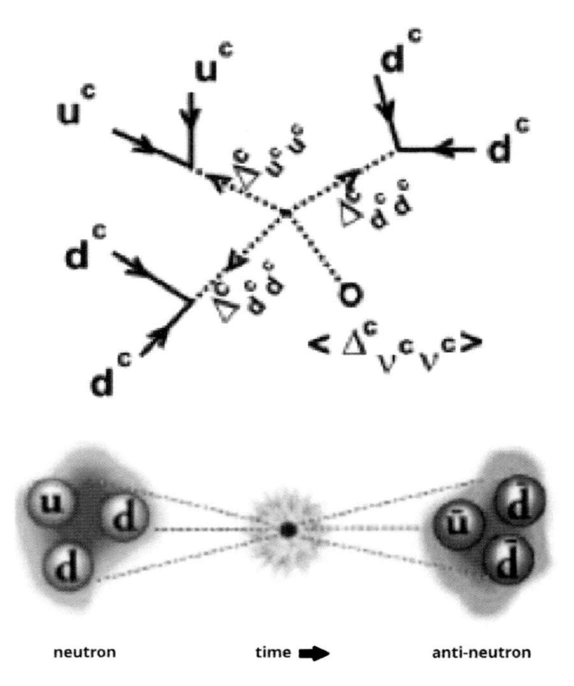

Fig. 3. Feynman diagram for neutron oscillation and Neutrino mass $n\bar{n}$ connection in a quark–lepton unified left–right seesaw model.

7. Role of $B - L$ in Baryon Number Violation

We saw above two extensions of standard model which lead to two specific examples of $B-L$ violation: one for neutrino mass and another for neutron–antineutron oscillation. One can ask the question as to whether we can say in a model independent way about the $B - L$ violation in SM extensions to high scale without detailed specification. One way to explore this would be to consider higher-dimensional B and L violating operators that are invariant under the SM gauge group. This discussion was carried out many years ago by Weinberg[36] and Wilczek and Zee.[37] They pointed out that there is one $d + 5$ operator invariant under the SM involving only the SM fields, i.e. $\mathcal{O}_1 = LHLH$, which leads to the Majorana mass for neutrinos once electroweak symmetry is broken down by the Higgs VEV and changes L by two units. The left–right model is a ultraviolet (UV) complete model that leads to this operator below the right-handed scale. At the level of $d = 6$, there are five operators: $\mathcal{O}_2 = QQQL$, $\mathcal{O}_3 = Q^TC^{-1}\boldsymbol{\tau}Q \cdot Q^TC^{-1}\boldsymbol{\tau}L$, $\mathcal{O}_4 = QL(u^cd^c)^*$ and $\mathcal{O}_5 = QQ(u^ce^c)^*$, $\mathcal{O}_6 = u^cu^cd^ce^c$. The interesting property is that they all conserve the $B - L$ quantum number and proton decay of type $p \to e^+\pi^0$. It is interesting that operators at the level of $d = 7$ break baryon number in such a way that they lead to $B - L = 2$ proton decay models, e.g. $n \to e^-\pi^+$:[38–41,43]

$$\begin{aligned}
\tilde{\mathcal{O}}_1 &= (d^cu^c)^*(d^cL_i)^*H_j^*\epsilon_{ij}\,,\\
\tilde{\mathcal{O}}_2 &= (d^cd^c)^*(u^cL_i)^*H_j^*\epsilon_{ij}\,,\\
\tilde{\mathcal{O}}_3 &= (Q_iQ_j)(d^cL_k)^*H_l^*\epsilon_{ij}\epsilon_{kl}\,,\\
\tilde{\mathcal{O}}_4 &= (Q_iQ_j)(d^cL_k)^*H_l^*(\boldsymbol{\tau}\epsilon)_{ij}\cdot(\boldsymbol{\tau}\epsilon)_{kl}\,,\\
\tilde{\mathcal{O}}_5 &= (Q_ie^c)(d^cd^c)^*H_i^*\,,\\
\tilde{\mathcal{O}}_6 &= (d^cd^c)^*(d^cL_i)^*H_i\,,\\
\tilde{\mathcal{O}}_7 &= (d^cD_\mu d^c)^*(\bar{L}_i\gamma^\mu Q_i)\,,\\
\tilde{\mathcal{O}}_8 &= (d^cD_\mu L_i)^*(\overline{d^c}\gamma^\mu Q_i)\,,\\
\tilde{\mathcal{O}}_9 &= (d^cD_\mu d^c)^*(\overline{d^c}\gamma^\mu e^c)\,.
\end{aligned} \tag{15}$$

A well known example of a UV complete theory where the above $d = 6$ operators emerge is the minimal $SU(5)$ model,[42] where after symmetry breaking to the standard model, $B-L$ remains a good symmetry. Examples of UV complete theories where the above $d = 7$ as well as the $d = 6$ operators arise have also been recently discussed[43] based on $SO(10)$ grand unified theories with the seesaw mechanism. These models lead to both $B - L$ conserving and violating nucleon decay.

8. $B - L$ Breaking and How the Presence of Nucleon Decay and $n\bar{n}$ Together Imply Majorana Neutrinos

A key question in physics beyond the standard model (BSM) is whether neutrinos are Dirac or Majorana type fermions. The answer to this question will dictate the path of BSM physics. The most direct experimental way is to settle this question is by searching for neutrinoless double beta decay ($\beta\beta_{o\nu}$) of certain nuclei. Rightly therefore, there is intense activity in this field at the moment at various laboratories around the world. The current round of experiments, however, are sensitive enough to probe only a small region of neutrino masses and that too provided the neutrino mass ordering is inverted type.[44,45] Even a large bulk of the inverted mass hierarchy region cannot be reached by the current experiments. On the other hand, if neutrino mass hierarchy is normal, it is indeed very unlikely that we will know the answer to this very important question from searches for neutrinoless double beta decay for a long time. It is therefore not without interest to search for alternative experimental strategies to answer this question. Below we suggest that one such strategy is to put renewed effort on the search for baryon number violation.[46]

We saw above that $B - L$ breaking connects Majorana neutrinos and baryon number violation via the electric charge formula if it is a local symmetry. Pursuing this line of thinking, it should be possible to use only B-violating processes, to experimentally resolve the above key questions of the nature of neutrino masses. As suggested recently,[46] consider two different B-violating processes such as $p \to e^+\pi^0$ and neutron oscillation or two B-violating nucleon decay modes, one of which conserves $B - L$ and another that breaks it such as $p \to e^+\pi^0$ (which is of $B - L = 0$ type) and $n \to e^-\pi^+$ (which has $B - L = 2$).[46] Simultaneous discovery of any pair of these processes will imply that there must be neutrinoless double beta decay via a typical Feynman diagram of type in Fig. 4 and once neutrinoless double beta decay has a nonzero amplitude (whether obtained directly or indirectly), no matter how small it is, it will imply a Majorana mass for the neutrino.[47] This can provide an alternative way to experimentally answer the Majorana or Dirac nature of the neutrino regardless of how small the effective neutrino mass contributing to $\beta\beta_{0\nu}$ decay is.

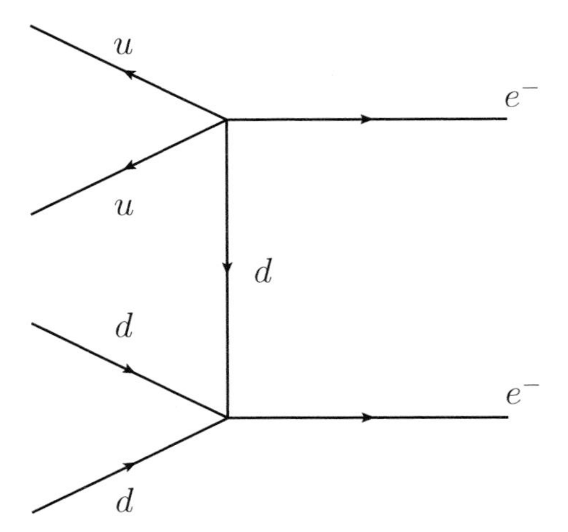

Fig. 4. How discovery of $n \to e^-\pi^+$ and $p \to e^+\pi^0$ implies a nonzero amplitude for neutrinoless double beta decay.

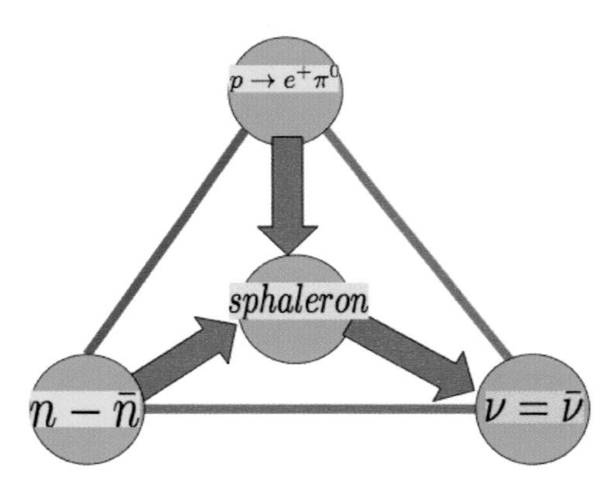

Fig. 5. "$B - L$ triangle" explains how discovering neutron oscillation and proton decay implies neutrinos are Majorana fermions.

Another way to settle the same issue is to invoke the sphaleron operator of the standard model and combine it with proton decay and $n - \bar{n}$ oscillation. To see how this argument goes, note that the nonperturbative effects of the standard model lead to B-violation given by the operator $QQQQQQQQQLLL$. Here Q and L are the $SU(2)_L$ doublets. We can formally rewrite this operator as product of three operators, i.e. $QQQQQQ$, $QQQL$ and LL. Explicitly rewriting this by expanding $Q \equiv (u, d)$ and $L \equiv (\nu_e, e)$ and using quark mixings to change generations, we get for one of the pieces of the sphaleron operator to be $uddudd \cdot uude \cdot \nu\nu$. The strength of this operator is of course highly suppressed. However, as a matter of principle, note that the first part is the piece that contributes to $n\bar{n}$ oscillation, second part to the $p \to e^+\pi^0$ decay and the last part to Majorana mass for the neutrinos. One can represent this in terms a triangle which I call "$B-L$ triangle" (see Fig. 5). The advantage here is that this combination directly gives Majorana neutrino mass without the intermediary of neutrinoless double beta decay.

9. $B-L$, Supersymmetry and Neutralino as a Dark Matter

A final application of local $B-L$ symmetry concerns an understanding of the widely discussed suggestion that in supersymmetric models, the lightest superpartner (LSP) of the standard model particles may play the role of dark matter of the universe. While this suggestion has led to a great deal of dark matter related activity in both theory and experiment, a key question is that in the minimal supersymmetric standard model (MSSM), there are interactions, the so-called R-parity violating ones, which imply that the LSP is actually unstable. So is there a natural extension of MSSM that would lead to a stable dark matter. It was pointed out in mid-eighties and early 90s that if MSSM is extended to include a local $B-L$ symmetry and if $B-L$ symmetry is broken by a Higgs field that has $B-L=2$, then R-parity is indeed automatic.[55–57] A simple way to see this is to realize that R-parity symmetry can be written as: $R=(-1)^{3(B-L)+2S}$ and when R-parity is violated by a Higgs field with $B-L=2$, R remains unbroken. It is interesting that the same Higgs field that gives Majorana mass to the right-handed neutrinos also preserves R-parity symmetry leading to a stable neutralino dark matter.

10. Summary

In summary, we have provided a broad brush overview of the history of $B-L$ as a new symmetry in particle physics and how in recent years following the discovery of neutrino masses, interest in this possible new symmetry has grown enormously. In particular, its connection to both neutrino mass and baryon number violation have provided new insights into physics beyond the standard model. All these have to be confirmed experimentally. At the same time, there are phenomenological studies of many different aspects of this symmetry. To summarize the efforts to unravel the degree of freedom corresponding to local $B-L$ symmetry experimentally, I mention only a few topics. Deciphering whether neutrinos are Majorana fermions is a direct confirmation of whether $B-L$ symmetry is broken or not. This does not say whether it is a global or local symmetry. Furthermore, by itself, discovery of $\beta\beta_{0\nu}$ decay cannot tell where the scale of $B-L$ symmetry breaking is. Supplemented by a discovery (or nondiscovery) of neutron oscillation, one can get an idea about a possible range (or exclude a possible range) of this scale but not the actual scale. The most definitive way to discover the scale of $B-L$ symmetry is to directly search for the gauge boson associated with this in the collider such as the LHC.[b] The same could also be inferred from a discovery of the W_R combined with a Majorana right-handed neutrino. Such searches are currently under way at the LHC.[53,54]

Acknowledgments

This work is supported by the National Science Foundation grant No. PHY-1315155. I am grateful to Tom Ferbel for a careful reading of the manuscript.

References

1. M. Gell-Mann, *Phys. Lett.* **8**, 214 (1964).
2. G. Zweig, CERN, Geneva – TH. 401 (rec. Jan. 64), 24pp.
3. A. Gamba, R. E. Marshak and S. Okubo, *Proc. Natl. Acad. Sci.* (*USA*) **45**, 881 (1959).
4. R. E. Marshak, *Prog. Theor. Phys.* **85**, 61 (1985).
5. S. L. Glashow, *Nucl. Phys.* **22**, 579 (1961).
6. S. Weinberg, *Phys. Rev. Lett.* **19**, 1264 (1967).
7. A. Salam, *Proceedings of the Nobel Symposium*, eds. N. Svartholm*et al.* (1968).
8. J. D. Bjorken and S. L. Glashow, *Phys. Lett.* **11**, 255 (1964).
9. F. Englert and R. Brout, *Phys. Rev. Lett.* **13**, 321 (1964).
10. P. W. Higgs, *Phys. Rev. Lett.* **13**, 508 (1964).
11. G. S. Guralnik, C. R. Hagen and T. W. B. Kibble, *Phys. Rev. Lett.* **13**, 585 (1964).
12. S. L. Glashow, J. Iliopoulos and L. Maiani, *Phys. Rev. D* **2**, 1285 (1970).
13. S. L. Adler, *Phys. Rev.* **177**, 2426 (1969).
14. J. S. Bell and R. Jackiw, *Nuovo Cimento A* **60**, 47 (1969).
15. S. L. Adler and W. A. Bardeen, *Phys. Rev.* **182**, 1517 (1969).
16. W. Bardeen, *Phys. Rev.* **184**, 1848 (1969).
17. G. 't Hooft, *Phys. Rev. Lett.* **37**, 8 (1976).
18. J. C. Pati and A. Salam, *Phys. Rev. D* **10**, 425 (1974).
19. R. N. Mohapatra and J. C. Pati, *Phys. Rev. D* **11**, 566, 2558 (1975).
20. G. Senjanović and R. N. Mohapatra, *Phys. Rev. D* **12**, 1502 (1975).
21. R. E. Marshak and R. N. Mohapatra, *Phys. Lett. B* **91**, 222 (1980).
22. A. Davidson, *Phys. Rev. D* **20**, 776 (1979).
23. P. Minkowski, *Phys. Lett. B* **77**, 421 (1977).

[b]For a small sample of recent papers on $B-L$ gauge boson searches, see Refs. 48–51 and for a review of general Z' searches, see Ref. 52.

24. M. Gell-Mann, P. Ramond and R. Slansky, *Supergravity*, eds. D. Freedman *et al.* (North-Holland, Amsterdam, 1980).
25. T. Yanagida, *Proc. KEK Workshop*, 1979 (unpublished).
26. S. L. Glashow, Cargese Lectures (1979).
27. R. N. Mohapatra and G. Senjanović, *Phys. Rev. Lett.* **44**, 912 (1980).
28. L. Maiani, Notes from the Ettore Majorana Lectures (for recent pedagogical reviews of the developments in neutrino physics), arXiv:1406.5503; S. M. Bilenky, arXiv:1408.1432 [hep-ph].
29. Y. Chikashige, R. N. Mohapatra and R. D. Peccei, *Phys. Lett. B* **98**, 265 (1981).
30. R. N. Mohapatra and R. E. Marshak, *Phys. Rev. Lett.* **44**, 1316 (1980) [Erratum: *ibid.* **44**, 1643 (1980)].
31. R. N. Mohapatra, *J. Phys. G* **36**, 104006 (2009), arXiv:0902.0834 [hep-ph].
32. K. Babu *et al.*, arXiv:1310.8593 [hep-ex].
33. J. C. Pati and A. Salam, Ref. 18.
34. H. Georgi, *Particles and Fields*, ed. C. E. Carlson (AIP, 1975).
35. H. Fritzsch and P. Minkowski, *Ann. Phys.* **93**, 193 (1975).
36. S. Weinberg, *Phys. Rev. Lett.* **43**, 1566 (1979).
37. F. Wilczek and A. Zee, *Phys. Rev. Lett.* **43**, 1571 (1979).
38. H. A. Weldon and A. Zee, *Nucl. Phys. B* **173**, 269 (1980).
39. R. E. Marshak and R. N. Mohapatra, *Coral Gables 1980, Proceedings, Recent Developments in High-Energy Physics*, 277-287 and Virginia Polytech. Blacksburg – VPI-HEP-80-02 (80, rec. Feb), 15pp. (002309).
40. K. S. Babu and R. N. Mohapatra, *Phys. Rev. Lett.* **109**, 091803 (2012).
41. S. M. Barr and X. Calmet, *Phys. Rev. D* **86**, 116010 (2012).
42. H. Georgi and S. L. Glashow, *Phys. Rev. Lett.* **32**, 438 (1974).
43. K. S. Babu and R. N. Mohapatra, *Phys. Rev. Lett.* **109**, 091803 (2012).
44. S. M. Bilenky, S. Pascoli and S. T. Petcov, *Phys. Rev. D* **64**, 053010 (2001).
45. F. Feruglio, A. Strumia and F. Vissani, *Nucl. Phys. B* **637**, 345 (2002) [Addendum: *ibid.* **659**, 359 (2003)]
46. K. S. Babu and R. N. Mohapatra, arXiv:1408.0803 [hep-ph].
47. J. Schechter and J. W. F. Valle, *Phys. Rev. D* **25**, 2951 (1982).
48. W. Emam and S. Khalil, *Eur. Phys. J. C* **52**, 625 (2007), arXiv:0704.1395 [hep-ph].
49. L. Basso, arXiv:1106.4462 [hep-ph].
50. L. Basso, A. Belyaev, S. Moretti and G. M. Pruna, *Nuovo Cimento C* **33**, 171 (2010), arXiv:1002.1214 [hep-ph].
51. K. Huitu, S. Khalil, H. Okada and S. K. Rai, *Phys. Rev. Lett.* **101**, 181802 (2008), arXiv:0803.2799 [hep-ph].
52. P. Langacker, *Rev. Mod. Phys.* **81**, 1199 (2009).
53. CMS Collab. (V. Khachatryan *et al.*), arXiv:1407.3683 [hep-ex].
54. ATLAS Collab. (G. Aad *et al.*), *Eur. Phys. J. C* **72**, 2056 (2012), arXiv:1203.5420 [hep-ex].
55. R. N. Mohapatra, *Phys. Rev. D* **34**, 3457 (1986).
56. A. Font, L. E. Ibanez and F. Quevedo, *Phys. Lett. B* **228**, 79 (1989).
57. S. P. Martin, *Phys. Rev. D* **46**, 2769 (1992).

Quark Mass Hierarchy and Flavor Mixing Puzzles

Zhi-Zhong Xing
Institute of High Energy Physics, Chinese Academy of Sciences, Beijing 100049, China
Center for High Energy Physics, Peking University, Beijing 100080, China
xingzz@ihep.ac.cn

The fact that quarks of the same electric charge possess a mass hierarchy is a big puzzle in particle physics, and it must be highly correlated with the hierarchy of quark flavor mixing. This chapter is intended to provide a brief description of some important issues regarding quark masses, flavor mixing and CP-violation. A comparison between the salient features of quark and lepton flavor mixing structures is also made.

1. A Brief History of Flavors

In the subatomic world, the fundamental building blocks of matter are known as "flavors," including both quarks and leptons. Fifty years ago, the quark model was born thanks to the seminal work done by Murray Gell-Mann[1] and George Zweig[2] independently; and it turned out to be a great milestone in the history of particle physics. The phrase "flavor physics" was first coined by Harald Fritzsch and Murray Gell-Mann in 1971, when they were trying different flavors of ice cream at one of the Baskin Robbins stores in Pasadena. Since then quarks and leptons have been *flavored.*

There are a total of twelve different flavors within the Standard Model (SM): six quarks and six leptons. Table 1 is an incomplete list of the important discoveries in flavor physics, which can give one a ball-park feeling of a century of developments in particle physics. The SM contains thirteen free flavor parameters in its electroweak sector: three charged-lepton masses, six quark masses, three quark flavor mixing angles and one CP-violating phase. Since the three neutrinos must be massive beyond the SM, one has to introduce seven (or nine) extra free parameters to describe their flavor properties: three neutrino masses, three lepton flavor mixing angles and one (or three) CP-violating phase(s), corresponding to their Dirac (or Majorana) nature.[a]

[a]In this connection we have assumed the 3×3 lepton flavor mixing matrix U to be unitary for the sake of simplicity. Whether U is unitary depends on the mechanism of neutrino mass generation.

Table 1. An incomplete list of some important discoveries in the 100-year developments of flavor (quark and lepton) physics.[3]

	Discoveries of lepton flavors, quark flavors and weak CP violation
1897	electron (Thomson[4])
1919	proton (up and down quarks) (Rutherford[5])
1932	neutron (up and down quarks) (Chadwick[6])
1933	positron (Anderson[7])
1937	muon (Neddermeyer and Anderson[8])
1947	Kaon (strange quark) (Rochester and Butler[9])
1956	electron antineutrino (Cowan *et al.*[10])
1962	muon neutrino (Danby *et al.*[11])
1964	CP-violation in s-quark decays (Christenson *et al.*[12])
1974	charm quark (Aubert *et al.*[13] and Abrams *et al.*[14])
1975	tau (Perl *et al.*[15])
1977	bottom quark (Herb *et al.*[16])
1995	top quark (Abe *et al.*[17] and Abachi *et al.*[18])
2001	tau neutrino (Kodama *et al.*[19])
2001	CP-violation in b-quark decays (Aubert *et al.*[20] and Abe *et al.*[21])

So there are at least twenty flavor parameters at low energies. Why is the number of degrees of freedom so big in the flavor sector? What is the fundamental physics behind these parameters? Such puzzles constitute the flavor problems in modern particle physics.

2. Quark Mass Hierarchy

Quarks are always confined inside hadrons, and hence the values of their masses cannot be directly measured. The only way to determine the masses of six quarks is to study their impact on hadron properties based on quantum chromodynamics (QCD). Quark mass parameters in the QCD and electroweak Lagrangians depend both on the renormalization scheme adopted to define the theory and on the scale parameter μ — this dependence reflects the fact that a bare quark is surrounded by a cloud of gluons and quark–antiquark pairs. In the limit where all the quark masses vanish, the QCD Lagrangian possesses an $\mathrm{SU(3)_L \times SU(3)_R}$ chiral symmetry, under which the left- and right-handed quarks transform independently. The scale of dynamical chiral symmetry breaking (i.e. $\Lambda_\chi \simeq 1$ GeV) can therefore be used to distinguish between light quarks ($m_q < \Lambda_\chi$) from heavy quarks ($m_q > \Lambda_\chi$).[22]

One may make use of the chiral perturbation theory, the lattice gauge theory and QCD sum rules to determine the *current* masses of the three light quarks u, d and s.[23] Their values can be rescaled to $\mu = 2$ GeV in

Table 2. Running quark masses at some typical energy scales in the SM, including the Higgs mass $M_H \simeq 125$ GeV and the vacuum-stability cutoff scale $\Lambda_{\rm VS} \simeq 4 \times 10^{12}$ GeV.[24]

μ	m_u (MeV)	m_d (MeV)	m_s (MeV)	m_c (GeV)	m_b (GeV)	m_t (GeV)
$m_c(m_c)$	$2.79^{+0.83}_{-0.82}$	$5.69^{+0.96}_{-0.95}$	116^{+36}_{-24}	$1.29^{+0.05}_{-0.11}$	$5.95^{+0.37}_{-0.15}$	$385.7^{+8.1}_{-7.8}$
2 GeV	$2.4^{+0.7}_{-0.7}$	4.9 ± 0.8	100^{+30}_{-20}	$1.11^{+0.07}_{-0.14}$	$5.06^{+0.29}_{-0.11}$	$322.2^{+5.0}_{-4.9}$
$m_b(m_b)$	$2.02^{+0.60}_{-0.60}$	$4.12^{+0.69}_{-0.68}$	84^{+26}_{-17}	$0.934^{+0.058}_{-0.120}$	$4.19^{+0.18}_{-0.16}$	$261.8^{+3.0}_{-2.9}$
M_W	$1.39^{+0.42}_{-0.41}$	$2.85^{+0.49}_{-0.48}$	58^{+18}_{-12}	$0.645^{+0.043}_{-0.085}$	$2.90^{+0.16}_{-0.06}$	174.2 ± 1.2
M_Z	$1.38^{+0.42}_{-0.41}$	2.82 ± 0.48	57^{+18}_{-12}	$0.638^{+0.043}_{-0.084}$	$2.86^{+0.16}_{-0.06}$	172.1 ± 1.2
M_H	$1.34^{+0.40}_{-0.40}$	$2.74^{+0.47}_{-0.47}$	56^{+17}_{-12}	$0.621^{+0.041}_{-0.082}$	$2.79^{+0.15}_{-0.06}$	$167.0^{+1.2}_{-1.2}$
$m_t(m_t)$	$1.31^{+0.40}_{-0.39}$	2.68 ± 0.46	55^{+17}_{-11}	$0.608^{+0.041}_{-0.080}$	$2.73^{+0.15}_{-0.06}$	163.3 ± 1.1
1 TeV	1.17 ± 0.35	$2.40^{+0.42}_{-0.41}$	49^{+15}_{-10}	$0.543^{+0.037}_{-0.072}$	$2.41^{+0.14}_{-0.05}$	148.1 ± 1.3
$\Lambda_{\rm VS}$	$0.61^{+0.19}_{-0.18}$	1.27 ± 0.22	26^{+8}_{-5}	$0.281^{+0.02}_{-0.04}$	$1.16^{+0.07}_{-0.02}$	82.6 ± 1.4
M_q	—	—	—	$1.84^{+0.07}_{-0.13}$	$4.92^{+0.21}_{-0.08}$	172.9 ± 1.1
$m_q(M_q)$	—	—	—	$1.14^{+0.06}_{-0.12}$	$4.07^{+0.18}_{-0.06}$	162.5 ± 1.1

the modified minimal subtraction ($\overline{\rm MS}$) scheme, as shown in Table 2. On the other hand, heavy quark effective theory, lattice gauge theory and QCD sum rules allow us to determine the *pole* masses M_c and M_b of the charm and bottom quarks. The pole mass M_t of the top quark can directly be measured. The relation between the pole mass M_q and the $\overline{\rm MS}$ running mass $m_q(\mu)$ can be established by taking account of the perturbative QCD corrections.[23] Given the observed mass of the Higgs boson $M_H \simeq 125$ GeV, we have calculated the running masses of six quarks at a number of typical energy scales and have listed their values in Table 2,[24] where $\Lambda_{\rm VS} \simeq 4 \times 10^{12}$ GeV denotes the cutoff scale of vacuum stability in the SM. Note that the values of the pole masses M_q and running masses $m_q(M_q)$ themselves, rather than the running masses $m_q(\mu)$ at these mass scales, are given in the last two rows of Table 2 for the sake of comparison. But the pole masses of the three light quarks are not listed simply because the perturbative QCD calculation is not reliable in that energy region.[24]

The quark mass values shown in Table 2 indicate the existence of a strong hierarchy either in the (u, c, t) sector or in the (d, s, b) sector. We find that it is instructive to consider the quark mass spectrum at the reference scale $\mu = M_Z$ by adopting the $\overline{\rm MS}$ scheme. The reason is simply that an extension

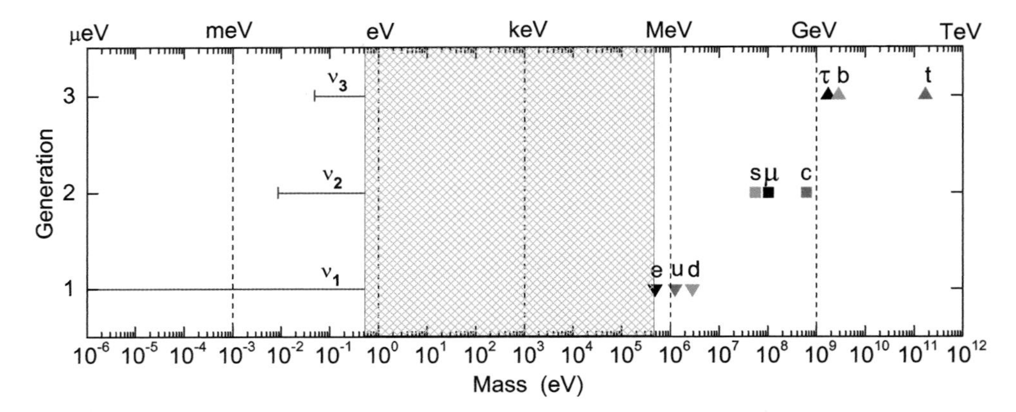

Fig. 1. A schematic illustration of the flavor "hierarchy" and "desert" problems in the SM fermion mass spectrum at the electroweak scale.

of the SM with new physics should be highly necessary far above M_Z, and the strong coupling constant $\alpha_{\rm s}$ becomes sizable far below M_Z. Quantitatively,

$$\begin{aligned} Q_q = +2/3: \quad & \frac{m_u}{m_c} \simeq \frac{m_c}{m_t} \simeq \lambda^4 \,, \\ Q_q = -1/3: \quad & \frac{m_d}{m_s} \simeq \frac{m_s}{m_b} \simeq \lambda^2 \,, \end{aligned} \tag{1}$$

hold to an acceptable degree of accuracy, where $\lambda \equiv \sin\theta_{\rm C} \approx 0.225$ with $\theta_{\rm C}$ being the famous Cabibbo angle of quark flavor mixing.[25] The three charged leptons have a similar mass hierarchy.

To be more intuitive, we present a schematic plot for the mass spectrum of six quarks and six leptons at the electroweak scale in Fig. 1,[26] where a normal neutrino mass ordering has been assumed. One can see that the span between the neutrino masses m_i and the top quark mass m_t is at least twelve orders of magnitude. Furthermore, the "desert" between the heaviest neutral fermion (e.g. ν_3) and the lightest charged fermion (i.e. e^-) spans at least six orders of magnitude. Why do the twelve fermions have such a strange mass pattern and a remarkable hierarchy and desert? A convincing answer to this fundamental question is yet to be seen.

3. Flavor Mixing Pattern

In a straightforward extension of the SM which allows its three neutrinos to be massive, a nontrivial mismatch between the mass and flavor eigenstates of leptons or quarks arises from the fact that lepton or quark fields can interact with both scalar and gauge fields, leading to the puzzling phenomena of flavor mixing and CP-violation. The 3×3 lepton and quark flavor mixing matrices that appear in the weak charged-current interactions are referred to, respectively, as the Pontecorvo–Maki–Nakagawa–Sakata (PMNS) matrix U[27] and the Cabibbo–Kobayashi–Maskawa (CKM) matrix V:[28]

$$-\mathcal{L}^{\ell}_{\rm cc} = \frac{g}{\sqrt{2}} \overline{(e \quad \mu \quad \tau)_{\rm L}} \gamma^\mu U \begin{pmatrix} \nu_1 \\ \nu_2 \\ \nu_3 \end{pmatrix}_{\rm L} W^-_\mu + {\rm h.c.}\,, \tag{2}$$
$$-\mathcal{L}^{q}_{\rm cc} = \frac{g}{\sqrt{2}} \overline{(u \quad c \quad t)_{\rm L}} \gamma^\mu V \begin{pmatrix} d \\ s \\ b \end{pmatrix}_{\rm L} W^+_\mu + {\rm h.c.}\,,$$

in which all the fermion fields are the mass eigenstates. By convention, U and V are defined to be associated with W^- and W^+, respectively. Note that V is unitary as dictated by the SM itself, but whether U is unitary or not depends on the mechanism responsible for the origin of neutrino masses.

In $\mathcal{L}^{\ell}_{\rm cc}$ and $\mathcal{L}^{q}_{\rm cc}$, the charged leptons and quarks with the same electric charges all have normal mass hierarchies (namely, $m_e \ll m_\mu \ll m_\tau$, $m_u \ll m_c \ll m_t$ and $m_d \ll m_s \ll m_b$, as shown in Fig. 1 or Table 2). Yet it remains unclear whether the three neutrinos also have a normal mass ordering ($m_1 < m_2 < m_3$) or not. Now that $m_1 < m_2$ has been fixed from the solar neutrino oscillations, the only likely "abnormal" mass ordering is $m_3 < m_1 < m_2$. The neutrino mass ordering is one of the central concerns in flavor physics, and it will be determined in the foreseeable future with the help of either an accelerator-based neutrino oscillation experiment or a reactor-based antineutrino oscillation experiment, or both of them.

Up to now, the moduli of the nine elements of the CKM matrix V have been determined from current experimental data to a good degree of accuracy:[23]

$$|V| = \begin{pmatrix} 0.97427 \pm 0.00015 & 0.22534 \pm 0.00065 & 0.00351^{+0.00015}_{-0.00014} \\ 0.22520 \pm 0.00065 & 0.97344 \pm 0.00016 & 0.0412^{+0.0011}_{-0.0005} \\ 0.00867^{+0.00029}_{-0.00031} & 0.0404^{+0.0011}_{-0.0005} & 0.999146^{+0.000021}_{-0.000046} \end{pmatrix} . \tag{3}$$

We see that V has a clear hierarchy: $|V_{tb}| > |V_{ud}| > |V_{cs}| \gg |V_{us}| > |V_{cd}| \gg |V_{cb}| > |V_{ts}| \gg |V_{td}| > |V_{ub}|$, which must have something to do with the strong hierarchy of quark masses. Figure 2 illustrates the salient structural features of V, as compared with the more or less "anarchical" structure of the PMNS matrix U. There exists at least two open questions:[29] (a) Is there any intrinsic relationship between the flavor mixing parameters of leptons

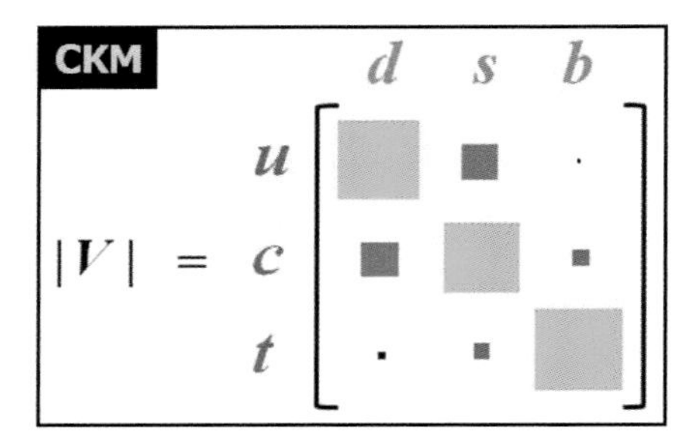

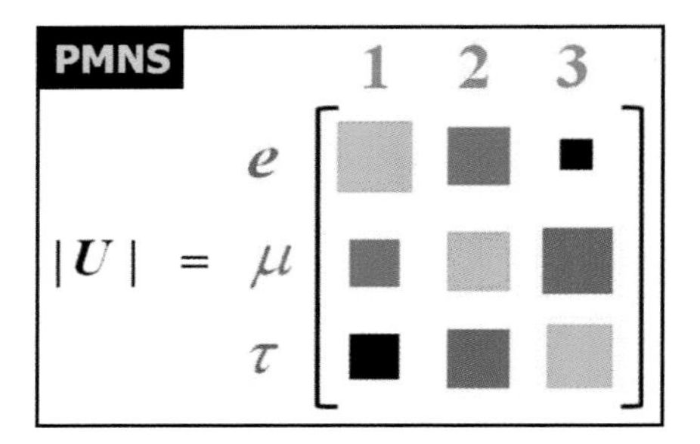

Fig. 2. A schematic illustration of the different flavor mixing structures of six quarks and six leptons at the electroweak scale.

and quarks in a certain Grand Unified Theory? (b) If yes, does this kind of relationship hold between U and V or between U and $V^\dagger$ (or between $U^\dagger$ and V) which are both associated with W^- (or W^+)?

The CKM matrix V can be parametrized in terms of three flavor mixing angles and a nontrivial CP-violating phase in nine different ways.[30] Among them, the most popular one is the so-called "standard" parametrization as advocated by the Particle Data Group:[23]

$$V = \begin{pmatrix} c_{12}c_{13} & s_{12}c_{13} & s_{13}e^{-i\delta_q} \\ -s_{12}c_{23} - c_{12}s_{13}s_{23}e^{i\delta_q} & c_{12}c_{23} - s_{12}s_{13}s_{23}e^{i\delta_q} & c_{13}s_{23} \\ s_{12}s_{23} - c_{12}s_{13}c_{23}e^{i\delta_q} & -c_{12}s_{23} - s_{12}s_{13}c_{23}e^{i\delta_q} & c_{13}c_{23} \end{pmatrix} , \quad (4)$$

in which $c_{ij} \equiv \cos\vartheta_{ij}$ and $s_{ij} \equiv \sin\vartheta_{ij}$ (for $ij = 12, 13, 23$) are defined. The present experimental data lead us to

$$\vartheta_{12} = 13.023^\circ \pm 0.038^\circ , \quad \vartheta_{13} = 0.201^{+0.009^\circ}_{-0.008^\circ} , \quad \vartheta_{23} = 2.361^{+0.063^\circ}_{-0.028^\circ} , \quad (5)$$

and $\delta_q = 69.21^{+2.55^\circ}_{-4.59^\circ}$. In comparison, the similar parameters of the PMNS lepton flavor mixing matrix U lie in the following 3σ ranges as obtained from a global analysis of current neutrino oscillation data:[31]

$$\theta_{12} = 30.6^\circ \to 36.8^\circ , \quad \theta_{13} = 7.6^\circ \to 9.9^\circ , \quad \theta_{23} = 37.7^\circ \to 52.3^\circ , \quad (6)$$

and $\delta_\ell = 0^\circ \to 360^\circ$ provided the neutrino mass ordering is normal (i.e. $m_1 < m_2 < m_3$); or

$$\theta_{12} = 30.6^\circ \to 36.8^\circ , \quad \theta_{13} = 7.7^\circ \to 9.9^\circ , \quad \theta_{23} = 38.1^\circ \to 53.2^\circ , \quad (7)$$

and $\delta_\ell = 0^\circ \to 360^\circ$ provided the neutrino mass ordering is inverted (i.e. $m_3 < m_1 < m_2$). In either case, U exhibits an anarchical pattern as shown in Fig. 2 (left panel). In literature, the possibilities of $\theta_{12} + \vartheta_{12} = 45^\circ$ and $\theta_{23} \pm \vartheta_{23} = 45^\circ$ have been discussed, although such relations depend on both the chosen parametrization and the chosen energy scale.

It is worth mentioning the off-diagonal asymmetries of the CKM matrix V in modulus,[32] which provide another measure of the structure of V about its V_{ud}–V_{cs}–V_{tb} and V_{ub}–V_{cs}–V_{td} axes, respectively:

$$\begin{aligned} \Delta^q_{\rm L} &\equiv |V_{us}|^2 - |V_{cd}|^2 = |V_{cb}|^2 - |V_{ts}|^2 = |V_{td}|^2 - |V_{ub}|^2 \simeq A^2\lambda^6(1-2\rho)\,, \\ \Delta^q_{\rm R} &\equiv |V_{us}|^2 - |V_{cb}|^2 = |V_{cd}|^2 - |V_{ts}|^2 = |V_{tb}|^2 - |V_{ud}|^2 \simeq \lambda^2\,, \end{aligned} \tag{8}$$

where $A \simeq 0.811$, $\lambda \simeq 0.225$ and $\rho \simeq 0.134$ denote the so-called Wolfenstein parameters.[33] It becomes obvious that $\Delta^q_{\rm L} \simeq 6.3 \times 10^{-5}$ and $\Delta^q_{\rm R} \simeq 5.1 \times 10^{-2}$ hold, implying that the CKM matrix V is symmetric about its V_{ud}–V_{cs}–V_{tb} axis to a high degree of accuracy. In comparison, the PMNS matrix U is not that symmetric about of its either axes, but it may possess an approximate or partial μ–τ permutation symmetry;[34] i.e. $|U_{\mu i}| \simeq |U_{\tau i}|$ (for $i = 1, 2, 3$). Such an interesting lepton flavor mixing structure at low energies might originate, via the renormalization-group running effects, from a super high energy PMNS matrix with the exact μ–τ symmetry.[35]

4. The Unitarity Triangles

Thanks to the six orthogonality relations of the unitary CKM matrix V, one may define six unitarity triangles in the complex plane:

$$\begin{aligned} \triangle_\alpha &: V_{\beta d}V^*_{\gamma d} + V_{\beta s}V^*_{\gamma s} + V_{\beta b}V^*_{\gamma b} = 0\,, \\ \triangle_i &: \ V_{uj}V^*_{uk} + V_{cj}V^*_{ck} + V_{tj}V^*_{tk} = 0\,, \end{aligned} \tag{9}$$

where α, β and γ co-cyclically run over the up-type quarks u, c and t, while i, j and k co-cyclically run over the down-type quarks d, s and b. The inner angles of triangles $\triangle_\alpha$ and $\triangle_i$ are universally defined as

$$\Phi_{\alpha i} \equiv \arg\left(-\frac{V_{\beta j}V^*_{\gamma j}}{V_{\beta k}V^*_{\gamma k}}\right) = \arg\left(-\frac{V_{\beta j}V^*_{\beta k}}{V_{\gamma j}V^*_{\gamma k}}\right)\,, \tag{10}$$

where the Greek and Latin subscripts keep them separately co-cyclically running. So $\triangle_\alpha$ and $\triangle_i$ share a common inner angle $\Phi_{\alpha i}$, as shown in Fig. 3.

Let us proceed to define the Jarlskog invariant of CP-violation $\mathcal{J}_q$ for the CKM matrix V through the equation[36]

$$\mathrm{Im}\big(V_{\alpha i}V_{\beta j}V^*_{\alpha j}V^*_{\beta i}\big) = \mathcal{J}_q \sum_\gamma \epsilon_{\alpha\beta\gamma} \sum_k \epsilon_{ijk}\,, \tag{11}$$

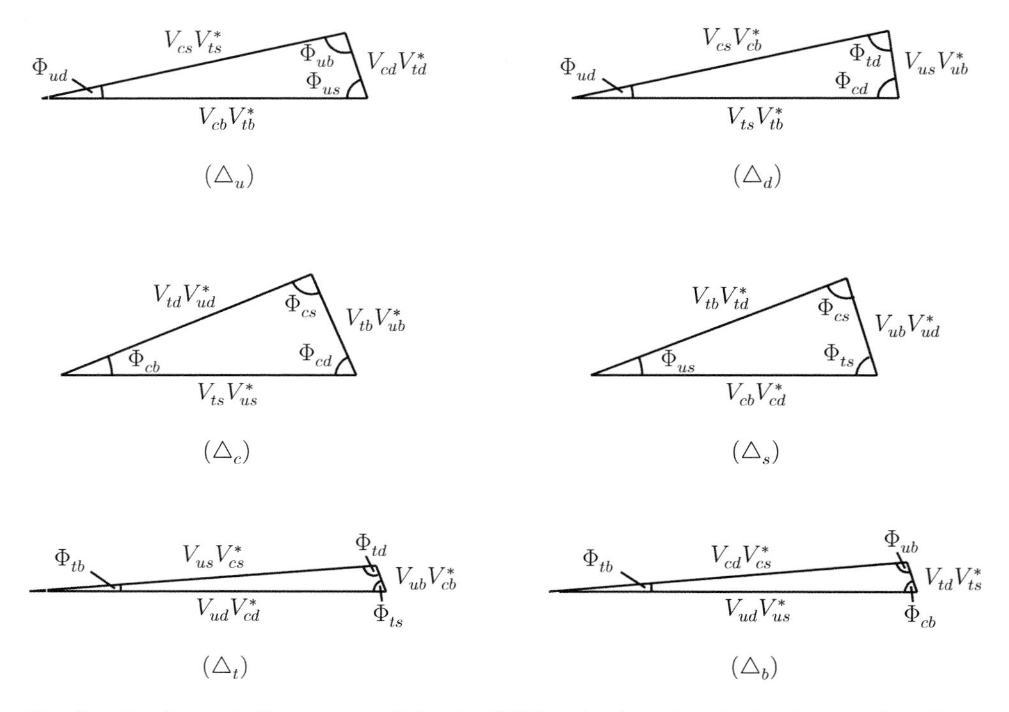

Fig. 3. A schematic illustration of the six CKM unitarity triangles in the complex plane, where each triangle is named by the flavor index that does not manifest itself in the sides.

where the relevant Greek and Latin subscripts run over (u, c, t) and (d, s, b), respectively. Given the standard (or Wolfenstein) parametrization of V,

$$\mathcal{J}_q = c_{12}s_{12}c_{13}^2 s_{13}c_{23}s_{23}\sin\delta_q \simeq A^2\lambda^6\eta\,, \tag{12}$$

where $\eta \simeq 0.354$ is the fourth Wolfenstein parameter.[33] We are therefore left with $\mathcal{J}_q \simeq 3.0\times 10^{-5}$, comparable with $\Delta_{\rm L}^q$ in magnitude. The six CKM unitarity triangles have the same area, equal to $\mathcal{J}_q/2$. If $\Delta_{\rm L}^q$ or $\Delta_{\rm R}^q$ were vanishing, there would be the congruence between two unitarity triangles:[37]

$$\begin{aligned} \Delta_{\rm L}^q = 0 &\Longrightarrow \triangle_u \cong \triangle_d\,, \quad \triangle_c \cong \triangle_s\,, \quad \triangle_t \cong \triangle_b\,, \\ \Delta_{\rm R}^q = 0 &\Longrightarrow \triangle_u \cong \triangle_b\,, \quad \triangle_c \cong \triangle_s\,, \quad \triangle_t \cong \triangle_d\,. \end{aligned} \tag{13}$$

Figure 3 clearly shows that $\Delta_{\rm L}^q \simeq 0$ is actually a rather good approximation.

Note that triangle $\triangle_s$ has been well studied in B-meson physics, and its three inner angles are usually denoted as $\alpha = \Phi_{cs}$, $\beta = \Phi_{us}$ and $\gamma = \Phi_{ts}$. A very striking result is $\alpha = 89.0^{+4.4^\circ}_{-4.2^\circ}$, as reported by the Particle Data Group,[23] implying that triangles $\triangle_s$ and $\triangle_c$ are almost the *right* triangles. In fact, $\alpha = 90^\circ$ leads us to the parameter correlation $\mathcal{J}_{\rm q} = |V_{ud}|\cdot|V_{ub}|\cdot|V_{td}|\cdot|V_{tb}|$, or equivalently $\cos\delta_q = \sin\vartheta_{13}/(\tan\vartheta_{12}\tan\vartheta_{23})$ or $\eta \simeq \sqrt{\rho(1-\rho)}$. If the CKM matrix V is parametrized as[38]

$$V = \begin{pmatrix} s_{\rm u}s_{\rm d}c_{\rm h} + c_{\rm u}c_{\rm d}e^{-{\rm i}\phi} & s_{\rm u}c_{\rm d}c_{\rm h} - c_{\rm u}s_{\rm d}e^{-{\rm i}\phi} & s_{\rm u}s_{\rm h} \\ c_{\rm u}s_{\rm d}c_{\rm h} - s_{\rm u}c_{\rm d}e^{-{\rm i}\phi} & c_{\rm u}c_{\rm d}c_{\rm h} + s_{\rm u}s_{\rm d}e^{-{\rm i}\phi} & c_{\rm u}s_{\rm h} \\ -s_{\rm d}s_{\rm h} & -c_{\rm d}s_{\rm h} & c_{\rm h} \end{pmatrix} , \quad (14)$$

where $c_x \equiv \cos\vartheta_x$ and $s_x \equiv \sin\vartheta_x$ with $x = $ u (up), d (down) or h (heavy), then one can easily obtain the following relationship:[39]

$$\frac{\sin\alpha}{\sin\phi} \simeq 1 - \tan\vartheta_{\rm u}\tan\vartheta_{\rm d}\cos\vartheta_{\rm h}\cos\phi - \frac{1}{2}\tan^2\vartheta_{\rm u}\tan^2\vartheta_{\rm d}\cos^2\vartheta_{\rm h} , \quad (15)$$

where the higher-order terms of $\tan\vartheta_{\rm u}$ and $\tan\vartheta_{\rm d}$ have been omitted. So $\alpha \simeq \phi$ holds to an excellent degree of accuracy. Taking account of

$$\begin{aligned} \vartheta_{\rm u} &= \arctan\left(\left|\frac{V_{ub}}{V_{cb}}\right|\right) \simeq 4.87^\circ , \\ \vartheta_{\rm d} &= \arctan\left(\left|\frac{V_{td}}{V_{ts}}\right|\right) \simeq 12.11^\circ , \\ \vartheta_{\rm h} &= \arcsin\left(\sqrt{|V_{ub}|^2 + |V_{cb}|^2}\right) \simeq 2.37^\circ , \end{aligned} \quad (16)$$

we obtain $\alpha \simeq 88.95^\circ$ from $\phi = 90^\circ$ according to Eq. (15). The numerical result $\phi - \alpha \simeq 1.05^\circ$ is extremely interesting in the sense that our current experimental data strongly hint at $\phi = 90^\circ$ either at the electroweak scale or at a super high energy scale. Such a conclusion holds because it has been proved that both ϕ and α are completely insensitive to the renormalization-group running effects.[39,40] Note also that ϕ essentially measures the phase difference between the up-type quark mass matrix and the down-type quark mass matrix, and thus $\phi = 90^\circ$ might have very profound meaning with respect to the origin of CP-violation.

Finally let us make a brief comment on the so-called CKM phase matrix, whose elements are just the nine inner angles of the CKM unitarity triangles as defined in Eq. (10):[40,41]

$$\Phi = \begin{pmatrix} \Phi_{ud} & \Phi_{us} & \Phi_{ub} \\ \Phi_{cd} & \Phi_{cs} & \Phi_{cb} \\ \Phi_{td} & \Phi_{ts} & \Phi_{tb} \end{pmatrix} \simeq \begin{pmatrix} 1.05^\circ & 21.38^\circ & 157.57^\circ \\ 68.65^\circ & 88.95^\circ & 22.4^\circ \\ 110.3^\circ & 69.67^\circ & 0.034^\circ \end{pmatrix} . \quad (17)$$

Each row or column of Φ corresponds to an explicit unitarity triangle as illustrated in Fig. 3, and thus its three matrix elements must satisfy the following six sum rules:

$$\sum_\alpha \Phi_{\alpha i} = \sum_i \Phi_{\alpha i} = 180^\circ \, . \tag{18}$$

So one may similarly define two off-diagonal asymmetries of Φ about its two axes.[40] If one of the two asymmetries were vanishing, we would be left with a result which is analogous to the one in Eq. (13). Of course, one may easily extend the same language to describe the PMNS phase matrix and discuss its evolution with the energy scales.[42]

5. Two Important Limits

Now let us speculate whether the observed pattern of the CKM matrix V can be partly understood the reasonable limits of quark masses. This idea is more or less motivated by two useful working symmetries in understanding the strong interactions of quarks and hadrons by means of QCD or an effective field theory based on QCD:[43] the chiral quark symmetry (i.e. $m_u, m_d, m_s \to 0$) and the heavy quark symmetry (i.e. $m_c, m_b, m_t \to \infty$). The reason for the usefulness of these two symmetries is simply that the masses of the light quarks are far below the typical QCD scale $\Lambda_{\rm QCD} \sim 0.2$ GeV, whereas the masses of the heavy quarks are far above it. Because the elements of V are dimensionless and their magnitudes lie in the range of 0 to 1, they can only depend on the mass ratios of lighter quarks to heavier quarks. The mass limits corresponding to the chiral and heavy quark symmetries are therefore equivalent to setting the relevant mass ratios to zero, and it is possible that they can help to reveal a part of the salient features of V. In this spirit, some preliminary attempts have been made to look at the quark flavor mixing pattern in the $m_u, m_d \to 0$ or $m_t, m_b \to \infty$ limits.[44]

We shall show that it is possible to gain an insight into the observed pattern of quark flavor mixing in the chiral and heavy quark mass limits. This model-independent access to the underlying quark flavor structure can explain why $|V_{us}| \simeq |V_{cd}|$ and $|V_{cb}| \simeq |V_{ts}|$ hold to a good degree of accuracy, why $|V_{cd}/V_{td}| \simeq |V_{cs}/V_{ts}| \simeq |V_{tb}/V_{cb}|$ is a reasonable approximation, and why $|V_{ub}/V_{cb}|$ should be smaller than $|V_{td}/V_{ts}|$. Furthermore, the empirical relations $|V_{ub}/V_{cb}| \sim \sqrt{m_u/m_c}$ and $|V_{td}/V_{ts}| \simeq \sqrt{m_d/m_s}$ can be reasonably conjectured in the heavy quark mass limits.[45]

Let us begin with the CKM matrix $V = O_{\rm u}^\dagger O_{\rm d}$ with $O_{\rm u}$ and $O_{\rm d}$ being the unitary transformations responsible for the diagonalizations of the up- and down-type quark mass matrices in the flavor basis. Namely,

$$
\begin{aligned}
O_{\rm u}^\dagger H_{\rm u} O_{\rm u} &= O_{\rm u}^\dagger M_{\rm u} M_{\rm u}^\dagger O_{\rm u} = {\rm Diag}\left\{m_u^2, m_c^2, m_t^2\right\} , \\
O_{\rm d}^\dagger H_{\rm d} O_{\rm d} &= O_{\rm d}^\dagger M_{\rm d} M_{\rm d}^\dagger O_{\rm d} = {\rm Diag}\left\{m_d^2, m_s^2, m_b^2\right\} ,
\end{aligned}
\tag{19}
$$

where $H_{\rm u}$ and $H_{\rm d}$ are defined to be Hermitian. To be more explicit, the nine matrix elements of V read

$$
V_{\alpha i} = \sum_{k=1}^{3} (O_{\rm u})^*_{k\alpha} (O_{\rm d})_{ki} , \tag{20}
$$

where α and i run over (u, c, t) and (d, s, b), respectively. In general, the mass limit $m_u \to 0$ (or $m_d \to 0$) does not correspond to a unique form of $H_{\rm u}$ (or $H_{\rm d}$). The reason is simply that the form of a fermion mass matrix is always basis-dependent. Without loss of any generality, one may choose a particular flavor basis such that $H_{\rm u}$ and $H_{\rm d}$ can be written as

$$
\begin{aligned}
\lim_{m_u \to 0} H_{\rm u} &= \begin{pmatrix} 0 & 0 & 0 \\ 0 & \times & \times \\ 0 & \times & \times \end{pmatrix} , \\
\lim_{m_d \to 0} H_{\rm d} &= \begin{pmatrix} 0 & 0 & 0 \\ 0 & \times & \times \\ 0 & \times & \times \end{pmatrix} ,
\end{aligned}
\tag{21}
$$

in which "$\times$" denotes an arbitrary nonzero element. Note that Eq. (21) is the result of a basis choice instead of an assumption.[45] When the mass of a given quark goes to infinity, we argue that it becomes decoupled from the masses of other quarks. In this case, one may also choose a specific flavor basis where $H_{\rm u}$ and $H_{\rm d}$ can be written as

$$
\begin{aligned}
\lim_{m_t \to \infty} H_{\rm u} &= \begin{pmatrix} \times & \times & 0 \\ \times & \times & 0 \\ 0 & 0 & \infty \end{pmatrix} , \\
\lim_{m_b \to \infty} H_{\rm d} &= \begin{pmatrix} \times & \times & 0 \\ \times & \times & 0 \\ 0 & 0 & \infty \end{pmatrix} .
\end{aligned}
\tag{22}
$$

In other words, the 3×3 Hermitian matrices $H_{\rm u}$ and $H_{\rm d}$ can be simplified to the effective 2×2 Hermitian matrices in either the chiral quark mass limit or the heavy quark mass limit. In view of the fact that $m_u \ll m_c \ll m_t$ and $m_d \ll m_s \ll m_b$ hold at an arbitrary energy scale, as shown in Table 2, we believe that Eqs. (21) and (22) are phenomenologically reasonable and can help explain some of the observed properties of quark flavor mixing in a model-independent way. Let us go into details.

(1) *Why does* $|V_{us}| \simeq |V_{cd}|$ *and* $|V_{cb}| \simeq |V_{ts}|$ *hold?* — A glance at Eq. (3) tells us that $|V_{us}| \simeq |V_{cd}|$ is an excellent approximation. It can be well understood in the heavy quark mass limits, where Hermitian $H_{\rm u}$ and $H_{\rm d}$ may take the form of Eq. (22). In this case the unitary matrices $O_{\rm u}$ and $O_{\rm d}$ used to diagonalize $H_{\rm u}$ and $H_{\rm d}$ can be expressed as

$$\lim_{m_t \to \infty} O_{\rm u} = P_{12} \begin{pmatrix} c_{12} & s_{12} & 0 \\ -s_{12} & c_{12} & 0 \\ 0 & 0 & 1 \end{pmatrix} ,$$
$$\lim_{m_b \to \infty} O_{\rm d} = P'_{12} \begin{pmatrix} c'_{12} & s'_{12} & 0 \\ -s'_{12} & c'_{12} & 0 \\ 0 & 0 & 1 \end{pmatrix} , \tag{23}$$

where $c^{(\prime)}_{12} \equiv \cos\vartheta^{(\prime)}_{12}$, $s^{(\prime)}_{12} \equiv \sin\vartheta^{(\prime)}_{12}$, and $P^{(\prime)}_{12} = {\rm Diag}\{e^{i\phi^{(\prime)}_{12}}, 1, 1\}$. Therefore, we immediately arrive at

$$|V_{us}| = \left| c_{12} s'_{12} - s_{12} c'_{12} e^{i\Delta_{12}} \right| = |V_{cd}| \tag{24}$$

in the $m_t \to \infty$ and $m_b \to \infty$ limits, where $\Delta_{12} \equiv \phi'_{12} - \phi_{12}$ denotes the nontrivial phase difference between the up- and down-quark sectors. Since $m_u/m_c \sim m_c/m_t \sim \lambda^4$ and $m_d/m_s \sim m_s/m_b \sim \lambda^2$ hold, the mass limits taken above are surely a good approximation. So the approximate equality $|V_{us}| \simeq |V_{cd}|$ is naturally attributed to the fact that both $m_t \gg m_u$, m_c and $m_b \gg m_d$, m_s hold.[b]

One may similarly consider the chiral quark mass limits $m_u \to 0$ and $m_d \to 0$ so as to understand why $|V_{ts}| \simeq |V_{cb}|$ holds. Equation (21) leads us to

$$\lim_{m_u \to 0} O_{\rm u} = P_{23} \begin{pmatrix} 1 & 0 & 0 \\ 0 & c_{23} & s_{23} \\ 0 & -s_{23} & c_{23} \end{pmatrix} ,$$
$$\lim_{m_d \to 0} O_{\rm d} = P'_{23} \begin{pmatrix} 1 & 0 & 0 \\ 0 & c'_{23} & s'_{23} \\ 0 & -s'_{23} & c'_{23} \end{pmatrix} , \tag{25}$$

[b]Quantitatively, $|V_{us}| \simeq |V_{cd}| \simeq \lambda$ holds. Hence $s_{12} \simeq \sqrt{m_u/m_c} \simeq \lambda^2$ and $s'_{12} \simeq \sqrt{m_d/m_s} \simeq \lambda$ are often conjectured and can easily be derived from some ansätze of quark mass matrices.[46]

in which $c^{(\prime)}_{23} \equiv \cos\vartheta^{(\prime)}_{23}$, $s^{(\prime)}_{23} \equiv \sin\vartheta^{(\prime)}_{23}$, and $P^{(\prime)}_{23} = \mathrm{Diag}\{1, 1, e^{i\phi^{(\prime)}_{23}}\}$. We are therefore left with

$$|V_{cb}| = \left|c_{23}s'_{23} - s_{23}c'_{23}e^{i\Delta_{23}}\right| = |V_{ts}| \qquad (26)$$

in the $m_u \to 0$ and $m_d \to 0$ limits, where $\Delta_{23} \equiv \phi'_{23} - \phi_{23}$ stands for the nontrivial phase difference between the up- and down-quark sectors. This model-independent result is also in good agreement with the experimental data $|V_{cb}| \simeq |V_{ts}|$ as given in Eq. (3). Namely, the approximate equality $|V_{cb}| \simeq |V_{ts}|$ is a natural consequence of $m_u \ll m_c$, m_t and $m_d \ll m_s$, m_b with no need for any specific assumptions.[c]

(2) *Why does* $|V_{cd}/V_{td}| \simeq |V_{cs}/V_{ts}| \simeq |V_{tb}/V_{cb}|$ *hold?* — Given the magnitudes of the CKM matrix elements in Eq. (3), it is easy to get $|V_{cd}/V_{td}| \simeq 26.0$, $|V_{cs}/V_{ts}| \simeq 24.1$ and $|V_{tb}/V_{cb}| \simeq 24.3$. Thus $|V_{cd}/V_{td}| \simeq |V_{cs}/V_{ts}| \simeq |V_{tb}/V_{cb}|$ holds as a reasonably good approximation. We find that such an approximate relation becomes exact in the mass limits $m_u \to 0$ and $m_b \to \infty$. To be much more explicit, we arrive at

$$V = \lim_{m_u \to 0} O^\dagger_{\rm u} \lim_{m_b \to \infty} O_{\rm d} = P'_{12} \begin{pmatrix} c'_{12} & s'_{12} & 0 \\ -c_{23}s'_{12} & c_{23}c'_{12} & -s_{23} \\ -s_{23}s'_{12} & s_{23}c'_{12} & c_{23} \end{pmatrix} P^\dagger_{23}, \qquad (27)$$

where Eqs. (23) and (25) have been used. Therefore,

$$\left|\frac{V_{cd}}{V_{td}}\right| = \left|\frac{V_{cs}}{V_{ts}}\right| = \left|\frac{V_{tb}}{V_{cb}}\right| = |\cot\vartheta_{23}| \qquad (28)$$

holds in the chosen quark mass limits, which assures the smallest CKM matrix element V_{ub} vanishes. This simple result is essentially consistent with the experimental data if $\vartheta_{23} \simeq 2.35°$ is taken.[d] Note that the quark mass limits $m_t \to \infty$ and $m_d \to 0$ are less favored because they predict both $|V_{td}| = 0$ and $|V_{us}/V_{ub}| = |V_{cs}/V_{cb}| = |V_{tb}/V_{ts}|$, which are in conflict with current experimental data. In particular, the limit $|V_{ub}| = 0$ is apparently closer to reality than the limit $|V_{td}| = 0$. But why V_{ub} is smaller in magnitude than all the other CKM matrix elements remains a puzzle since it is difficult for us to judge that the quark mass limits $m_u \to \infty$ and $m_b \to 0$ should make more sense than the quark mass limits $m_t \to \infty$ and $m_d \to 0$ from a phenomenological point of view. The

[c] It is possible to obtain the quantitative relationship $|V_{cb}| \simeq |V_{ts}| \simeq \lambda^2$ through $s_{23} \simeq m_c/m_t \simeq \lambda^4$ and $s'_{23} \simeq m_s/m_b \simeq \lambda^2$ from a number of ansätze of quark mass matrices.[37]

[d] This numerical estimate implies $\tan\vartheta_{23} \simeq \lambda^2 \simeq \sqrt{m_c/m_t}$, which can easily be derived from the Fritzsch ansatz of quark mass matrices.[47]

experimental data in Eq. (3) indicate $|V_{td}| \gtrsim 2|V_{ub}|$ and $|V_{ts}| \simeq |V_{cb}|$. So a comparison between the ratios $|V_{ub}/V_{cb}|$ and $|V_{td}/V_{ts}|$ might be able to tell us an acceptable reason for $|V_{td}| > |V_{ub}|$.

(3) *Why is $|V_{ub}/V_{cb}|$ smaller than $|V_{td}/V_{ts}|$?* — Given Eqs. (21)–(23), we can calculate the ratios $|V_{ub}/V_{cb}|$ and $|V_{td}/V_{ts}|$ in the respective heavy quark mass limits:[45]

$$\lim_{m_b \to \infty} \left| \frac{V_{ub}}{V_{cb}} \right| = \left| \frac{(O_{\rm u})_{3u}}{(O_{\rm u})_{3c}} \right| , \qquad \lim_{m_t \to \infty} \left| \frac{V_{td}}{V_{ts}} \right| = \left| \frac{(O_{\rm d})_{3d}}{(O_{\rm d})_{3s}} \right| . \tag{29}$$

This result is quite nontrivial in the sense that $|V_{ub}/V_{cb}|$ turns out to be independent of the mass ratios of three down-type quarks in the $m_b \to \infty$ limit, and $|V_{td}/V_{ts}|$ has nothing to do with the mass ratios of three up-type quarks in the $m_t \to \infty$ limit. In particular, the flavor indices showing up on the right-hand side of Eq. (29) is rather suggestive: $|V_{ub}/V_{cb}|$ is relevant to u and c quarks, and $|V_{td}/V_{ts}|$ depends on d and s quarks. We are therefore encouraged to conjecture that $|V_{ub}/V_{cb}|$ (or $|V_{td}/V_{ts}|$) should be a simple function of the mass ratio m_u/m_c (or m_d/m_s) in the $m_t \to \infty$ (or $m_b \to \infty$) limit. If the values of m_u, m_d, m_s and m_c in Table 2 are taken into account, the simplest phenomenological conjectures should be

$$\lim_{m_b \to \infty} \left| \frac{V_{ub}}{V_{cb}} \right| \simeq c_1 \sqrt{\frac{m_u}{m_c}} , \qquad \lim_{m_t \to \infty} \left| \frac{V_{td}}{V_{ts}} \right| \simeq c_2 \sqrt{\frac{m_d}{m_s}} , \tag{30}$$

where c_1 and c_2 are the coefficients of $\mathcal{O}(1)$. In view of $\sqrt{m_u/m_c} \simeq \lambda^2$ and $\sqrt{m_d/m_s} \simeq \lambda$, we expect that $|V_{ub}/V_{cb}|$ is naturally smaller than $|V_{td}/V_{ts}|$ in the heavy quark mass limits. Taking $c_1 = 2$ and $c_2 = 1$ for example, we obtain $|V_{ub}/V_{cb}| \simeq 0.093$ and $|V_{td}/V_{ts}| \simeq 0.222$ from Eq. (30), consistent with current data $|V_{ub}/V_{cb}| \simeq 0.085$ and $|V_{td}/V_{ts}| \simeq 0.214$ in Eq. (3). Given $m_t \simeq 172$ GeV and $m_b \simeq 2.9$ GeV at M_Z, one may argue that $m_t \to \infty$ is a much better limit and thus the relation $|V_{td}/V_{ts}| \simeq \sqrt{m_d/m_s}$ has a good chance to be true. In comparison, $|V_{ub}/V_{cb}| \simeq 2\sqrt{m_u/m_c}$ suffers from much bigger uncertainties associated with the values of m_u and m_c, and even its coefficient "2" is questionable.

6. On the Texture Zeros

In the lack of a quantitatively convincing flavor theory, one has to make use of possible flavor symmetries or assume possible texture zeros to reduce the number of free parameters associated with the fermion mass matrices so as to achieve some phenomenological predictions for flavor mixing and CP-violation. Note that the texture zeros of a given fermion mass matrix mean that the corresponding matrix elements are either exactly vanishing or sufficiently suppressed compared with their neighboring counterparts. There are usually two types of texture zeros:

- They may just originate from a proper choice of the flavor basis, and thus have no definite physical meaning;
- They originate as a natural or contrived consequence of an underlying discrete or continuous flavor symmetry.

A typical example of this kind is the famous Fritzsch mass matrices with six texture zeros,[47,e] in which three of them come from the basis transformation and the others arise from either a phenomenological assumption or a flavor model (e.g. based on the Froggatt–Nielsen mechanism[48]). Such zeros allow one to establish a few simple and testable relations between flavor mixing angles and fermion mass ratios. If such relations are in good agreement with the relevant experimental data, they may have a good chance to be close to the truth — namely, the same or similar relations should be predicted by a more fundamental flavor model with far fewer free parameters. Hence a study of possible texture zeros of fermion mass matrices *does* make some sense in order to get useful hints about flavor dynamics which are responsible for the generation of fermion masses and the origin of CP-violation.

The original six-zero Fritzsch quark mass matrices were ruled out in the late 1980's, because it failed in making the smallness of V_{cb} compatible with the largeness of m_t. A straightforward extension of the Fritzsch ansatz with five or four texture zeros has been discussed by a number of authors.[37] Given current experimental data on quark flavor mixing and CP-violation, it is found that only the following five five-zero Hermitian textures of quark mass matrices are still allowed at the 2σ level:[49]

$$M_{\rm u} = \begin{pmatrix} 0 & \times & 0 \\ \times & \times & \times \\ 0 & \times & \times \end{pmatrix}, \quad M_{\rm d} = \begin{pmatrix} 0 & \times & 0 \\ \times & \times & 0 \\ 0 & 0 & \times \end{pmatrix}; \tag{31}$$

[e] Given a Hermitian or symmetric mass matrix, a pair of off-diagonal texture zeros have been counted as one zero in the literature.

or

$$M_{\rm u} = \begin{pmatrix} 0 & \times & 0 \\ \times & \times & 0 \\ 0 & 0 & \times \end{pmatrix} , \quad M_{\rm d} = \begin{pmatrix} 0 & \times & 0 \\ \times & \times & \times \\ 0 & \times & \times \end{pmatrix} ; \qquad (32)$$

or

$$M_{\rm u} = \begin{pmatrix} 0 & \times & 0 \\ \times & 0 & \times \\ 0 & \times & \times \end{pmatrix} , \quad M_{\rm d} = \begin{pmatrix} 0 & \times & 0 \\ \times & \times & \times \\ 0 & \times & \times \end{pmatrix} ; \qquad (33)$$

or

$$M_{\rm u} = \begin{pmatrix} 0 & 0 & \times \\ 0 & \times & \times \\ \times & \times & \times \end{pmatrix} , \quad M_{\rm d} = \begin{pmatrix} 0 & \times & 0 \\ \times & \times & 0 \\ 0 & 0 & \times \end{pmatrix} ; \qquad (34)$$

or

$$M_{\rm u} = \begin{pmatrix} 0 & 0 & \times \\ 0 & \times & 0 \\ \times & 0 & \times \end{pmatrix} , \quad M_{\rm d} = \begin{pmatrix} 0 & \times & 0 \\ \times & \times & \times \\ 0 & \times & \times \end{pmatrix} . \qquad (35)$$

In comparison, the Hermitian $M_{\rm u}$ and $M_{\rm d}$ may also have a parallel structure and contain four texture zeros:[50]

$$M_{\rm u} = \begin{pmatrix} 0 & \times & 0 \\ \times & \times & \times \\ 0 & \times & \times \end{pmatrix} , \quad M_{\rm d} = \begin{pmatrix} 0 & \times & 0 \\ \times & \times & \times \\ 0 & \times & \times \end{pmatrix} , \qquad (36)$$

where only a single zero does not originate from the basis transformation. But it has been found that a finite $(1,1)$ matrix element of $M_{\rm u}$ or $M_{\rm d}$ does not significantly affect the main phenomenological consequences of Eq. (36), if its magnitude is naturally small (i.e. $\lesssim m_u$ or $\lesssim m_d$).[51]

The hierarchical structures of four-zero quark mass matrices in Eq. (36) can be approximately illustrated as follows:[52]

$$\begin{aligned} M_{\rm u} &\sim m_t \begin{pmatrix} 0 & \vartheta_{\rm u}^3 & 0 \\ \vartheta_{\rm u}^3 & \epsilon_{\rm u}^2 & \epsilon_{\rm u} \\ 0 & \epsilon_{\rm u} & 1 \end{pmatrix} , \\ M_{\rm d} &\sim m_b \begin{pmatrix} 0 & \vartheta_{\rm d}^3 & 0 \\ \vartheta_{\rm d}^3 & \epsilon_{\rm d}^2 & \epsilon_{\rm d} \\ 0 & \epsilon_{\rm d} & 1 \end{pmatrix} , \end{aligned} \qquad (37)$$

where $\vartheta_{\rm u}$ and $\vartheta_{\rm d}$ essentially correspond to the definitions in Eq. (14), and they are related to $\epsilon_{\rm u}$ and $\epsilon_{\rm d}$ in the following way:

$$\begin{aligned} \vartheta_{\rm u}^2 &\sim \epsilon_{\rm u}^6 \sim \frac{m_u}{m_c} \sim 2.2 \times 10^{-3} \,, \\ \vartheta_{\rm d}^2 &\sim \epsilon_{\rm d}^6 \sim \frac{m_d}{m_s} \sim 4.9 \times 10^{-2} \,. \end{aligned} \tag{38}$$

If the phase difference between $M_{\rm u}$ and $M_{\rm d}$ is 90°, then it will be straightforward to obtain $|V_{us}| \simeq |V_{cd}| \sim \sqrt{\vartheta_{\rm u}^2 + \vartheta_{\rm d}^2}$ and $|V_{cb}| \simeq |V_{ts}| \sim |\epsilon_{\rm u} - \epsilon_{\rm d}|$. While $|V_{td}/V_{ts}| \sim \vartheta_{\rm d}$ is apparently expected, $|V_{ub}/V_{cb}| \sim \vartheta_{\rm u}$ must get modified due to the non-negligible contribution of $\mathcal{O}(\vartheta_{\rm d}^2) \sim \mathcal{O}(\vartheta_{\rm u})$ from the down quark sector. Note that $\epsilon_{\rm u}$ and $\epsilon_{\rm d}$ are not very small,[52] and their partial cancellation results in a small $|V_{cb}|$ or $|V_{ts}|$.

It is worth pointing out that one may also relax the Hermiticity of quark mass matrices with a number of texture zeros, such that they can fit current experimental data very well.[53] On the other hand, some texture zeros of quark mass matrices are not preserved to all orders or at any energy scales in a given flavor model. If the model is built at a super high energy scale, where a proper flavor symmetry can be used to constrain the structures of quark mass matrices, one has to take account of the renormalization-group running effects in order to compare its phenomenological results with the experimental data at the electroweak scale.[37]

7. On the Strong CP Problem

So far we have discussed weak CP-violation based on the CKM matrix V in the SM. Now let us make a brief comment on the strong CP problem, because it is closely related to the overall phase of quark mass matrices and may naturally disappear if one of the six quark masses vanishes. It is well known that there exists a P- and T-violating term $\mathcal{L}_\theta$, which originates from the instanton solution to the $\mathrm{U(1)_A}$ problem,[54] in the Lagrangian of QCD for strong interactions of quarks and gluons.[55] This CP-violating term can be compared with the mass term of six quarks, $\mathcal{L}_{\rm m}$, as follows:

$$\mathcal{L}_\theta = \theta \frac{\alpha_{\rm s}}{8\pi} G^a_{\mu\nu} \tilde{G}^{a\mu\nu} \,, \quad \mathcal{L}_{\rm m} = \overline{(u \; c \; t \; d \; s \; b)_{\rm L}} \, \mathcal{M} \begin{pmatrix} u \\ c \\ t \\ d \\ s \\ b \end{pmatrix}_{\rm R} + {\rm h.c.} \,, \tag{39}$$

where θ is a free dimensionless parameter characterizing the presence of CP-violation, α_s is the strong fine-structure constant, $G^a_{\mu\nu}$ (for $a = 1, 2, \ldots, 8$) represents the SU(3)$_c$ gauge fields, $\tilde{G}^{a\mu\nu} \equiv \epsilon^{\mu\nu\alpha\beta} G^a_{\mu\nu}/2$, and $\mathcal{M}$ stands for the overall 6×6 quark mass matrix. The chiral transformation of the quark fields $q \to \exp(i\phi_q \gamma_5) q$ (for $q = u,\ c,\ t$ and $d,\ s,\ b$) leads to the changes

$$\theta \longrightarrow \theta - 2\sum_q \phi_q \,, \quad \arg(\det \mathcal{M}) \longrightarrow \arg(\det \mathcal{M}) + 2\sum_q \phi_q \,, \tag{40}$$

where the change of θ follows from the chiral anomaly[56] in the chiral currents

$$\partial_\mu (\bar{q}\gamma^\mu \gamma_5 q) = 2im_q \bar{q}\gamma_5 q + \frac{\alpha_s}{4\pi} G^a_{\mu\nu} \tilde{G}^{a\mu\nu} \,. \tag{41}$$

Then the effective CP-violating term in QCD, which is invariant under the above chiral transformation, turns out to be

$$\mathcal{L}_{\bar{\theta}} = \bar{\theta} \frac{\alpha_s}{8\pi} \, G^a_{\mu\nu} \tilde{G}^{a\mu\nu} \,, \tag{42}$$

in which $\bar{\theta} = \theta + \arg(\det \mathcal{M})$ is a sum of both the QCD contribution and the electroweak contribution.[57] The latter depends on the phase structure of the quark mass matrix $\mathcal{M}$. Because of

$$|\det \mathcal{M}| = m_u m_c m_t m_d m_s m_b \,, \tag{43}$$

the determinant of $\mathcal{M}$ becoming vanishing in the $m_u \to 0$ (or $m_d \to 0$) limit. In this case, the phase of $\det \mathcal{M}$ is arbitrary, and thus it can be arranged to cancel out θ such that $\bar{\theta} \to 0$. Namely, QCD would be a CP-conserving theory if one of the six quarks were massless. But current experimental data have definitely ruled out the possibility of $m_u = 0$ or $m_d = 0$. Moreover, the experimental upper limit on the neutron electric dipole moment yields $\bar{\theta} < 10^{-10}$.[58] The strong CP problem is therefore a theoretical problem of how to explain why $\bar{\theta}$ appears but is so small.[59]

A comparison between weak and strong CP-violating effects might make sense, but it is difficult to find out a proper measure for either of them. The issue involves the reference scale and flavor parameters which may directly or indirectly determine the strength of CP-violation. To illustrate,[45,f]

$$\begin{aligned} \mathrm{CP}_{\text{weak}} &\sim \frac{1}{\Lambda^6_{\text{EW}}} (m_u - m_c)(m_c - m_t)(m_t - m_u)(m_d - m_s) \\ &\quad \times (m_s - m_b)(m_b - m_d) \mathcal{J}_{\text{q}} \sim 10^{-13} \,, \\ \mathrm{CP}_{\text{strong}} &\sim \frac{1}{\Lambda^6_{\text{QCD}}} m_u m_c m_t m_d m_s m_b \sin\bar{\theta} \sim 10^4 \sin\bar{\theta} < 10^{-6} \,, \end{aligned} \tag{44}$$

[f] We admit that running the heavy quark masses m_c, m_b and m_t down to the QCD scale might not make sense.[60] One may only consider the masses of up and down quarks[61] and then propose $\mathrm{CP}_{\text{strong}} \sim m_u m_d \sin\bar{\theta}/\Lambda^2_{\text{QCD}}$ as an alternative measure of strong CP-violation.

where $\Lambda_{\rm EW} \sim 10^2$ GeV, $\Lambda_{\rm QCD} \sim 0.2$ GeV, and the sine function of $\bar{\theta}$ has been adopted to take account of the periodicity in its values. So the effect of weak CP-violation would vanish if the masses of any two quarks in the same (up or down) sector were equal,[g] and the effect of strong CP-violation would vanish if $m_u \to 0$ or $\sin\bar{\theta} \to 0$ held. The remarkable suppression of CP-violation in the SM implies that an interpretation of the observed matter–antimatter asymmetry of the universe[23] requests for a new source of CP-violation beyond the SM, such as leptonic CP-violation in the decays of heavy Majorana neutrinos based on the seesaw and leptogenesis mechanisms.[63]

8. Concluding Remarks

Let us make some concluding remarks with the help of the Fritzsch–Xing "pizza" plot as shown in Fig. 4. It offers a summary of the 28 free parameters associated with the SM itself and neutrino masses, lepton flavor mixing angles and CP-violating phases. Here, our focus is on the five parameters of strong and weak CP-violation. In the quark sector, the strong CP-violating phase $\bar{\theta}$ remains unknown, but the weak CP-violating phase δ_q has been determined to a good degree of accuracy. In the lepton sector, however, none of the CP-violating phases have been measured. While the Dirac CP-violating phase δ_ℓ can be determined in the future long-baseline neutrino oscillation experiments, how to probe or constrain the Majorana CP-violating phases ρ and σ is still an open question.

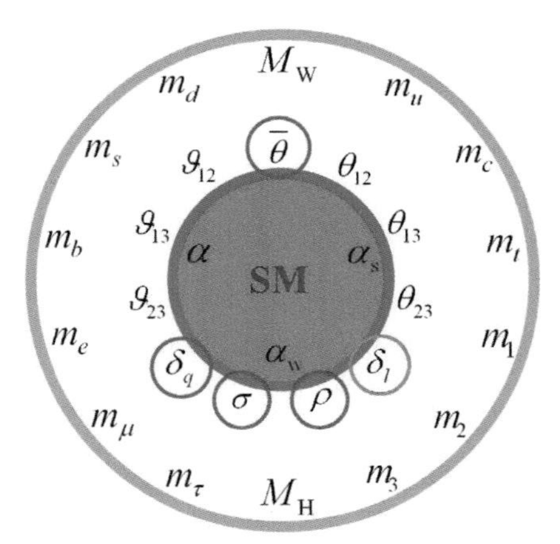

Fig. 4. The Fritzsch–Xing "pizza" plot of the 28 parameters associated with the SM itself and neutrino masses, lepton flavor mixing angles and CP-violating phases.

[g]In this special case one of the three mixing angles of V must vanish, leading to $\mathcal{J}_{\rm q} = 0$ too.[62]

Perhaps some of the flavor puzzles cannot be resolved unless we finally find out the fundamental flavor theory. But the latter cannot be achieved without a lot of phenomenological and experimental attempts. As Leonardo da Vinci emphasized, "Although nature commences with reason and ends in experience, it is necessary for us to do the opposite. That is, to commence with experience and from this to proceed to investigate the reason."

Of course, we have learnt a lot about flavor physics from the quark sector, and are learning much more in the lepton sector. We find that the flavors of that big pizza in Fig. 4 are very appealing to us.

Acknowledgments

I am deeply indebted to Harald Fritzsch for inviting me to write this brief review article and for fruitful collaboration in flavor physics from which I have benefitted a lot. This work is supported in part by the National Natural Science Foundation of China under grant No. 11375207 and the National Key Basic Research Program of China under contract No. 2015CB856700.

References

1. M. Gell-Mann, *Phys. Lett.* **8**, 214 (1964).
2. G. Zweig, CERN-TH-401 and CERN-TH-412 (1964).
3. Z. Z. Xing and S. Zhou, *Neutrinos in Particle Physics, Astronomy and Cosmology* (Zhejiang University Press and Springer-Verlag, 2011).
4. J. J. Thomson, *Phil. Mag.* **44**, 293 (1897).
5. E. Rutherford, *Phil. Mag.* **37**, 149 (1919).
6. J. Chadwick, *Nature* **129**, 312 (1932).
7. C. D. Anderson, *Phys. Rev.* **43**, 491 (1933).
8. S. H. Neddermeyer and C. D. Anderson, *Phys. Rev.* **51**, 884 (1937).
9. G. D. Rochester and C. C. Butler, *Nature* **106**, 885 (1947).
10. C. L. Cowan, F. Reines, F. B. Harrison, H. W. Kruse and A. D. McGuire, *Science* **124**, 103 (1956).
11. G. Danby *et al.*, *Phys. Rev. Lett.* **9**, 36 (1962).
12. J. H. Christenson, J. W. Cronin, V. L. Fitch and R. Turlay, *Phys. Rev. Lett.* **13**, 138 (1964).
13. E598 Collab. (J. J. Aubert *et al.*), *Phys. Rev. Lett.* **33**, 1404 (1974).
14. SLAC-SP-017 Collab. (G. S. Abrams *et al.*), *Phys. Rev. Lett.* **33**, 1406 (1974).
15. M. L. Perl *et al.*, *Phys. Rev. Lett.* **35**, 1489 (1975).
16. S. W. Herb *et al.*, *Phys. Rev. Lett.* **39**, 252 (1977).
17. CDF Collab. (F. Abe *et al.*), *Phys. Rev. Lett.* **74**, 2626 (1995).
18. D0 Collab. (S. Abachi *et al.*), *Phys. Rev. Lett.* **74**, 2422 (1995).
19. DONUT Collab. (K. Kodama *et al.*), *Phys. Rev. Lett.* **504**, 218 (2001).
20. BABAR Collab. (B. Aubert *et al.*), *Phys. Rev. Lett.* **87**, 091801 (2001).

21. Belle Collab. (K. Abe *et al.*), *Phys. Rev. Lett.* **87**, 091802 (2001).
22. J. Gasser and H. Leutwyler, *Phys. Rep.* **87**, 77 (1982).
23. Particle Data Group (J. Beringer *et al.*), *Phys. Rev. D* **86**, 010001 (2012).
24. Z. Z. Xing, H. Zhang and S. Zhou, *Phys. Rev. D* **77**, 113016 (2008); *Phys. Rev. D* **86**, 013013 (2012).
25. N. Cabibbo, *Phys. Rev. Lett.* **10**, 531 (1963).
26. Y. F. Li and Z. Z. Xing, *Phys. Lett. B* **695**, 205 (2011).
27. Z. Maki, M. Nakagawa and S. Sakata, *Prog. Theor. Phys.* **28**, 870 (1962); B. Pontecorvo, *Sov. Phys. JETP* **26**, 984 (1968).
28. M. Kobayashi and T. Maskawa, *Prog. Theor. Phys.* **49**, 652 (1973).
29. Z. Z. Xing, arXiv:1309.2102.
30. H. Fritzsch and Z. Z. Xing, *Phys. Rev. D* **57**, 594 (1998).
31. F. Capozzi, G. L. Fogli, E. Lisi, A. Marrone, D. Montanino and A. Palazzo, *Phys. Rev. D* **89**, 093018 (2014).
32. Z. Z. Xing, *Nuovo Cimento A* **109**, 115 (1996); *J. Phys. G* **23**, 717 (1997).
33. L. Wolfenstein, *Phys. Rev. Lett.* **51**, 1945 (1983).
34. Z. Z. Xing and S. Zhou, *Phys. Lett. B* **737**, 196 (2014).
35. S. Luo and Z. Z. Xing, *Phys. Rev. D* **90**, 073005 (2014); Y. L. Zhou, arXiv:1409.8600.
36. C. Jarlskog, *Phys. Rev. Lett.* **55**, 1039 (1985); *Z. Phys. C* **29**, 491 (1985).
37. H. Fritzsch and Z. Z. Xing, *Prog. Part. Nucl. Phys.* **45**, 1 (2000).
38. H. Fritzsch and Z. Z. Xing, *Phys. Lett. B* **413**, 396 (1997).
39. Z. Z. Xing, *Phys. Lett. B* **679**, 111 (2009).
40. S. Luo and Z. Z. Xing, *J. Phys. G* **37**, 075018 (2010).
41. P. F. Harrison, S. Dallison and W. G. Scott, *Phys. Lett. B* **680**, 328 (2009).
42. S. Luo, *Phys. Rev. D* **85**, 013006 (2012).
43. F. Wilczek, arXiv:1206.7114.
44. H. Fritzsch, *Phys. Lett. B* **184**, 391 (1987); Z. Z. Xing, *J. Phys. G* **23**, 1563 (1997); H. Fritzsch and Z. Z. Xing, *Nucl. Phys. B* **556**, 49 (1999).
45. Z. Z. Xing, *Phys. Rev. D* **86**, 113006 (2012).
46. S. Weinberg, *Trans. New York Acad. Sci.* **38**, 185 (1977); F. Wilczek and A. Zee, *Phys. Lett. B* **70**, 418 (1970); H. Fritzsch, *Phys. Lett. B* **70**, 436 (1977).
47. H. Fritzsch, *Phys. Lett. B* **73**, 317 (1978); *Nucl. Phys. B* **155**, 189 (1979).
48. C. D. Froggatt and H. B. Nielsen, *Nucl. Phys. B* **147**, 277 (1979).
49. P. Ramond, R. G. Roberts and G. G. Ross, *Nucl. Phys. B* **406**, 19 (1993); L. Ibanez and G. G. Ross, *Phys. Lett. B* **332**, 100 (1994); B. R. Desai and A. R. Vaucher, *Phys. Rev. D* **63**, 113001 (2001); H. D. Kim, S. Raby and L. Schradin, *Phys. Rev. D* **69**, 092002 (2004); W. A. Ponce, J. D. Gomez and R. H. Benavides, *Phys. Rev. D* **87**, 053016 (2013).
50. D. Du and Z. Z. Xing, *Phys. Rev. D* **48**, 2349 (1993); P. S. Gill and M. Gupta, *J. Phys. G* **21**, 1 (1995); *Phys. Rev. D* **56**, 3143 (1997); H. Lehmann, C. Newton and T. T. Wu, *Phys. Lett. B* **384**, 249 (1996); Z. Z. Xing, *J. Phys. G* **23**, 1563 (1997); K. Kang and S. K. Kang, *Phys. Rev. D* **56**, 1511 (1997); T. Kobayashi and Z. Z. Xing, *Int. J. Mod. Phys. A* **13**, 2201 (1998); J. L. Chkareuli and C. D. Froggatt, *Phys. Lett. B* **450**, 158 (1999); A. Mondragon and E. Rodriguez-Jauregui, *Phys. Rev. D* **59**, 093009 (1999); H. Nishiura, K. Matsuda and

T. Fukuyama, *Phys. Rev. D* **60**, 013006 (1999); G. C. Branco, D. Emmanuel-Costa and R. G. Felipe, *Phys. Lett. B* **477**, 147 (2000); R. Rosenfeld and J. L. Rosner, *Phys. Lett. B* **516**, 408 (2001).

51. R. Verma, *J. Phys. G* **40**, 125003 (2013).
52. H. Fritzsch and Z. Z. Xing, *Phys. Lett. B* **555**, 63 (2003); Z. Z. Xing and H. Zhang, *J. Phys. G* **30**, 129 (2004); R. Verma, G. Ahuja, N. Mahajan, M. Gupta and M. Randhawa, *J. Phys. G* **37**, 075020 (2010).
53. H. Fritzsch, Z. Z. Xing and Y. L. Zhou, *Phys. Lett. B* **697**, 357 (2011).
54. S. Weinberg, *Phys. Rev. D* **11**, 3583 (1975).
55. G. 't Hooft, *Phys. Rev. Lett.* **37**, 8 (1976); R. Jackiw and C. Rebbi, *Phys. Rev. Lett.* **37**, 172 (1976); C. G. Callan, R. F. Dashen and D. J. Gross, *Phys. Lett. B* **63**, 334 (1976).
56. S. Adler, *Phys. Rev.* **177**, 2426 (1969); J. S. Bell and R. Jackiw, *Nuovo Cimento A* **60**, 47 (1969).
57. J. E. Kim, *Phys. Rep.* **150**, 1 (1987); H. Y. Cheng, *Phys. Rep.* **158**, 1 (1988).
58. C. A. Baker *et al.*, *Phys. Rev. Lett.* **97**, 131801 (2006).
59. R. D. Peccei, arXiv:hep-ph/9807516; P. Sikivie, *Comptes Rendus Physique* **13**, 176 (2012); Y. H. Ahn, arXiv:1410.1634.
60. H. Fusaoka and Y. Koide, *Phys. Rev. D* **57**, 3986 (1998).
61. Z. Huang, *Phys. Rev. D* **48**, 270 (1993).
62. J. W. Mei and Z. Z. Xing, *J. Phys. G* **30**, 1243 (2004).
63. M. Fukugita and T. Yanagida, *Phys. Lett. B* **174**, 45 (1986).

Analytical Determination of the QCD Quark Masses

C. A. Dominguez

Centre for Theoretical and Mathematical Physics and Department of Physics, University of Cape Town, Rondebosch 7700, South Africa

The current status of determinations of the QCD running quark masses is reviewed. Emphasis is on recent progress on analytical precision determinations based on finite energy QCD sum rules. A critical discussion of the merits of this approach over other alternative QCD sum rules is provided. Systematic uncertainties from both the hadronic and the QCD sector have been recently identified and dealt with successfully, thus leading to values of the quark masses with unprecedented accuracy. Results currently rival in precision with lattice QCD determinations.

1. Introduction

Quark and gluon confinement in Quantum Chromodynamics (QCD) precludes direct experimental measurements of the fundamental QCD parameters, i.e. the strong interaction coupling and the quark masses. Hence, in order to determine these parameters analytically, one needs to relate them to experimentally measurable quantities. Alternatively, simulations of QCD on a lattice (LQCD) provide increasingly accurate numerical values for these parameters, but little if any insight into their origin. The first approach relies on the intimate relation between QCD Green functions, in particular, their Operator Product Expansion (OPE) beyond perturbation theory, and their hadronic counterparts. This relation follows from Cauchy's theorem in the complex energy plane, and is known as the finite energy QCD sum rule (FESR) technique.[1] In addition to producing numerical values for the QCD parameters, this method provides a detailed breakdown of the relative impact of the various dynamical contributions. For instance, the strong coupling at the scale of the τ-lepton mass essentially follows from the relation between the experimentally measured τ ratio, R_τ, and a contour integral involving the perturbative QCD (PQCD) expression of the $V + A$ correlator, a classic example of a FESR. This is the cleanest, most transparent, and model-independent determination of the strong coupling.[2,3] It also allows one to gauge the impact of each individual term in PQCD, up to the currently known five-loop order. Similarly, in the case of the quark masses, one considers a QCD correlation function which, on the one hand, involves the quark masses and other QCD parameters and on the other hand, it involves a measurable (hadronic) spectral function. Using Cauchy's theorem to relate both representations, the quark masses become a function of QCD parameters, e.g. the strong coupling, some vacuum condensates reflecting confinement, etc., and measurable hadronic parameters. The virtue of this approach is that it provides a breakdown of each contribution to the final value of the quark masses. More importantly, it allows one to tune the relative weight of each of these contributions by introducing suitable integration kernels. This last feature has been used recently in the case of the charm- and bottom-quark masses leading to very accurate values. It has also been employed to unveil the hadronic systematic uncertainties affecting light-quark mass determinations, to wit. In this case, the ideal Green function is the light-quark pseudoscalar current correlator. This contains the square of the quark masses as an overall factor multiplying the PQCD expansion, and the leading power corrections in the OPE. Unfortunately, this correlator is not realistically accessible experimentally beyond the pseudoscalar meson pole. While the existence of at least two radial excitations of the pion and the kaon are known from hadronic interaction data, this information is hardly enough to reconstruct the full spectral functions. In spite of many attempts over the years to model them, there remained an unknown systematic uncertainty that has plagued light quark mass determinations from QCD sum rules (QCDSR). The use of the vector current correlator, for which there is plenty of experimental data from τ decays and e^+e^- annihilation, is not a realistic option for the light quarks as their masses enter as sub-leading terms in the OPE. The scalar correlator, involving the square of quark mass differences, at some stage offered some promise for determining the strange quark mass with reduced systematic uncertainties. This was due to the availability of data on $K - \pi$ phase shifts. Unfortunately, these data do not fully determine the hadronic spectral function. The latter can be reconstructed from phase shift data only after substantial theoretical manipulations, implying a large unknown systematic uncertainty. A breakthrough has been made recently by introducing an integration kernel in the contour integral in the complex energy plane. This allows one to suppress substantially the unknown hadronic resonance contribution to the pseudoscalar current correlator. As it follows from Cauchy's theorem, this suppression implies that the quark masses are determined essentially from the well-known pseudoscalar meson pole and PQCD (well known up to five-loop level). In this way it has been possible to reduce the hadronic resonance contribution to the 1% level, allowing for an unprecedented accuracy of some 8–10% in the values of the up-, down-, and strange-quark masses. Nevertheless, there still remained a well-known shortcoming in the PQCD sector due to the poor convergence of the pseudoscalar correlator. In fact, the contribution to the quark mass from each perturbative term, up to five-loop level, is essentially identical. While this problem was well known, it remained unresolved for decades. A breakthrough has finally been made recently by using Padè approximants to accelerate efficiently the perturbative convergence. When used for the strange-quark mass, this procedure unveils a systematic uncertainty of some 30% in all previous determinations based on the original perturbative expansion. The strange-quark mass is now known with a 10% error, but essentially free from systematics from the hadronic and the QCD sector. Further improvement on this accuracy will be possible with further reduction of the uncertainty in the strong coupling, now the main source of error.

The determination of the charm- and bottom-quark masses has been free of systematic uncertainties due to the hadronic resonance sector, as there is plenty of experimental information in the vector channel from e^+e^- annihilation into hadrons. One problem, though, is that the massive vector current correlator is not known in PQCD to the same level as the light pseudoscalar correlation function. Nevertheless, substantial theoretical progress has been made over the years leading to extremely accurate charm- and bottom-quark masses. The novel idea of introducing suitable integration kernels in Cauchy's contour integrals, as described above, has also been used recently as a way of improving accuracy in the heavy-quark sector. For instance, kernels can be used to suppress regions where the data is either not as accurate, or simply unavailable. This will also be reported here.

This paper is organized as follows. First, determinations of quark-mass ratios from the various hadronic data, as well as from the chiral perturbation theory (CHPT), will be reviewed in Sec. 2. These ratios are quite useful as consistency checks for results from QCDSR. Section 3 describes the OPE beyond perturbation theory, one of the two pillars of QCDSR. Section 4 discusses quark–hadron duality and FESR, while Sec. 5 provides a critical discussion of Laplace sum rules, as originally proposed and applied to a large number of issues in hadronic physics. With precision determinations being the name of the game at present, these Laplace sum rules cannot compete in accuracy with, e.g. FESR, thus falling out of favor. Several conceptual flaws affecting these Laplace sum rules are pointed out and discussed, paving the way for FESR as the preferred method to determine QCD as well as hadronic parameters. FESR weighted by suitable integration kernels will be analyzed in the light-quark sector in Sec. 6. In particular, it will be shown how this technique unveils the underlying hadronic systematic uncertainty plaguing light-quark mass determinations for the past thirty years. In Sec. 7, recent progress on charm- and bottom-quark mass determinations will be reported. Comparison with LQCD results for all quark masses will also be made. Finally, Sec. 8 provides a very brief summary of this report.

As an important disclaimer, this paper is not a comprehensive review of all past quark mass determinations from QCDSR. It is, rather, a report on very recent progress on the subject. Given that past determinations of light-quark masses were affected by unknown systematic uncertainties, both from the hadronic resonance sector as well as the QCD sector, it makes little sense to review them. Any agreement between values affected by these uncertainties and current results, free of them, would only be fortuitous. For instance, once the hadronic resonance uncertainty is removed, the values of all three light-quark masses get reduced by some 15–20%, with a similar situation in the QCD sector (due to the solution of the well-known problem with the poor convergence of the pseudoscalar correlator). This is a clear sign of a systematic uncertainty acting in only one direction. In addition, light-quark masses from QCDSR before 2006 employed correlators up to at most four-loop level in PQCD, together with superseded values of the strong coupling. Last but not least, quark masses determined from Laplace QCDSR are affected by very large, mostly unacknowledged, systematic uncertainties from the hadronic as well as the QCD sector, as discussed in Sec. 5.

2. Quark Mass Ratios

Quark masses actually precede QCD by a number of years, albeit under the guise of *current algebra quark masses*, which clearly lacked today's detailed understanding of quark-mass renormalization. In fact, the study of global $SU(3) \times SU(3)$ chiral symmetry realized *á la* Nambu–Goldstone, and its breaking down to $SU(2) \times SU(2)$, followed by a breaking down to $SU(2)$, and finally to $U(1)$ was first done using the strong interaction Hamiltonian[4–7]

$$H(x) = H_0(x) + \epsilon_0 u_0(x) + \epsilon_3 u_3(x) + \epsilon_8 u_8(x)\,. \tag{1}$$

The term $H_0(x)$ above is $SU(3) \times SU(3)$ invariant, the $\epsilon_{0,3,8}$ are symmetry breaking parameters, and the scalar densities $u_{0,3,8}(x)$ transform according to the $3\bar{3} + \bar{3}3$ representation of $SU(3) \times SU(3)$. In modern language, ϵ_8 is related to the strange quark mass m_s, and ϵ_3 to the difference between the down- and the up-quark masses $m_d - m_u$, while the scalar densities are related to products of quark–antiquark field operators. For instance, the ratio of SU(3) breaking to SU(2) breaking is given by

$$R \equiv \frac{m_s - m_{ud}}{m_d - m_u} = \frac{\sqrt{3}}{2}\frac{\epsilon_8}{\epsilon_3}, \tag{2}$$

where $m_{ud} \equiv (m_u + m_d)/2$. In the pre-QCD era many relations for quark-mass ratios were obtained from hadron mass ratios, as well as from other hadronic information, e.g. $\eta \to 3\pi$, K_{l3} decay, etc.[7] To mention a pioneering determination of the ratio R above, from a solution to the $\eta \to 3\pi$ puzzle proposed in Ref. 8 it followed[9] $R^{-1} = 0.020 \pm 0.002$, in remarkable agreement with a later determination based on baryon mass splitting[10] $R^{-1} = 0.021 \pm 0.003$, and with the most recent value[11] $R^{-1} = 0.025 \pm 0.003$. With the advent of CHPT,[6,7,11–14] certain quark mass ratios turned out to be renormalization scale independent to leading order, and could be expressed in terms of pseudoscalar meson mass ratios,[7,12,13] e.g.

$$\frac{m_u}{m_d} = \frac{M_{K^+}^2 - M_{K^0}^2 + 2M_{\pi^0}^2 - M_{\pi^+}^2}{M_{K^0}^2 - M_{K^+}^2 + M_{\pi^+}^2} = 0.56\,, \tag{3}$$

$$\frac{m_s}{m_d} = \frac{M_{K^+}^2 + M_{K^0}^2 - M_{\pi^+}^2}{M_{K^0}^2 - M_{K^+}^2 + M_{\pi^+}^2} = 20.2\,, \tag{4}$$

where the numerical results follow after some subtle corrections due to electromagnetic self-energies.[11] Beyond leading order in CHPT, things become complicated. At next-to-leading order (NLO) the only parameter-free relation is

$$Q^2 \equiv \frac{m_s^2 - m_{ud}^2}{m_d^2 - m_u^2} = \frac{M_K^2 - M_\pi^2}{M_{K^0}^2 - M_{K^+}^2}\frac{M_K^2}{M_\pi^2}\,. \tag{5}$$

Other quark mass ratios at NLO and beyond depend on the renormalization scale as well as on some CHPT low energy constants which need to be determined independently.[11,12] After taking into account electromagnetic self-energies, Eq. (5) gives[12] $Q = 24.3$, a recent analysis[12,14] of $\eta \to 3\pi$ gives $Q = 22.3 \pm 0.8$, and the most recent value from the FLAG Collaboration[11] is

$$Q = 22.6 \pm 0.7 \pm 0.6\,. \tag{6}$$

The ratios R, Eq. (2), and Q, Eq. (5), together with the leading-order ratios Eqs. (3) and (4), will prove useful for comparisons with QCD sum rule results.

An additional useful quark mass ratio involving the ratios Eqs. (3) and (4) is

$$r_s \equiv \frac{m_s}{m_{ud}} = \frac{2m_s/m_d}{1 + m_u/m_d} = 28.1 \pm 1.3 \,, \tag{7}$$

where the numerical value follows from the NLO CHPT relation,[12] to be compared with the LO result from Eqs. (3) and (4), $r_s = 25.9$, and a large N_c estimate[15] $r_s = 26.6 \pm 1.6$. The most recent FLAG Collaboration result is[11]

$$r_s = 27.46 \pm 0.15 \pm 0.41 \,. \tag{8}$$

3. Operator Product Expansion Beyond Perturbation Theory

The OPE beyond perturbation theory in QCD, one of the two pillars of the sum rule technique, is an effective tool to introduce quark–gluon confinement dynamics. It is not a model, but rather a parametrization of quark and gluon propagator corrections due to confinement, done in a rigorous renormalizable quantum field theory framework. Let us consider a typical object in QCD in the form of the two-point function, or current correlator

$$\Pi(q^2) = i \int d^4x \, e^{iqx} \langle 0|T(J(x)J(0))|0\rangle \,, \tag{9}$$

where the local current $J(x)$ is built from the quark and gluon fields entering the QCD Lagrangian. Equivalently, this current can also be written in terms of hadronic fields with the same quantum numbers. A relation between the two representations follows from Cauchy's theorem in the complex energy (squared) plane. This is often referred to as quark–hadron duality, the second pillar of the QCDSR method to be discussed in the next section. The QCD correlator, Eq. (9), contains a perturbative piece (PQCD), and a nonperturbative one mostly reflecting quark–gluon confinement. The leading order in PQCD is shown in Fig. 1. Since confinement has not been proven analytically in QCD, its effects can only be introduced effectively, e.g. by

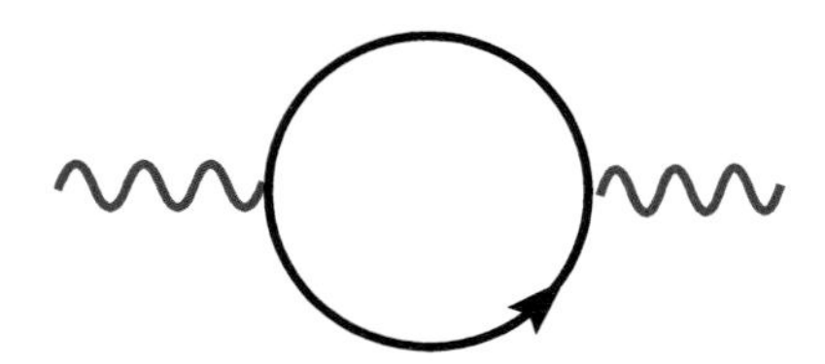

Fig. 1. Leading order PQCD correlator. All values of the four-momentum of the quarks in the loop are allowed. The wiggly line represents the current of momentum q ($-q^2 \gg 0$).

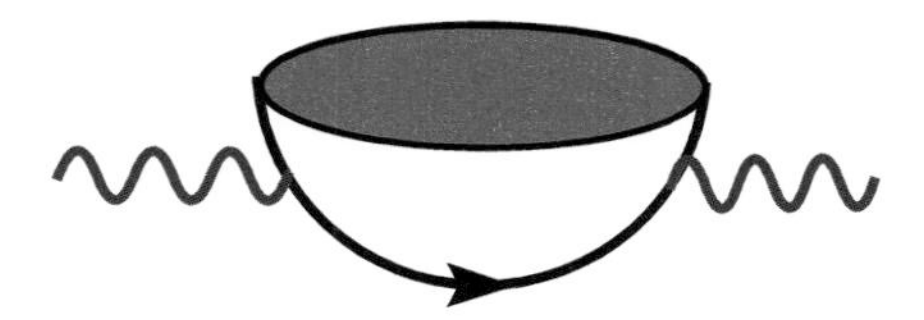

Fig. 2. Quark propagator modification due to (infrared) quarks interacting with the physical QCD vacuum, and involving the quark condensate. Large momentum flows through the bottom propagator.

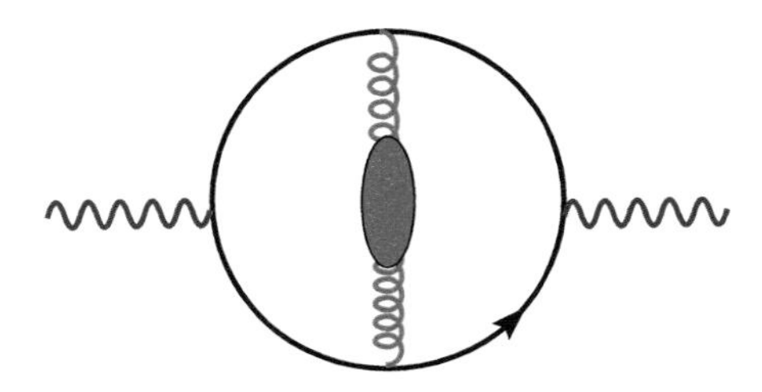

Fig. 3. Gluon propagator modification due to (infrared) gluons interacting with the physical QCD vacuum, and involving the gluon condensate. Large momentum flows through the quark propagators.

parametrizing quark and gluon propagator corrections in terms of vacuum condensates. This is done as follows. In the case of the quark propagator

$$S_F(p) = \frac{i}{\not p - m} \Rightarrow \frac{i}{\not p - m + \Sigma(p^2)}, \tag{10}$$

the propagator correction $\Sigma(p^2)$ contains the information on confinement, a purely nonperturbative effect. One expects this correction to peak at and near the quark mass-shell, e.g. for $p \simeq 0$ in the case of light quarks. Effectively, this can be viewed as in Fig. 2, where the (infrared) quarks in the loop have zero momentum and interact strongly with the physical QCD vacuum. This effect is then parametrized in terms of the quark condensate $\langle 0|\bar{q}(0)q(0)|0\rangle$.

Similarly, in the case of the gluon propagator

$$D_F(k) = \frac{i}{k^2} \Rightarrow \frac{i}{k^2 + \Lambda(k^2)}, \tag{11}$$

the propagator correction $\Lambda(k^2)$ will peak at $k \simeq 0$, and the effect of confinement in this case can be parametrized by the gluon condensate $\langle 0|\alpha_s \vec{G}^{\mu\nu} \cdot \vec{G}_{\mu\nu}|0\rangle$ (see Fig. 3). In addition to the quark and the gluon condensate, there is a plethora of higher-order condensates entering the OPE of the current correlator at short distances, i.e.

$$\Pi(q^2)|_{\mathrm{QCD}} = C_0\hat{I} + \sum_{N=0} C_{2N+2}(q^2,\mu^2)\langle 0|\hat{O}_{2N+2}(\mu^2)|0\rangle\,, \tag{12}$$

where μ^2 is the renormalization scale, and where the Wilson coefficients in this expansion, $C_{2N+2}(q^2,\mu^2)$, depend on the Lorentz indices and quantum numbers of $J(x)$ and of the local gauge invariant operators $\hat{O}_N$ built from the quark and gluon fields. These operators are ordered by increasing dimensionality and the Wilson coefficients, calculable in PQCD, fall off by corresponding powers of $-q^2$. In other words, this OPE achieves a factorization of short distance effects encapsulated in the Wilson coefficients, and long distance dynamics present in the vacuum condensates. Since there are no gauge invariant operators of dimension $d = 2$ involving the quark and gluon fields in QCD, it is normally assumed that the OPE starts at dimension $d = 4$. This is supported by results from QCD sum rule analyses of τ-lepton decay data, which show no evidence of $d = 2$ operators.[16,17] The unit operator $\hat{I}$ in Eq. (12) has dimension $d = 0$ and $C_0\hat{I}$ stands for the purely perturbative contribution. The Wilson coefficients as well as the vacuum condensates depend on the renormalization scale. For light quarks, and for the leading $d = 4$ terms in Eq. (12), the μ^2 dependence of the quark mass cancels the corresponding dependence of the quark condensate, so that this contribution is a renormalization group (RG) invariant. Similarly, the gluon condensate is also a RG invariant, hence once determined in some channel these condensates can be used throughout. The numerical values of the vacuum condensates cannot be calculated analytically from first principles as this would be tantamount to solving QCD exactly. One exception is that of the quark condensate which enters in the Gell-Mann–Oakes–Renner relation,[5] a QCD low energy theorem following from the global chiral symmetry of the QCD Lagrangian.[18] Otherwise, it is possible to extract values for the leading vacuum condensates using QCDSR together with experimental data, e.g. e^+e^- annihilation into hadrons, and hadronic decays of the τ-lepton. Alternatively, as LQCD improves in accuracy it should become a valuable source of information on these condensates.

4. Quark–Hadron Duality and Finite Energy QCD Sum Rules

Turning to the hadronic sector, bound states and resonances appear in the complex energy (squared) plane (s-plane) as poles on the real axis, and singularities in the second Riemann sheet, respectively. All these singularities lead to a discontinuity across the positive real axis. Choosing an integration

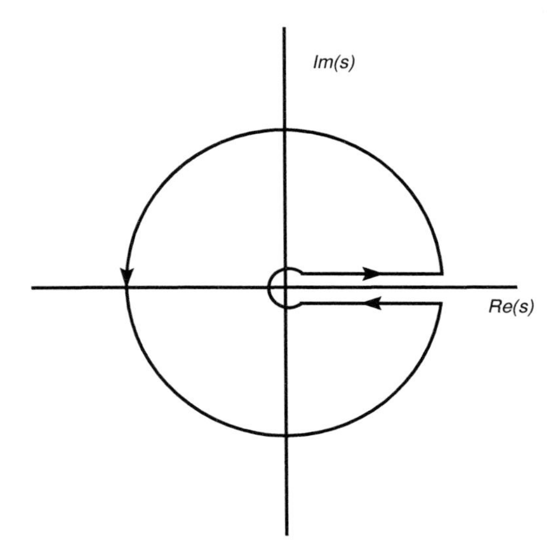

Fig. 4. Integration contour in the complex s-plane. The discontinuity across the real axis brings in the hadronic spectral function, while integration around the circle involves the QCD correlator. The radius of the circle is s_0, the onset of QCD.

contour as shown in Fig. 4, and given that there are no other singularities in the complex s-plane, Cauchy's theorem leads to the FESR

$$\int_{s_{\rm th}}^{s_0} ds \frac{1}{\pi} f(s) \, {\rm Im}\, \Pi(s) \Big|_{\rm HAD} = -\frac{1}{2\pi i} \oint_{C(|s_0|)} ds f(s) \Pi(s) \Big|_{\rm QCD} + {\rm Res}[\Pi(s) f(s), s = 0] \,, \tag{13}$$

where $f(s)$ is an arbitrary function, s_{th} is the hadronic threshold, and the finite radius of the circle, s_0, is large enough for QCD and the OPE to be used on the circle. Depending on the particular form of the integration kernel, $f(s)$, the last term above may or may not be present. Physical observables determined from FESR should be independent of s_0. In practice, though, this is not exact, and there is usually a region of stability where observables are fairly independent of s_0, typically somewhere inside the range $s_0 \simeq 1$–4 GeV^2. Since $f(s)$ is often a polynomial, the existence of a wide stability region is a highly nontrivial feature. Equation (13) is the mathematical statement of what is usually referred to as quark–hadron duality. Since QCD is not valid in the time-like region ($s \geq 0$), in principle there is a possibility of problems on the circle near the real axis (duality violations). This issue was identified very early in Ref. 19 long before the present formulation of QCDSR, and is currently referred to as (quark–hadron) duality violations. First attempts at identifying and quantifying this problem were made in Ref. 20 using data on hadronic decays of the tau-lepton, the pseudoscalar (pionic) channel and the strangeness changing scalar channel, and in Ref. 21 by considering chiral sum rules. It appears that the size of these duality violations might be channel dependent, as an analysis of tau-decay data extended beyond the kinematical end point finds no effect.[22] For recent work on this problem, see e.g. Refs. 23 and 24, and references therein.

The right-hand side of this FESR involves the QCD correlator, which is expressed in terms of the OPE as in Eq. (12). The left-hand side involves the hadronic spectral function which, in principle, is written as

$$\mathrm{Im}\,\Pi(s)|_{\mathrm{HAD}} = \mathrm{Im}\,\Pi(s)|_{\mathrm{POLE}} + \mathrm{Im}\,\Pi(s)|_{\mathrm{RES}}\theta(s_0 - s) + \mathrm{Im}\,\Pi(s)|_{\mathrm{PQCD}}\theta(s - s_0)\,, \tag{14}$$

where the ground state pole (if present) is followed by the resonances which merge smoothly into the hadronic continuum above some threshold s_0. This continuum is expected to be well represented by PQCD if s_0 is large enough. Hence, if one were to consider an integration contour in Eq. (13) extending to infinity, the cancellation between the hadronic continuum on the left-hand side and the PQCD contribution on the right-hand side, would render the sum rule a FESR. In practice, though, there is a finite value s_0 beyond which this cancellation takes place, and s_0 is identified with the onset of PQCD. In this case, the last term in Eq. (14) is obviously redundant. The integration in the complex s-plane of the QCD correlator is usually carried out in two different ways, Fixed Order Perturbation Theory (FOPT) and Contour Improved Perturbation Theory (CIPT). The first method treats running quark masses and the strong coupling as fixed at a given value of s_0. After integrating all logarithmic terms ($\ln(-s/\mu^2)$) the RG improvement is achieved by setting the renormalization scale to $\mu^2 = -s_0$. In CIPT, the RG improvement is performed before integration, thus eliminating logarithmic terms, and the running quark masses and strong coupling are integrated around the circle. This requires solving numerically the RGE for the quark masses and the coupling at each point on the circle. The FESR Eq. (13) with $f(s) = 1$ and in FOPT can be written as

$$(-)^N C_{2N+2}\langle 0|\hat{O}_{2N+2}|0\rangle = \int_0^{s_0} ds\, s^N \frac{1}{\pi}\,\mathrm{Im}\,\Pi(s)\Big|_{\mathrm{HAD}} - s_0^{N+1} M_{2N+2}(s_0), \tag{15}$$

where the dimensionless PQCD moments $M_{2N+2}(s_0)$ are given by

$$M_{2N+2}(s_0) = \frac{1}{s_0^{(N+1)}} \int_0^{s_0} ds\, s^N \frac{1}{\pi}\,\mathrm{Im}\,\Pi(s)\Big|_{\mathrm{PQCD}}\,. \tag{16}$$

If the hadronic spectral function is known in some channel from experiment, e.g. from τ-decay into hadrons, then $\mathrm{Im}\,\Pi(s)|_{\mathrm{HAD}} \equiv \mathrm{Im}\,\Pi(s)|_{\mathrm{DATA}}$, and Eq. (15) can be used to determine the values of the vacuum condensates. Subsequently, Eq. (15) can be used in a different channel for a different application. It is important to mention that the correlator $\Pi(q^2)$ is generally not a physical observable. However, this has no effect in FOPT as the unphysical quantities (polynomials) in the correlator do not contribute to the integrals. In the case of CIPT, though, this requires modified sum rules involving as many derivatives of the correlator as necessary to render it physical.

5. Laplace Transform QCD Sum Rules

The original QCD sum rule method proposed in Ref. 25, had as a starting point the well-known dispersion relation, or Hilbert transform, which follows from Cauchy's theorem in the complex squared energy s-plane

$$\frac{1}{N!}\left(-\frac{d}{dQ^2}\right)^N \Pi(Q^2)\bigg|_{Q^2=Q_0^2} = \frac{1}{\pi}\int_0^\infty \frac{\mathrm{Im}\,\Pi(s)}{(s+Q_0^2)^{N+1}}ds\,, \tag{17}$$

where N is the number of derivatives required for the integral to converge asymptotically, and $Q^2 \equiv -q^2 > 0$. As it stands, the dispersion relation, Eq. (17), is a tautology. Next, a specific asymptotic limit process in the parameters N and Q^2 was performed, i.e. $\lim Q^2 \to \infty$ and $\lim N \to \infty$, with $Q^2/N \equiv M^2$ fixed, leading to Laplace transform QCDSR

$$\begin{aligned}\hat{L}_M[\Pi(Q^2)] &\equiv \lim_{\substack{Q^2,N\to\infty \\ Q^2/N\equiv M^2}} \frac{(-)^N}{(N-1)!}(Q^2)^N\left(\frac{d}{dQ^2}\right)^N \Pi(Q^2) \equiv \Pi(M^2) \\ &= \frac{1}{M^2}\int_0^\infty \frac{1}{\pi}\,\mathrm{Im}\,\Pi(s)e^{-s/M^2}ds\,. \end{aligned}\tag{18}$$

This equation is still a tautology. In order to turn it into something with useful content, one needs to invoke Eq. (14). This procedure makes no explicit use of the concept of quark–hadron duality, thus not relying on Cauchy's theorem in the complex s-plane, other than initially at the level of Eq. (17). In applications of these sum rules, $\Pi(M^2)$ was computed in QCD by applying the Laplace operator $\hat{L}_M$ to the OPE expression of $\Pi(Q^2)$, and the spectral function on the right-hand side was parametrized as in Eq. (14). The function $\Pi(M^2)$ in PQCD involves the transcendental function $\mu(t,\beta,\alpha)$,[26] as first discussed in Ref. 27. This novel method had an enormous impact, as witnessed by the several thousand publications to date on analytic solutions to QCD in the nonperturbative domain.[1] However, in the past decade, and as the subject moved towards high precision determinations to compete with LQCD, this particular sum rule has fallen out of favor for a variety of reasons as detailed next.

The first thing to notice in Eq. (18) is the introduction of an *ad hoc* new parameter, M^2, the Laplace variable, which determines the squared energy regions where the exponential kernel would have a minor/major impact. It has regularly been advertised in the literature that a judicious choice of M^2 would lead to an exponential suppression of the often experimentally unknown resonance region beyond the ground state, as well as to a factorial suppression of higher-order condensates in the OPE. In practice, though, this was hardly factually achieved, thus becoming an oracular statement. Indeed, since the parameter M^2 has no physical significance, other than being a mathematical artifact, results from these QCDSR would have to be independent of M^2 in a hopefully broad region. This so-called stability window is often unacceptably narrow, and the expected exponential suppression of the unknown resonance region does not materialize. Furthermore, the factorial suppression of higher-order condensates only starts at dimension $d = 6$ with a mild suppression by a factor $1/\Gamma(3) = 1/2$. But beyond $d = 6$ little, if anything, is numerically known about the vacuum condensates to profit from this feature. Another serious shortcoming of these QCDSR is that the role of the threshold for PQCD in the complex s-plane, s_0, i.e. the radius of the circular contour in Fig. 4, is often exponentially suppressed, or at best reduced in importance compared with its role in FESR. This is unfortunate, as s_0 is a parameter which, unlike M^2, has a clear physical interpretation, and which can be easily determined from data in some instances. On a separate issue, Laplace sum rules, unlike FESR, do not facilitate the insertion of nontrivial integration kernels. These kernels have been proved essential in the followings:

(a) Modern determinations of the light-quark masses, to quench significantly the experimentally unknown resonance region beyond the pion and kaon poles, and thus reducing considerably systematic uncertainties from this sector (see Sec. 6). The Laplace exponential kernel is unsuited for this purpose, thus making it close to impossible to eliminate this systematic uncertainty. Hence, quark-mass determinations in this framework are all affected by some 20–30% error, a fact hardly acknowledged in the literature.

(b) In tuning the contribution of data in the charm- and bottom-quark regions, thus allowing for very high precision determinations of these quark masses (see Sec. 7).

(c) Allowing for a purely theoretical determination of the charm- and bottom-quark region contributions to the hadronic part of the $g-2$ of the muon,[28] with an excellent agreement with later LQCD results.[29]
(d) Similar to (c) but in relation to the hadronic contribution to the QED running coupling at the scale of the Z boson.[30]

Last, but not least, Laplace sum rule results are often too dependent on the renormalization scale μ^2. In fact, in some applications results are linearly dependent on μ^2, with no plateau in sight, i.e. straight lines with large slopes.

Many of these shortcomings of the Laplace QCDSR can be traced back to the way Cauchy's theorem in the complex s-plane is being invoked. This is done trivially, and only initially at the level of the dispersion relation, Eq. (17). In contrast, FESR are derived directly from Cauchy's theorem, with the upper limit of the integration range, i.e. the radius of the contour s_0, being finite on account of the quark–hadron duality assumption. If one were to invoke this assumption in the Laplace sum rule, Eq. (18), it would lead to a serious mismatch between the PQCD contribution to right-hand side and its contribution to the left-hand side. In fact, in the integration range $s_0 - \infty$ the integral of the PQCD imaginary part has no counterpart in $\Pi(M^2)$ entering the left-hand side. The latter involves the transcendental functions $\mu(t, \beta, \alpha)$, while the former does not. For this reason a power series expansion of the exponential in the Laplace sum rules cannot strictly lead to FESR. Clearly, it is still possible to choose the integration kernel $f(s)$ in Eq. (13) to be a negative exponential, thus leading to a different version of Laplace sum rules. However, this would be very different from Eq. (18), plus it would still lead to the rest of the shortcomings mentioned above.

6. Light-Quark Masses

Traditionally, the light-quark masses have been determined using the correlator, Eq. (9), involving the pseudoscalar currents $J(x) \equiv \partial_\mu A^\mu(x)|^i_j = [\bar{m}_i(\mu) + \bar{m}_j(\mu)] :\bar{q}_j(x) i\gamma_5 q_i(x):$, where $A_\mu(x)$ is the axial vector current of flavors i and j, $\bar{m}_i(\mu)$ the quark mass in the $\overline{\text{MS}}$ scheme, μ the renormalization scale and $q_i(x)$ are the quark fields. An issue of major concern in the past was the presence of logarithmic quark-mass singularities in these correlators. This problem has been satisfactorily resolved some time ago in Refs. 31 and 32. These correlators are now known to five-loop order in PQCD,[33] and free of logarithmic quark mass singularities. The Wilson coefficients of the leading power corrections, i.e. the gluon and the quark condensates, are also known up to two-loop level.[34] Higher-dimensional condensates, as well as quark mass corrections of order $\mathcal{O}(m_i^4)$ (with respect to the one-loop term) and higher turn out to be negligible. From Cauchy's theorem, Eq. (13), the FESR to determine the quark masses can be written as

$$\delta_5^{\rm QCD}(s_0) \equiv -\frac{1}{2\pi i}\oint_{C(|s_0|)} ds\, \psi_5^{\rm QCD}(s)\Delta_5(s) = \delta_5^{\rm HAD}$$
$$\equiv 2f_P^2 M_P^4 \Delta_5(M_P^2) + \int_{s_{\rm th}}^{s_0} ds\, \frac{1}{\pi}\, {\rm Im}\, \psi_5(s)\bigg|_{\rm RES} \Delta_5(s)\,, \qquad (19)$$

where $\Delta_5(s)$ is an (analytic) integration kernel to be introduced shortly, the first term on the right-hand side is the pseudoscalar meson pole contribution ($P = \pi, K$), $s_{\rm th}$ is the hadronic threshold, and ${\rm Im}\,\psi_5(s)|_{\rm RES}$ is the hadronic resonance spectral function. The radius of integration s_0 is assumed to be large enough for QCD to be valid on the circle. For later convenience this FESR can be rewritten as

$$\delta_5(s_0)|_{\rm QCD} = \delta_5|_{\rm POLE} + \delta_5(s_0)|_{\rm RES}\,, \qquad (20)$$

where the meaning of each term is self-evident. Historically, the problem with the pseudoscalar correlator has been the lack of direct experimental information on the hadronic resonance spectral functions. Two radial excitations of the pion and of the kaon, with known masses and widths, have been observed in hadronic interactions.[35] However, this information is hardly enough to reconstruct the full spectral function. In fact, inelasticity, nonresonant background and resonance interference are impossible to guess, leaving no choice but to model these functions. This introduces an unknown systematic uncertainty which has been present in all past QCD sum rule determinations of the light-quark masses. Since the FESR Eq. (19) is valid for any analytic $\Delta_5(s)$ one can choose this kernel in such a way as to suppress $\delta_5(s_0)|_{\rm RES}$ as much as possible. An example of such a function is the second degree polynomial[36–39]

$$\Delta_5(s)|_{\rm RES} = 1 - a_0 s - a_1 s^2\,, \qquad (21)$$

where a_0 and a_1 are constants fixed by the requirement $\Delta_5(M_1^2) = \Delta_5(M_2^2) = 0$, where $M_{1,2}$ are the masses of the first two radial excitations of the pion or kaon. This simple kernel suppresses enormously the resonance contribution, which becomes only a couple of a percent of the pole contribution, and well below the current uncertainty due to the strong coupling. This welcome feature is essentially independent of the model chosen to parametrize the resonances. A practical parametrization consists of two Breit–Wigner forms normalized at threshold according to chiral perturbation theory, as first proposed in Ref. 40 for the pionic channel, and in Ref. 41 for the kaonic channel (for an alternative parametrization see Ref. 42 and references therein). Detailed results for $\delta_5(s_0)|_{\rm QCD}$, to five-loop order in PQCD and up to dimension $d = 4$ in the OPE, after integrating in FOPT may be found in Ref. 37. In the case of CIPT, the FESR must be written in terms of the second derivative of the current correlator. This is in order to eliminate the unphysical first degree polynomial present in $\psi_5(s)$, which unlike the case of FOPT would otherwise contribute to the FESR which then becomes

$$-\frac{1}{2\pi i}\oint_{C(|s_0|)} ds\, \psi_5''^{\rm QCD}(s)[F(s) - F(s_0)]$$
$$= 2f_P^2 M_P^4 \Delta_5(M_P^2) + \frac{1}{\pi}\int_{s_{th}}^{s_0} ds\, {\rm Im}\,\psi_5(s)\Big|_{\rm RES} \Delta_5(s)\,, \tag{22}$$

where

$$F(s) = -s\left(s_0 - a_0\frac{s_0^2}{2} - a_1\frac{s_0^3}{3}\right) + \frac{s^2}{2} - a_0\frac{s^3}{6} - a_1\frac{s^4}{12}\,, \tag{23}$$

and

$$F(s_0) = -\frac{s_0^2}{2} + a_0\frac{s_0^3}{3} + a_1\frac{s_0^4}{4}\,. \tag{24}$$

The RG improvement is used before integration, so that all logarithmic terms vanish. The running coupling as well as the running quark masses are no longer frozen as in FOPT, but must be integrated. This can be done by solving numerically the respective RG equations at each point on the integration circle in the complex s-plane. Detailed expressions are given in Refs. 37 and 38.

The parameters of the integration kernel, Eq. (21), are $a_0 = 0.897$ GeV^{-2} and $a_1 = -0.1806$ GeV^{-4} for the pionic channel, and $a_0 = 0.768$ GeV^{-2} and $a_1 = -0.140$ GeV^{-4} for the kaonic channel. These values correspond to the radial excitations $\pi(1300)$, $\pi(1800)$, $K(1460)$ and $K(1830)$. The pion and kaon decay constants are[35] $f_\pi = 92.21 \pm 0.14$ MeV, and $f_K = (1.22 \pm 0.01) f_\pi$. In the QCD sector it is best to use the value of the strong coupling determined at the scale of the τ-mass, as this is close to the scale in current use for the light-quark masses, i.e. $\mu = 2$ GeV. The extraction of $\alpha_s(M_\tau)$ from the R_τ ratio involves an integral with a natural kinematical integration kernel that eliminates the contribution of the $d = 4$ term in the OPE. This welcome feature improves the accuracy of the determination, and it makes little sense to introduce additional spurious integration kernels which would artificially recover this $d = 4$ contribution. The different values obtained from τ decay using CIPT are all in agreement with each other, i.e. $\alpha_s(M_\tau) = 0.338 \pm 0.012$,[3] $\alpha_s(M_\tau) = 0.341 \pm 0.008$,[44] $\alpha_s(M_\tau) = 0.344 \pm 0.009$,[45] $\alpha_s(M_\tau) = 0.332 \pm 0.016$,[46] and the most recent result $\alpha_s(M_\tau) = 0.331 \pm 0.013$.[47] These determinations are model independent and extremely transparent, with α_s obtained essentially by confronting PQCD with the single experimental number R_τ. The $d = 4$ gluon condensate has been extracted from τ decays,[17] but one can conservatively consider the wide range $\langle \alpha_s G^2 \rangle = 0.01$–$0.12$ GeV^4. The impact of the light-quark condensate is at the level of 1% in the quark masses. A $\pm 30\%$ uncertainty in the resonance contribution $\delta_5(s_0)|_{\text{RES}}$ in Eq. (20) translates into a safe 1% change in the quark masses. Finally, it has been assumed that the unknown six-loop PQCD contribution is equal to the five-loop result, a supposedly conservative estimate of higher orders in PQCD. Nevertheless, there is a convergence problem with the PQCD expansion to be discussed later.

Beginning with the strange quark mass, Fig. 5 shows the results for $\bar{m}_s$ (2 GeV)$|_{\overline{\text{MS}}}$ with no integration kernel, $\Delta_5(s) = 1$, and taking into account only the kaon pole, curve (a), and the kaon pole plus a two Breit–Wigner resonance model with a threshold constraint from CHPT,[41] curve (b) (a misprint in the formula for the spectral function in Ref. 41 has been corrected

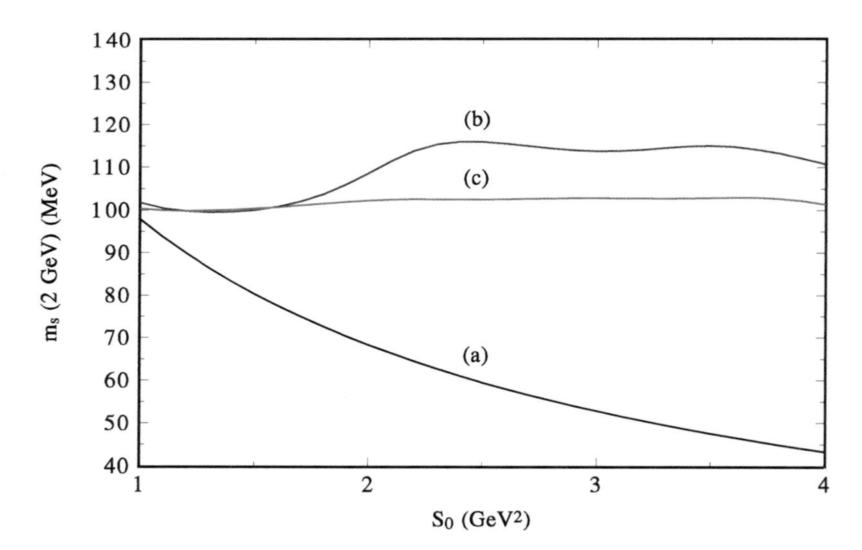

Fig. 5. The strange quark mass $\bar{m}_s$(2 GeV) in the $\overline{\text{MS}}$ scheme taking into account only the kaon pole with $\Delta_5(s) = 1$ (curve (a)), and the two Breit–Wigner resonance spectral function with a threshold constraint from CHPT,[41] with $\Delta_5(s) = 1$ (curve (b)), and $\Delta_5(s)$ as in Eq. (21) (curve (c)). A systematic uncertainty of some 20% due to the resonance sector is dramatically unveiled.

in Ref. 43). These curves are for the central value of $\alpha_s(M_\tau)$ whose uncertainties will be considered afterwards. The latter result is reasonably stable in the wide region $s_0 = 2$–4 GeV^2, so that it could lead us to conclude that $\bar{m}_s$ (2 GeV)$|_{\overline{\rm MS}} \simeq 100$–$120$ MeV, albeit with a yet unknown systematic uncertainty arising from the resonance sector. Introducing the kernel, Eq. (21), leads to curve (c) and to a dramatic unveiling of this systematic uncertainty. In fact, the *real* value of the quark mass is $\bar{m}_s$ (2 GeV)$|_{\overline{\rm MS}} = 102 \pm 8$ MeV, or some 20% below the former result (this error now includes the uncertainty in α_s). In addition, and as a bonus, the systematic uncertainty-free result is remarkably stable in the unusually wide region $s_0 \simeq 1$–4 GeV^2 (typical stability regions are only half as wide).

It must be recalled that the pseudoscalar correlator involves the overall factor $(m_s + m_{ud})^2$. Hence, in order to determine m_s an input value for the ratio m_s/m_{ud} is needed in the result from the sum rule, which is

$$\bar{m}_s(2\ \text{GeV})|_{\overline{\rm MS}} = \frac{105.5 \pm 8.2\ \text{MeV}}{1 + m_{ud}/m_s}\,. \tag{25}$$

Using the wide range $m_s/m_{ud} = 24 - 29$ leads to $\bar{m}_s$ (2 GeV)$|_{\overline{\rm MS}} = 102 \pm 8$ MeV. In this case, the impact of the uncertainty in the quark mass ratio is small. However, in the case of the up- and down-quark masses, the corresponding ratio m_u/m_d plays a more important role in the result from the sum rule, which is

$$\bar{m}_d\ (2\ \text{GeV})|_{\overline{\rm MS}} = \frac{8.2 \pm 0.6\ \text{MeV}}{1 + m_u/m_d}\,. \tag{26}$$

The input used in Ref. 38 for the mass ratio in Eq. (26) is $m_u/m_d = 0.553$ from CHPT.[48] Once m_d is determined from Eq. (26), m_u follows. Using these results for the individual masses, one obtains the ratios m_u/m_d and m_s/m_{ud} shown in Table 1. Using instead the ratio $m_u/m_d = 0.50 \pm 0.04$ from Ref. 11 gives $m_u = 2.7 \pm 0.3$ MeV, and $m_d = 5.5 \pm 0.4$ MeV. These quark masses $\bar{m}_u$ (2 GeV)$|_{\overline{\rm MS}}$ and $\bar{m}_d$ (2 GeV)$|_{\overline{\rm MS}}$ also exhibit a remarkably wide stability region $s_0 \simeq 1$–4 GeV^2.[38,39]

I now return to the problem with the PQCD convergence of the pseudoscalar correlator, and its impact on the strange-quark mass[49] (the impact on the up- and down-quark masses, leading to improved results, is currently under investigation[50]). For the integration kernel, Eq. (21), and $s_0 = 4.2$ GeV^2, the FOPT result for $\delta_5^{\rm QCD}(s_0)$ in PQCD is given by

$$\delta_5^{\rm PQCD} = 0.23\ \text{GeV}^8[1 + 2.2\alpha_s + 6.7\alpha_s^2 + 19.5\alpha_s^3 + 56.5\alpha_s^4]\,, \tag{27}$$

Table 1. The running quark masses in the $\overline{\text{MS}}$ scheme at a scale $\mu = 2$ GeV in units of MeV from QCDSR (first three rows), and from the FLAG lattice QCD analysis.[11] The ratios $\bar{m}_u/\bar{m}_d$ and $\bar{m}_s/\bar{m}_{ud}$ are an input in the QCDSR (see text). The ratios R and Q are defined on the left-hand sides of Eqs. (2) and (3).

Source	$\bar{m}_u$	$\bar{m}_d$	$\bar{m}_s$	$\bar{m}_{ud}$	$\bar{m}_u/\bar{m}_d$	$\bar{m}_s/\bar{m}_{ud}$	R	Q
QCDSR[38]	2.9 ± 0.2	5.3 ± 0.4	—	4.1 ± 0.2	0.553 (input)	—	—	—
QCDSR[37]	—	—	102 ± 8	—	—	24.9 ± 2.7	33 ± 6	21 ± 3
QCDSR[49]	—	—	94 ± 9	—	—	27 ± 1 (input)	30 ± 6	19 ± 3
FLAG[11]	2.40 ± 0.23	4.80 ± 0.23	101 ± 3	3.6 ± 0.2	0.50 ± 0.04	28.1 ± 1.2	40.7 ± 4.3	24.3 ± 1.5

which after replacing a typical value of α_s leads to all terms beyond the leading order to be roughly the same, e.g. for $\alpha_s = 0.3$ the result is

$$\delta_5^{\text{PQCD}} = 0.23 \text{ GeV}^8[1 + 0.65 + 0.60 + 0.53 + 0.46] , \tag{28}$$

which is hardly (if at all) convergent. In fact, judging from the first five terms, this expansion is worse behaved than the nonconvergent harmonic series. An integration kernel[49] shown to be better suited than Eq. (21) is

$$\Delta_5(s) = (s - a)(s - s_0) , \tag{29}$$

with $a = 2.8$ GeV^2. In this case, the perturbative expansion for the strange-quark mass, with $m_s/m_{ud} = 27 \pm 1$, becomes

$$\bar{m}_s \text{ (2 GeV)} = 248.3 \text{ MeV } \left(1 + 2.59\alpha_s + 8.60\alpha_s^2 + 26.50\alpha_s^3 + 75.47\alpha_s^4\right)^{-1/2} , \tag{30}$$

with all terms being roughly of the same size for α_s (2 GeV) $\simeq 0.3$. This result implies an obvious systematic uncertainty in the QCD sector, which was not exposed and dealt with in Ref. 37. The ideal tool to deal with this problem is that of Padè approximants[51]

$$f(z) \approx [m/n] \equiv \frac{a_0 + a_1 z + \cdots + a_m z^m}{1 + b_1 z + \cdots + b_n z^n} , \quad m + n = k , \tag{31}$$

with $[m/0]$ being the standard Taylor series expansion of $f(z)$. This particularly simple Padè approximant already accelerates the PQCD convergence

$$\bar{m}_s \text{ (2 GeV)} = 248.3 \text{ MeV } \left(1 - 1.30\alpha_s + 1.80\alpha_s^2 - 1.95\alpha_s^3 - 0.34\alpha_s^4\right)^{-1/2} , \tag{32}$$

leading to

$$\bar{m}_s \text{ (2 GeV)}|_{\overline{\text{MS}}} = \frac{97.48 \pm 10.6 \text{ MeV}}{1 + m_{ud}/m_s} = 94 \pm 9 \text{ MeV} . \tag{33}$$

In contrast, the original expression, Eq. (30) would give $\bar{m}_s$ (2 GeV) $\simeq$ 125 MeV. A systematic uncertainty in the QCD sector of roughly 30% has thus been exposed and eliminated. For a more detailed discussion of this procedure see Ref. 52.

In Table 1 one finds a summary of the results for the light-quark masses, and the ratios R and Q defined on the left-hand sides of Eqs. (2) and (5), together with the results of the Flag group.[11] The values of the up- and down-quark masses and their ratios in Table 1 are slightly different from those in Ref. 39 due to the input value for the ratio m_u/m_d. Using the FLAG ratio[11] instead, gives similar values within errors as mentioned earlier after Eq. (26). In either case, there seems to be some tension between these results and those from Ref. 11. Perhaps once the QCD systematic uncertainty is dealt with in this channel the tension might be resolved.

The various sources of errors in the quark masses discussed earlier combine into the final values given in Table 1. Having all but eliminated the systematic uncertainty from the hadronic resonance sector, the main source of error is now due to the strong coupling, and the PQCD sector for the up- and down-quark masses, i.e. the poor convergence of the perturbative series. Improved accuracy in the determination of α_s would then allow for a reduction of the uncertainties in the light-quark mass sector.

7. Heavy-Quark Masses

Determinations of the charm- and bottom-quark masses are not affected by a lack of data, as there is plenty of experimental information from e^+e^- annihilation into hadrons at high energies,[35] except for a gap in the region 25 GeV$^2 \lesssim s \lesssim 50$ GeV^2. On the theoretical side, there has been very good progress on PQCD up to four-loop level.[53–72] The leading power correction in the OPE is due to the gluon condensate with its Wilson coefficient known at the two-loop level.[70]

The correlator, Eq. (9), involves the vector current $J(x) \equiv V_\mu(x) = \bar{Q}(x)\gamma_\mu Q(x)$, where $Q(x)$ is the charm- or bottom-quark field. The experimental data is in the form of the R_Q-ratio for charm (bottom) production, which determines the hadronic spectral function. Modern determinations of the heavy-quark masses have been based on inverse moment (Hilbert-type) QCDSR, e.g. Eq. (13) with $f(s) = 1/s^n$, in which case Eq. (13) requires the additional term on the right-hand side, i.e. the residue at the pole: $\mathrm{Res}[\Pi(s)f(s), s = 0]$. These sum rules require QCD knowledge of the vector correlator in the low energy region, around the open charm (bottom) threshold, as well as in the high energy region. A recent update[71] of earlier determinations[53–55,57–59] reports a charm-quark mass in the $\overline{\text{MS}}$ scheme accurate to 1%, and half this uncertainty for the bottom-quark mass. However, the analysis of Ref. 72 claims an error a factor two larger for the charm-quark mass. It appears that the discrepancy arises from the treatment of PQCD. In fact, in Ref. 72 two different renormalization scales were used, one for the strong coupling and another one for the quark mass. This unconventional choice results in an artificially larger error in the charm-quark mass obtained from inverse (Hilbert) moment QCDSR. It does not affect, though, sum rules involving positive powers of s. In any case, the philosophy in current use is to choose the result from the method leading to the smallest uncertainty.

Beginning with the charm-quark mass, an alternative procedure was proposed some years ago based only on the high energy expansion of the heavy-quark vector correlator.[73,74] This method was followed recently,[75] but with updated PQCD information and the inclusion of integration kernels in the FESR, Eq. (13), tuned to enhance/suppress contributions from data in certain regions. One such kernel is the so-called *pinched* kernel[20,21]

$$f(s) = 1 - \frac{s}{s_0}\,, \tag{34}$$

which is supposed to suppress potential duality violations close to the real $s-$axis in the complex s-plane. In connection with the charm-quark mass application, this kernel enhances the contribution from the first two narrow resonances, J/ψ and $\psi(2S)$, and reduces the weight of the broad resonance region, particularly near the onset of the continuum. The latter feature is better achieved with the alternative kernel[76]

$$f(s) = 1 - \left(\frac{s_0}{s}\right)^2, \tag{35}$$

which produces an obvious larger enhancement of the narrow resonances, and a higher quenching of the broad resonance region. This kernel has been used together with both the high and the low energy expansion of the vector correlator in Ref. 76. The results for $\bar{m}_c$ (3 GeV) in the $\overline{\text{MS}}$ scheme using two different integration kernels are listed in Table 2, and the related uncertainties are shown in Table 3. The kernel $f(s) = 1/s^2$ is from Ref. 77. The merits of each kernel may be judged by its ability to minimize these uncertainties, in particular those that might be most affected by systematic errors, such as e.g. the experimental data. The kernel Eq. (35) appears to be optimal as it produces the smallest uncertainty due to the data, and is very stable against changes in s_0. Some recent determinations of the charm-quark mass are based on Hilbert moments with no s_0-dependent kernel, such as that in

Table 2. Results for the charm-quark mass at different orders in PQCD, and for two integration kernels from Ref. 76. The result for $f(s) = 1/s^2$ is obtained using slightly different values of the QCD parameters, and a different integration procedure as in Ref. 77.

	$\bar{m}_c$ (3 GeV) (in MeV)			
Kernel	$\bar{m}_c^{(0)}$	$\bar{m}_c^{(1)}$	$\bar{m}_c^{(2)}$	$\bar{m}_c^{(3)}$
s^{-2}	1129	1021	998	995
$1-(s_0/s)^2$	1146	1019	991	987

Table 3. The various uncertainties due to the data (EXP), the value of α_s ($\Delta\alpha_s$), changes of $\pm 35\%$ in the renormalization scale around $\mu = 3$ GeV ($\Delta\mu$), and the value of the gluon condensate (NP).[76]

		Uncertainties (in MeV)				
Kernel	$\bar{m}_c$ (3 GeV)	Exp.	$\Delta\alpha_s$	$\Delta\mu$	NP	Total
s^{-2}	995	9	3	1	1	9.6
$1-(s_0/s)^2$	987	7	4	1	1	8.2

Table 2. While there is no explicit s_0-dependence in Hilbert moments (the integrals extend to infinity), there is definitely a residual dependence when choosing the threshold for the onset of PQCD. From a FESR perspective, the major drawback of the kernel $1/s^2$ is clearly the poor stability against changes in s_0. An important remark is in order concerning the uncertainty due to changes in s_0. From current data it is not totally clear where does PQCD actually start. This problem not only affects FESR, with their explicitly obvious s_0-dependence, but also Hilbert moments with an implicit s_0-dependence, as there is no data all the way up to infinity.

Two of the most recent results for $\bar{m}_c$ (3 GeV),[71,72] together with the weighted FESR value[76] are

$$\bar{m}_c\,(3\text{ GeV}) = \begin{cases} 986 \pm 13\text{ MeV} & (\text{Ref. } 71)\,, \\ 998 \pm 29\text{ MeV} & (\text{Ref. } 72)\,, \\ 987 \pm 9\text{ MeV} & (\text{Ref. } 76)\,, \end{cases} \tag{36}$$

in very good agreement with each other, except for the errors. The small uncertainty from Ref. 76 is due in part to improved quenching of the data in the broad resonance region, but mostly due to a strong reduction in the sensitivity to s_0, i.e. the onset of PQCD. For comparison, a recent LQCD determination gives[78]

$$\bar{m}_c\ (3\ \text{GeV}) = 986 \pm 6\ \text{MeV}\,, \tag{37}$$

in excellent agreement in magnitude and uncertainty with Ref. 76.

Turning to the bottom-quark mass, the most recent precision determination[79] is based on e^+e^- data from the *BABAR* Collaboration,[80] and integration kernels related to Legendre-type Laurent polynomials, as used e.g. in the charm-quark case,[75] to wit

$$f(s) \equiv \mathcal{P}_3^{(i,j,k)}(s, s_0) = A(s^i + Bs^j + Cs^k)\,, \tag{38}$$

subject to the global constraint

$$\int_{s^*}^{s_0} \mathcal{P}_3^{(i,j,k)}(s, s_0) s^{-n}\, ds = 0\,, \tag{39}$$

where $n \in \{0, 1\}$, $i, j, k \in \{-3, -2, -1, 0, 1\}$, and i, j, k are all different. The above constraint determines the constants B and C. The constant A is an arbitrary overall normalization which cancels out in the sum rule Eq. (13). The reason for the presence of the integrand s^{-n} above is that the behaviour of $R_b(s)$ in the region to be quenched resembles a monotonically decreasing logarithmic function. Hence, an inverse power of s optimizes the quenching. As an example, taking $s_0 = (16\ \text{GeV})^2$ (and $A = 1$) one finds

$$\mathcal{P}_3^{(-3,-1,0)}(s, s_0) = s^{-3} - (1.02 \times 10^{-4}\ \text{GeV}^{-4}) s^{-1} + 3.70 \times 10^{-7}\ \text{GeV}^{-6}\,, \tag{40}$$

with s in units of GeV^2. There are five different kernels $\mathcal{P}_3^{(i,j,k)}$, and the spread of values obtained for $\bar{m}_b$ using this set of different kernels was used as a consistency check on the method. The fourth-order Laurent polynomial $\mathcal{P}_4^{(i,j,k,r)(s,s_0)}$ is also defined by the constraint Eq. (39), but with $n \in 0, 1, 2$, and there are also five different kernels of this type.

Table 4 summarizes the results obtained in Ref. 79. Options **A**, **B**, and **C** refer to different options for considering the data (for details, see Ref. 79). In Ref. 79, a total of 15 different kernels, $f(s)$ in Eq. (13), were considered, i.e. 10 from the class $\mathcal{P}_3^{(i,j,k)}(s, s_0)$, and 5 from the class $\mathcal{P}_4^{(i,j,k,r)}(s, s_0)$. Figure 6 shows the range of values for $\bar{m}_b$ (10 GeV) obtained using all of the 10 kernels in the class $\mathcal{P}_3^{(i,j,k)}(s, s_0)$, as a function of s_0. Remarkably, between 12 GeV $< \sqrt{s_0} <$ 28 GeV, all of the masses obtained using all 10 kernels from the class $\mathcal{P}_3^{(-3,-1,0)}(s, s_0)$ fall in the range 3621 MeV $\leq \bar{m}_b$ (10 GeV) $\leq$ 3625 MeV. The method gives a consistent result even in the region $\sqrt{s_0} < 4\bar{m}_b(\mu) \approx 15$ GeV where the high-energy expansion used in the contour

Table 4. Results from Ref. 79 for $\overline{m}_b$ (10 GeV) using kernels $f(s)$ selected for producing the lowest uncertainty. Results from the kernels $f(s) = s^{-3}$ and $f(s) = s^{-4}$ used in Refs. 71 and 81 are given here for comparison. The errors are from experiment,[80] (ΔEXP), from the strong coupling ($\Delta\alpha_s$) and from variation of the renormalization scale by ± 5 GeV around $\mu = 10$ GeV ($\Delta\mu$). These sources were added in quadrature to give the total uncertainty (ΔTOTAL). Options **A**, **B** and **C** refer to different ways of treating the data (see Ref. 79). The option uncertainties Δ**A**, Δ**B** and Δ**C** are the differences between $\overline{m}_b$ (10 GeV) obtained with and without Option **A**, **B** or **C**. As in Refs. 71 and 81, these are not added to the total uncertainty, and are listed only for comparison purposes.

			Uncertainties (MeV)				Options **A**, **B**, **C** (MeV)		
$f(s)$	$\overline{m}_b$ (10 GeV)	$\sqrt{s_0}$ (GeV)	ΔEXP	$\Delta\alpha_s$	$\Delta\mu$	ΔTOTAL	Δ**A**	Δ**B**	Δ**C**
s^{-3}	3612	∞	9	4	1	10	20	-17	16
s^{-4}	3622	∞	7	5	10	13	12	-12	8
$\mathcal{P}_3^{(-3,-1,0)}(s_0, s)$	3623	16	6	6	2	9	1	-6	0
$\mathcal{P}_3^{(-3,-1,1)}(s_0, s)$	3623	16	6	6	2	9	2	-7	0
$\mathcal{P}_3^{(-3,0,1)}(s_0, s)$	3624	16	7	6	2	9	2	-7	0
$\mathcal{P}_3^{(-1,0,1)}(s_0, s)$	3625	16	8	5	4	10	4	-12	0
$\mathcal{P}_4^{(-3,-1,0,1)}(s_0, s)$	3623	20	6	6	3	9	0	-4	0

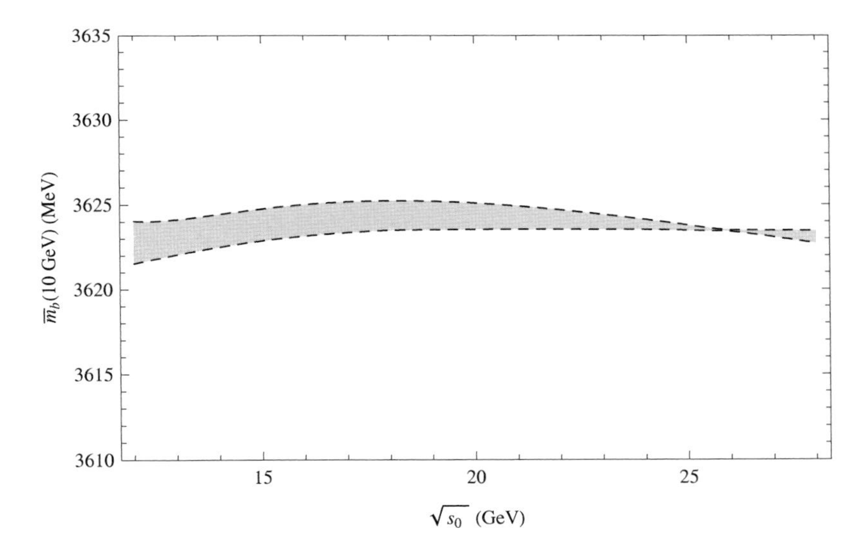

Fig. 6. The values of $\overline{m}_b$ (10 GeV), obtained for different values of s_0 and using the 10 different kernels in the class $\mathcal{P}_3^{(i,j,k)}(s_0, s)$. All results lie within the shaded region.

integral in Eq. (13) is not guaranteed to converge. Using, rather, the 5 kernels in the class $\mathcal{P}_4^{(i,j,k,r)}(s, s_0)$, and varying s_0 in the range 18 GeV $< \sqrt{s_0} <$ 70 GeV, all of the masses thus obtained lie in the interval 3620 MeV $\leq \overline{m}_b$ (10 GeV) $\leq$ 3626 MeV. These results show a great insensitivity of this method on the parameter s_0, and also on which powers of s are used to construct $\mathcal{P}_3^{(i,j,k)}(s,s_0)$ and $\mathcal{P}_4^{(i,j,k,r)}(s,s_0)$. This in turn demonstrates the consistency between the high and low energy expansions of PQCD. The final result chosen in Ref. 79 from the optimal kernel $\mathcal{P}_3^{(-3,-1,0)}(s_0,s)$ is

$$\bar{m}_b\ (10\ \text{GeV}) = 3623(9)\ \text{MeV}\,, \tag{41}$$

$$\bar{m}_b(\bar{m}_b) = 4171(9)\ \text{MeV}\,. \tag{42}$$

This value is fully consistent with the latest lattice determination $\bar{m}_b$ (10 GeV) = 3617(25) MeV.[78] It is also consistent with a previous QCD sum rule precision determination[81] giving $\bar{m}_b$ (10 GeV) = 3610(16) MeV.

8. Conclusions

After a short review of quark mass ratios, the method of QCDSR was discussed in connection with determinations of individual values of the quark masses. The historical (unknown) hadronic systematic uncertainty affecting light-quark mass determinations for over thirty years was highlighted. Details of the recent breakthrough in strongly reducing this uncertainty were provided. In addition, it was explained how the previously unknown systematic QCD uncertainty, due to the poor convergence of the light-quark pseudoscalar correlator, was exposed and essentially eliminated for the strange-quark mass. The up- and down-quark cases are currently under investigation, which should lead to a similar successful result. Future improvement in accuracy is now possible and it depends essentially on more accurate determinations of the strong coupling, the remaining main source of error. In the heavy-quark sector, recent high precision determinations of the charm- and bottom-quark masses were reported. While these values are all in agreement, there is some disagreement on the size of the errors. The use of suitable multi-purpose integration kernels in FESR allows one to tune the weight of the various contributions to the quark masses. This in turn allows one to minimize the error due to the data, as well as to the uncertainty in the onset of PQCD. The latter uncertainty impacts FESR as well as Hilbert moment sum rules, as there is no data all the way up to infinity. If no kernel other than simple (nonpinched) inverse powers of s are used, this uncertainty would be much larger than normally reported, as may be appreciated from Table 3, and especially Table 4. However, according to the current philosophy one chooses the determination having the smallest error. Marginal improvement of the current total error in this framework should be possible with improved accuracy in the data and in the strong coupling.

Acknowledgments

The author wishes to thank his collaborators in the various projects on quark masses reported here: S. Bodenstein, J. Bordes, N. Nasrallah, J. Peñarrocha, R. Röntsch and K. Schilcher. This work was supported in part by NRF (South Africa).

References

1. P. Colangelo and A. Khodjamirian, *At the Frontier of Particle Physics: Handbook of QCD*, Vol. 3, ed. M. A. Shifman (World Scientific, Singapore, 2001), p. 1495.
2. A. Pich, *Acta Phys. Polon. Suppl.* **3**, 165 (2010).
3. A. Pich, arXiv:1310.7922.
4. S. Glashow and S. Weinberg, *Phys. Rev. Lett.* **20**, 224 (1968).
5. M. Gell-Mann, R. Oakes and B. Renner, *Phys. Rev.* **175**, 2195 (1968).
6. H. Pagels, *Phys. Rep. C* **16**, 219 (1975).
7. J. Gasser and H. Leutwyler, *Phys. Rep. C* **87**, 77 (1982).
8. C. A. Dominguez and A. Zepeda, *Phys. Rev. D* **18**, 884 (1978).
9. C. A. Dominguez, *Phys. Lett. B* **86**, 171 (1979).
10. P. Minkowski and A. Zepeda, *Nucl. Phys. B* **164**, 25 (1980).
11. FLAG Working Group, S. Aoki *et al.*, arXiv:1310.8555.
12. H. Leutwyler, *PoS* **CD09**, 005 (2009).
13. S. Weinberg, *Trans. New York Acad. Sci.* **38**, 185 (1977).
14. G. Colangelo, S. Lanz and E. Passemar, *PoS* **CD09**, 047 (2009).
15. H. Leutwyler, *Nucl. Phys. B* (*Proc. Suppl.*) **64**, 223 (1998).
16. C. A. Dominguez and K. Schilcher, *Phys. Rev. D* **61**, 114020 (2000).
17. C. A. Dominguez and K. Schilcher, *J. High Energy Phys.* **01**, 093 (2007).
18. J. Bordes, C. A. Dominguez, P. Moodley, J. Peñarrocha and K. Schilcher, *J. High Energy Phys.* **05**, 064 (2010); *ibid.* **10**, 102 (2012).
19. R. Shankar, *Phys. Rev. D* **15**, 755 (1977).
20. K. Maltman, *Phys. Lett. B* **440**, 367 (1998).
21. C. A. Dominguez and K. Schilcher, *Phys. Lett. B* **448**, 93 (1999); *ibid.* **581**, 193 (2004).
22. C. A. Dominguez, N. F. Nasrallah and K. Schilcher, *Phys. Rev. D* **80**, 054014 (2009).
23. M. Gonzalez-Alonso, A. Pich and J. Prades, *Phys. Rev. D* **82**, 014019 (2010).
24. I. Caprini, M. Golterman and S. Peris, arXiv:1407.2577.
25. M. A. Shifman, A. I. Vainshtein and V. I. Zakharov, *Nucl. Phys. B* **147**, 385 (1979).
26. A. Erdelyi *et al.*, *Higher Transcendental Functions* (McGraw-Hill, New York, 1955).
27. E. de Rafael, Centre de Physique Theorique, Marseille, Report, CPT-81/P.1344 (1981) and *Proc. 1981 French-American Seminar: Theoretical Aspects of Quantum Chromodynamics*, J. W. Dash, editor, CPT-81/P.1345 (1981).

28. S. Bodenstein, C. A. Dominguez and K. Schilcher, *Phys. Rev. D* **85**, 014029 (2012).
29. ETM Collab. (F. Burger *et al.*), *J. High Energy Phys.* **1402**, 099 (2014); K. Jansen (private communication); HPQCD Collab. (B. Chakraborty *et al.*), arXiv:1403.177.
30. S. Groote, J. G. Körner, K. Schilcher and N. F. Nasrallah, *Phys. Lett. B* **440**, 375 (1998); J. H. Kühn and M. Steinhauser, *Phys. Lett. B* **437**, 425 (1998); S. Bodenstein, C. A. Dominguez, K. Schilcher and H. Spiesberger, *Phys. Rev. D* **86**, 093013 (2012).
31. K. G. Chetyrkin, C. A. Dominguez, D. Pirjol and K. Schilcher, *Phys. Rev. D* **51**, 5090 (1995).
32. M. Jamin and M. Münz, *Z. Phys. C* **66**, 633 (1995).
33. P. A. Baikov, K. G. Chetyrkin and J. H. Kühn, *Phys. Rev. Lett.* **96**, 012003 (2006).
34. L. V. Lanin, V. P. Spidorov and K. G. Chetyrkin, *Sov. J. Nucl. Phys.* **44**, 892 (1986); D. J. Broadhurst and S. C. Generalis, Open University Report, OUT-4102-22, 1988 (unpublished); S. C. Generalis, *J. Phys. G* **15**, L225 (1989); *ibid.* **16**, L117 (1990); L. R. Surguladze and F. V. Tkachov, *Nucl. Phys. B* **331**, 35 (1990); D. J. Broadhurst, P. A. Baikov, V. A. Ilyin, J. Fleischer, O. V. Tarasov and V. A. Smirnov, *Phys. Lett. B* **329**, 103 (1994).
35. Particle Data Group (J. Beringer *et al.*), *Phys. Rev. D* **86**, 010001 (2012).
36. C. A. Dominguez, N. F. Nasrallah and K. Schilcher, *J. High Energy Phys.* **02**, 072 (2008).
37. C. A. Dominguez, N. F. Nasrallah, R. Röntsch and K. Schilcher, *J. High Energy Phys.* **05**, 020 (2008).
38. C. A. Dominguez, N. F. Nasrallah, R. Röntsch and K. Schilcher, *Phys. Rev. D* **79**, 014009 (2009).
39. C. A. Dominguez, *Int. J. Mod. Phys. A* **25**, 5223 (2010).
40. C. A. Dominguez, *Z. Phys. C* **26**, 269 (1984).
41. C. A. Dominguez, L. Pirovano and K. Schilcher, *Phys. Lett. B* **425**, 193 (1998).
42. K. Maltman and J. Kambor, *Phys. Rev. D* **65**, 074013 (2002).
43. C. A. Dominguez, A. Ramlakan and K. Schilcher, *Phys. Lett. B* **511**, 59 (1998).
44. G. Cvetic *et al.*, *Phys. Rev. D* **82**, 093007 (2010).
45. M. Davier *et al.*, *Eur. Phys. J. C* **56**, 305 (2006).
46. P. A. Baikov, K. G. Chetyrkin and J. H. Kühn, *Phys. Rev. Lett.* **101**, 012002 (2008).
47. A. Pich, *Prog. Part. Nucl. Phys.* **75**, 41 (2014).
48. H. Leutwyler, *Phys. Lett. B* **378**, 313 (1996).
49. S. Bodenstein, C. A. Dominguez and K. Schilcher, *J. High Energy Phys.* **07**, 138 (2013).
50. C. A. Dominguez and K. Schilcher, work in progress.
51. J. R. Ellis, E. Gardi, M. Karliner and M. A. Samuel, *Phys. Lett. B* **366**, 268 (1996); *Phys. Rev. D* **54**, 6986 (1996); E. Gardi, *Phys. Rev. D* **56**, 68 (1997).
52. S. Bodenstein, Precision determination of QCD fundamental parameters from sum rules, Ph.D. thesis, University of Cape Town (2014).

53. K. G. Chetyrkin, R. Harlander, J. H. Kühn and M. Steinhauser, *Nucl. Phys. B* **503**, 339 (1997).
54. P. A. Baikov, K. G. Chetyrkin and J. H. Kühn, *Nucl. Phys. B* (*Proc. Suppl.*) **189**, 49 (2009).
55. K. G. Chetyrkin, R. Harlander and J. H. Kühn, *Nucl. Phys. B* **586**, 56 (2000).
56. Y. Kiyo, A. Maier, P. Maierhöfer and P. Marquard, *Nucl. Phys. B* **823**, 269 (2009).
57. P. A. Baikov, K. G. Chetyrkin and J. H. Kühn, *Phys. Rev. Lett.* **101**, 012002 (2008).
58. P. A. Baikov, K. G. Chetyrkin and J. H. Kühn, *Nucl. Phys. B* (*Proc. Suppl.*) **135**, 243 (2004).
59. K. G. Chetyrkin, J. H. Kühn and M. Steinhauser, *Phys. Lett. B* **371**, 93 (1996).
60. K. G. Chetyrkin, J. H. Kühn and M. Steinhauser, *Nucl. Phys. B* **482**, 213 (1996).
61. K. G. Chetyrkin, J. H. Kühn and M. Steinhauser, *Nucl. Phys. B* **505**, 40 (1997).
62. R. Boughezal, M. Czakon and T. Schutzmeier, *Phys. Rev. D* **74**, 074006 (2006).
63. R. Boughezal, M. Czakon and T. Schutzmeier, *Nucl. Phys. B* (*Proc. Suppl.*) **160**, 164 (2006).
64. A. Maier, P. Maieröfer and P. Marquard, *Nucl. Phys. B* **797**, 218 (2008).
65. A. Maier, P. Maieröfer and P. Marquard, *Phys. Lett. B* **669**, 88 (2008).
66. K. G. Chetyrkin, J. H. Kühn and C. Sturm, *Eur. Phys. J. C* **48**, 107 (2006).
67. A. Maier, P. Maierhöfer, P. Marquard and A. V. Smirnov, *Nucl. Phys. B* **824**, 1 (2010).
68. A. H. Hoang, V. Mateu and S. M. Zebarjad, *Nucl. Phys. B* **813**, 349 (2009).
69. J. Hoff and M. Steinhauser, arXiv:1103.1481.
70. D. J. Broadhurst *et al.*, *Phys. Lett. B* **329**, 103 (1994).
71. K. G. Chetyrkin *et al.*, *Theor. Math. Phys.* **170**, 217 (2012).
72. B. Dehnadi, A. H. Hoang, V. Mateu and S. M. Zebarjad, *J. High Energy Phys.* **09**, 103 (2013).
73. J. Peñarrocha and K. Schilcher, *Phys. Lett. B* **515**, 291 (2001).
74. J. Bordes, J. Peñarrocha and K. Schilcher, *Phys. Lett. B* **562**, 81 (2003).
75. S. Bodenstein, J. Bordes, C. A. Dominguez, J. Peñarrocha and K. Schilcher, *Phys. Rev. D* **82**, 114013 (2010).
76. S. Bodenstein, J. Bordes, C. A. Dominguez, J. Peñarrocha and K. Schilcher, *Phys. Rev. D* **83**, 074014 (2011).
77. J. H. Kühn, M. Steimhauser and C. Sturm, *Nucl. Phys. B* **778**, 192 (2007); K. G. Chetyrkin *et al.*, *Phys. Rev. D* **80**, 074010 (2009).
78. C. McNeile, C. T. H. Davies, E. Follana, K. Hornbostel and G. P. Lepage, *Phys. Rev. D* **82**, 034512 (2010).
79. S. Bodenstein, J. Bordes, C. A. Dominguez, J. Peñarrocha and K. Schilcher, *Phys. Rev. D* **85**, 034003 (2012).
80. B. Aubert *et al.*, *Phys. Rev. Lett.* **102**, 012001 (2009).
81. K. G. Chetyrkin *et al.*, *Phys. Rev. D* **80**, 074010 (2009).

CP Violation in Six Quark Scheme — Legacy of Sakata Model

Makoto Kobayashi
KEK, 1-1 Oho, Tsukuba, Ibaraki 305 0801, Japan
kobayath@post.kek.jp

After a short review of the activities of Shoichi Sakata and his group, how the six-quark model explains CP violation is described. Experimental verification of the model at the B-factories is also briefly discussed.

1. Introduction

CP violation was discovered in the same year as the quark model was proposed.[1,2] The number of quark flavors was three at first, but it turned out that it is closely related to CP violation. This issue was discussed by Maskawa and myself in 1972.[3] At the time, the discovery of renormalizability of the non-Abelian gauge theory opened the possibility that all the interactions are described by the gauge theory without difficulty of divergence. What we investigated then was whether CP violation can be explained in this new framework.

I started my research work under the supervision of Professor Shoichi Sakata in Nagoya University. He is known for the Sakata Model,[4] which is an important precursor of the quark model. He passed away when I was still a graduate student, but his group continued active research from the viewpoint of composite model even after his demise. Since I was much influenced by Prof. Sakata and his group, I begin this article with an overview of the Sakata Model and its development.

2. Sakata Model

The number of hadron species substantially increased due to the discovery of strange particles. In the Sakata Model, Sakata considered that only three hadrons, the proton (p), the neutron (n) and the lambda-particle (Λ) are the fundamental particles and all other hadrons are composite states. If we denote the fundamental particles as t, mesons and baryons other than p, n and Λ have structures of $t\bar{t}$ and $tt\bar{t}$, respectively, in this model. Some examples are

$$\text{Meson}: \quad \pi^+ = (p\bar{n})\,, \qquad K^+ = (p\bar{\Lambda})\,,$$

$$\text{Baryon}: \quad \Sigma^+ = (p\Lambda\bar{n})\,, \quad \Xi^- = (\Lambda\Lambda\bar{p})\,.$$

The Sakata Model was presented at the autumn meeting of the physical society of Japan in 1955, although the paper was published in the next year issue of the Progress of Theoretical Physics. Sakata graduated from Kyoto University in 1933. In the previous year of his graduation, the neutron was discovered and it resolved the mystery of the structure of atomic nuclei. He was strongly impressed by this fact. He overlaid the role of the lambda particle on the role of the neutron in the atomic nuclei. In his paper,[4] he says "It seems to me that the present state of the theory of new particles is very similar to that of the atomic nuclei 25 years ago" and "Supposing that the similar situation is realized at present, I proposed a compound hypothesis for unstable particles to account for Nishijima–Gell-Mann's rule."

Although the Sakata Model changed the viewpoint on the structure of hadrons fundamentally, it lacked a concrete theory describing dynamics of the compound system, so that direct outcomes from the model were not so many. It is remarkable, however, Ikeda, Ogawa and Ohnuki developed group theoretical approach focusing on symmetry among the fundamental triplets.[5] They discussed classification of the hadrons according to the representation of the $U(3)$ group. Certainly their work was going ahead of the subsequent boom of the group theory.

One of the advantageous points of the Sakata Model is that the weak interaction of the hadrons can be explained by the following two types of transitions among the fundamental triplets,

$$\begin{array}{cc} p & \\ \updownarrow & \nwarrow\!\!\searrow \\ n & \Lambda \end{array} \,.$$

This patten of the transition is very similar to that of the lepton,

$$\begin{array}{cc} \nu & \\ \updownarrow & \nwarrow\!\!\searrow \\ e & \mu \end{array} \,.$$

This similarity was called $B-L$ symmetry by Gamba, Marshak and Okubo.[6]

Focusing on this similarity, Maki, Nakagawa, Ohnuki and Sakata proposed an ambitious model, which is called Nagoya Model.[7] In this model, the fundamental triplet, p, n and Λ, are regarded as the compound system of the following form,

$$p = \langle \nu B^+ \rangle\,, \quad n = \langle e^- B^+ \rangle\,, \quad \Lambda = \langle \mu B^+ \rangle\,,$$

where B^+ denotes a hypothetical object called B-matter. The idea is that strong and weak interactions of the fundamental triplet are attributed to the B-matter and the constituent leptons, respectively. At that time, neutrino was still thought to consist of a single species. It seems that Sakata used the word "B-matter" attempting to express something transcending an ordinary particle from his sentiment on the field theory. He was thinking that success of the renormalization theory in the quantum electrodynamics is tentative and the field theory would be replaced by a more fundamental theory soon or later.

The Nagoya Model also lacked a proper description of dynamics of the compound system, but it showed an interesting development. In 1962, it was found that ν_e and ν_μ are different particles. When this news was brought to Japan, two interesting papers were published, one by Katayama, Matsumoto, Tanaka and Yamada[8] and the other by Maki, Nakagawa and Sakata.[9] Both papers discussed the modification of the Nagoya Model to accommodate two neutrinos:

$$p = \langle \nu_1 B^+ \rangle\,, \quad n = \langle e^- B^+ \rangle\,, \quad \Lambda = \langle \mu B^+ \rangle\,, \quad p' = \langle \nu_2 B^+ \rangle\,,$$

where

$$\nu_1 = \cos\theta \nu_e + \sin\theta \nu_\mu\,,$$

$$\nu_2 = -\sin\theta \nu_e + \cos\theta \nu_\mu\,.$$

Since the transitions of p to n and p to Λ take place through the constituent ν_e and ν_μ of the proton, respectively, this model naturally explains the difference of strength of two transitions. As for the fourth element, p', both groups took a stance that it may or may not exist. At that time, it was still thought that the fundamental particles are baryons, but the structure of the weak interaction is essentially the same as that of Glashow, Iliopoulos and Maiani.[10]

In the course of the above argument, Maki *et al.*,[9] noticed that ν_e and ν_μ are not the mass eigenstates in general and precisely formulated the flavor mixing in the lepton sector. They identified the mass eigenstates with ν_1 and ν_2 as the constituent of the fundamental baryons. But this is an extra assumption specific to the Nagoya Model, so that it is irrelevant to the lepton flavor mixing itself. However, they were aware of the implication of their arguments on the pure lepton sector. They correctly pointed out that if the mass difference is very small, the mixing effect will be difficult to see in the usual setup of the experiment, because the oscillation length is too long. Today, we call the mixing angles in the lepton sector "MNS angles."

Meanwhile, it turned out that the Sakata Model is incompatible with experimental observations. It treats the triplet baryons, p, n and Λ differently from the rest of baryons. In terms of representation of the $SU(3)$ group, p, n and Λ belong to **3**, while Σ and Ξ belong to either **6** or **15**. However, experiments were indicating that the baryons belong to the octet representation. Eventually, the Sakata Model was abandoned and gave way to the quark model.

3. Six-Quark Model and CP Violation

CP violation was discovered by Cronin, Fitch *et al.* in 1964.[1] They found that a small fraction of K_L decays into two pions. This decay would not take place if CP is conserved, so that it was an evidence of CP violation. CP violation was seen only in the neutral K meson system for long time. Accordingly, it was hard to elucidate the mechanism of CP violation from such limited experimental information at first.

However, situation had changed when renormalizability of the non-Abelian guage theory opened the possibility of describing weak interactions in a renormalizable manner. In 1972, I obtained Ph.D. from Nagoya University and moved to Kyoto, where I resumed collaboration with Toshihide Maskawa, who also graduated from Nagoya University and moved to Kyoto some time before. The subject we chose at that time was how we can explain CP violation in a renormalizable guage theory. Renormalizability imposes stringent constrains on possible interactions, so that simple systems consisting of a small number of particles become automatically CP symmetric. Accordingly, the system must be complex to some extent to accommodate CP violation.

We discussed this problem in the Glashow–Weinberg–Salam theory.[11–13] In this scheme, ordinary three-quark model has a problem of strangeness changing neutral current. This problem can be avoided by considering a four-quark model of the GIM type. However, as is shown below, this system is CP symmetric, provided that there are no additional flavor changing interactions.

The quark sector of this model consists of two left-handed doublets

$$\begin{pmatrix} u_{1L}^0 \\ d_{1L}^0 \end{pmatrix}, \quad \begin{pmatrix} u_{2L}^0 \\ d_{2L}^0 \end{pmatrix},$$

and four right-handed singlets

$$u_{1R}^0\,,\ u_{2R}^0\,,\ d_{1R}^0\,,\ d_{2R}^0\,.$$

The most general form of their Yukawa couplings with the Higgs doublet φ is

$$L_Y = -\sum_{i,j=1,2}\left\{\left(\bar{u}_{iL}^0, \bar{d}_{iL}^0\right) M_{ij}^{(d)} \varphi d_{jR}^0 + \left(\bar{u}_{iL}^0, \bar{d}_{iL}^0\right) M_{ij}^{(u)} \tilde{\varphi} u_{jR}^0\right\} + \text{h.c.}\,,$$

where $M^{(d)}$ and $M^{(u)}$ are arbitrary 2×2 complex matrices and

$$\varphi = \begin{pmatrix} \varphi^+ \\ \varphi^0 \end{pmatrix}, \quad \tilde{\varphi} = \begin{pmatrix} \varphi^{0*} \\ -\varphi^- \end{pmatrix}.$$

These Yukawa couplings yield the following mass terms through the vacuum expectation value v of φ^0:

$$L_{\text{mass}} = -v\left\{\left(\bar{d}_{1L}^0, \bar{d}_{2L}^0\right) M^{(d)} \begin{pmatrix} d_{1R}^0 \\ d_{2R}^0 \end{pmatrix} + \left(\bar{u}_{1L}^0, \bar{u}_{2L}^0\right) M^{(u)} \begin{pmatrix} u_{1R}^0 \\ u_{2R}^0 \end{pmatrix}\right\} + \text{h.c.}$$

The mass matrices $vM^{(d)}$ and $vM^{(u)}$ are not necessarily Hermitian but they are diagonalized by the bi-unitary transformations

$$\begin{pmatrix} d_{1L}^0 \\ d_{2L}^0 \end{pmatrix} = V^{(d)} \begin{pmatrix} d_L \\ s_L \end{pmatrix}, \quad \begin{pmatrix} d_{1R}^0 \\ d_{2R}^0 \end{pmatrix} = W^{(d)} \begin{pmatrix} d_R \\ s_R \end{pmatrix},$$

and

$$\begin{pmatrix} u_{1L}^0 \\ u_{2L}^0 \end{pmatrix} = V^{(u)} \begin{pmatrix} u_L \\ c_L \end{pmatrix}, \quad \begin{pmatrix} u_{1R}^0 \\ u_{2R}^0 \end{pmatrix} = W^{(u)} \begin{pmatrix} u_R \\ c_R \end{pmatrix},$$

where d, s, u and c are the mass eigenstates, which are the physical quark states. Then the interaction Lagrangian of the charged current weak interactions can be expressed as

$$L_{\text{int}} = -\frac{g}{2\sqrt{2}} W_\mu^+ (\bar{u}, \bar{c}) V_{(2)} \gamma^\mu (1-\gamma^5) \begin{pmatrix} d \\ s \end{pmatrix} + \text{h.c.},$$

with

$$V_{(2)} = V^{(u)\dagger} V^{(d)}\,.$$

From the arbitrariness of $M^{(d)}$ and $M^{(u)}$, there is no limitation on $V_{(2)}$ except that it is a 2×2 unitary matrix.

The left-handed doublets can be expressed as

$$\begin{pmatrix} u_L \\ d'_L \end{pmatrix}, \quad \begin{pmatrix} c_L \\ s'_L \end{pmatrix},$$

where d' and s' are given by

$$\begin{pmatrix} d' \\ s' \end{pmatrix} = V_{(2)} \begin{pmatrix} d \\ s \end{pmatrix} .$$

All the matrix elements of $V_{(2)}$ are not necessarily physical parameters. If we rewrite the Lagrangian in terms of the quark fields with different phase convention,

$$\tilde{u} = e^{i\delta_u} u \,, \quad \tilde{d} = e^{i\delta_d} d \,, \quad \tilde{c} = e^{i\delta_c} c \,, \quad \tilde{s} = e^{i\delta_s} s \,,$$

then $V_{(2)}$ is transformed to

$$\tilde{V}_{(2)} = \begin{pmatrix} e^{i\delta_u} & 0 \\ 0 & e^{i\delta_c} \end{pmatrix} V_{(2)} \begin{pmatrix} e^{-i\delta_d} & 0 \\ 0 & e^{-i\delta_b} \end{pmatrix} .$$

It is easy to see that, by choosing the phase factors properly, we can make $\tilde{V}_{(2)}$ a real matrix of the GIM type,

$$\tilde{V}_{(2)} = \begin{pmatrix} \cos\theta & \sin\theta \\ -\sin\theta & \cos\theta \end{pmatrix} .$$

In the case that the charged current interactions are only source of the flavor change, this phase redefinition does not make any change in the rest of the Lagrangian. Therefore, if the Higgs sector consists of a single doublet, this type of the four-quark system is CP symmetric. However, this is not necessarily the case, if there is a flavor changing interaction other than the charged current weak interaction. This is because the phase redefinition of the quark fields may generate a new phase factor to the coupling constants of the flavor changing interaction. The multi-Higgs model is one such example.

Now, we consider a six-quark system consisting of three left-handed doublets. In this case, the doublets can be written as

$$\begin{pmatrix} u_L \\ d'_L \end{pmatrix} , \quad \begin{pmatrix} c_L \\ s'_L \end{pmatrix} , \quad \begin{pmatrix} t_L \\ b'_L \end{pmatrix} ,$$

with

$$\begin{pmatrix} d' \\ s' \\ b' \end{pmatrix} = V_{(3)} \begin{pmatrix} d \\ s \\ b \end{pmatrix} ,$$

where $V_{(3)}$ is a 3×3 unitary matrix.

Simple parameter counting shows that we cannot make this matrix real in general, by the redefinition of the phase convention of the quark fields. The general form of a 3×3 unitary matrix is described by nine parameters. On the other hand, the number of adjustable phases is six, corresponding to six-quark fields. However, the overall common phase does not affect to $V_{(3)}$, so that only five phases can be used to reduce the number of parameters of $V_{(3)}$. As a result, $V_{(3)}$ contains four physical parameters, while a real form of $V_{(3)}$ is parametrized by three Euler angles. This implies that V cannot be transformed to a real form in general. Therefore, CP violation can be accommodated in the charged current weak interaction of this type. This is an essence of the arguments we made in our paper.

As for the actual representation of the matrix, there are many ways of parametrization. We adopted the following form in our paper,

$$V_{(3)} = \begin{pmatrix} c_1 & -s_1c_2 & -s_1s_2 \\ s_1c_2 & c_1c_2c_3 - s_2s_3e^{i\delta} & c_1c_2s_3 + s_2c_3e^{i\delta} \\ s_1s_2 & c_1s_2c_3 + c_2s_3e^{i\delta} & c_1s_2s_3 - c_2c_3e^{i\delta} \end{pmatrix},$$

where $c_i = \cos\theta_i$, $s_i = \sin\theta_i$ $(i = 1, 2, 3)$ and θ_1, θ_2, θ_3 and δ are four physical parameters in this representation. Unless $\sin\delta = 0$, the system is violating CP symmetry.

More popular and convenient for the practical use is the following parametrization proposed by Wolfenstein,[14]

$$V_{(3)} = \begin{pmatrix} 1 - \lambda^2/2 & \lambda & A\lambda^3(\rho - i\eta) \\ -\lambda & 1 - \lambda^2/2 & A\lambda^2 \\ A\lambda^3(1 - \rho - i\eta) & -A\lambda^2 & 1 \end{pmatrix},$$

where the physical parameters are λ, A, ρ and η. This parametrization is based on the peculiar pattern existing in the actual values of the matrix elements, which can be expressed by the power of λ with $\lambda \approx 0.22$. This expression is not exactly unitary, but the deviation is $O(\lambda^4)$.

After our paper was published, experiments revealed the existence of new flavors of the quark and lepton one after another. Discovery of the J/ψ particle and the charmed particles proved the existence of the fourth quark c. Although this discovery had a great importance to convince the reality of quarks, it had nothing to do with CP violation. However, the situation has changed, when the τ lepton was discovered in 1975. It was suggesting the existence of the third generation in the quark sector too. Since then, our paper began to attract attention. Soon after that, the existence of the fifth quark b was revealed by the discovery of the Υ particle. The discovery of the t quark was as late as 1995, but the six-quark scheme became a standard picture well before that.

4. CP Violation in the B-Meson System

The existence of six kinds of quarks does not necessarily imply that observed CP violation phenomena are caused through the mixing angles mentioned in the previous section. It was known that the ϵ parameter of CP violation in the neutral K-meson system can be explained by the six-quark model with a certain range of mixing parameters,[15] but this is not a decisive test. In particular, it does not differentiate the six-quark model from the super weak model.[16] For this reason, great effort has been poured into the determination of the ϵ' parameter of the two-pion decay of the neutral K-meson, both experimentally and theoretically. Experimental results are now available[17,18] but theoretical estimate based on the lattice calculation seems not yet finalized.

Meanwhile, a new method of measuring CP violation in the B-meson system was developed.[19–22] To see this, we consider a state oscillating between B_d and $\bar{B}_d$ and its decay into a final state f to which both B_d and $\bar{B}_d$ can decay directly. In this case, the decay amplitudes through B_d and $\bar{B}_d$ interfere, and if CP is violated, the time profile of the decay probability of the oscillating state shows an oscillating behavior. Then the asymmetry defined by

$$A(t) = \frac{\Gamma(B_d(t) \to f) - \Gamma(\bar{B}_d(t) \to f)}{\Gamma(B_d(t) \to f) + \Gamma(\bar{B}_d(t) \to f)} = a \sin(\Delta m t + \delta)$$

is a measure of CP violation. Here, $B_d(t)$ and $\bar{B}_d(t)$ are the oscillating states which are B_d and $\bar{B}_d$ at $t = 0$, respectively, and Δm is a difference of the mass eigenvalues of $B_d - \bar{B}_d$ system. It is assumed that the total width of two eigenstates is the same.

The case that f is a $J/\psi + K_S$ state is particularly important from both theoretical and experimental points of view. In this case, the asymmetry $A(t)$ is predicted from the six-quark model in good approximation as

$$A(t) = -\sin(2\phi_1)\sin(\Delta m t)\,,$$

where

$$\phi_1 = \arg\left(-\frac{V_{cd}V_{cb}^*}{V_{td}V_{tb}^*}\right).$$

This implies that $A(t)$ is directly given by the fundamental parameters of weak interaction without the strong interaction effect. Experimentally, the signal of $J/\psi + K_S$ is very clear.

It is interesting to note that ϕ_1 is related to the so-called unitarity triangle. The orthogonality of the first and third column vectors of V is expressed as

$$V_{ud}V_{ub}^* + V_{cd}V_{cb}^* + V_{td}V_{tb}^* = 0\,.$$

Regarding each term of the left-hand side of this equation as a two-dimensional vector in the complex plane, we can draw a triangle. Then ϕ_1 is nothing but the angle between two sides corresponding to the second and third terms. In a similar manner, various observable quantities can be understood in relation to this triangle. The orthogonality relations of other combination of the column vectors also define triangles. Their shapes are different to each other but their areas are common. The common area is a measure of CP violation of this model. If the CP violating parameter is zero, all the triangles collapse and the area vanishes.

The measurement of the asymmetry $A(t)$ is no easy task, because the lifetime of B_d is around 10^{-12} seconds. To overcome the difficulty, specially designed asymmetric B-factories were build at SLAC and KEK. They are e^+e^- colliders with the center-of-mass energy tuned to the $\Upsilon(4S)$ resonance, but colliding electron and positron have different energies. The mass of $\Upsilon(4S)$ is just above the $B_d+\bar{B}_d$ threshold and about a half of $\Upsilon(4S)$ decays into $B_d+\bar{B}_d$. The produced B-mesons are boosted due to the asymmetric energy. This makes possible measurement of the decay time through the decay position.

The experiments at SLAC and KEK discovered CP violation in the B-meson system, for the first time outside of the K-meson system.[23,24] Many results so far obtained in the B-factory experiments are consistent with the six-quark model. It is commonly believed that the major source of the CP violation observed in the laboratory experiments is the flavor mixing among the six quarks.

5. Remarks

The fact that this universe is made of matter is apparent asymmetry between particle and antiparticle. Sakharov discussed conditions for the asymmetry to be generated in the course of the evolution of the universe.[25] The conditions include (1) baryon number nonconservation, (2) C and CP violation and (3) out of thermal equilibrium. The Standard Model satisfies all of the three conditions, provided that electroweak symmetry breaking is the first-order phase transition. However, the detailed study shows that the Standard Model does not generate enough asymmetry between matter and antimatter. This implies that there must be other source of CP violation than the quark flavor mixing.

In this connection, flavor mixing in the lepton sector is attracting attention. Discovery of the neutrino oscillation revealed that neutrinos are massive and there must be mixing among them.[26] This opened an interesting possibility that CP symmetry is violated in the lepton sector, too. At the moment, CP violation in the lepton sector is yet to be observed.

We do not know the fundamental principle that determines the masses and the mixing parameters of quarks and leptons. This degrades understanding of the real origin of CP violation in the quark and lepton sectors. It is conceivable that the origin of CP violation is on a more fundamental level and it comes out only through the flavor mixing. At any rate, the whole picture of CP violation is still veiled in mystery.

References

1. J. H. Christenson, J. W. Cronin, V. L. Fitch and R. Turlay, *Phys. Rev. Lett.* **13**, 138 (1964).
2. M. Gell-Mann, *Phys. Lett.* **8**, 214 (1964).
3. M. Kobayashi and T. Maskawa, *Prog. Theor. Phys.* **49**, 652 (1973).
4. S. Sakata, *Prog. Theor. Phys.* **16**, 686 (1956).
5. M. Ikeda, S. Ogawa and Y. Ohnuki, *Prog. Theor. Phys.* **22**, 715 (1959).
6. A. Gamba, R. E. Marshak and S. Okubo, *Proc. Natl. Acad. Sci.* **45**, 881 (1959).
7. Z. Maki, M. Nakagawa, Y. Ohnuki and S. Sakata, *Prog. Theor. Phys.* **23**, 1174 (1960).
8. Y. Katayama, K. Matsumoto, S. Tanaka and E. Yamada, *Prog. Theor. Phys.* **28**, 675 (1962).
9. Z. Maki, M. Nakagawa and S. Sakata, *Prog. Theor. Phys.* **28**, 870 (1962).
10. S. L. Glashow, J. Iliopoulos and L. Maiani, *Phys. Rev. D* **2**, 1285 (1970).
11. S. L. Glashow, *Nucl. Phys.* **22**, 579 (1961).
12. S. Weinberg, *Phys. Rev. Lett.* **19**, 1264 (1967).
13. A. Salam and N. Svartholm (eds.), *Proc. 8th Nobel Symposium* (Almquist and Wiksell, Stockholm, 1968), p. 367.
14. L. Wolfenstein, *Phys. Rev. Lett.* **51**, 1945 (1983).
15. J. R. Ellis, M. K. Gaillard and D. V. Nanopoulos, *Nucl. Phys. B* **109**, 213 (1976).
16. L. Wolfenstein, *Phys. Rev. Lett.* **13**, 562 (1964).
17. J. R. Batley *et al.*, *Phys. Rev. Lett. B* **544**, 97 (2001).
18. E. Abouzaid *et al.*, *Phys. Rev. D* **83**, 092001 (2011).
19. A. B. Carter and A. I. Sanda, *Phys. Rev. Lett.* **45**, 952 (1980).
20. A. B. Carter and A. I. Sanda, *Phys. Rev. D* **23**, 1567 (1981).
21. I. I. Bigi and A. I. Sanda, *Nucl. Phys. B* **193**, 85 (1981).
22. I. I. Bigi and A. I. Sanda, *Nucl. Phys. B* **291**, 41 (1987).
23. B. Aubert *et al.*, *Phys. Rev. Lett.* **87**, 091801 (2001).
24. K. Abe *et al.*, *Phys. Rev. Lett.* **87**, 091802 (2001), see also http://ckmfitter.in203.fr.
25. A. D. Sakharov, *JETP Lett.* **5**, 24 (1967).
26. Super-Kamiokande Collab. (Y. Fukuda *et al.*), *Phys. Rev. Lett.* **81**, 1562 (1988).

The Constituent-Quark Model — Nowadays

W. Plessas

Theoretical Physics, Institute of Physics, University of Graz,
Universitätsplatz 5, A-8010 Graz, Austria

The present performance of the constituent-quark model as an effective tool to describe low-energy hadrons on the basis of quantum chromodynamics is exemplified along the relativistic constituent-quark model with quark dynamics relying on a realistic confinement and a Goldstone-boson-exchange hyperfine interaction. In particular, the spectroscopy of all known baryons is covered within a universal model in good agreement with phenomenology. The structure of the nucleons as probed under electromagnetic, weak, and gravitational interactions is produced in a correct manner. Likewise, the electroweak form factors of the baryons with u, d, and s flavor contents, calculated so far, result reasonably. Shortcomings still remain with regard to hadronic resonance decays.

1. Introductory Remarks

Since the notion of quarks was introduced 50 years ago[1] there has ever been the quest for their proper interaction and how to account for it.[2] Of course, the quark dynamics was formulated by a gauge theory in late 1972/early 1973,[3,4] however, over more than four decades quantum chromodynamics (QCD) has hitherto not yet been amenable to a complete solution, with severe problems prevailing especially at low energies. In order to describe the wealth of hadron phenomena in this energy domain, one still has to resort to simulations, approximations, and/or models.

The constituent-quark model has a history almost as long as QCD itself. For quite some time, however, it was formulated within an inadequate framework, namely, along non-relativistic quantum mechanics, employed a simplistic confinement (e.g., harmonic-oscillator forces), and relied on deficient hyperfine interactions, mostly one-gluon exchange (OGE). Only over the recent decades or so it has become manifest that most importantly

- a relativistically invariant formulation is mandatory,[5,6]
- the confinement needs to be taken strictly according to QCD (e.g., congruent with lattice-QCD results),[7] and

- the hyperfine interaction must take into account the transformation of the active degrees of freedom due to the spontaneous breaking of chiral symmetry (SBχS) of QCD at low energies.[8]

All of these shortcomings were remedied by the so-called Goldstone-boson-exchange (GBE) relativistic constituent-quark model (RCQM).[9,10] While its first version covered only $SU(3)_F$ baryons, the same model was subsequently extended to c and b flavors,[11] and was recently adapted to comprise all known baryons on a universal basis.[12,13]

In this article we demonstrate how far and in which quality baryon properties and reactions can nowadays be described by the GBE RCQM.

2. The Relativistic Constituent-Quark Model

We first explain the definition of the universal Goldstone-boson-exchange (UGBE) RCQM comprising all known baryons of flavors u, d, s, c, and b within a single framework. It is formulated in a Poincaré-invariant Hamiltonian quantum theory and is based on a relativistically invariant mass operator

$$\hat{M} = \hat{M}_{\text{free}} + \hat{M}_{\text{int}} , \tag{1}$$

where the free part corresponds to the total kinetic energy of the three-quark system and the interaction part contains the dynamics of the constituent quarks Q.

In the rest frame of the baryon, where its three-momentum $\vec{P} = \sum_i^3 \vec{k}_i^2 = 0$, we may express the terms as

$$\hat{M}_{\text{free}} = \sum_{i=1}^{3} \sqrt{\hat{m}_i^2 + \hat{\vec{k}}_i^2} , \tag{2}$$

$$\hat{M}_{\text{int}} = \sum_{i<j}^{3} \hat{V}_{ij} = \sum_{i<j}^{3} \left(\hat{V}_{ij}^{\text{conf}} + \hat{V}_{ij}^{\text{hf}} \right) . \tag{3}$$

Here, the $\hat{\vec{k}}_i$ correspond to the three-momentum operators of the individual quarks with rest masses m_i and the Q–Q potential operators $\hat{V}_{ij}$ are composed of confinement and hyperfine interactions.

The confinement depending linearly on the Q–Q distance r_{ij} is adopted as

$$V_{ij}^{\text{conf}}(\vec{r}_{ij}) = V_0 + C r_{ij} \tag{4}$$

Table 1. Free parameters of the UGBE RCQM.

Singlet coupling constant $(g_0/g_{24})^2$	Cutoff parameters Λ_{24} [fm^{-1}]	Λ_0 [fm^{-1}]
1.5	3.55	7.52

with the strength $C = 2.33$ fm^{-2}, corresponding to the string tension of QCD. The parameter $V_0 = -402$ MeV is only necessary to set the ground state of the whole baryon spectrum, i.e., the proton mass to $M_p = 939$ MeV; it is irrelevant, if one considers only level spacings.

The hyperfine interaction is deduced from the $SU(5)_F$ GBE leading to the instantaneous potential

$$V_{\mathrm{hf}}(\vec{r}_{ij}) = \left[V_{24}(\vec{r}_{ij}) \sum_{a=1}^{24} \lambda_i^a \lambda_j^a + V_0(\vec{r}_{ij}) \lambda_i^0 \lambda_j^0\right] \vec{\sigma}_i \cdot \vec{\sigma}_j \,. \tag{5}$$

Here, we take into account only its spin–spin component, which produces the most important hyperfine forces for the baryon spectra. While $\vec{\sigma}_i$ represent the Pauli spin matrices of $SU(2)_S$, the λ_i^a are the generalized Gell-Mann flavor matrices of $SU(5)_F$ for quark i. In addition to the exchange of the pseudoscalar 24-plet also the flavor-singlet is included because of the $U(1)$ anomaly. The radial form of the GBE potential resembles the one of the pseudoscalar meson exchange

$$V_\beta(\vec{r}_{ij}) = \frac{g_\beta^2}{4\pi} \frac{1}{12 m_i m_j} \left[\mu_\beta^2 \frac{e^{-\mu_\beta r_{ij}}}{r_{ij}} - 4\pi \delta(\vec{r}_{ij})\right] \tag{6}$$

for $\beta = 24$ and $\beta = 0$. Herein the δ-function must be smeared out leading to[10,14]

$$V_\beta(\vec{r}_{ij}) = \frac{g_\beta^2}{4\pi} \frac{1}{12 m_i m_j} \left[\mu_\beta^2 \frac{e^{-\mu_\beta r_{ij}}}{r_{ij}} - \Lambda_\beta^2 \frac{e^{-\Lambda_\beta r_{ij}}}{r_{ij}}\right] . \tag{7}$$

The UGBE RCQM uses only one GBE mass μ_{24} and a single cutoff Λ_{24} for the 24-plet of $SU(5)_F$. The singlet exchange comes with another mass μ_0 and another cutoff Λ_0 with a separate coupling constant g_0. Consequently the number of open parameters in the hyperfine interaction, determined by a best fit to the baryon spectra, could be kept as low as only three (see Table 1).

All other parameters entering the model have judiciously been predetermined by existing phenomenological insights. In this way the constituent quark masses have been set to the values as given in Table 2. The 24-plet

Table 2. Predetermined parameters of the UGBE RCQM.

Constituent-quark masses [MeV]				Exchange masses [MeV]		24-plet coupling constant
$m_u = m_d$	m_s	m_c	m_b	μ_{24}	μ_0	$g_{24}^2/4\pi$
340	480	1675	5055	139	958	0.7

Goldstone-boson mass has been assumed as the value of the π mass and similarly the singlet mass as the one of the η'. The universal coupling constant of the 24-plet has been chosen according to the value for the π-quark coupling constant, derived from the phenomenologically known π–N coupling constant via the Goldberger–Treiman relation.

We note that the earlier GBE RCQM,[9] relies on the exchange of the $SU(3)_F$ pseudoscalar (ps) octet and singlet mesons (π, K, η_8, and η_0, respectively). For each of these mesons it uses different exchange masses μ_γ and different cutoffs Λ_γ, corresponding to $\gamma = \pi$, K, $\eta = \eta_8$, and $\eta' = \eta_0$ mesons. However, for both the ps GBE and the UGBE RCQMs the fit to the u-, d-, and s-flavored baryons is practically of the same quality. It is certainly remarkable that for a unified description of all baryon spectra a single 24-plet Goldstone-boson mass as employed in the UGBE RCQM turns out to be sufficient.

3. Baryon Spectroscopy

We now demonstrate for the case of the UGBE RCQM the reproduction of a representative ensemble of baryon spectra, usually up to the first one or two excitations above the ground state in each J^P multiplet of fixed spin (total angular momentum) J and parity P. Of course, the CQM in principle produces an infinity of excited states according to the ever rising (linear) confinement interaction. However, one must not consider higher than the first few resonances to be realistically described by a quark-model hyperfine dynamics that rests on SBχS. It can only be expected to provide an effective description for the ground states and excitations not too high in the spectra.

In Fig. 1 a comparison of the theoretical levels for the N and Δ states as produced by the UGBE RCQM with experimental data as compiled by the PDG[15] is given. A notable achievement of the GBE RCQM over all other quark models consists in the right level orderings of positive- and negative-parity N excitations. In particular, the Roper resonance $N^*(1440)$ falls well below the negative-parity $N^*(1535)$ resonance as required by experiment. At the same time the usual ordering of positive- and negative-parity states is

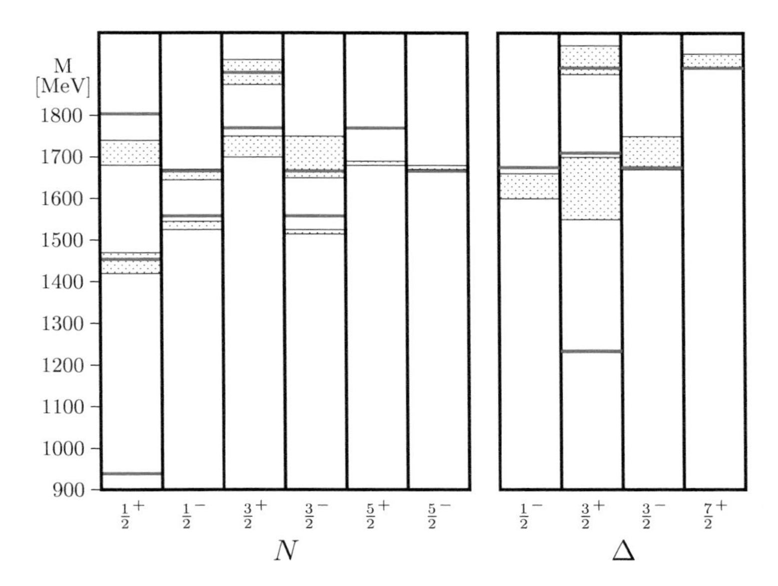

Fig. 1. Nucleon and Δ excitation spectra (solid/red levels) as produced by the UGBE RCQM in comparison to phenomenological data[15] (the gray/blue lines and shadowed/blue boxes show the experimental masses and their uncertainties).

maintained in the Λ spectrum, see Fig. 2 below. This behaviour can only be accounted for, if the hyperfine interaction is flavor dependent, like the GBE. A OGE hyperfine interaction, being flavor independent, cannot provide such a pattern.

In Fig. 2 the same comparison is given for the strange baryons. Here, there is only the serious shortcoming of reproducing the Λ(1405) resonance.

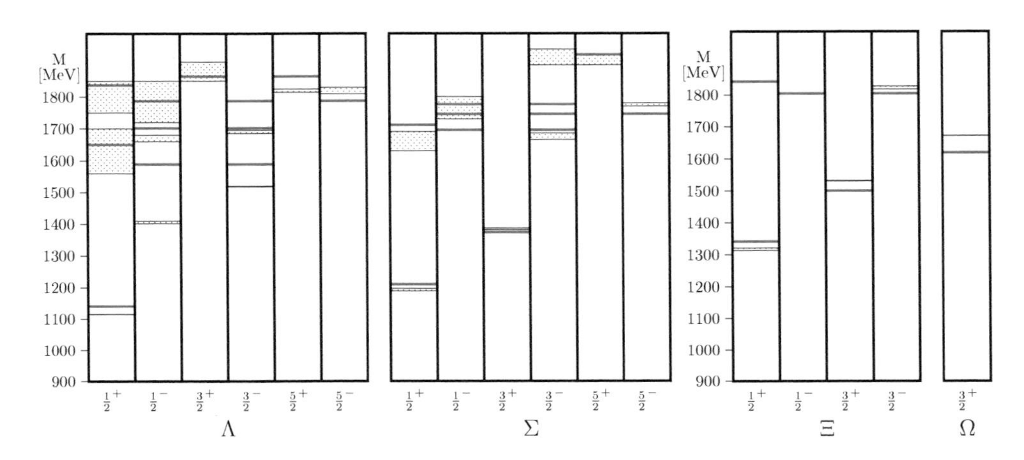

Fig. 2. Same as Fig. 1 but for the strange baryons.

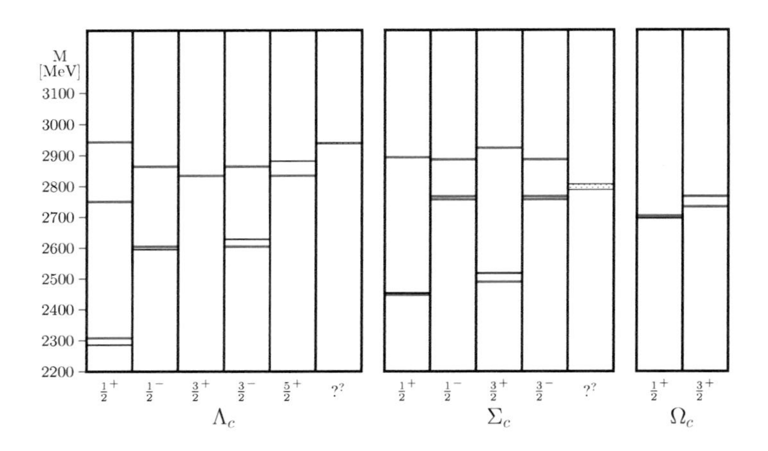

Fig. 3. Same as Fig. 1 but for charm baryons.

The theoretical level comes out much too high, a common problem to all CQMs and other approaches relying only on $\{QQQ\}$ configurations. This state lying so close to the threshold for the K–N decay, is presumably much affected by coupling to the decay channel.

The predictions of the UGBE RCQM for the charm baryons are shown in Fig. 3 in comparison to available experimental data. For some of the Λ_c and Σ_c excitations the J^P quantum numbers are not yet definitely determined from experiment, however, the UGBE RCQM can easily provide theoretical levels in the required energy range.

Finally the spectra of the one-b flavor baryons are shown in Fig. 4 and compared to the known experimental levels.

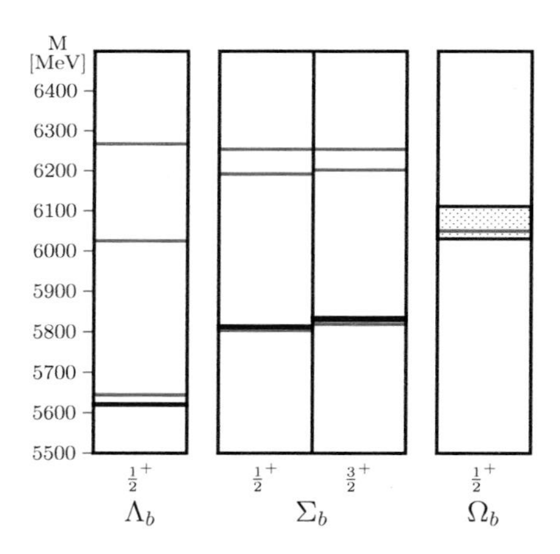

Fig. 4. Same as Fig. 1 but for bottom baryons.

Of course, there are many more states in the spectra than shown here that are tentatively well reproduced by the UGBE RCQM. However, for them we cannot compare to experiment but only to results from lattice-QCD. As far as we have seen there are no discrepancies of the predictions by the UGBE RCQM with the ones from lattice-QCD. Such detailed comparisons are given in Ref. 16. Due to limited space they are left out here.

In summary, we find a very reasonable description of all baryon spectra by the UGBE RCQM. A single notorious problem remains with regard to the $\Lambda(1405)$ state and some minor improvements would be welcome for a few higher excitations, e.g., in the N or Δ spectra.

4. Nucleon Structure

In this section we demonstrate, how the structures of the nucleons as observed with electromagnetic, weak, and gravitational probes are predicted by the GBE RCQM. We have performed relativistically invariant calculations using as input the baryon eigenstates produced by the invariant mass operator $\hat{M}$ of the pseudoscalar GBE RCQM. No further parameters are introduced in any place.

In all cases the transformations of the Poincaré group have been performed following the point form of relativistic quantum mechanics. In particular, this allows to carry out the Lorentz transformation exactly, since the corresponding generators are interaction-free. Similarly, the usual angular-momentum addition can be applied, since the generators of rotations are also interaction-free. Only the generators of spatial and time translations contain interactions, but the problem of translational invariance can be avoided by imposing momentum conservation. The whole formalism employed in our approach is summarized in Refs. 17–19.

4.1. *Elastic electromagnetic form factors of the nucleons*

The predictions of the ps GBE RCQM were obtained already long ago.[20,21] We repeat them here for completeness and because of the discussion of their flavor decompositions in the following subsection. Figure 5 shows the proton electromagnetic form factors, Fig. 6 the ones of the neutron in the range of low momentum transfers, up to $Q^2 = 4$ GeV^2. The covariant predictions are immediately found in line with the experimental data, with small deviations appearing only in a few instances, such as the subtle observables of electric radii and magnetic moments, which do not exactly coincide with the measured values (see Tables 3 and 4). If any relativistic contributions

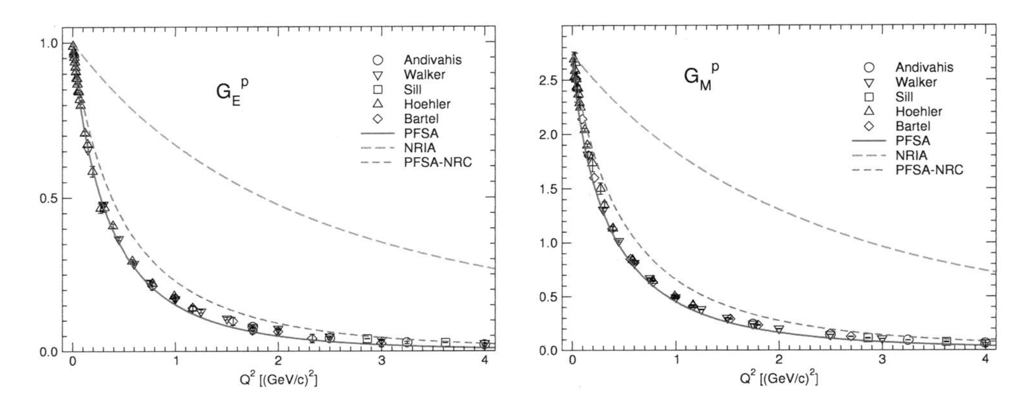

Fig. 5. Elastic electromagnetic form factors of the proton as predicted by the ps GBE RCQM[9] in comparison to the experimental data base. The solid/red curves are the covariant results obtained along the PFSM.[20] The short-dashed/magenta curves show the results with a non-relativistic electromagnetic current operator. The long-dashed/green curves represent the NRIA.

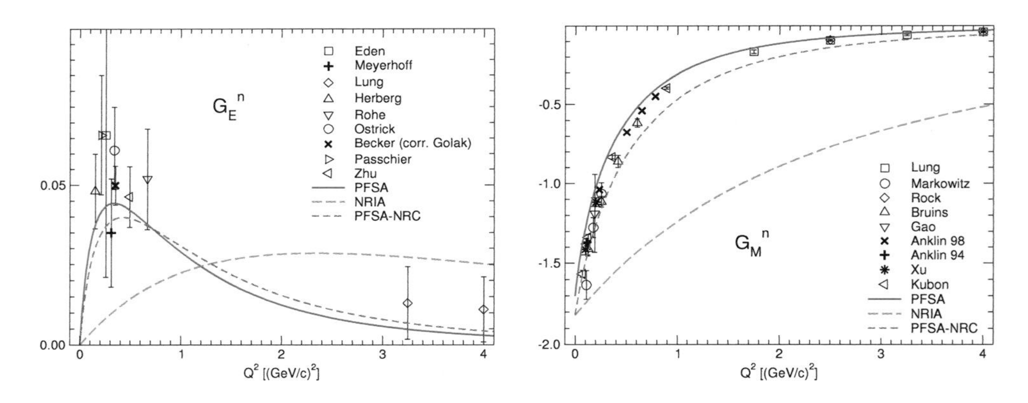

Fig. 6. Elastic electromagnetic form factors of the neutron as predicted by the ps GBE RCQM[9] in comparison to the experimental data base. Same description of the curves as in Fig. 5.

Table 3. Electric radii r_E^2 [fm^2] of the nucleons.

Nucleon	GBE RCQM	NRIA	Experiment
p	0.82	0.10	0.7700 ± 0.0090[15]
			0.70706 ± 0.00066[22]
n	-0.13	-0.01	-0.1161 ± 0.0022[15]

Table 4. Magnetic moments of the nucleons.

Nucleon	GBE RCQM	NRIA	Experiment
p	2.70	2.74	2.792847356(23)[15]
n	−1.70	−1.82	−1.9130427(5)[15]

are left out, the results fall short, with the non-relativistic impulse approximation (NRIA) failing completely. We emphasize that this is not only true with the form factors as a whole but also with the electric radii and magnetic moments; both of them contain considerable relativistic effects (see also the corresponding detailed discussions in Ref. 23, where various relativistic contributions to these observables are considered quantitatively).

4.2. *Flavor analysis of the N electromagnetic form factors*

Over the recent years phenomenological data on the flavor contents in the electromagnetic nucleon form factors have become available by analyses of the world data of electron-scattering experiments.[24–26] They provide stringent tests for any model or calculation of the electromagnetic nucleon structures. We have exposed the ps GBE RCQM to these tests.[27,28] Here we refer to some of the pertinent results, with the complete set of observables to be found in Ref. 29.

The phenomenological analysis were made under the assumption of charge symmetry, which is also observed in the GBE RCQM. Again a rather accurate reproduction of the various flavor contributions to all the proton and neutron electric as well as magnetic form factors is found, see Figs. 7 and 8. It should be emphasized that these results are produced from a RCQM relying only on $\{QQQ\}$ configurations with no contributions from other than u and d flavors. Obviously, these degrees of freedom suffice, and no contributions from either s flavors or higher Fock configurations, such as $\{QQQQ\bar{Q}\}$, need to be advocated; there is also no hint to a quark–diquark clustering in the nucleons (for quantities suspected to hint to such configurations see Refs. 27–29).

4.3. *Axial form factors of the nucleons*

The structure of the nucleons as revealed under weakly interacting probes has also been explored quite some time ago. In particular, the axial and induced pseudoscalar form factors, G_A and G_P, respectively, were calculated

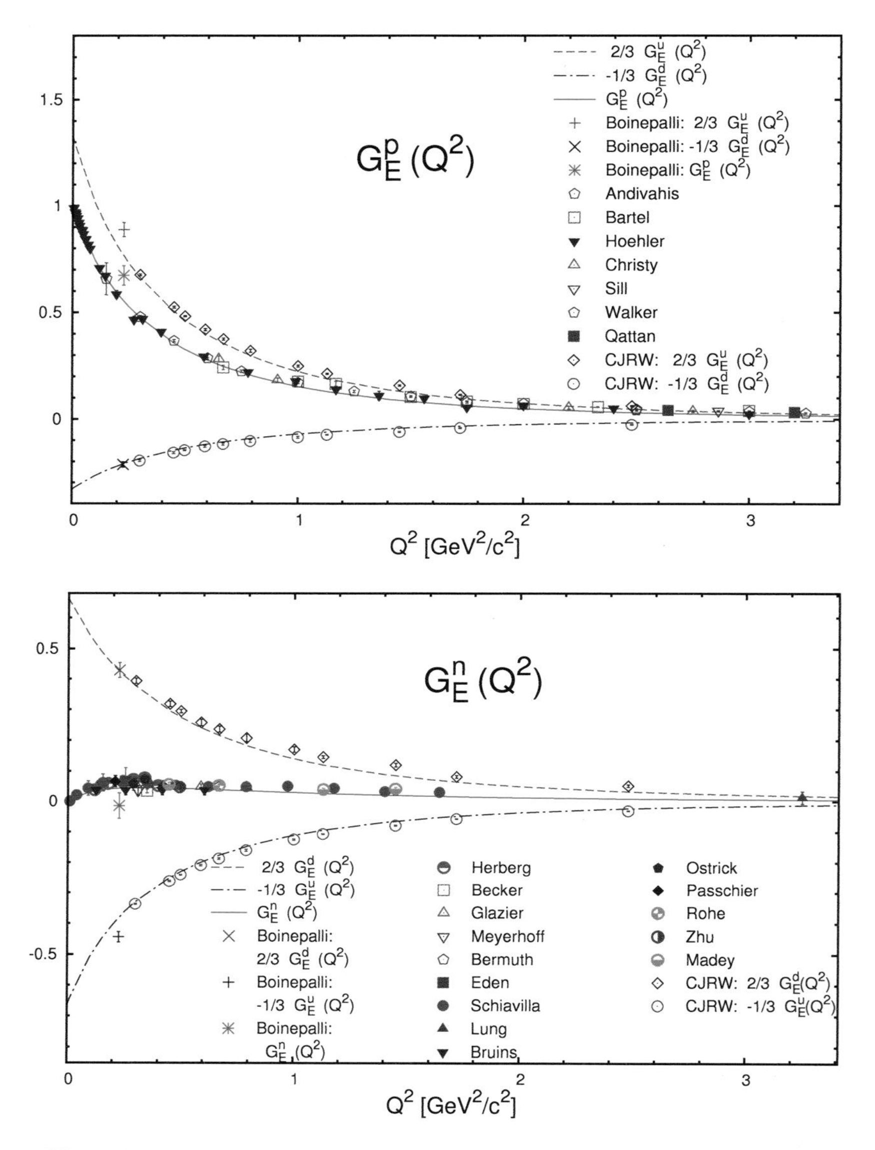

Fig. 7. Flavor decompositions of the proton and neutron electric form factors. The curves represent the predictions of the ps GBE RCQM as denoted in the inserts. The total result is compared to the world data base, the flavor contributions to the phenomenological data published in Ref. 24. A comparison is also made to points from a lattice-QCD calculation by Boinepalli *et al.*[30]

in a covariant way, using the point form in a manner consistent with the electromagnetic form factors in the previous subsection.[21,31] For completeness we show them in Sec. 4.3.

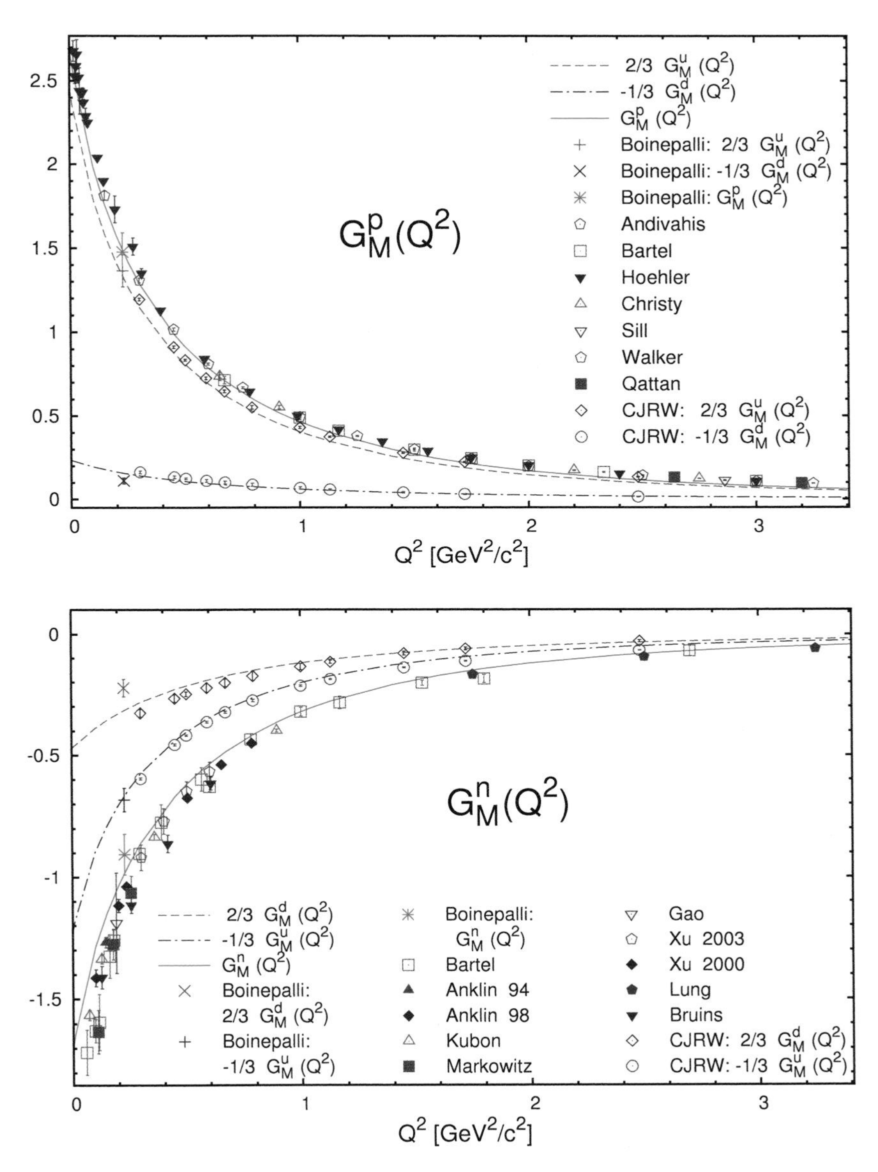

Fig. 8. Flavor decompositions of the proton and neutron magnetic form factors. Same description as in Fig. 7.

Again a rather accurate description of the experimental data is achieved by the ps GBE RCQM up to momentum transfers of ~ 4 GeV^2. The nucleon axial charge is found to be $g_A = 1.15$, i.e. the phenomenological value[15] of 1.2723 ± 0.0023 is undershooted, like in most lattice-QCD calculations.

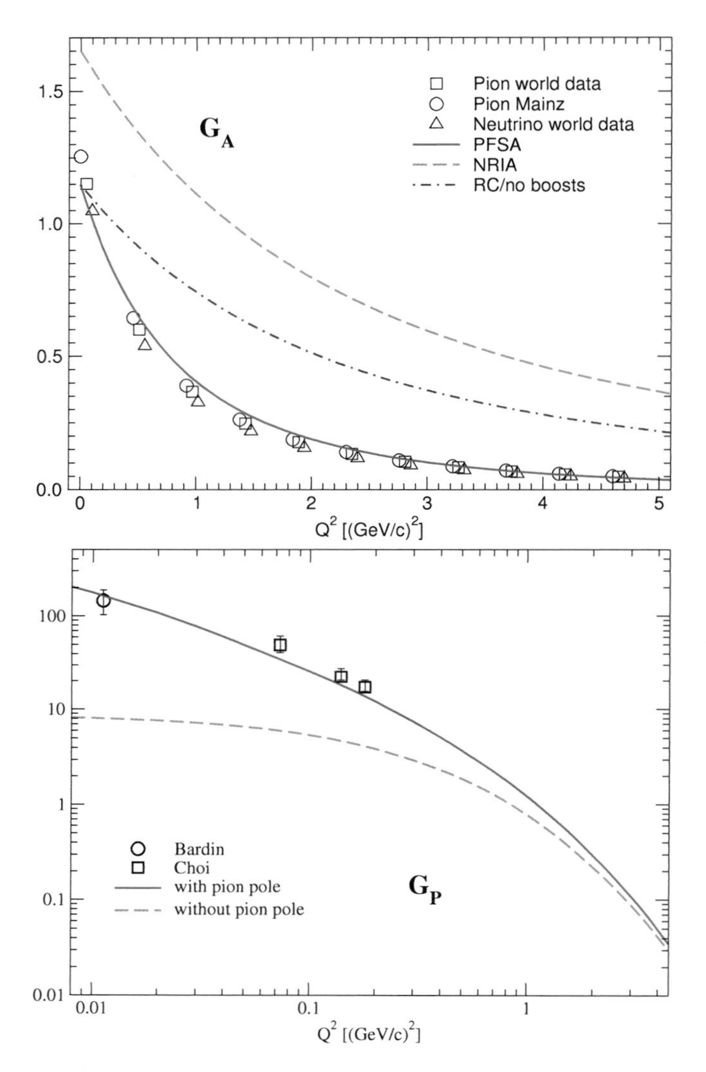

Fig. 9. Axial and induced pseudoscalar nucleon form factors as predicted by the ps GBE RCQM[9] in comparison to experimental data as indicated. The solid/red curves are the covariant results obtained along the PFSM.[21,31] The dashed-dotted/blue curve in the upper panel shows the result without boost effects, and the dashed/green curve is the NRIA. The dashed/green curve in the lower panel represents the result without inclusion of the π-pole term.

4.4. *Gravitational form factors of the nucleons*

Recently, we have also obtained the nucleon structure as revealed by gravitational forces.[16] In particular we have calculated the gravitational form factor $A(Q^2)$, which corresponds to the matrix element of the Θ^{00} component of the

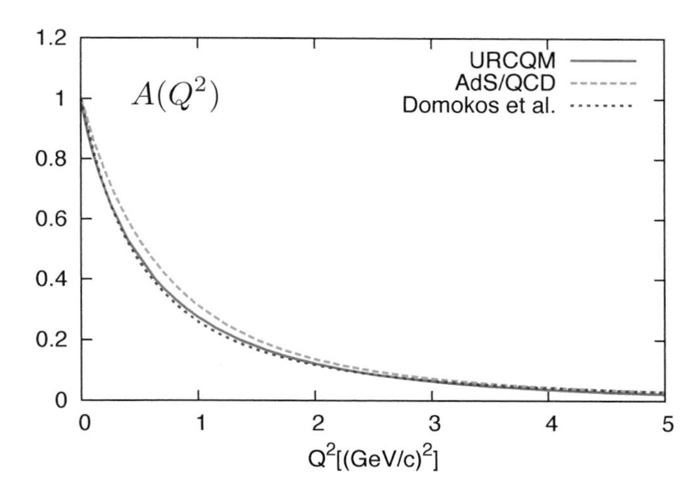

Fig. 10. Gravitational form factor $A(Q^2)$ as predicted by the UGBE RCQM[12] in comparison to results from AdS/QCD[32] and a pomeron model.[33]

energy–momentum tensor between the nucleon mass eigenstates. Of course, there is no experimental evidence, but the result of the GBE RCQM is well congruent with findings from other approaches, as is evident from Fig. 10. There all form factor results have been normalized to 1 at $Q^2 = 0$. The momentum dependences are then found very similar for all models.

5. Electroweak Structures of the Δ's and the Hyperons

The studies discussed in the previous section for the nucleon ground states have also been extended to the Δ's and all the octet as well as decuplet hyperons both for electromagnetic and axial form factors. Of course, the comparison to experiments is rather restricted in this domain. However, one can also compare with other approaches such as lattice QCD and chiral perturbation theory. Here we give only a selected insight into the performance of the GBE RCQM in describing these observables. More can be found in the literature,[34,35] and a complete account of all observables is given in Ref. 36.

5.1. *Electroweak form factors of the Δ's*

In Fig. 11 we show the electromagnetic form factors of the Δ's. In case of the Δ^+ we can compare the predictions of the ps GBE RCQM to lattice-QCD calculations by Alexandrou *et al.*,[37] and we observe a satisfying agreement.

The axial form factor of the Δ is shown in Fig. 12, where again a nice agreement is found with the most advanced lattice-QCD calculations.

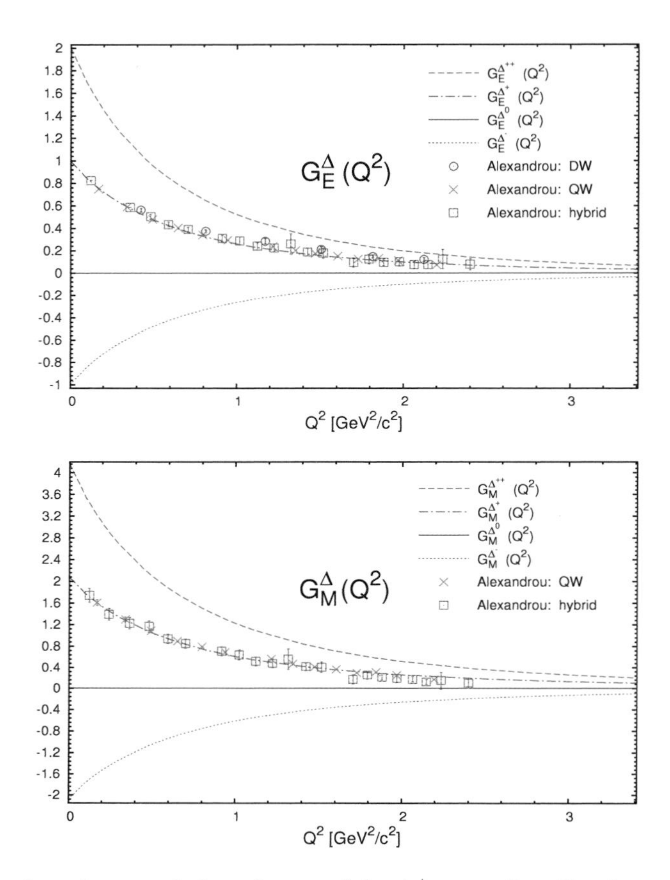

Fig. 11. Electric and magnetic form factors of the Δ^+ as predicted by the ps GBE RCQM[9] in comparison to results from lattice-QCD calculations.[37]

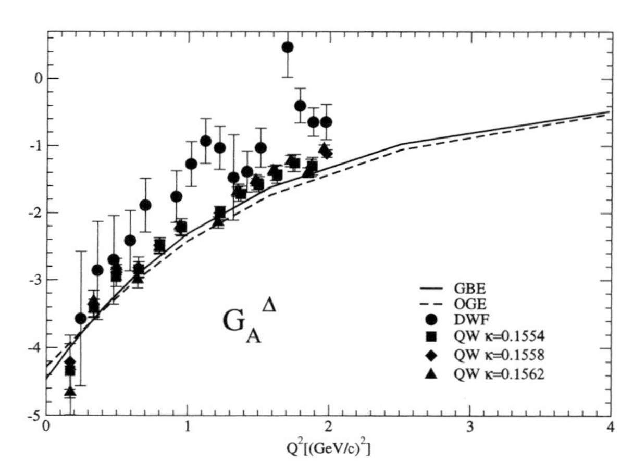

Fig. 12. Axial form factor of the Δ as predicted by the ps GBE RCQM[9] in comparison to results from lattice-QCD calculations.[38–40]

5.2. *Electroweak form factors of the hyperons*

For the hyperons we restrict ourselves to showing only the electromagnetic form factors of the Ω^-. In this case we can again compare to lattice-QCD results by Alexandrou *et al.*,[41] see Fig. 13. Again a good agreement is found, only with the magnetic form factors exhibiting indeed a similar momentum dependence but having different absolute values.

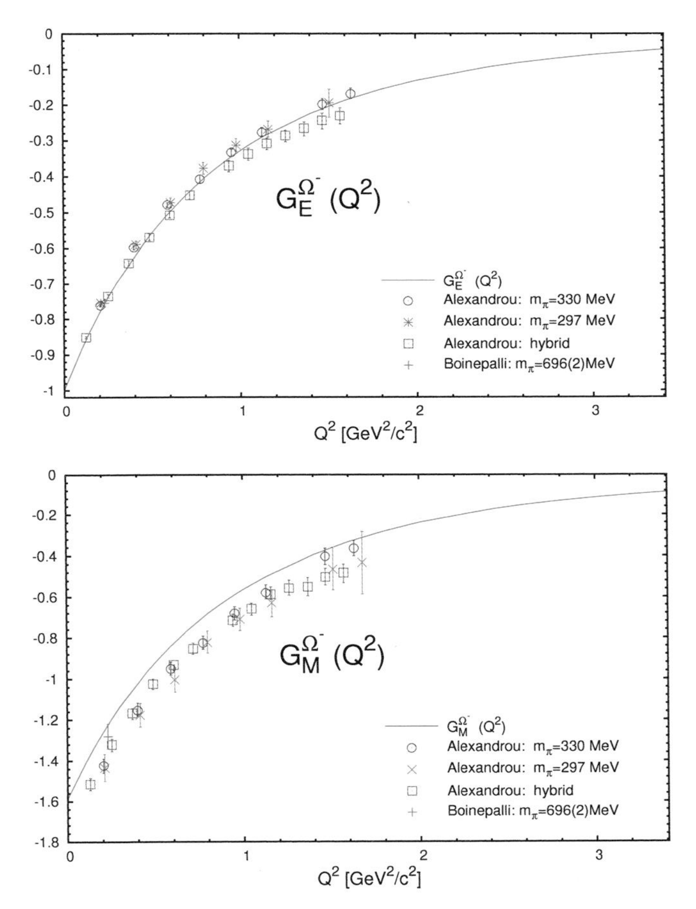

Fig. 13. Electric and magnetic form factors of the Ω^- as predicted by the ps GBE RCQM[9] in comparison to results from lattice-QCD calculations.[41]

6. πNN and $\pi N\Delta$ Vertex Form Factors

Finally, in this article we address the microscopic description of meson–baryon vertex form factors. We have applied the ps GBE RCQM to the calculation of the momentum dependences of the $G_{\pi NN}$ and $G_{\pi N\Delta}$ vertex

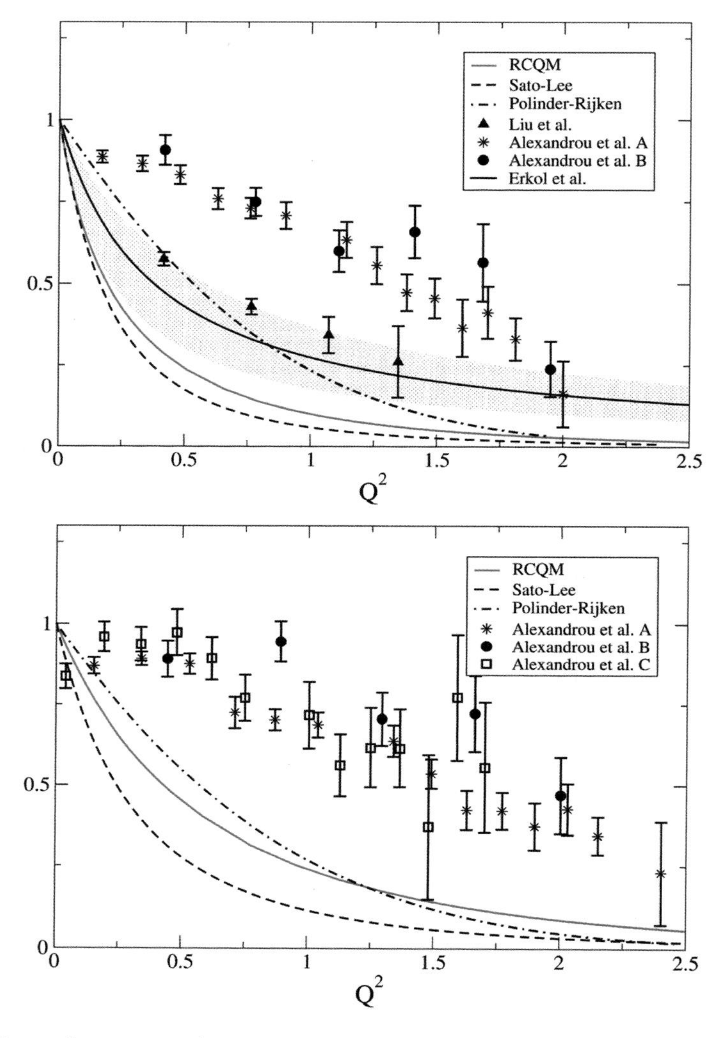

Fig. 14. $G_{\pi NN}$ (upper panel) and $G_{\pi N\Delta}$ (lower panel) vertex form factors as produced by the ps GBE RCQM[9] (solid/red curves). The dashed and dashed-dotted curves show the phenomenological form factors as employed in the meson–baryon models by Sato and Lee[43] as well as Polinder and Rijken,[44] respectively. The solid line in the shaded area represents the lattice-QCD results of Erkol *et al.*,[45] with their theoretical uncertainties, the solid triangles represent the lattice-QCD results by Liu *et al.*,[46,47] and all the other points various lattice-QCD predictions by Alexandrou *et al.*[48]

form factors.[42] The results we have obtained provide a very realistic description of these vertex form factors, since the GBE RCQM produces momentum dependences just as employed in macroscopic meson–baryon interaction models, where they were parametrized by fitting phenomenological data, see Fig. 14. In addition, the coupling constants turn out to be very similar to the ones used in these meson–baryon models (see the vertex parametrizations provided in Ref. 42). Here, however, we do not find agreement with lattice-QCD results, where the latter also differ drastically among each other. We are tempted to conclude that the GBE RCQM provides a more realistic account of the structure of meson–baryon interaction vertices.

7. Summary and Discussion

We have collected into this article a number of baryon observables produced by the GBE RCQM. Thereby we have demonstrated that a RCQM can nowadays indeed produce a very realistic effective description of low-energy baryons. Having reviewed the spectroscopy of all measured baryons, the nucleon structure as revealed under electromagnetic, weak, and gravitational probes, as well as the Δ and hyperon electroweak form factors, we have not found a single place, where the RCQM fails or appears inadequate. The same is true with observables not shown here explicitly due to space limitations. When experiments are missing but lattice-QCD results are available, most of the results obtained with the GBE RCQM coincide with them. Except for a few minor deviations here and there, discrepancies to lattice-QCD calculations occur in the πNN and $\pi N\Delta$ vertex form factors. To date the reason for that remains unclear.

Due to limited space in the present article we have not addressed baryon resonance decays. For this type of problems the situation is not so pleasant. First covariant results of decay widths usually fall below the experimental data.[49–52] We suspect that a more realistic description of resonant states is required, providing an explicit inclusion of the various decay channels. In other words, a coupled-channels RCQM should be set up taking into account also the mesonic decay channels. Exploratory studies in this direction have already turned out to be promising.[53]

Half a century after the concept of quarks was introduced, we find the constituent-quark model nowadays as a reliable effective tool to describe baryon properties and reactions. According to our experience the most important ingredients consist in

- a strict observation of relativistic invariance and
- the inclusion of the Q–Q dynamics according to the SBχS of low-energy QCD.

Acknowledgments

The results discussed in the present article originated from a number of joint efforts with diploma and doctoral students as well as collaborators at several foreign institutions. I am grateful to all of them for having set the various aims through in elaborating on many aspects of the RCQM. The most recent results have been worked out with Ki-Seok Choi, Joseph P. Day, and M. Rohrmoser, under constant advice by Robert F. Wagenbrunn, and I want to express my particular thanks to them.

References

1. M. Gell-Mann, *Phys. Lett.* **8**, 214 (1964).
2. M. Gell-Mann, *Acta Phys. Austr. Suppl.* **9**, 733 (1972).
3. H. Fritzsch and M. Gell-Mann, *eConf* **C720906V2**, 135 (1972), arXiv:hep-ph/0208010.
4. H. Fritzsch, M. Gell-Mann and H. Leutwyler, *Phys. Lett. B* **47**, 365 (1973).
5. S. Capstick, S. Godfrey, N. Isgur and J. E. Paton, *Phys. Lett. B* **175**, 457 (1986).
6. S. Capstick and N. Isgur, *Phys. Rev. D* **34**, 2809 (1986).
7. G. S. Bali and K. Schilling, *Phys. Rev. D* **46**, 2636 (1992).
8. L. Y. Glozman and D. O. Riska, *Phys. Rept.* **268**, 263 (1996).
9. L. Y. Glozman, W. Plessas, K. Varga and R. F. Wagenbrunn, *Phys. Rev. D* **58**, 094030 (1998).
10. L. Y. Glozman, Z. Papp, W. Plessas, K. Varga and R. F. Wagenbrunn, *Phys. Rev. C* **57**, 3406 (1998).
11. L. Y. Glozman and D. O. Riska, *Nucl. Phys. A* **603**, 326 (1996) [Erratum: *ibid.* **620**, 510 (1997)].
12. J. P. Day, W. Plessas and K. S. Choi, arXiv:1205.6918 [hep-ph].
13. J. P. Day, K. S. Choi and W. Plessas, *Few-Body Syst.* **54**, 329 (2013).
14. L. Ya. Glozman, Z. Papp, W. Plessas, K. Varga and R. F. Wagenbrunn, *Phys. Rev. C* **61**, 019804 (2000).
15. Particle Data Group Collab. (K. A. Olive *et al.*), *Chin. Phys. C* **38**, 090001 (2014).
16. J. P. Day, PhD Thesis, University of Graz (2013).
17. T. Melde, L. Canton, W. Plessas and R. F. Wagenbrunn, *Eur. Phys. J. A* **25**, 97 (2005).
18. T. Melde, K. Berger, L. Canton, W. Plessas and R. F. Wagenbrunn, *Phys. Rev. D* **76**, 074020 (2007).
19. W. Plessas, *PoS* **LC2010**, 017 (2010), arXiv:1011.0156 [hep-ph].
20. R. F. Wagenbrunn, S. Boffi, W. Klink, W. Plessas and M. Radici, *Phys. Lett. B* **511**, 33 (2001).
21. S. Boffi, L. Y. Glozman, W. Klink, W. Plessas, M. Radici and R. F. Wagenbrunn, *Eur. Phys. J. A* **14**, 17 (2002).
22. A. Antognini *et al.*, *Science* **339**, 417 (2013).

23. K. Berger, R. F. Wagenbrunn and W. Plessas, *Phys. Rev. D* **70**, 094027 (2004).
24. G. D. Cates, C. W. de Jager, S. Riordan and B. Wojtsekhowski, *Phys. Rev. Lett.* **106**, 252003 (2011).
25. I. A. Qattan and J. Arrington, *Phys. Rev. C* **86**, 065210 (2012).
26. M. Diehl and P. Kroll, *Eur. Phys. J. C* **73**, 2397 (2013).
27. M. Rohrmoser, K.-S. Choi and W. Plessas, arXiv:1110.3665 [hep-ph].
28. M. Rohrmoser, K. S. Choi and W. Plessas, *Acta Phys. Polon. Suppl.* **6**, 371 (2013).
29. M. Rohrmoser, Master Thesis, University of Graz (2013).
30. S. Boinepalli, D. B. Leinweber, A. G. Williams, J. M. Zanotti and J. B. Zhang, *Phys. Rev. D* **74**, 093005 (2006).
31. L. Y. Glozman, M. Radici, R. F. Wagenbrunn, S. Boffi, W. Klink and W. Plessas, *Phys. Lett. B* **516**, 183 (2001).
32. S. J. Brodsky and G. F. de Teramond, *Phys. Rev. D* **78**, 025032 (2008).
33. S. K. Domokos, J. A. Harvey and N. Mann, *Phys. Rev. D* **80**, 126015 (2009).
34. K.-S. Choi, W. Plessas and R. F. Wagenbrunn, *Phys. Rev. D* **82**, 014007 (2010); *ibid.* 039901 (2010).
35. K.-S. Choi, W. Plessas and R. F. Wagenbrunn, *Few-Body Syst.* **50**, 203 (2011).
36. K.-S. Choi, PhD Thesis, University of Graz (2011).
37. C. Alexandrou *et al.*, *Phys. Rev. D* **79**, 014507 (2009).
38. C. Alexandrou, E. B. Gregory, T. Korzec, G. Koutsou, J. Negele, T. Sato and A. Tsapalis, *PoS* **Lattice2010**, 141 (2010).
39. C. Alexandrou, E. B. Gregory, T. Korzec, G. Koutsou, J. W. Negele, T. Sato and A. Tsapalis, *Phys. Rev. Lett.* **107**, 141601 (2011).
40. C. Alexandrou, E. B. Gregory, T. Korzec, G. Koutsou, J. W. Negele, T. Sato and A. Tsapalis, *Phys. Rev. D* **87**, 114513 (2013).
41. C. Alexandrou, T. Korzec, G. Koutsou, J. W. Negele and Y. Proestos, *Phys. Rev. D* **82**, 034504 (2010).
42. T. Melde, L. Canton and W. Plessas, *Phys. Rev. Lett.* **102**, 132002 (2009).
43. T. Sato and T. S. H. Lee, *Phys. Rev. C* **54**, 2660 (1996).
44. H. Polinder and T. A. Rijken, *Phys. Rev. C* **72**, 065210 (2005); *ibid.* 065211 (2005).
45. G. Erkol, M. Oka and T. T. Takahashi, *Phys. Rev. D* **79**, 074509 (2009).
46. K. F. Liu, S. J. Dong, T. Draper and W. Wilcox, *Phys. Rev. Lett.* **74**, 2172 (1995).
47. K. F. Liu *et al.*, *Phys. Rev. D* **59**, 112001 (1999).
48. C. Alexandrou, G. Koutsou, T. Leontiou, J. W. Negele and A. Tsapalis, *Phys. Rev. D* **76**, 094511 (2007).
49. T. Melde, W. Plessas and R. F. Wagenbrunn, *Phys. Rev. C* **72**, 015207 (2005); *ibid.* **74**, 069901 (2006).
50. B. Sengl, T. Melde and W. Plessas, *Phys. Rev. D* **76**, 054008 (2007).
51. T. Melde, W. Plessas and B. Sengl, *Phys. Rev. C* **76**, 025204 (2007).
52. T. Melde, W. Plessas and B. Sengl, *Phys. Rev. D* **77**, 114002 (2008).
53. R. Kleinhappel, W. Plessas and W. Schweiger, *Few-Body Syst.* **54**, 339 (2013).

From Ω^- to Ω_b, Doubly Heavy Baryons and Exotics

Marek Karliner

Raymond and Beverly Sackler School of Physics and Astronomy, Tel Aviv University, Tel Aviv, Israel
marek@proton.tau.ac.il

I discuss accurate theoretical predictions for masses of baryons containing the b quark. I point out an approximate effective supersymmetry between heavy quark baryons and mesons and provide predictions for the magnetic moments of Λ_c and Λ_b. Proper treatment of the color-magnetic hyperfine interaction in QCD is crucial for obtaining these results. Closely related methods are then applied to doubly-heavy hadrons: the recently observed exotic $\bar{Q}Q\bar{q}q$ mesons and QQq baryons. Predictions are given for the masses and decay modes of additional $\bar{Q}Q\bar{q}q$ mesons and of QQq baryons: $\Xi_{cc}(ccq)$, $\Xi_{bb}(bbq)$ and $\Xi_{bc}(bcq)$, which might soon be seen experimentally, as indicated by the large number of B_c mesons observed by LHCb.

1. Introduction

QCD describes hadrons as valence quarks in a sea of gluons and $\bar{q}q$ pairs. At distances above ~ 1 GeV^{-1} quarks acquire an effective *constituent mass* due to chiral symmetry breaking. A hadron can then be thought of as a bound state of constituent quarks. In the zeroth-order approximation the hadron mass M is then given by the sum of the masses of its constituent quarks m_i,

$$M = \sum_i m_i \,.$$

The binding and kinetic energies are "swallowed" by the constituent quarks masses. The first and most important correction comes from the color hyperfine (HF) chromomagnetic interaction,

$$\begin{aligned} M &= \sum_i m_i + V_{i<j}^{\mathrm{HF(QCD)}} \,, \\ V_{ij}^{\mathrm{HF(QCD)}} &= v_0(\lambda_i \cdot \lambda_j)\frac{\sigma_i \cdot \sigma_j}{m_i m_j}\langle\psi|\delta(r_i - r_j)|\psi\rangle \,, \end{aligned} \tag{1}$$

where v_0 gives the overall strength of the hyperfine interaction, $\lambda_{i,j}$ are the $SU(3)$ color matrices, $\sigma_{i,j}$ are the quark spin operators and $|\psi\rangle$ is the hadron wave function. This is a contact spin–spin interaction, analogous to the EM hyperfine interaction, which is a product of the magnetic moments,

$$V_{ij}^{\mathrm{HF(QED)}} \propto \mu_i \cdot \mu_j = e^2 \frac{\sigma_i \cdot \sigma_j}{m_i m_j} \tag{2}$$

in QCD, the $SU(3)_c$ generators take place of the electric charge. From Eq. (1) many very accurate results have been obtained for the masses of the ground-state hadrons. Nevertheless, several caveats are in order. First, this is a low-energy phenomenological model, still awaiting a rigorous derivation from QCD. It is far from providing a complete description of the hadronic spectrum, but it provides excellent predictions for mass splittings and magnetic moments. The crucial assumptions of the model are:

(a) hyperfine interaction is considered as a perturbation which does not change the wave function;
(b) effective masses of quarks are the same inside mesons and baryons;
(c) there are no 3-body effects.

2. Quark Masses

The first example of the application of Eq. (1) is Table 1, showing the quark mass differences obtained from mesons and baryons.[1] The mass difference between two quarks of different flavors denoted by i and j are seen to have the same value to a good approximation when they are bound to a "spectator" quark of a given flavor.

On the other hand, Table 1 shows clearly that *constituent quark mass differences depend strongly on the flavor of the spectator quark.* For example, $m_s - m_d \approx 180$ MeV when the spectator is a light quark but the same mass difference is only about 90 MeV when the spectator is a b quark.

Since these are *effective masses*, we should not be surprised that their difference is affected by the environment, but the large size of the shift is quite surprising and its quantitative derivation from QCD is an outstanding challenge for theory.

A second example shows how we can extract the ratio of the constituent quark masses from the ratio of the the hyperfine splittings in the corresponding mesons:

$$\frac{M(K^*) - M(K)}{M(D^*) - M(D)} = \frac{4v_0 \frac{\lambda_u \cdot \lambda_s}{m_u m_s} \langle\psi|\delta(r)|\psi\rangle}{4v_0 \frac{\lambda_u \cdot \lambda_c}{m_u m_c} \langle\psi|\delta(r)|\psi\rangle} \approx \frac{m_c}{m_s} \,. \tag{3}$$

Table 1. Quark mass differences from baryons and mesons.

Observable	Baryons		Mesons				$\Delta m_{\rm Bar}$ MeV	$\Delta m_{\rm Mes}$ MeV
			$J=1$		$J=0$			
	B_i	B_j	$\mathcal{V}_i$	$\mathcal{V}_j$	$\mathcal{P}_i$	$\mathcal{P}_j$		
$\langle m_s - m_u\rangle_d$	sud	uud	$s\bar{d}$	$u\bar{d}$	$s\bar{d}$	$u\bar{d}$	177	179
	Λ	N	K^*	ρ	K	π		
$\langle m_s - m_u\rangle_c$			$c\bar{s}$	$c\bar{u}$	$c\bar{s}$	$c\bar{u}$		103
			D_s^*	D_s^*	D_s	D_s		
$\langle m_s - m_u\rangle_b$			$b\bar{s}$	$b\bar{u}$	$b\bar{s}$	$b\bar{u}$		91
			B_s^*	B_s^*	B_s	B_s		
$\langle m_c - m_u\rangle_d$	cud	uud	$c\bar{d}$	$u\bar{d}$	$c\bar{d}$	$u\bar{d}$	1346	1360
	Λ_c	N	D^*	ρ	D	π		
$\langle m_c - m_u\rangle_c$			$c\bar{c}$	$u\bar{c}$	$c\bar{c}$	$u\bar{c}$		1095
			ψ	D^*	η_c	D		
$\langle m_c - m_s\rangle_d$	cud	sud	$c\bar{d}$	$s\bar{d}$	$c\bar{d}$	$s\bar{d}$	1169	1180
	Λ_c	Λ	D^*	K^*	D	K		
$\langle m_c - m_s\rangle_c$			$c\bar{c}$	$s\bar{c}$	$c\bar{c}$	$s\bar{c}$		991
			ψ	D_s^*	η_c	D_s		
$\langle m_b - m_u\rangle_d$	bud	uud	$b\bar{d}$	$u\bar{d}$	$b\bar{d}$	$u\bar{d}$	4685	4700
	Λ_b	N	B^*	ρ	B	π		
$\langle m_b - m_u\rangle_s$			$b\bar{s}$	$u\bar{s}$	$b\bar{s}$	$u\bar{s}$		4613
			B_s^*	K^*	B_s	K		
$\langle m_b - m_s\rangle_d$	bud	sud	$b\bar{d}$	$s\bar{d}$	$b\bar{d}$	$s\bar{d}$	4508	4521
	Λ_b	Λ	B^*	K^*	B	K		
$\langle m_b - m_c\rangle_d$	bud	sud	$b\bar{d}$	$c\bar{d}$	$b\bar{d}$	$c\bar{d}$	3339	3341
	Λ_b	Λ_c	B^*	D^*	B	D		
$\langle m_b - m_c\rangle_s$			$b\bar{s}$	$c\bar{s}$	$b\bar{s}$	$c\bar{s}$		3328
			B_s^*	D_s^*	B_s	D_s		

2.1. *Color hyperfine splitting in baryons*

Analogously to Eq. (3) we can obtain the quark mass ratio from baryons, and can then compare the two sets of results:

$$\left(\frac{m_c}{m_s}\right)_{\rm Bar} = \frac{M_{\Sigma^*} - M_{\Sigma}}{M_{\Sigma_c^*} - M_{\Sigma_c}} = 2.84\,, \quad \left(\frac{m_c}{m_s}\right)_{\rm Mes} = \frac{M_{K^*} - M_K}{M_{D^*} - M_D} = 2.81\,, \quad (4)$$

$$\left(\frac{m_c}{m_u}\right)_{\rm Bar} = \frac{M_{\Delta} - M_p}{M_{\Sigma_c^*} - M_{\Sigma_c}} = 4.36\,, \quad \left(\frac{m_c}{m_u}\right)_{\rm Mes} = \frac{M_{\rho} - M_{\pi}}{M_{D^*} - M_D} = 4.46\,. \quad (5)$$

We find the same value from mesons and baryons $\pm 2\%$.

The presence of a fourth flavor gives us the possibility of obtaining a new type of mass relation between mesons and baryons. The $\Sigma - \Lambda$ mass difference is believed to be due to the difference between the $u-d$ and $u-s$ hyperfine interactions. Similarly, the $\Sigma_c - \Lambda_c$ mass difference is believed to be due to the difference between the $u-d$ and $u-c$ hyperfine interactions. We therefore obtain the relation

$$\begin{aligned} \left(\frac{\frac{1}{m_u^2}-\frac{1}{m_u m_c}}{\frac{1}{m_u^2}-\frac{1}{m_u m_s}}\right)_{\text{Bar}} &= \frac{M_{\Sigma_c}-M_{\Lambda_c}}{M_\Sigma - M_\Lambda} = 2.16\,, \\ \left(\frac{\frac{1}{m_u^2}-\frac{1}{m_u m_c}}{\frac{1}{m_u^2}-\frac{1}{m_u m_s}}\right)_{\text{Mes}} &= \frac{(M_\rho - M_\pi)-(M_{D^*}-M_D)}{(M_\rho - M_\pi)-(M_{K^*}-M_K)} = 2.10\,. \end{aligned} \tag{6}$$

The meson and baryon relations agree to $\pm 3\%$.

We can write down an analogous relation for hadrons containing the b quark instead of the s quark, obtaining the prediction for splitting between Σ_b and Λ_b:

$$\frac{M_{\Sigma_b}-M_{\Lambda_b}}{M_\Sigma - M_\Lambda} = \frac{(M_\rho - M_\pi)-(M_{B^*}-M_B)}{(M_\rho - M_\pi)-(M_{K^*}-M_K)} = 2.51 \tag{7}$$

yielding $M(\Sigma_b) - M(\Lambda_b) = 194$ MeV.[1,2]

This splitting was subsequently measured by CDF.[3] They obtained the masses of the Σ_b^- and Σ_b^+ from the decay $\Sigma_b \to \Lambda_b + \pi$ by measuring the corresponding mass differences in MeV

$$\begin{aligned} M(\Sigma_b^-) - M(\Lambda_b) &= 195.5^{+1.0}_{-1.0}\,(\text{stat.}) \pm 0.1\,(\text{syst.})\,, \\ M(\Sigma_b^+) - M(\Lambda_b) &= 188.0^{+2.0}_{-2.3}\,(\text{stat.}) \pm 0.1\,(\text{syst.}) \end{aligned} \tag{8}$$

with isospin-averaged mass difference $M(\Sigma_b) - M(\Lambda_b) = 192$ MeV.

There is also the prediction for the spin splittings, good to 5%

$$M(\Sigma_b^*) - M(\Sigma_b) = \frac{M(B^*)-M(B)}{M(K^*)-M(K)} \cdot [M(\Sigma^*) - M(\Sigma)] = 22 \text{ MeV} \tag{9}$$

to be compared with 21 MeV from the isospin-average of CDF measurements.[3]

The challenge is to understand how and under what assumptions one can derive from QCD the very simple model of hadronic structure at low energies which leads to such accurate predictions.

3. Effective Meson–Baryon SUSY

Some of the results described above can be understood[2] by observing that in the hadronic spectrum there is an approximate effective supersymmetry between mesons and baryons related by replacing a light antiquark by a light diquark.

Fig. 1. A heavy quark coupled to "brown muck" color antitriplet.

This supersymmetry transformation goes beyond the simple constituent quark model. It assumes only a valence quark of flavor i with a model independent structure bound to "light quark brown muck color antitriplet" of model-independent structure carrying the quantum numbers of a light antiquark or a light diquark, cf. Fig. 1. Since it assumes no model for the valence quark, nor the brown muck antitriplet coupled to the valence quark, it holds also for the quark–parton model in which the valence is carried by a current quark and the rest of the hadron is a complicated mixture of quarks and antiquarks.

This light quark supersymmetry transformation, denoted here by T^S_{LS}, connects a meson denoted by $|\mathcal{M}(\bar{q}Q_i)\rangle$ and a baryon denoted by $|\mathcal{B}([qq]_S Q_i)\rangle$ both containing the same valence quark of some fixed flavor Q_i, $i = (u, s, c, b)$ and a light color-antitriplet "brown muck" state with the flavor and baryon quantum numbers respectively of an antiquark $\bar{q}$ (u or d) and two light quarks coupled to a diquark of spin S,

$$T^S_{LS}|\mathcal{M}(\bar{q}Q_i)\rangle \equiv |\mathcal{B}([qq]_S Q_i)\rangle \,. \tag{10}$$

The mass difference between the meson and baryon related by this T^S_{LS} transformation has been shown[4] to be independent of the quark flavor i for all four flavors (u, s, c, b) when the contribution of the hyperfine interaction energies is removed. For the two cases of spin-0[4] $S = 0$ and spin-1 $S = 1$ diquarks,

$$\begin{aligned} M(N) - \tilde{M}(\rho) &= 323 \text{ MeV} \approx \\ \approx M(\Lambda) - \tilde{M}(K^*) &= 321 \text{ MeV} \approx \\ \approx M(\Lambda_c) - \tilde{M}(D^*) &= 312 \text{ MeV} \approx \\ \approx M(\Lambda_b) - \tilde{M}(B^*) &= 310 \text{ MeV} \,, \end{aligned} \tag{11}$$

$$
\begin{aligned}
\tilde{M}(\Delta) - \tilde{M}(\rho) &= 517.56 \text{ MeV} \approx \\
\approx \tilde{M}(\Sigma) - \tilde{M}(K^*) &= 526.43 \text{ MeV} \approx \\
\approx \tilde{M}(\Sigma_c) - \tilde{M}(D^*) &= 523.95 \text{ MeV} \approx \\
\approx \tilde{M}(\Sigma_b) - \tilde{M}(B^*) &= 512.45 \text{ MeV}\,,
\end{aligned} \tag{12}
$$

where

$$
\tilde{M}(V_i) \equiv \frac{3M_{\mathcal{V}_i} + M_{\mathcal{P}_i}}{4}\,; \tag{13}
$$

are the weighted averages of vector and pseudoscalar meson masses, denoted respectively by $M_{\mathcal{V}_i}$ and $M_{\mathcal{P}_i}$, which cancel their hyperfine contribution, and

$$
\tilde{M}(\Sigma_i) \equiv \frac{2M_{\Sigma_i^*} + M_{\Sigma_i}}{3}\,, \qquad \tilde{M}(\Delta) \equiv \frac{2M_\Delta + M_N}{3} \tag{14}
$$

are the analogous weighted averages of baryon masses which cancel the hyperfine contribution between the diquark and the additional quark.

4. Magnetic Moments of Heavy Quark Baryons

In Λ, Λ_c and Λ_b baryons the light quarks are coupled to spin 0. Therefore the magnetic moments of these baryons are determined by the magnetic moments of the s, c and b quarks, respectively. The latter are proportional to the chromomagnetic moments which determine the hyperfine splitting in baryon spectra. We can use this fact to predict the Λ_c and Λ_b baryon magnetic moments by relating them to the hyperfine splittings in the same way as given in the original prediction[5] of the Λ magnetic moment,

$$
\mu_\Lambda = -\frac{\mu_p}{3} \cdot \frac{M_{\Sigma^*} - M_\Sigma}{M_\Delta - M_N} = -0.61 \text{ n.m.} \quad (\text{EXP} = -0.61 \text{ n.m.})\,. \tag{15}
$$

We obtain

$$
\begin{aligned}
\mu_{\Lambda_c} &= -2\mu_\Lambda \cdot \frac{M_{\Sigma_c^*} - M_{\Sigma_c}}{M_{\Sigma^*} - M_\Sigma} = 0.43 \text{ n.m.}\,, \\
\mu_{\Lambda_b} &= \mu_\Lambda \cdot \frac{M_{\Sigma_b^*} - M_{\Sigma_b}}{M_{\Sigma^*} - M_\Sigma} = -0.067 \text{ n.m.}
\end{aligned} \tag{16}
$$

We hope these observables can be measured in foreseeable future and view the predictions (16) as a challenge for the experimental community.

Table 2. Ratio of the hyperfine splittings in mesons and baryons, for different potentials.

	Δ_K/Δ_Σ	$\Delta_D/\Delta_{\Sigma_c}$	$\Delta_B/\Delta_{\Sigma_b}$
m_Q/m_q	1.33	4.75	14
EXP	2.08 ± 0.01	2.18 ± 0.08	2.15 ± 0.20
Harmonic	1.65	1.62	1.59
Coulomb	5.07 ± 0.08	5.62 ± 0.02	5.75 ± 0.01
Linear	1.88 ± 0.06	1.88 ± 0.08	1.86 ± 0.08
Log	2.38 ± 0.02	2.43 ± 0.02	2.43 ± 0.01
Cornell ($k = 0.28$)	2.10 ± 0.05	2.16 ± 0.07	2.17 ± 0.08

5. Testing Confining Potentials Through Meson/Baryon hyperfine Splitting Ratio

The ratio of color hyperfine splitting in mesons and baryons is a sensitive probe of the details of the confining potential. This is because this ratio depends only on the value of the wave function at the origin, which in turn is determined by the confining potential and by the ratio of quark masses, together with the fact that the color quark–antiquark interaction in mesons is twice as strong as the quark–quark interaction in baryons, $(\lambda_u \cdot \lambda_s)_{\text{meson}} = 2(\lambda_u \cdot \lambda_s)_{\text{baryon}}$. We then have

$$\frac{M_(K^*) - M(K)}{M(\Sigma^*) - M(\Sigma)} = \frac{4}{3} \frac{\langle\psi|\delta(\mathbf{r}_u - \mathbf{r}_s)|\psi\rangle_{\text{meson}}}{\langle\psi|\delta(\mathbf{r}_u - \mathbf{r}_s)|\psi\rangle_{\text{baryon}}} \tag{17}$$

and analogous expressions with the s quark replaced by another heavy quark Q. From the experiment we have three data points for this ratio, with $Q = s, c, b$. We can then compute the ratio (17) for five different representative confining potentials and compare with experiment. The five potentials are

- harmonic oscillator,
- Coulomb interaction,
- linear potential,
- linear + Coulomb, i.e. Cornell potential,
- logarithmic.

The results are shown in Fig. 2 and Table 2.[6]

For all potentials which contain one coupling constant the coupling strength cancels in the meson–baryon ratio. The Cornell potential which is a combination of a Coulomb and linear potential contains two couplings, one of which cancels in the meson–baryon ratio. The remaining coupling is denoted by k. The gray band corresponds to the range of values $0.2 < k < 0.5$ of

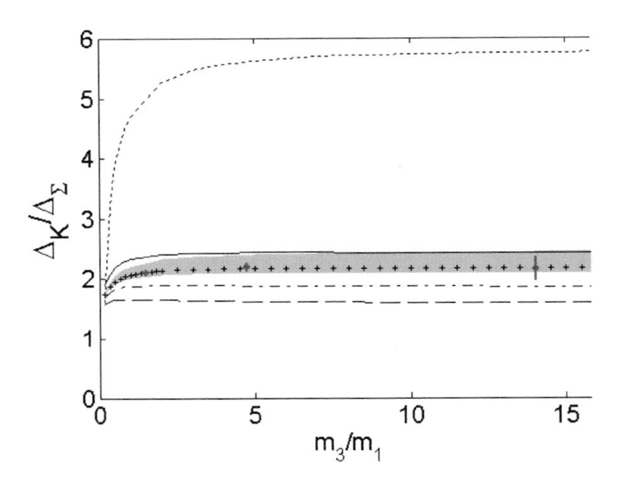

Fig. 2. Ratio of the hyperfine splittings in mesons and baryons, as function of the quark mass ratio. Shaded region: Cornell potential for $0.2 < k < 0.5$; crosses: Cornell, $k = 0.28$; long dashes: harmonic oscillator; short dashes: Coulomb; dot-dash: linear; continuous: logarithmic; thick dots: experimental data.

the Cornell potential. The crosses correspond to $k = 0.28$ which is the value previously used to fit the charmonium data. Clearly the Cornell potential with $k = 0.28$ provides the best fit to the experiment.

6. Predicting the Mass of *b*-Baryons

On top of the already discussed Σ_b with quark content bqq, there are two additional ground-state b-baryons, Ξ_b and Ω_b. We will now discuss the theoretical prediction of their masses and compare it with experiment.

6.1. Ξ_b

The Ξ_Q baryons quark content is Qsd or Qsu. They can be obtained from "ordinary" Ξ (ssd or ssu) by replacing one of the s quarks by a heavier quark $Q = c, b$. There is one important difference, however. In the ordinary Ξ, Fermi statistics dictates that two s quarks must couple to spin 1, while in the ground state of Ξ_Q the (sd) and (su) diquarks have spin 0. Consequently, the Ξ_b mass is given by the expression: $\Xi_q = m_q + m_s + m_u - 3v\langle\delta(r_{us})\rangle/m_u m_s$. The Ξ_b mass can thus be predicted using the known Ξ_c baryon mass as a starting point and adding the corrections due to mass differences and hyperfine interactions:

$$\Xi_b = \Xi_c + (m_b - m_c) - 3v(\langle\delta(r_{us})\rangle_{\Xi_b} - \langle\delta(r_{us})\rangle_{\Xi_c})/(m_u m_s)\,. \qquad (18)$$

Since the Ξ_Q baryon contains a strange quark, and the effective constituent quark masses depend on the spectator quark, the optimal way to estimate the mass difference $(m_b - m_c)$ is from mesons which contain both s and Q quarks:

$$m_b - m_c = \frac{1}{4}(3B_s^* + B_s) - \frac{1}{4}(3D_s^* + D_s) = 3324.6 \pm 1.4\,. \tag{19}$$

On the basis of these results we predicted[7] $M(\Xi_b) = 5795 \pm 5$ MeV. Our paper was submitted on June 14, 2007. The next day CDF announced the result,[9] $M(\Xi_b) = 5792.9 \pm 2.5 \pm 1.7$ MeV, following up on an earlier D0 measurement, $M(\Xi_b) = 5774 \pm 11 \pm 15$ MeV,[8] as shown in Fig. 3.

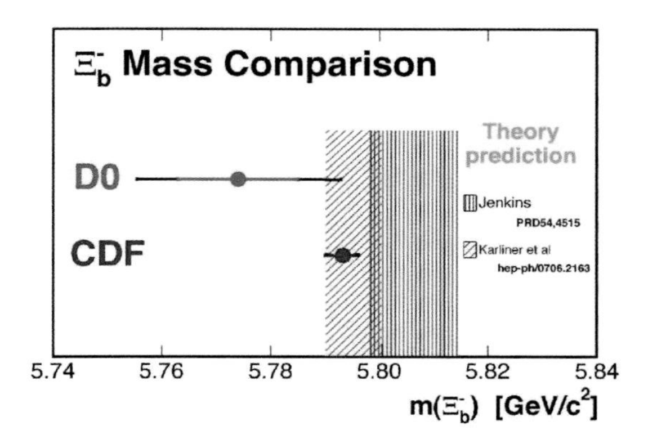

Fig. 3. Ξ_b mass: comparison of theoretical predictions with CDF and X0 data (from a CDF talk by D. Litvinstev).

6.2. *Mass of the* Ω_b

For the spin-averaged Ω_b mass we have

$$\begin{aligned}\frac{1}{3}(2M(\Omega_b^*) + M(\Omega_b)) &= \frac{1}{3}(2M(\Omega_c^*) + M(\Omega_c)) + (m_b - m_c)_{B_s - D_s}\\ &= 6068.9 \pm 2.4 \text{ MeV}\,.\end{aligned} \tag{20}$$

For the hyperfine splitting we obtain

$$\begin{aligned}M(\Omega_b^*) - M(\Omega_b) &= (M(\Omega_c^*) - M(\Omega_c))\frac{m_c}{m_b}\frac{\langle\delta(r_{bs})\rangle_{\Omega_b}}{\langle\delta(r_{cs})\rangle_{\Omega_c}}\\ &= 30.7 \pm 1.3 \text{ MeV}\end{aligned} \tag{21}$$

leading to the following predictions:

$$\Omega_b = 6052.1 \pm 5.6 \text{ MeV}\,, \quad \Omega_b^* = 6082.8 \pm 5.6 \text{ MeV}\,. \tag{22}$$

About four months after our prediction (22) for Ω_b mass was published,[10] D0 collaboration published the first measurement of Ω_b mass:[11]

$$M(\Omega_b)_{D0} = 6165 \pm 10\,(\text{stat.}) \pm 13\,(\text{syst.})\ \text{MeV}\,.$$

The deviation from the central value of our prediction was huge, 113 MeV. Understandably, we were very eager to see the CDF result. CDF published their result about nine months later, in May 2009:[12]

$$M(\Omega_b)_{\text{CDF}} = 6054 \pm 6.8\,(\text{stat.}) \pm 0.9\,(\text{syst.})\ \text{MeV}\,.$$

The central values of theoretical prediction and CDF agree to within 2 MeV, or about 1/3 standard deviation.

Figure 4 shows a comparison of our predictions for the masses of Σ_b, Ξ_b and Ω_b baryons with the CDF experimental data.

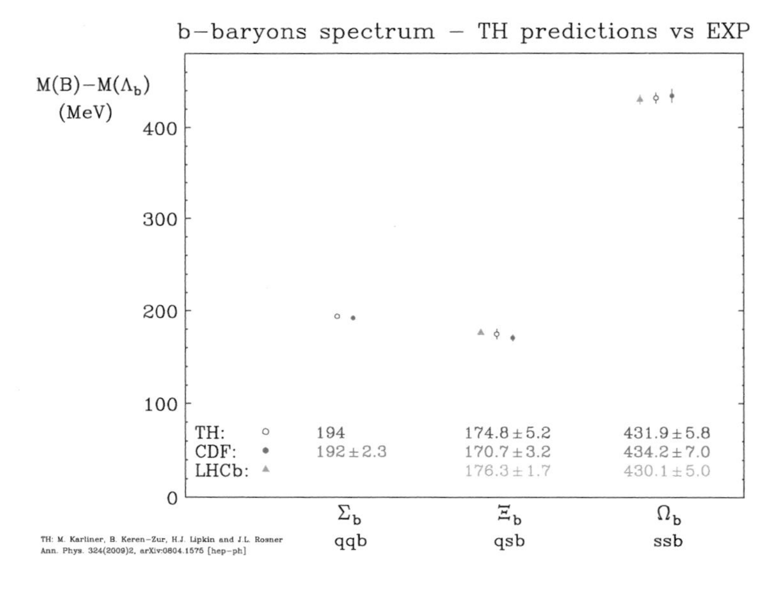

Fig. 4. Masses of b-baryons: comparison of theoretical predictions[7,10] with experiment.

The sign in our prediction

$$M(\Sigma_b^*) - M(\Sigma_b) < M(\Omega_b^*) - M(\Omega_b) \tag{23}$$

appears to be counterintuitive, since the color hyperfine interaction is inversely proportional to the quark mass. The expectation value of the interaction with the same wave function for Σ_b and Ω_b violates our inequality. When wave function effects are included, the inequality is still violated if the potential is linear, but is satisfied in predictions which use the Cornell potential.[6] This reversed inequality is not predicted by other recent

Table 3. Comparison of predictions for b baryons with those of some other recent approaches[13–15] and with experiment. Masses quoted are isospin averages unless otherwise noted. Our predictions are those based on the Cornell potential.

	Value in MeV				
Quantity	Ref. 13	Ref. 14	Ref. 15	This work	Experiment
$M(\Lambda_b)$	5622	5612	Input	Input	5619.7 ± 1.7
$M(\Sigma_b)$	5805	5833	Input	—	5811.5 ± 2
$M(\Sigma_b^*)$	5834	5858	Input	—	5832.7 ± 2
$M(\Sigma_b^*) - M(\Sigma_b)$	29	25	Input	20.0 ± 0.3	$21.2^{+2.2}_{-2.1}$
$M(\Xi_b)$	5812	$5806^{\rm a}$	Input	5790–5800	$5792.9 \pm 3.0^{\rm b}$
$M(\Xi_b')$	5937	$5970^{\rm a}$	5929.7 ± 4.4	5930 ± 5	—
$\Delta M(\Xi^b)^{\rm c}$	—	—	—	6.4 ± 1.6	—
$M(\Xi_b^*)$	5963	$5980^{\rm a}$	5950.3 ± 4.2	5959 ± 4	—
$M(\Xi_b^*) - M(\Xi_b')$	26	$10^{\rm a}$	20.6 ± 1.9	29 ± 6	—
$M(\Omega_b)$	6065	6081	6039.1 ± 8.3	6052.1 ± 5.6	$6054.4 \pm 7^{\rm d}$
$M(\Omega_b^*)$	6088	6102	6058.9 ± 8.1	6082.8 ± 5.6	—
$M(\Omega_b^*) - M(\Omega_b)$	23	21	19.8 ± 3.1	30.7 ± 1.3	—
$M(\Lambda^*_{b[1/2]})$	5930	5939	—	5929 ± 2	—
$M(\Lambda^*_{b[3/2]})$	5947	5941	—	5940 ± 2	—
$M(\Xi^*_{b[1/2]})$	6119	6090	—	6106 ± 4	—
$M(\Xi^*_{b[3/2]})$	6130	6093	—	6115 ± 4	—

[a] Value with configuration mixing taken into account; slightly higher without mixing.
[b] CDF[9] value of $M(\Xi_b^-)$.
[c] $M(bsd) - M(bsu)$.
[d] CDF[12] value of $M(\Omega_b)$.

approaches[13–15] which all predict an Ω_b splitting smaller than a Σ_b splitting. However the reversed inequality is also seen in the corresponding charm experimental data,

$$M(\Sigma_c^*) - M(\Sigma_c) < M(\Omega_c^*) - M(\Omega_c)$$

$$64.3 \pm 0.5 \text{ MeV} \qquad 70.8 \pm 1.5 \text{ MeV}\,. \tag{24}$$

This suggests that the sign of the $SU(3)$ symmetry breaking gives information about the form of the potential. It is of interest to follow this clue theoretically and experimentally.

We have made additional predictions[7,10] for some excited states of b-baryons. Our results are summarized in Table 3.

In addition to mesons and baryons containing a single heavy quark, I want to discuss here two important types of doubly-heavy hadrons which have attracted a lot of attention recently: (a) baryons containing two heavy- and one light quark; (b) *exotic* mesons containing a heavy quark–antiquark pair and a light quark–antiquark pair. Since the latter have recently been seen by experiments, we discuss them first, in the next section.

7. First Observation of Manifestly Exotic Hadrons

In late 2007 the Belle Collaboration reported[16] anomalously large partial widths of $\Upsilon(5S) \to \Upsilon(2S)$ and $\Upsilon(5S) \to \Upsilon(1S)$, two orders of magnitude larger than the analogous decays of $\Upsilon(3S)$. Soon afterward Harry Lipkin and I proposed[17] that a four-quark exotic resonance $[b\bar{b}u\bar{d}]$ might mediate these decays through the cascade $\Upsilon(mS) \to [b\bar{b}u\bar{d}]\pi^- \to \Upsilon(nS)\pi^+\pi^-$. We suggested looking for the $[b\bar{b}u\bar{d}]$ resonance in these decays as peaks in the invariant mass of $\Upsilon(1S)\pi$ or $\Upsilon(2S)\pi$ systems.

More recently Belle collaboration confirmed this prediction, announcing[18,19] the observation of two charged bottomonium-like resonances Z_b as narrow structures in $\pi^\pm\Upsilon(nS)$ $(n = 1, 2, 3)$ and $\pi^\pm h_b(mP)$ $(m = 1, 2)$ (h_b-s are spin-singlet, P-wave bottomonia with $J = 1$) mass spectra that are produced in association with a single charged pion in $\Upsilon(5S)$ decays.

Since these states decay into bottomonium and a charged pion, they must contain both a $\bar{b}b$ heavy quark–antiquark pair *and* a $\bar{d}u$ light quark–antiquark pair. Thus their minimal quark content is $\bar{b}b\bar{d}u$. They are *manifestly exotic.* Their discovery by Belle was the first time such manifestly exotic hadron resonances have been unambiguously observed experimentally.

The measured masses of the two structures averaged over the five final states are $M_1 = 10608.4 \pm 2.0$ MeV, $M_2 = 10653.2 \pm 1.5$ MeV, both with a width of about 15 MeV. Interestingly enough, the two masses M_1 and M_2 are about 3 MeV above the respective $B^*\bar{B}$ and $B^*\bar{B}^*$ thresholds.

7.1. *Interpretation as deuteron-like quasi-bound states of two heavy mesons*

The most interesting theoretical question is *what are these states?* Their quantum numbers are those of a $\bar{b}b\bar{u}d$ tetraquark, but such quantum numbers can also be realized by a system consisting of $B^*\bar{B}$ and $B^*\bar{B}^*$ "hadronic molecules" loosely bound by pion exchange, as schematically shown in Fig. 5.

The difference between these two possibilities is subtle, because they have the same quantum numbers and therefore *in principle* they can can mix with each other. The extent of the mixing depends on the overlap between the respective wave functions. By a "tetraquark" I mean a state where all four

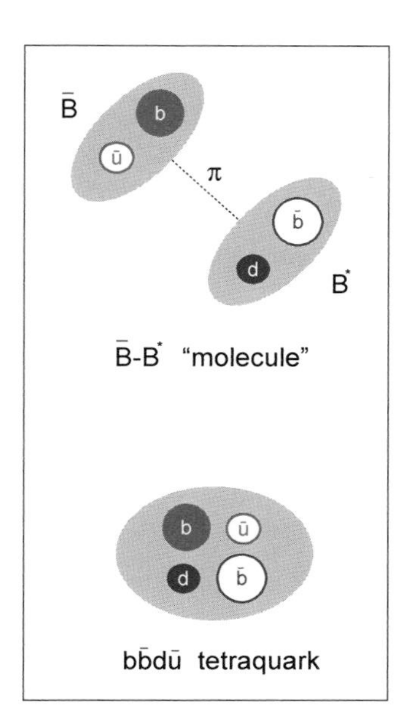

Fig. 5. A schematic depiction of a $\bar{B}$–B^* deuteron-like "hadronic molecule" versus a $b\bar{b}d\bar{u}$ tetraquark. Note that, unlike in the "molecule," in the tetraquark configuration the $b\bar{u}$ and $d\bar{b}$ diquarks in general won't be color singlets.

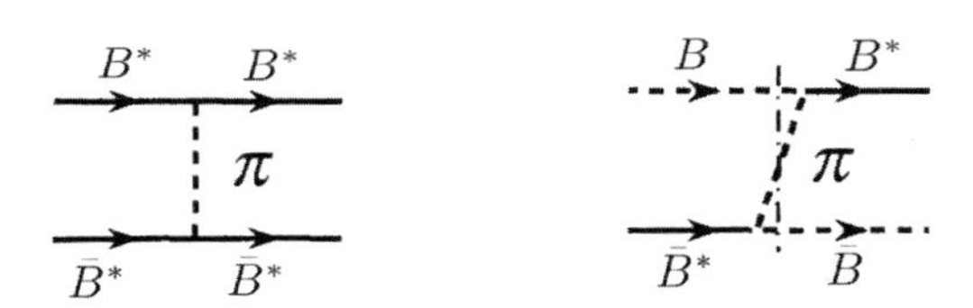

Fig. 6. Diagrams contributing to $B\bar{B}^*$ and $B^*\bar{B}^*$ binding through pion exchange. Analogous diagrams contribute to $D\bar{D}^*$ and $D^*\bar{D}^*$ binding, modulo the caveat that $D^* \to D\pi$, while $B^* \to B\pi$ is kinematically forbidden, as discussed in the text.

quarks are within the same "bag" or confinement volume, while by "hadronic molecule" I mean a state where there are two color-singlet heavy-light mesons attracting each other by exchange of pions and possibly other light mesons.

The proximity of the two resonances to the $B^*\bar{B}$ and $B^*\bar{B}^*$ thresholds strongly suggests a parallel with $X(3872)$, whose mass is almost exactly at the $D^*\bar{D}$ threshold.

It also provides strong support for the the possibility that these state indeed are deuteron-like "molecules" of two heavy mesons quasi-bound by pion exchange,[20–22] as schematically shown in Fig. 6. This is because it is very unlikely that two "genuine" tetraquarks just happened to sit at the respective two-meson thresholds.

The attraction due to π exchange is three times weaker in the $I = 1$ channel than in the $I = 0$ channel. Consequently, in the charm system the $I = 1$ state is expected to be well above the $D^*\bar{D}$ threshold and the $I = 0$ $X(3872)$ is at the threshold.[a] In the bottom system the attraction due to π exchange is essentially the same, but the kinetic energy is much smaller by a factor of $\sim m(B)/m(D) \approx 2.8$. Therefore the net binding is much stronger than in the charm system.

The recently discovered manifestly exotic charged resonances are surprisingly narrow. This is the case in both $\bar{b}b$ systems[18,19] — and in the exotic charmonium, namely the remarkable peak $Z_c(3900)$ at $3899.0\pm3.6\pm4.9$ MeV with $\Gamma = 46 \pm 10 \pm 20$ MeV reported by BESIII.[23]

The relatively slow decay of these exotic resonances implies that the dominant configuration of the $\bar{Q}Q\bar{q}q$ four-body system is *not* that of a low-lying $\bar{Q}Q$ quarkonium and pion(s). The latter have a much lower energy than the respective two-meson thresholds $\bar{M}M^*$ and $\bar{M}^*M^*$, $(M = D, B)$, but do readily fall apart into $(\bar{Q}Q)$ and pion(s) and would result in very large decay widths. Thus we should view these systems as loosely bound states and/or near threshold resonances in the two heavy-meson system.

Such "molecular" states, $\bar{D}D^*$, etc. were introduced in Ref. 20. They were later extensively discussed[21,22] in analogy with the deuteron which binds via exchange of pions and other light mesons, and were referred to as "*deusons*." The key observation is that the coupling to the heavy mesons of the light mesons exchanged (π, ρ, etc.) becomes universal and independent of M_Q for $M_Q \to \infty$, and so does the resulting potential in any given J^{CP} and isospin channel. In this limit the kinetic energy $\sim p^2/(M_Q)$ vanishes, and the two heavy mesons bind with a binding energy $\sim$ the maximal depth of the attractive meson-exchange potential.

For a long time it was an important open question whether these consideration apply in the real world with large but finite masses of the D and B mesons. The recent experimental results of Belle[18,19] and BESIII,[23] together with theoretical analysis in Refs. 24 and 25 strongly indicate that such exotic states do exist — some were already found and more are predicted below.

Due to parity conservation, the pion cannot be exchanged in the $\bar{M}M$ system, but it does contribute in the $\bar{M}M^*$ and $\bar{M}^*M^*$ channels. This fact provides additional support to the molecular interpretation, because resonances have been observed close to $\bar{D}D^*$, $\bar{D}^*D^*$, $\bar{B}B^*$ and $\bar{B}^*B^*$ thresholds, but no resonances have been seen at $\bar{D}D$ nor at $\bar{B}B$ thresholds.

[a]For simplicity we treat $X(3872)$ as an isoscalar, since it has no charged partners, and we ignore here the issue of isospin breaking in its decays. A more refined treatment results in the same conclusions.

The $\tau_1 \cdot \tau_s$ isospin nature of the exchange implies that the the binding is three times stronger in the isoscalar channel. It was estimated[24,25] that in the bottomonium system this difference in the binding potentials raises the $I = 1$ exotics well above the $I = 0$ exotics. In the charmonium system this splitting is expected to be slightly smaller, because the $\bar{D}D^*/\bar{D}^*D^*$ states are larger than $\bar{B}B^*/\bar{B}^*B^*$. This is because the reduced mass in the $\bar{B}B^*$ system is approximately 2.5 times larger than in the $\bar{D}D^*$ system. On the other hand, the net attractive potential due to the light mesons exchanged between the heavy-light mesons is approximately the same, since $m_c, m_b \gg \Lambda_{\text{QCD}}$. As usual in quantum mechanics, for a given potential the radius of a bound state or a resonance gets smaller when the reduced mass grows, so the $\bar{D}D^*$ states are larger than the $\bar{B}B^*$ states. Because of this difference in size the attraction in both $I = 0$ and $I = 1$ charmonium channels is expected to be somewhat smaller. In addition, Ref. 26 suggested that the asymptotic coupling between the two heavy mesons and a pion, $g_{MM^*\pi}$ for $m_Q \to \infty$ is approached from below, so that $g_{BB^*\pi} > g_{DD^*\pi}$.

7.2. *Prediction of additional related exotic states*

Since the quarks are heavy, we can treat their kinetic energy as a perturbation depending linearly on a parameter inversely proportional to μ_{red}, the reduced mass of the two meson system, which scales like the mass of the heavy quark,[27] with the Hamiltonian $H = a \cdot p^2 + V$, where $a = 1/\mu_{\text{red}} \sim 1/m_Q$. We can use the existing data in order to make a very rough estimate of the isovector binding potential in the $m_Q \to \infty$ limit. We have two data points: $Z_c(3900)$ at $a(D)$ is approximately 27 MeV above $\bar{D}D^*$ threshold and $Z_b(10610)$ at $a(B)$ is approximately 3 MeV above $\bar{B}B^*$ threshold. Linear extrapolation to $a = 0$ yields $E_b^{I=1}(a = 0) \approx -11.7$ MeV. In view of the convexity of the binding energy in a, the actual binding energy is likely to slightly exceed this linear extrapolation.

We can then use this result for the isovector channel to estimate the $\bar{B}B^*$ binding in the isoscalar channel. Assuming that the isoscalar binding energy in the $m_Q \to \infty$ limit is three times larger than for the isovector, we have $E_b^{I=0}(a = 0) \approx 3 \cdot (-11.7) = -35$ MeV. The state $X(3872)$ is at $\bar{D}D^*$ threshold, providing an additional data point of $E_b^{I=0}(a(D)) \approx 0$ in the isoscalar channel. Linear extrapolation to $a(B)$ yields approximately -20 MeV as the $\bar{B}B^*$ binding energy in the isoscalar channel.

The upshot is that the newly discovered $Z_c(3900)$ isovector resonance confirms and refines the estimates in Refs. 24 and 25 for the mass of the putative $\bar{B}B^*$ isoscalar bound state. This immediately lead to several predictions:[27]

(a) two $I = 0$ narrow resonances X_b in the bottomonium system, about 23 MeV below $Z_b(106010)$ and $Z_b(10650)$, i.e. about 20 MeV below the corresponding $\bar{B}B^*$ and $\bar{B}^*B^*$ thresholds;
(b) an $I = 1$ resonance above $\bar{D}^*D^*$ threshold;
(c) an $I = 0$ resonance near $\bar{D}^*D^*$ threshold.

The X_b states can most likely be observed through the decays $X_b \to \Upsilon\omega$ or $X_b \to \chi_b\pi\pi$. G-parity prevents their decay to $\Upsilon\pi\pi$. The observed decay of their charmonium sector analogue $X(3872) \to J/\pi\pi$ is only possible because isospin is strongly broken between D^+ and D^0, and because $X(3872)$ is at the $\bar{D}D^*$ threshold. In contradistinction, in the bottomonium system isospin is almost perfectly conserved. *Thus the null result in CMS search*[28] *for $X_b \to \Upsilon(1S)\pi^+\pi^-$ does not tell us if X_b exists.*

Quite recently the BESIII collaboration reported observation in $e^+e^- \to (D^*\bar{D}^*)^{\pm}\pi^{\mp}$ of what looks just like (b) above, namely a new charmonium-like charged resonance $Z_c(4025)$, slightly above the $\bar{D}^*D^*$ threshold, at $\sqrt{s} = (4026.3 \pm 2.6 \pm 3.7)$ MeV, with width of $24.8 \pm 5.6 \pm 7.7$ MeV.[29] Shortly afterward, BESIII reported[30] observation of another charged charmonium-like structure $Z_c(4020)$ in $e^+e^- \to \pi^+\pi^-h_c$ at $(4022.9 \pm 0.8 \pm 2.7)$ MeV and width of $7.9 \pm 2.7 \pm 2.6$ MeV.

Figure 7 provides a concise summary of the experimental information about the masses of doubly-heavy exotics observed so far, together with our predictions for the masses of additional states, as discussed above. The observed and predicted decay channels of these states are shown as well.

7.3. *A $(\Sigma_b^+\Sigma_b^-)$ beauteron dibaryon?*

The discovery of the Z_b-s and their interpretation as quasi-bound $B^*\bar{B}$ and $B^*\bar{B}^*$, raises an interesting possibility of a *strongly bound $\Sigma_b^+\Sigma_b^-$ deuteron-like state, a beauteron.*[31] The Σ_b is about 500 MeV heavier than B^*. The $\Sigma_b\Sigma_b$ kinetic energy is therefore significantly smaller than that of $B\bar{B}^*$ or $B^*\bar{B}^*$. Moreover, Σ_b with $I = 1$ couples more strongly to pions than B and B^* with $I = \frac{1}{2}$. The opposite electric charges of Σ_b^+ and Σ_b^- provide additional 2–3 MeV of binding energy. The heavy dibaryon bound state might be sufficiently long-lived to be observed experimentally. A possible decay mode of the beauteron is $(\Sigma_b^+\Sigma_b^-) \to \Lambda_b\Lambda_b\pi^+\pi^-$, which might be observable in LHCb. It should also be seen in lattice QCD.

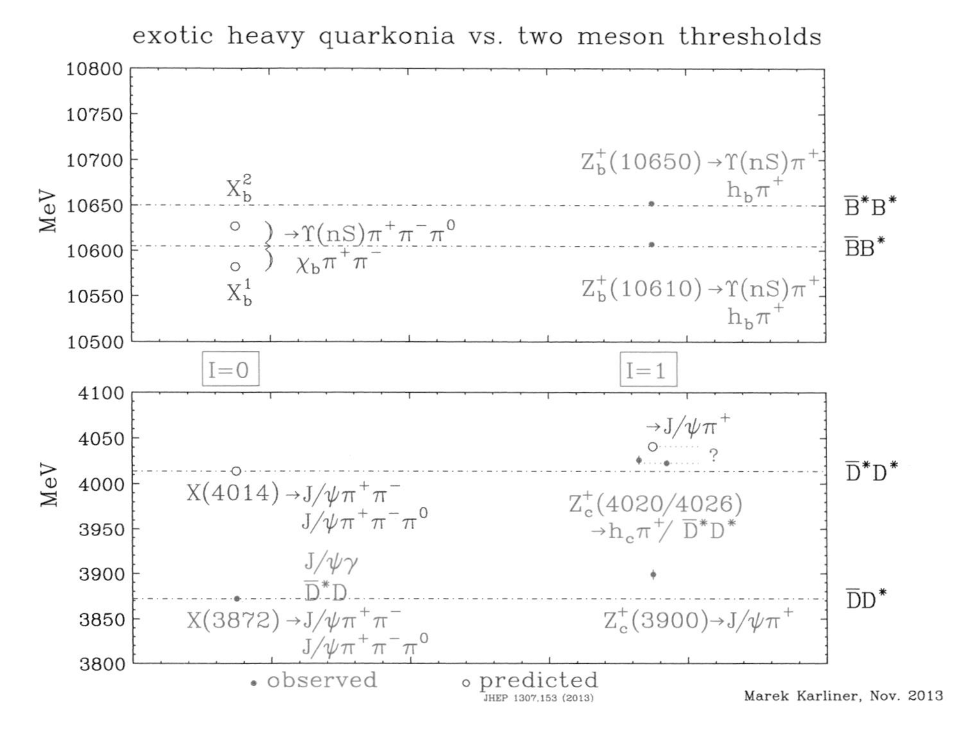

Fig. 7. Masses and decay channels of the doubly-heavy exotic quarkonia versus two-meson thresholds. The states observed so far are shown in red, the predicted states are shown in blue. $I = 0$ resonances are shown on the left and isovectors are shown on the right. Note the proximity of all the states to the corresponding two-meson thresholds.

8. Doubly-Heavy Baryons

From the point of view of QCD there is nothing exotic about baryons containing two heavy quarks (b or c, generically denoted by Q) and one light quark (u or d, generically denoted by q). Heavy quarks decay only by weak interaction, with a characteristic lifetime orders of magnitude larger than the typical QCD timescale, so from the point of view of strong interactions the QQq baryons are stable, just like protons, neutrons and hyperons. Thus these *doubly-heavy* baryons *must* exist.

On the other hand, producing and discovering them is an extraordinary challenge. One needs to produce *two* $\bar{Q}Q$ pairs and then rearrange them, so as to form QQ diquark. *Then* the QQ diquark needs to pick up a light quark q, to make a QQq baryon. At first it seems that the production cross section for such a process is too low for the current generation of accelerators.

This view is probably overly pessimistic. A substantial basis for optimism is the observation of a large number of the doubly-heavy $B_c = (b\bar{c})$ mesons by D0, CDF and especially LHCb,[33–35] indicating[27,32] that simultaneous production of $\bar{b}b$ and $\bar{c}c$ pairs which are close to each other in space and in rapidity and can coalesce to form doubly-heavy hadrons is not too rare.

There are interesting parallels between the doubly-heavy baryons QQq and the hypothetical $QQ\bar{q}\bar{q}$ tetraquarks. In both types of systems there is a light color triplet — a quark or an anti-diquark — bound to a heavy diquark. Because of this similarity, *experimental observation of doubly-heavy baryons is very important not just in its own right, but as source of extremely valuable information for deducing the properties of the more exotic $QQ\bar{q}\bar{q}$ tetraquarks.* Such deduction can carried out just as it was done for b-baryons.

As discussed above in the last few years it became possible to accurately predict at the level of 2–3 MeV the masses of heavy baryons containing the b-quark: $\Sigma_b(bqq)$, $\Xi_b(bsq)$ and $\Omega_b(bss)$,[1,7,10] as shown in Fig. 4. Similar approach can be used to predict the masses of doubly-heavy baryons,[32] as shown in Table 4. Using these one can then compute the corresponding lifetimes, as shown in Table 5 which also shows other authors' estimates.

The most important decay modes are those involving the most-favored Cabibbo–Kobayashi–Maskawa matrix elements, such as $c \to sW^{*+}$ and $b \to cW^{*-}$. Among the latter, we focus on those modes which can pass the trigger criteria in the collider experiments, such as

(a) $\Xi_{cc}^{++}(ccu) \to (csu)W^{*+} \to \Xi_c^+\pi^+ \to \Xi^-\pi^+\pi^+\pi^+,\ \Lambda_c^+K^-\pi^+\pi^+,$

(b) $\Xi_{cc}^{+}(ccd) \to (csd)W^{*+} \to \Xi_c^0\pi^+,\ \Lambda_c^+K^-\pi^+,$

(c) $\Xi_{bc}^{+}(bcu) \to \Xi_{cc}^{++}W^{*-},\ \Xi_b^0W^{*+},$

(d) $\Xi_{bc}^{0}(bcd) \to \Xi_{cc}^{+}W^{*-},\ \Xi_b^-W^{*+},\ (bsu)^*,$

(e) $\Xi_{bb}(bbq) \to (bcq)^*W^{*-}.$

An interesting decay involving the subprocess $b \to (J/\psi s)$ *twice* is the chain $\Xi_{bb} \to J/\psi\Xi_b^{(*)} \to J/\psi J/\psi\Xi^{(*)}$, where $\Xi_b^{(*)}$ denotes a (possibly excited) state with the minimum mass of $\Xi_b(5792)$, while $\Xi^{(*)}$ denotes a (possibly excited) state with the minimum mass of Ξ. Although this state is expected to be quite rare and one has to pay the penalty of two J/ψ leptonic branching fractions, it has a distinctive signature and is worth looking for.

In Ref. 32 we also estimated the hyperfine splitting between B_c^* and B_c mesons to be 68 MeV. P-wave excitations of the Ξ_{cc} with light-quark total angular angular momentum $j = 3/2$, the analog of those observed for D and B mesons, were estimated to lie around 420–470 MeV above the spin-weighted average of the Ξ_{cc} and Ξ_{cc}^* masses. Production rates could be as large as 50% of those for B_c, which also requires the production of two heavy

Table 4. Summary of our mass predictions[32] (in MeV) for lowest-lying baryons with two heavy quarks. States without a star have $J = 1/2$; states with a star are their $J = 3/2$ hyperfine partners. The quark q can be either u or d. The square or curved brackets around cq denote coupling to spin 0 or 1.

State	Quark content	$M(J = 1/2)$	$M(J = 3/2)$
$\Xi_{cc}^{(*)}$	ccq	3627 ± 12	3690 ± 12
$\Xi_{bc}^{(*)}$	$b[cq]$	6914 ± 13	6969 ± 14
Ξ'_{bc}	$b(cq)$	6933 ± 12	—
$\Xi_{bb}^{(*)}$	bbq	10162 ± 12	10184 ± 12

Table 5. Summary of lifetime predictions for baryons containing two heavy quarks. Values given are in fs.

Baryon	Our work[32]	Ref. 36	Ref. 37	Ref. 38	Ref. 39
$\Xi_{cc}^{++} = ccu$	185	430 ± 100	460 ± 50	500	~ 200
$\Xi_{cc}^{+} = ccd$	53	120 ± 100	160 ± 50	150	~ 100
$\Xi_{bc}^{+} = bcu$	244	330 ± 80	300 ± 30	200	—
$\Xi_{bc}^{0} = bcd$	93	280 ± 70	270 ± 30	150	—
$\Xi_{bb}^{0} = bbu$	370	—	790 ± 20	—	—
$\Xi_{bb}^{-} = bbd$	370	—	800 ± 20	—	—

quark pairs. We are optimistic that with the increased data samples soon to be available in hadronic and e^+e^- collisions, the first baryons with two heavy quarks will finally be seen.

Acknowledgments

The work described here was done in collaboration with B. Keren-Zur, H. J. Lipkin and J. Rosner. It was supported in part by a grant from the Israel Science Foundation. The work of J. Rosner was supported by the U.S. Department of Energy, Division of High Energy Physics, Grant No. DE-FG02-13ER41958.

References

1. M. Karliner and H. J. Lipkin, *Phys. Lett. B* **575**, 249 (2003), arXiv:hep-ph/0307243.
2. M. Karliner and H. J. Lipkin, *Phys. Lett. B* **660**, 539 (2008), arXiv:hep-ph/0611306.
3. CDF Collab. (T. Aaltonen *et al.*), *Phys. Rev. Lett.* **99**, 202001 (2007).
4. M. Karliner and H. J. Lipkin, *Phys. Lett. B* **650**, 185 (2007), arXiv:hep-ph/0608004.
5. A. De Rujula, H. Georgi and S. L. Glashow, *Phys. Rev. D* **12**, 147 (1975).
6. B. Keren-Zur, *Ann. Phys.* **323**, 631 (2008), arXiv:hep-ph/0703011.
7. M. Karliner, B. Keren-Zur, H. J. Lipkin and J. L. Rosner, arXiv:0706.2163v1 [hep-ph].
8. D0 Collab. (V. M. Abazov *et al.*), *Phys. Rev. Lett.* **99**, 052001 (2007).
9. CDF Collab. (T. Aaltonen *et al.*), *Phys. Rev. Lett.* **99**, 052002 (2007).
10. M. Karliner, B. Keren-Zur, H. J. Lipkin and J. L. Rosner, arXiv:0708.4027 [hep-ph], (unpublished); *Ann. Phys.* **324**, 2 (2009), arXiv:0804.1575 [hep-ph].
11. D0 Collab. (V. M. Abazov *et al.*), *Phys. Rev. Lett.* **101**, 232002 (2008), arXiv:0808.4142 [hep-ex].
12. CDF Collab. (T. Aaltonen *et al.*), *Phys. Rev. D* **80**, 072003 (2009), arXiv:0905.3123 [hep-ex].
13. D. Ebert *e al.*, *Phys. Rev. D* **72**, 034026 (2005); *Phys. Lett. B* **659**, 612 (2008).
14. W. Roberts and M. Pervin, *Int. J. Mod. Phys. A* **23**, 2817 (2008), arXiv:0711.2492 [nucl-th].
15. E. E. Jenkins, *Phys. Rev. D* **77**, 034012 (2008).
16. Belle Collab. (K. F. Chen *et al.*), *Phys. Rev. Lett.* **100**, 112001 (2008), arXiv:0710.2577 [hep-ex].
17. M. Karliner and H. J. Lipkin, arXiv:0802.0649 [hep-ph].
18. Belle Collab. (I. Adachi *et al.*), arXiv:1105.4583 [hep-ex].
19. Belle Collab. (A. Bondar *et al.*), *Phys. Rev. Lett.* **108**, 122001 (2012), arXiv:1110.2251 [hep-ex].
20. M. B. Voloshin and L. B. Okun, *JETP Lett.* **23**, 333 (1976) [*Pisma Zh. Eksp. Teor. Fiz.* **23**, 369 (1976)].
21. N. A. Tornqvist, *Phys. Rev. Lett.* **67**, 556 (1991); N. A. Tornqvist, *Z. Phys. C* **61**, 525 (1994), arXiv:hep-ph/9310247.
22. N. A. Tornqvist, *Phys. Lett. B* **590**, 209 (2004), arXiv:hep-ph/0402237.
23. BESIII Collab. (M. Ablikim *et al.*), arXiv:1303.5949 [hep-ex].
24. M. Karliner, H. J. Lipkin, N. A. Tornqvist, arXiv:1109.3472 [hep-ph].
25. M. Karliner, H. J. Lipkin and N. A. Tornqvist, *Nucl. Phys. B* (*Proc. Suppl.*) **225-227**, 102 (2012).
26. S. Nussinov, *Phys. Lett. B* **418**, 383 (1998).
27. M. Karliner and S. Nussinov, *J. High Energy Phys.* **1307**, 153 (2013), arXiv:1304.0345 [hep-ph].
28. CMS Collab. (S. Chatrchyan *et al.*), arXiv:1309.0250 [hep-ex].
29. BESIII Collab. (M. Ablikim *et al.*), arXiv:1308.2760 [hep-ex].
30. BESIII Collab. (M. Ablikim *et al.*), arXiv:1309.1896 [hep-ex].

31. M. Karliner, H. J. Lipkin, N. A. Törnqvist, unpublished; arXiv:1109.3472.
32. M. Karliner and J. L. Rosner, Baryons with two heavy quarks: Masses, production, decays, and detection, *Phys. Rev. D* **90**, 094007 (2014), arXiv:1408.5877 [hep-ph].
33. CDF Collab. (T. Aaltonen *et al.*), *Phys. Rev. Lett.* **100**, 182002 (2008), arXiv:0712.1506 [hep-ex].
34. D0 Collab. (V. M. Abazov *et al.*), *Phys. Rev. Lett.* **101**, 012001 (2008), arXiv:0802.4258 [hep-ex].
35. LHCb Collab. (R. Aaij *et al.*), *Phys. Rev. Lett.* **109**, 232001 (2012), arXiv:1209.5634 [hep-ex]; arXiv:1408.0971 [hep-ex]; arXiv:1407.2126 [hep-ex]; *Eur. Phys. J. C* **74**, 2839 (2014), arXiv:1401.6932 [hep-ex]; *Phys. Rev. Lett.* **111**, 181801 (2013), arXiv:1308.4544 [hep-ex]; *J. High Energy Phys.* **09**, 075 (2013), arXiv:1306.6723 [hep-ex]; *Phys. Rev. D* **87**, 112012 (2013), arXiv:1304.4530 [hep-ex].
36. K. Anikeev *et al.*, *B* physics at the Tevatron: Run II and beyond, in *Workshop on B Physics at Conferences C99-09-23.2 and C00-02-24*, arXiv:hep-ph/0201071.
37. V. V. Kiselev and A. K. Likhoded, *Phys. Usp.* **45**, 455 (2002) [*Usp. Fiz. Nauk* **172**, 497 (2002)], arXiv:hep-ph/0103169.
38. J. D. Bjorken, Fermilab preprint, Estimates of decay branching ratios for hadrons containing charm and bottom quarks, 1986, FERMILAB-PUB-86-189-T, `http://lss.fnal.gov/archive/1986/pub/fermilab-pub-86-189-t.pdf`.
39. M. A. Moinester, *Z. Phys. A* **355**, 349 (1996), arXiv:hep-ph/9506405.

Quark Elastic Scattering as a Source of High Transverse Momentum Mesons

Rick Field

Department of Physics, University of Florida,
Gainesville, Florida 32611, USA
rfield@phys.ufl.edu

From 7 GeV/c pions to 600 GeV/c jets, the wonderful journey from the "old days" of Feynman–Field collider phenomenology to the Tevatron and the LHC.

When I arrived as a postdoc at Caltech in 1973, it was already clear from SLAC deep inelastic scattering experiments that the proton was a composite particle made up of tiny hard pieces which were referred to as "partons." Also, there was mounting evidence that at least some of the partons were quarks. We knew that only about 50% of the protons momentum was carried by the quarks, but I do not think we knew that the other 50% was carried by point-like massless gluons. The ISR at CERN was studying proton–proton collisions at a center-of-mass energy of 53 GeV and Fermilab was colliding 200 GeV protons on fixed targets (i.e. $W = \sqrt{s} = 19.2$ GeV).

When two protons of equal and opposite momentum collide at high energy most of the time they simply fall apart producing a collection of hadrons moving along the direction of the two incoming protons and all of the outgoing particles have small transverse momenta relative to the beam direction (~ 300 MeV/c). However, it was noticed that occasionally a high transverse momentum, p_T, hadron (pion or kaon) would be produced. This did not happen often but it happened more often then one would expect if the proton was a "soft" object. In those days, high transverse momentum meant anything with $p_T > 2$ GeV/c and the highest transverse momentums observed were only around 7 GeV/c!

In 1974, R. P. Feynman and I were wondering about where these high transverse momentum hadrons came from. We did not believe that a pion traveling in the direction of one of the incoming protons could "turn the corner" and come out at high transverse momentum without falling apart into its constituent quarks. We believed that the high p_T particles came from a hard 2-to-2 scattering of the quarks within the incoming protons.

The two outgoing high transverse momentum quarks would then fragment into pions and kaons some of which would have high p_T. At that time, we did not know how to calculate the quark–quark elastic scattering differential cross-section. The theory of Quantum Chromodynamics (QCD) was just beginning to be understood and the perturbative 2-to-2 parton–parton differential cross-sections had not yet been calculated. People were just beginning to realize that QCD was an asymptotically free theory which allows perturbation theory to be applied at high p_T. Because we did not yet understand how to calculate anything, in the first Feynman–Field paper (FF1)[1] which we completed in 1975, but did not publish until 1977, we concocted the "quark–quark elastic scattering black-box" model which is illustrated in Fig. 1. We fit the SLAC deep inelastic scattering data to determine the

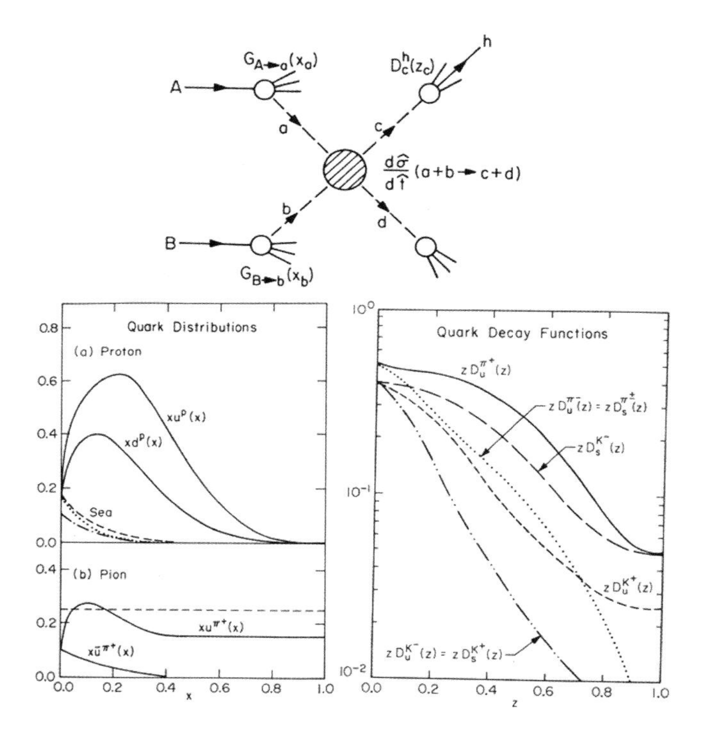

Fig. 1. Illustration of the Feynman–Field quark–quark elastic scattering "black-box" model for hadron–hadron collisions from FF1[1] (1977). (Top) The model assumed that high p_T particles arise from direct hard collisions between constituent quarks in the incoming particles, which fragment or cascade down into several hadrons. (Bottom left) The quark distribution functions were determined by fitting the SLAC deep inelastic scattering data. (Bottom right) The quark fragmentation functions were determined by fitting e^+e^- data and the 2-to-2 quark–quark elastic scattering cross-section, $d\sigma/dt$, was determined by fitting the data (i.e. "black-box").

probability of finding a quark of flavor f within a proton carrying a fraction, x, of the protons momentum, $G_{p\to f}(x)$. In addition, we fit e^+e^- data to determine the probability that a hadron, h, carrying fractional momentum, z, of an outgoing quark of flavor, f, is contained among the fragmentation products, $F_{f\to h}(z)$. The proton structure functions (we called them quark distribution functions) and quark fragmentation functions (we called them quark decay functions) were assumed to scale (i.e. were a function only of the fractional momentum x or z). We took the quark–quark elastic scattering differential cross-section to be a "black-box" and determined it by fitting the data.

I wrote the first draft of all the Feynman–Field papers and Feynman would come in and give me sentences or paragraphs that he would like to include in the paper. The following is a Feynman quote from FF1:[1]

> *"The model we shall choose is not a popular one, so that we will not duplicate too much of the work of others who are similarly analyzing various models (e.g. constituent interchange model, multiperipheral models, etc.). We shall assume that the high p_T particles arise from direct hard collisions between constituent quarks in the incoming particles, which fragment or cascade down into several hadrons."*

The "black-box" model was naïve, however, it convinced us we were on the right track. As shown in Fig. 2, we adjusted the quark–quark elastic differential cross-section to fit the experimentally measured high p_T meson cross-section at $W = 19.4$ GeV and then predicted it correctly at $W = 53$ GeV. The rise in the cross-section, of course, comes from the parton distribution function. We were amazed that we were able to use electron–proton and e^+e^- data to predict something about hadron–hadron collisions. The model also predicted the topology in high p_T hadron–hadron collisions that we are all familiar with today in which there is a "toward" side "jet" (i.e. collection of hadrons moving roughly in the same direction) and an "away" jet, together with the "beam–beam remnants" (we called them the beam and target jet). We studied this topology in more detail in FFF1.[2] The "beam–beam remnants" are part of the "underlying event" in hadron–hadron collisions which I am still working to understand and model.[3]

In FF1[1] we were able to predict particle ratios at high p_T. Actually, the reason we waited two years to publish the paper is that the model predicted the π^+/π^- ratio would increase at large p_T in proton–proton collisions and Feynman wanted to see some evidence of this before we published the paper. In July 1976 Feynman was at a meeting in Les Houches where he learned

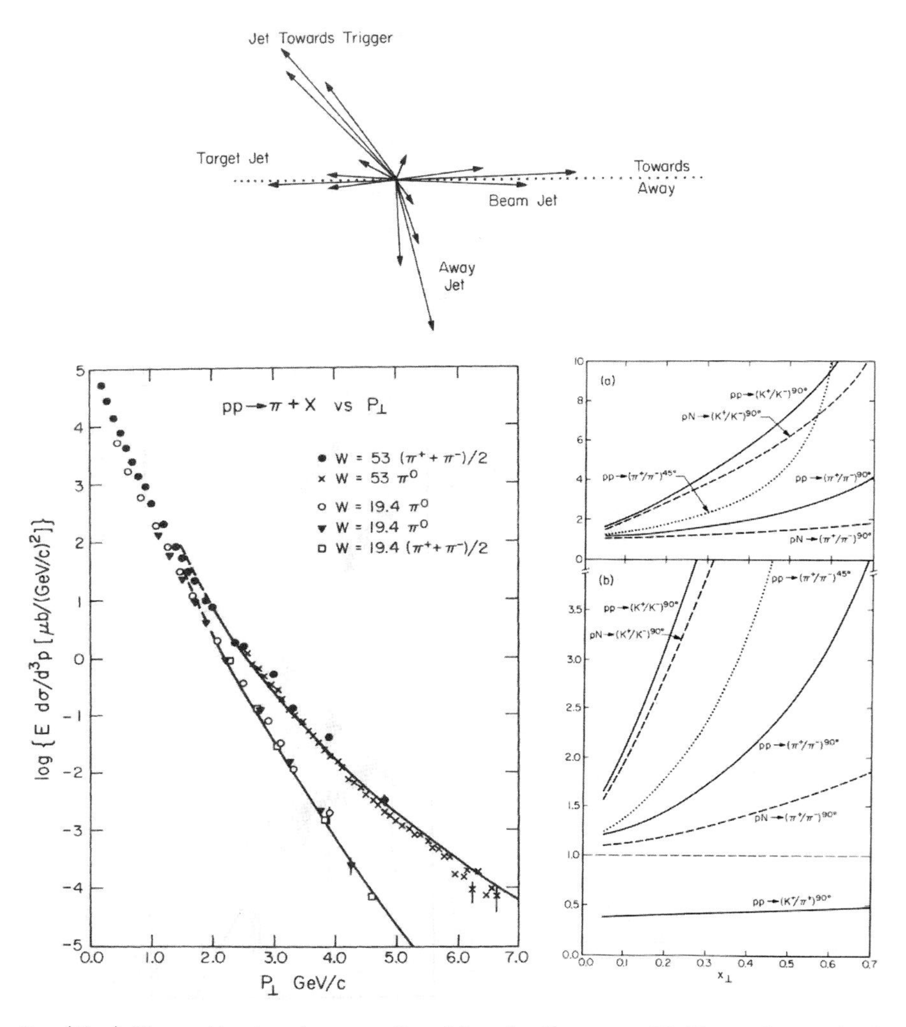

Fig. 2. (Top) Shows the topology predicted by the Feynman–Field quark–quark elastic scattering "black-box" model for hadron–hadron from FF1[1] (1977) in which there is a "toward" side "jet" (i.e. collection of hadrons moving roughly in the same direction) and an "away" jet, together with the beam and target jet (i.e. the "beam–beam" remnants). Also shows the predictions of the model for the inclusive meson cross-section at 19.4 GeV and 53 GeV (bottom left) and for the high p_T particle rations at 53 GeV (a) and 19.4 GeV (bottom right).

from Jim Cronin that the University of Chicago group did see the increase we expected in an experiment at Fermilab. The $x_T = 2p_T/W$ values at the ISR were too small to see much of an effect. As shown in Fig. 3, in July 1976 I received a telegram which Feynman sent from Les Houches which stated: "*Saw Cronin — Am now convinced were right track — Quick write — Feynman.*"

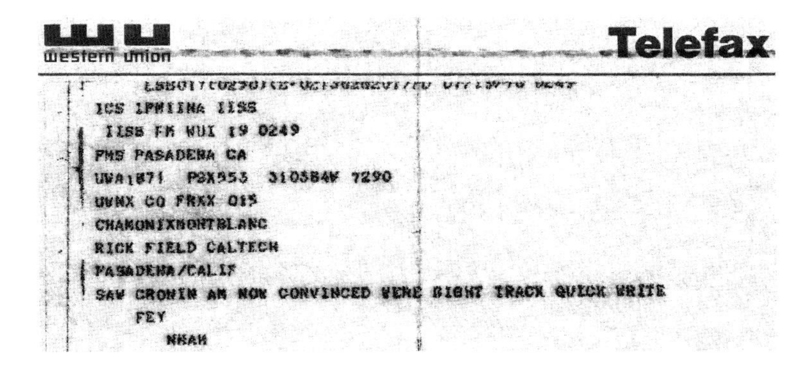

western union Telefax

ICS IPMIINA IISS
IISS FM WUI 19 0249
PMS PASADENA CA
UWA1871 P3X953 310384Y 7290
UWNX CO FRXX 015
CHAMONIXMONTBLANC
RICK FIELD CALTECH
PASADENA/CALIF
SAW CRONIN AM NOW CONVINCED WERE RIGHT TRACK QUICK WRITE
FEY
NMAN

Fig. 3. Telegram I received from Feynman in July 1976. Feynman was at the Les Houches conference in France and learned about the latest results on the hight p_T particle ratios from Jim Cronin. The message said, "*Saw Cronin am now convinced were right track quick write. Feynman.*"

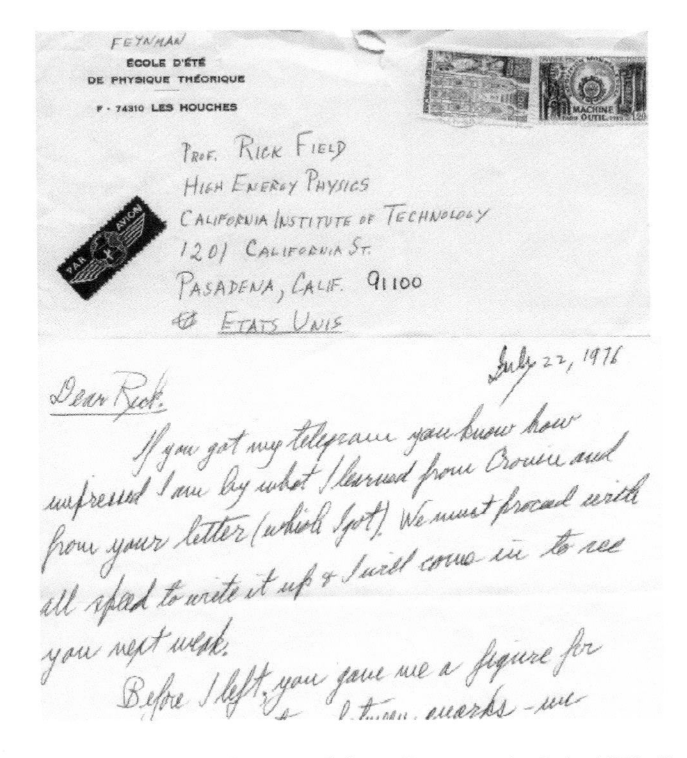

FEYNMAN
ÉCOLE D'ÉTÉ
DE PHYSIQUE THÉORIQUE
F - 74310 LES HOUCHES

PROF. RICK FIELD
HIGH ENERGY PHYSICS
CALIFORNIA INSTITUTE OF TECHNOLOGY
1201 CALIFORNIA ST.
PASADENA, CALIF. 91100
ETATS UNIS

July 22, 1976

Dear Rick,

If you got my telegram you know how impressed I am by what I learned from Cronin and from your letter (which I got) We must proceed with all speed to write it up & I will come in to see you next weak.

Before I left, you gave me a figure for

Fig. 4. First page of the letter I received from Feynman in July 1976. Feynman was at the Les Houches conference in France and learned about the latest results on the hight p_T particle ratios from Jim Cronin. The first paragraph said, "*If you got my telegram you know how impressed I am by what I learned from Cronin and from your letter (which I got). We must proceed with all speed to write it up and I will come to see you next weak*" (he mispelled week).

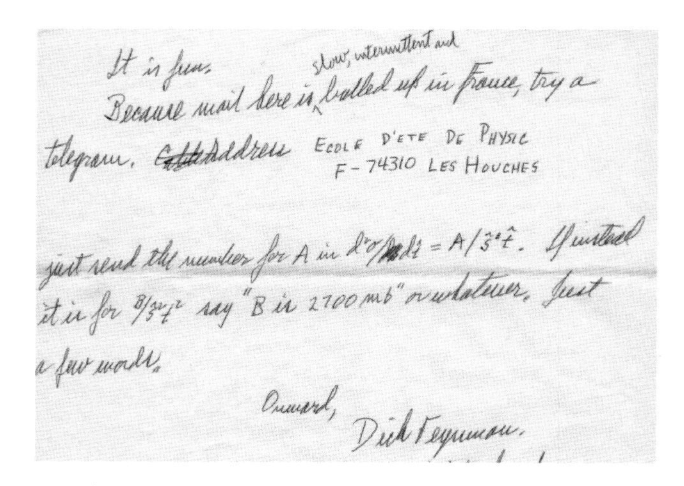

It is fun.
Because mail here is slow, intermittent and balled up in France, try a telegram. Address ECOLE D'ETE DE PHYSIC
F-74310 LES HOUCHES

just send the number for A in $d^2\sigma/d\hat{t} = A/\hat{s}^3\hat{t}$. If instead it is for $B/\hat{s}^3\hat{t}^2$ say "B is 2700 mb" or whatever. Just a few words.

Onward,
Dick Feynman.

Fig. 5. Last page of the letter I received from Feynman in July 1976. The line said, "*It is fun*" and at the end he said, "*Onward.*"

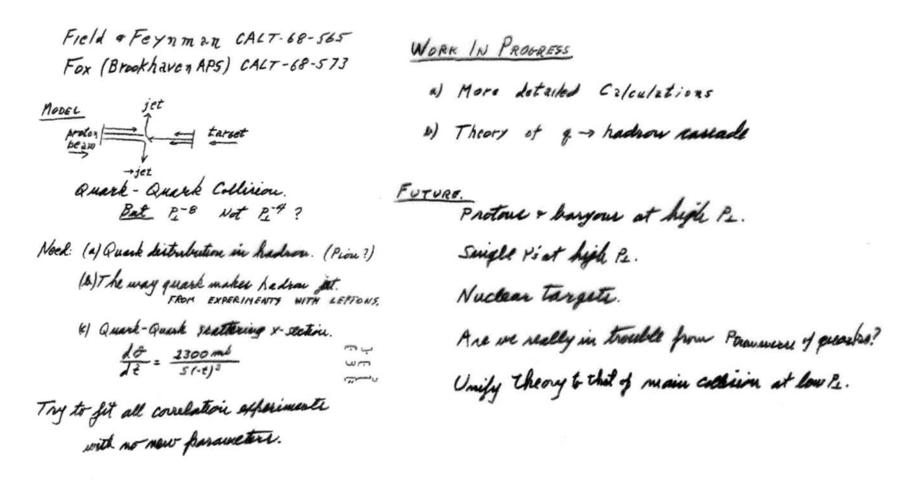

Fig. 6. Transparency from a talk Feynmen gave on our quark–quark elastic scattering "black-box" model at the Coral Gables Conference in December 1976. He listed as our work in progress: (a) More detailed calculations. (b) Theory for how quarks fragment into hadrons. He listed as our future studies: (1) Protons and baryons at high p_T. (2) Single photons at high p_T. (3) Nuclear targets. (4) Are we in trouble from premortial k_T of the quarks within the incoming hadrons? (5) Unified theory including the main collision at low p_T [in addition to the high p_T part].

Later in July 1976, I received a letter that Feynman sent from Les Houches shown in Figs. 4 and 5. He sent the letter to Professor Rick Field although I was only a postdoc at Caltech at that time. Figure 6 shows a transparency from a talk Feynmen gave on our quark–quark elastic scattering "black-box" model at the Coral Gables Conference in December 1976. He listed as our work in progress: (a) More detailed calculations. (b) Theory for how quarks fragment into hadrons. In addition, he listed as our future studies: (1) Protons and baryons at high p_T. (2) Single photons at high p_T. (3) Nuclear targets. (4) Are we in trouble from premortial k_T of the quarks within the incoming hadrons? (5) Unified theory including the main collision at low p_T [in addition to the high p_T part]. Many of his suggestions for future studies are still in progress. High p_T photon production has turned out to be an excellent way to study QCD. We found that in order to fit the data, we had to give the incoming partons a rather large intrensic (i.e. premordial) transverse momentum within the colliding hadrons, which is still the case today. Experiments at the LHC are currently studying protons and baryons at high p_T, and the current QCD Monte Carlo models are finding it difficult to fit this data.

The "black-box" model showed we were on the right track, but as one can see in retrospect there were many things we did not understand. For one, we thought the pion structure function went to a constant at high x and similarly we thought the quark fragmentation function to a pion went to a constant at large z (see Fig. 1). Of course, we all know now that there can be a constant term in these functions, but they are the so-called "higher twist" terms and fall off as a power of Q^2. Also, the "black-box" model did not include gluons. At that time we did not realize the gluon is a "hard" point-like parton just like the quark. We thought of it more like "glue."

The "black-box" model lasted less than a year. Things were happening fast. Even before the paper was published we were learning more about QCD. Once we realized it is an asymptotically free theory and that we could use perturbation theory to calculate high p_T phenomena we did everything over again, but this time using QCD as shown in Fig. 7. The parton distribution functions (PDF's) and the fragmentation functions now depended on the scale of the hard scattering (i.e. Q^2). Gluons were now included and all of the seven parton–parton scattering differential cross-sections were calculated by perturbation theory. Figure 8 shows some of the predictions of the QCD approach with $\Lambda = 400$ MeV from FFF2 (1978).[4,5] The following is a Feynman quote from FFF2:[5]

> *"We investigate whether the present experimental behavior of mesons with large transverse momentum in hadron–hadron collisions is consistent with the theory of quantum-chromodynamics (QCD) with asymptotic freedom, at least as the theory is now partially understood."*

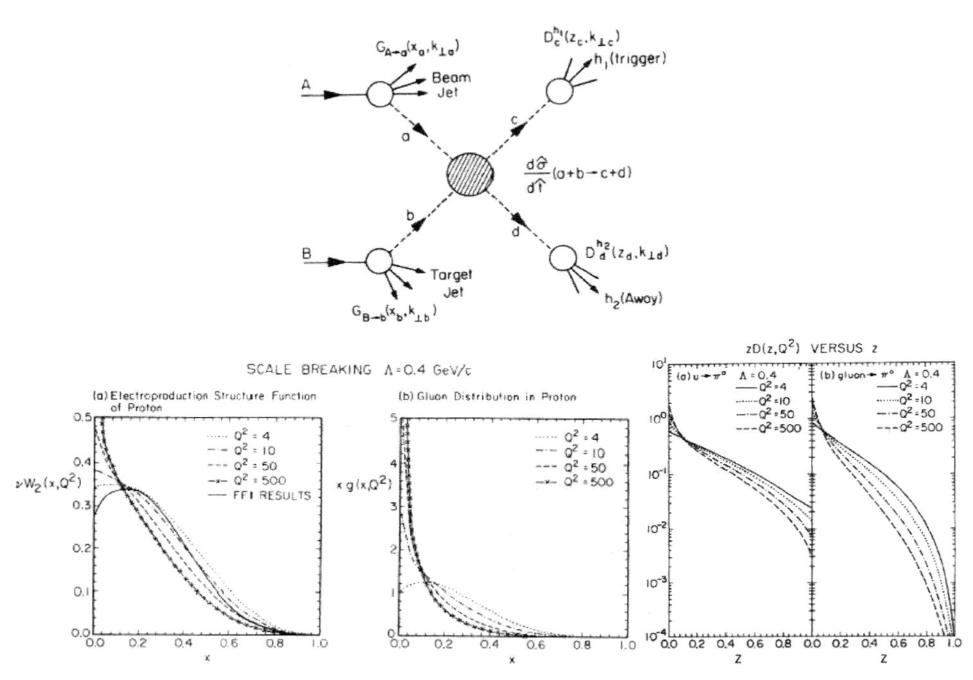

Fig. 7. Illustration of the QCD model for hadron–hadron collisions from FFF2[5] (1978). (Top) The model assumed that high p_T particles arise from direct hard collisions between constituent quarks and gluons in the incoming particles, which fragment into "jets" of hadrons. (Bottom left) The quark distribution functions were determined by fitting the SLAC deep inelastic scattering data at $Q^2 = 4$ GeV and determined at other values of Q^2 using QCD perturbation theory (bottom right) The quark fragmentation functions were determined by fitting e^+e^- data and the 2-to-2 quark–quark elastic scattering cross-section, $d\sigma/dt$, was determined from the data (i.e. "black-box").

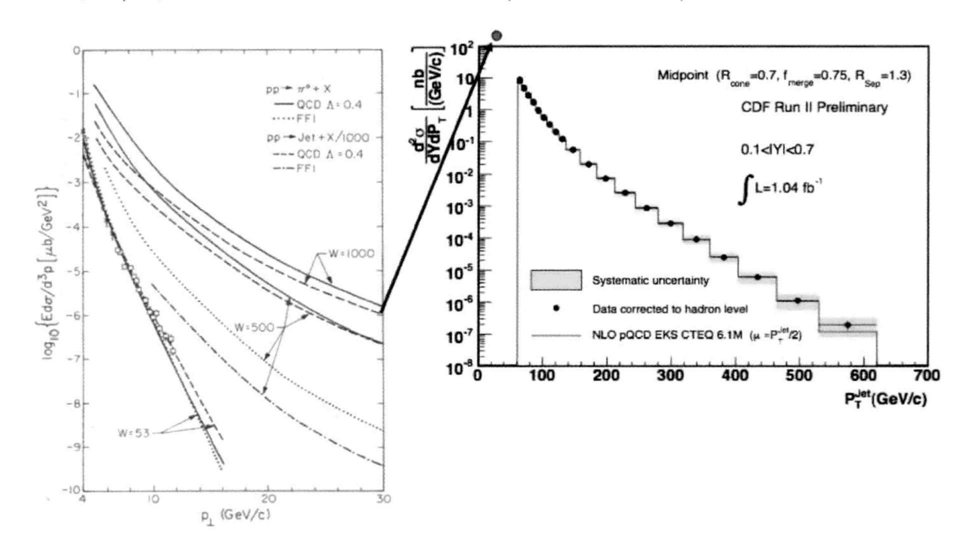

Fig. 8. (Left) Predictions of the QCD model for meson and "jet" production hadron–hadron collisions from FFF2[5] (1978). (Right) CDF Run 2 data on the inclusive "jet" cross-section at 1.96 TeV with an integrated luminosity of 1 fb^{-1}.

We realized from this study that the "jet" cross-section was much larger than the cross-section to produce a single charged hadron at the same p_T. We did not know if they would ever build a collider with a center-of-mass energy of 1 TeV, but as can be seen in Fig. 8, in 1978 we predicted the "jet" cross-section at $W = 1$ TeV. However, our transverse momentum scale only extended out to 30 GeV/c! Figure 8 shows our prediction at $p_T = 30$ GeV/c together with the inclusive jet cross-section measurment my graduate student, Craig Group, did at CDF in 2006.[6] Due to the resolution of the CDF calorimeter it is difficult to measure the jet cross-section below 60 GeV/c. What we thought in 1978 was a high p_T jet is too low of a p_T to be measured at the Tevatron! It has been a wonderful journey from the 7 GeV/c π^0's in Fig. 2 to the 600 GeV/c jets in Fig. 8! The CDF high p_T di-jet events are a bit "cleaner" than we would have thought back in 1978. This is because at that time we were using a QCD scale Λ of around 400 MeV and today we know that it is much smaller (around $\sim$200 MeV). Small Λ means a small QCD coupling α_s and hence less initial and final state gluon radiation, resulting in "cleaner" di-jet events. The following is another Feynman quote from FFF2:[5]

> *"At the time of this writing, there is still no sharp quantitative test of QCD. An important test will come in connection with the phenomena of high p_T discussed here."*

The calculations in FF1,[1] FFF1,[2] and FFF2[5] were done analytically by convoluting (i.e. integrating) over the parton distribution functions, fragmentation functions, and the parton cross-sections. We wanted to be able to simulate on an event-by-event bases hadron–hadron collisions (and e^+e^- annihilations) using Monte Carlo techniques, but to do so would require a model for the way the outgoing quarks and gluons fragment into hadrons. In FF2[7] (1978) Feynman and I proposed a simple model for parameterizing the properties of quark jets. The model assumes that quark jets can be analyzed on the basis of the recursive principle illustrated in Fig. 9. Our "chain decay" ansatz assumed that, if the rank 1 meson carries away momentum ξ, from a quark of flavor "a" and momentum P_0, the remaining cascade starts with a quark of flavor "b" and momentum $P_1 = P_0 - \xi$, and the remaining hadrons are described in precisely the same way as the hadrons which came from a jet originated by a quark of flavor "b" with momentum P_1. There is one generating function, $f(y)$, which gives the probability that the rank 1 meson leaves fractional momentum y to the remaining cascade. The generating function was determined by fitting the e^+e^- data. Additional parameters were

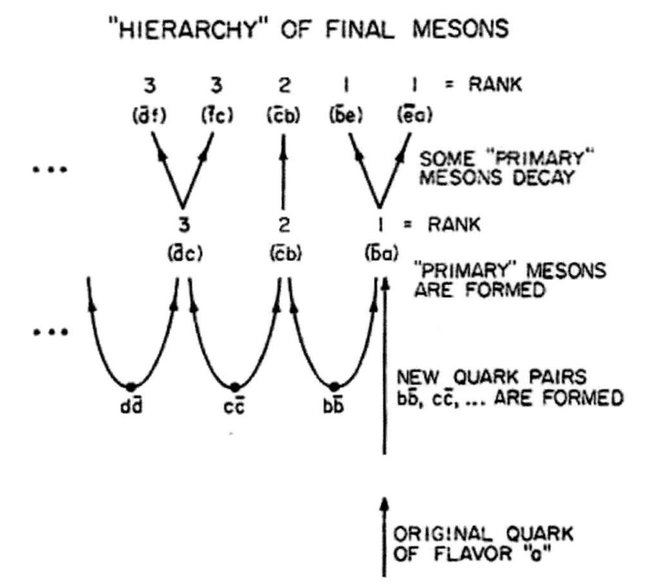

Fig. 9. Illustration of the hierarchy of mesons formed when a quark of flavor "a" fragments into hadrons from FF2[7] (1978). The initial quark of flavor "a" combines with an antiquark from a produced quark–antiquark pair, "bbbar," forming the meson for rank 1. The resulting quark of flavor "b" then combines with an antiquark from another produced quark–antiquark pair forming the meson of rank 2 and so on. These "primary" mesons are then allowed to decay into "secondary" mesons.

included to handle the flavor dependence of the fragmentation functions. We let β_u be the probability that the new quark–antiquark pair is a $u\bar{u}$ pair, and β_d be the probability that it is a $d\bar{d}$ pair, etc. We later generalized the model to include gluon jets. Figure 10 shows a transparency from a talk Feynmen gave on our model for how quarks fragment into hadrons at the International Symposium on Multiparticle Dynamics (ISMD), Kaysersberg, France, June 12, 1977. The following is a Feynman quote from FF2:[7]

> *"The predictions of the model are reasonable enough physically that we expect it may be close enough to reality to be useful in designing future experiments and to serve as a reasonable approximation to compare to data. We do not think of the model as a sound physical theory."*

The Feynman–Field jet model (FF fragmentation) was, of course, a naive scaling model and QCD certainly modifies the approach. Nevertheless, it was very easy to implement the model using Monte Carlo techniques and it allowed us, for the first time, to simulate hadron–hadron collisions (and

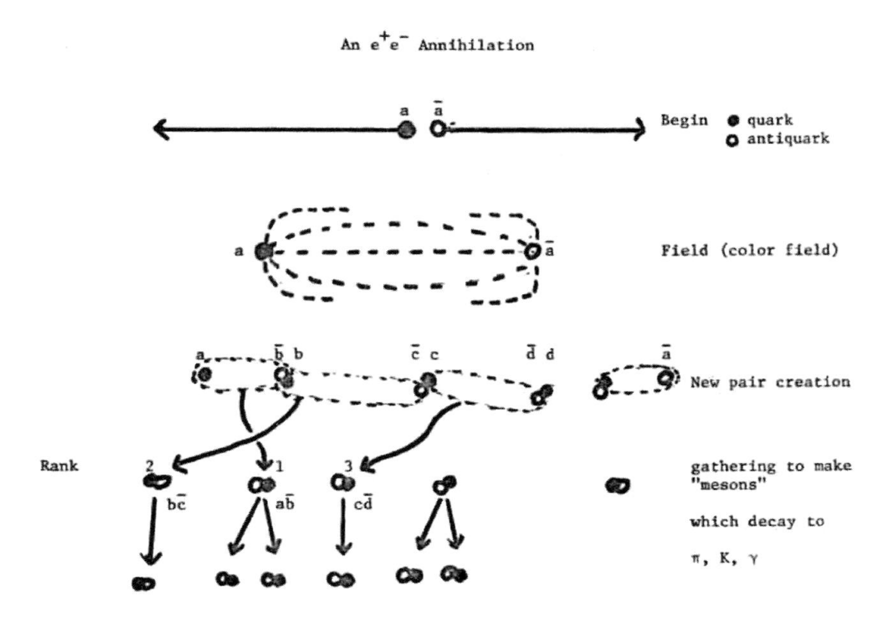

Fig. 10. Transparency from a talk Feynmen gave on our model for how quarks fragment into hadrons at the International Symposium on Multiparticle Dynamics (ISMD), Kaysersberg, France, June 12, 1977.

e^+e^- annihilations) on an event-by-event basis. In 1978 we constructed a QCD Monte Carlo event generator for hadron–hadron collisions (and e^+e^- annihilations) and began making predictions for experiment.

As illustrated in Fig. 11, quarks radiate gluons producing a "parton shower" which can be computed using perturbative QCD down to some small scale Q_0 at which time things become non-perturbative and hadrons are formed. It only makes sense to use the FF fragmentation model if one stops the parton shower at a rather high scale, say around 5 GeV. One needs another fragmentation approach if you want to generate the parton shower down to a smaller scale. With many partons of low momentum you cannot independently fragment each one of them using the FF model. In FW1[8] (1983) my first graduate student, Steven Wolfram, and I proposed a "cluster" fragmentation model which was more closely connected to perturbative QCD than the FF fragmentation model. Here one evolves the parton shower down to a rather low scale, follow the color flow, and forms low invariant mass color singlet clusters which are then allowed to decay into hadrons using two-body phase space. Our QCD Monte Carlo event generator had a switch which allowed us to run either of the two approaches. We could stop the parton shower evolution at a high scale and employ FF fragmentation

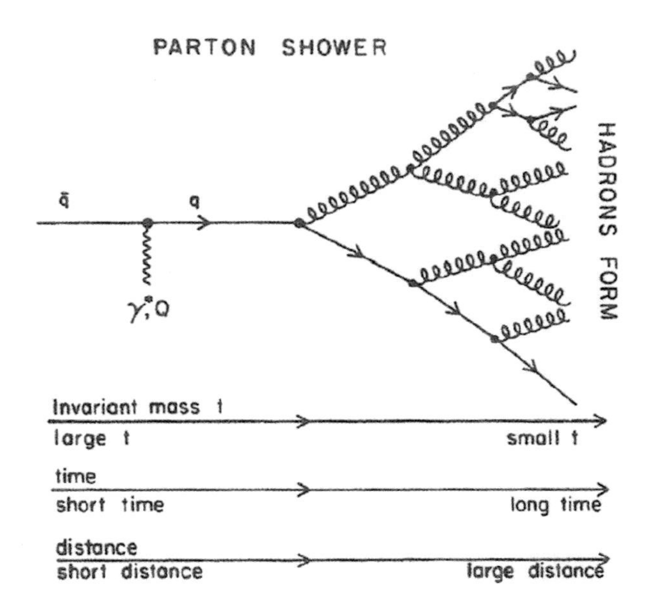

Fig. 11. Illustration of a "parton shower" in which a quark initially produced at a scale Q radiates gluons which in turn radiate additional gluons and quark–antiquark pairs. As time increases the shower progresses to larger distances from the point where in initial quark was produced and the invariant masses, t, of the partons decrease from a maximum of Q^2 to some small scale Q_0 where hadrons are formed.

to parameterize the rest of the hadronization process or we could evolve the parton-shower down to a small Q^2 and employ the simple two-body phase space decay model (i.e. FW fragmentation).

Feynman warned me not to release our QCD Monte Carlo event generator to the experimenters to use. He felt that it would be a "black box" to them and that they would take the predictions too seriously without understanding the subtleties involved in constructing the generator. In retrospect, I think Feynman was wrong and I should have given our QCD Monte Carlo event generator to the experimenters. Later Frank Paige constructed ISAJET[9] which uses FF fragmentation and Bryan Webber constructed HERWIG[10] which uses an improved version of FW fragmentation. PYTHIA[11] uses a "string fragmentation" model invented in 1983 by the Lund group.[12] PYTHIA, HERWIG, and ISAJET all employed different fragmentation models.

I left Caltech in 1980 to help build a high energy theory and experimental program at the University of Florida. In 1998 I joined the CDF Experimental Collaboration at Fermilab. I always wanted to produce my own data plot to compare with my theory predictions. At CDF I learned how to analyze data and worked to test and improve the QCD Monte Carlo models. The goal is to accurately simulate, on an event-by-event basis everything that occurs in a hadron–hadron collision. To do this one must, not only do a good job describing the hard scattering components of the collision, but in addition one must correctly model the "underlying event" (UE). The UE consists of the beam–beam remnants and the multiple parton interactions that accompany a hard scattering. Perhaps I am still trying to accomplish Feynman's future study item (5) in Fig. 6.

I have been a member of the CMS Collaboration at the LHC since Gena Mitselmaker formed the CMS group at the University of Florida in 1998. In 2006 Paolo Bartalini, Livio Fano, and I formed a group of people within CMS to study the UE at the LHC. In November 2010, we published the first LHC measurement of the UE at 900 GeV.[13] Wow! What a wonderful journey from the "old days" of Feynman–Field collider phenomenology to the Tevatron and the LHC. I am very greatful to everyone that has helped me along the way.

References

1. R. D. Field and R. P. Feynman, *Phys. Rev. D* **15**, 2590 (1977).
2. R. P. Feynman, R. D. Field and G. C. Fox, *Nucl. Phys. B* **128**, 1 (1977).
3. R. Field, *Annu. Rev. Nucl. Part. Sci.* **62**, 427 (2012).
4. R. D. Field, *Phys. Rev. Lett.* **40**, 997 (1978).
5. R. P. Feynman, R. D. Field and G. C. Fox, *Phys. Rev. D* **18**, 3320 (1978).
6. CDF Collab. (T. Aaltonen *et al.*), *Phys. Rev. D* **78**, 052006 (2008).
7. R. D. Field and R. P. Feynman, *Nucl. Phys. B* **136**, 1 (1978).
8. R. D. Field and S. Wolfram, *Nucl. Phys. B* **213**, 65 (1983).
9. F. Paige and S. Protopopescu, BNL Report, BNL38034, version 7.32, 1986 (unpublished).
10. G. Marchesini and B. R. Webber, *Nucl. Phys. B* **310**, 461 (1988); I. G. Knowles, *Nucl. Phys. B* **310**, 571 (1988); S. Catani, G. Marchesini and B. R. Webber, *Nucl. Phys. B* **349**, 635 (1991).
11. T. Sjostrand, *Phys. Lett. B* **157**, 321 (1985); M. Bengtsson, T. Sjostrand and M. van Zijl, *Z. Phys. C* **32**, 67 (1986); T. Sjostrand and M. van Zijl, *Phys. Rev. D* **36**, 2019 (1987).
12. B. Andersson, G. Gustafson, G. Ingelman and T. Sjostrand, *Phys. Rep.* **97**, 31 (1983).
13. CMS Collab., *Eur. Phys. J. C* **70**, 555 (2010), arXiv:1006.2083.

Exclusive Processes and the Fundamental Structure of Hadrons

Stanley J. Brodsky

SLAC National Accelerator Laboratory,
Stanford University, Stanford, California 94309, USA

I review the historical development of QCD predictions for exclusive hadronic processes, beginning with constituent counting rules and the quark interchange mechanism, phenomena which gave early validation for the quark structure of hadrons. The subsequent development of pQCD factorization theorems for hard exclusive amplitudes and the development of evolution equations for the hadron distribution amplitudes provided a rigorous framework for calculating hadronic form factors and hard scattering exclusive scattering processes at high momentum transfer. I also give a brief introduction to the field of "light-front holography" and the insights it brings to quark confinement, the behavior of the QCD coupling in the nonperturbative domain, as well as hadron spectroscopy and the dynamics of exclusive processes.

1. Introduction

One of the most important arenas for testing quantum chromodynamics are measurements of exclusive scattering reactions in which the kinematics of all of the initial and final-state particles are determined. Exclusive processes include hadronic spacelike and timelike hadronic form factors, two-photon reactions $\gamma\gamma \to H\bar{H}$, Compton scattering $\gamma p \to \gamma p$, deeply virtual Compton scattering $\gamma^* p \to \gamma p$; and two-body scattering reactions such as elastic proton–proton scatting $pp \to pp$ and pion photoproduction $\gamma^* p \to \pi^+ n$. In such reactions one is sensitive to the fundamental composition of hadrons in terms of their quark and gluon content, as well as the fundamental interactions and forces acting on their quark and gluonic constituents. Exclusive reactions are also sensitive to the dynamics of color confinement, and they illuminate the mechanisms underlying quark and gluon hadronization at the amplitude level in quantum chromodynamics (QCD).

A central feature of exclusive processes are the "*constituent counting rules*" which allow one to verify the underlying quark and gluon content of hadrons and enumerate their fundamental degrees of freedom. The counting rules were derived in the 1970s by Farrar and this author[1,2] and by Matveev, Muradyan and Tavkhelidze,[3] soon after the development of QCD by Fritzsch, Gell-Mann, and Leutwyler.[4]

This review begins with a summary of the main features of quark counting rules and quark interchange phenomena — which gave early validation of the quark structure of hadrons. I then review the development of perturbative QCD factorization theorems and the evolution equations for the hadron "*distribution amplitudes*" — the fundamental wave-functions which control form factors at large momentum transfer and other hard scattering exclusive scattering processes. I also give a short introduction to "light-front holography" and the insights it brings to quark confinement, the QCD coupling in the nonperturbative domain, light-quark hadron spectroscopy, and the dynamics of exclusive processes.

2. Quark Counting Rules

The physics principle underlying the counting rules is that at momentum transfers large compared to the QCD color confinement scale, where asymptotic freedom is applicable, the dominant internal interactions between quarks from gluon exchange within hadrons should be approximately scale independent — in fact, close to conformal.[5,6] The twist-dimension of the effective operator which creates the hadron at short distances[7,8] then controls the scaling of an exclusive amplitude. Here 'twist' refers to the dimension minus the spin of the interpolating operator; it also equals the number of quarks in the valence Fock state; e.g., the twist is 3 for baryons and the twist is 2 for mesons; higher Fock states will give "higher twist" power-law suppressed contributions. The result is that the power-law fall-off of any hard-scattering reaction can be predicted by simply counting the number of bound quarks in each of the interacting hadrons, thus providing a direct window to their constituent structure. Thus, according to the counting rules, an exclusive scattering amplitude at large momentum transfer such as elastic hadron–hadron scattering at fixed $\theta_{\rm CM}$ (or fixed t/s) will have the nominal power-law fall-off[1–3]

$$M(A+B \to C+D) = \frac{F_{A+B\to C+D}(\theta_{\rm CM})}{s^N},$$

where

$$N = n_A + n_B + n_C + n_D - 4\,.$$

Here n_i is the minimum number of fundamental ("valence") constituents in each hadron's wave-function; e.g., one counts $n_B = 3$ for baryons, $n_M = 2$

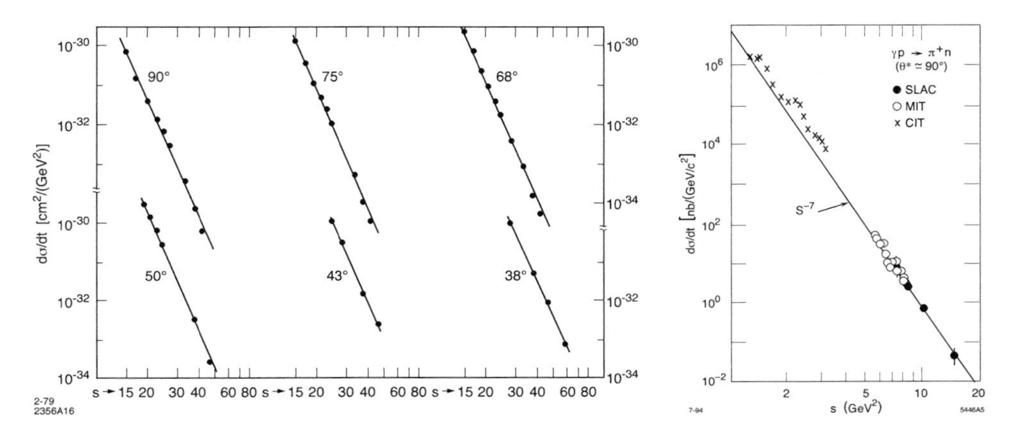

Fig. 1. Data for $\frac{d\sigma}{dt}(pp \to pp)$ and $\frac{d\sigma}{dt}(\gamma p \to \pi^0 p)$. The measured power-law fall-off at fixed $\theta_{\rm CM}$ are consistent with the quark counting rule predictions s^{-10} and s^{-7}, respectively. Further details and references can be found in Ref. 10.

for mesons, and $n = 1$ for interacting elementary particles such as the photon or lepton. Thus for any two-to-two scattering process the differential cross-section will fall as

$$\frac{d\sigma}{dt}(A + B \to C + D) \propto \frac{|M(A + B \to C + D)|^2}{s^2} \propto \frac{|F_{A+B\to C+D}(\theta_{\rm CM})|^2}{s^{2N-2}}$$

at large center-of-mass energy squared $s = E^2_{\rm CM}$ and fixed ratio of invariants such as fixed t/s or fixed center-of-mass angle $\theta_{\rm CM}$. A simple example is meson–baryon photoproduction:

$$\frac{d\sigma}{dt}(\gamma p \to K^+\Lambda) \propto \frac{|F(\theta_{\rm CM}|^2}{s^7}$$

since $N = 3 + 1 + 2 + 3 - 2 = 7$. The experimental test of the counting rule for proton–proton elastic scattering at fixed $\theta_{\rm CM}$ is shown in Fig. 1. The quark counting prediction is $N = 12 - 2 = 10$, and the best fit to the data is $N = 9.7 \pm 0.5$.[9]

In general, one finds that the fall-off predicted by quark counting is consistent with measurements when the momentum transfer p_T in the hard exclusive reaction exceeds a few GeV. Here $p_T^2 = tu/s$. The counting rule for elastic lepton–hadron scattering $\ell H \to \ell H$ also predicts the power-law fall-off of the dominant helicity-conserving hadron's form factor at large $Q^2 = -t$:

$$F_H(Q^2) \propto [1/Q^2]^{n_H - 1} ,$$

where n_H is the minimum number of elementary constituents in H; i.e. the "twist" of the leading interpolating operator of the hadron H at small

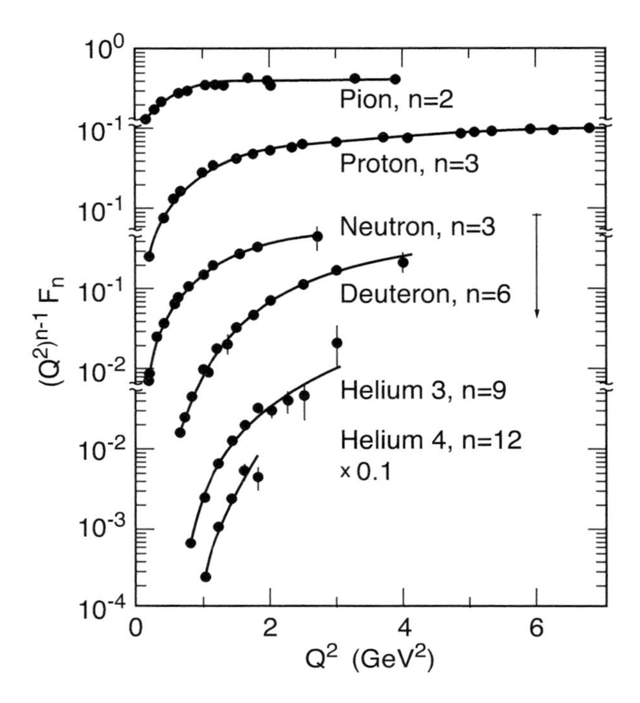

Fig. 2. Tests of quark counting for meson, nucleon and nuclear form factors at high momentum transfer: $[Q^2]^{n_H-1} \times F_H(Q^2) \to$ constant. Further details and references can be found in Ref. 11.

distances. Thus meson form factors fall-off as $1/Q^2$, the Dirac form factor falls at $1/Q^4$. A comparison with data is given in Fig. 2. Historically, the empirical success of the counting rules provided a primary check on the number of valence quarks underlying the bound-state structure of hadrons. Furthermore, they provided a test on the elementary, structureless property of photons, leptons, as well as the intermediate bosons exchanged in charged- or neutral-current exclusive processes. The counting rules can also be applied to large-angle real or virtual Compton scattering, hard exclusive nuclear reactions such as $\gamma d \to np$. They can also be extended to exclusive processes involving the production of multiple hadrons, such as $p\bar{p} \to K^+K^-\pi^0$. A detailed review is given in Ref. 11.

Because of the Drell–Yan exclusive–inclusive connection,[12] the quark counting rules for form factors at large momentum transfer also imply counting rules for the power-law fall-off of the quark and gluon structure functions in hadrons: $G_{a/H}(x)$ at $x \to 1$, the kinematic domain where the constituent a takes most of the momentum of the hadron H. The limiting power-law behavior at $x \sim 1$ of the distributions derived from minimally connected amplitudes is[13]

$$G_{a/H}(x) \propto (1-x)^p \,,$$

where

$$p = 2n - 1 + 2\Delta S_z \,.$$

Here n is the minimal number of spectator quark lines, and $\Delta S_z = |S_a^z - S_H^z|$. For example, $\Delta S_z = 0$ or 1 for parallel or anti-parallel quark and proton helicities, respectively. This counting rule reflects the fact that the valence Fock states with the minimum number of constituents give the leading contribution to structure functions when one quark carries nearly all of the light-cone momentum. Thus Fock states with a higher number of partons give structure functions which fall off faster as x approaches 1. The helicity dependence of the counting rule also reflects the helicity retention properties of the gauge couplings: a quark with a large momentum fraction of the hadron also tends to carry its helicity; the anti-parallel helicity quark is suppressed by an extra suppression of $(1-x)^2$. The counting rule for valence quarks can be combined with the splitting functions to predict the $x \sim 1$ behavior of gluon and non-valence quark distributions. In particular, the gluon distribution of non-exotic hadrons must fall by at least one power faster than the respective quark distributions. The counting rules for large x behavior of quark and gluon helicity distributions can also be derived from the continuity between the exclusive and inclusive channels at fixed invariant mass. For example, as shown by Drell and Yan,[12] a quark structure function $G_{q/H} \sim (1-x)^{2n-1}$ at $x \to 1$ if the corresponding form factor $F(Q^2) \propto (1/Q^2)^n$ at large Q^2. A similar analysis controls the $z \to 1$ behavior of the fragmentation functions $D_{H/a}(z)$. The results are constant with Gribov–Lipatov crossing relations. Higher twist corrections due to multiparton subprocesses are discussed in Ref. 14.

The quark counting rules were later shown to explicitly emerge from pQCD factorization theorems by Brodsky and Lepage,[15,16] and by Efremov and Radyushkin.[17] As we shall review below, one also predicts logarithmic corrections to the nominal power law fall-off which can be systematically derived from two sources: the evolution of the QCD coupling constant $\alpha_s(Q^2)$ and the evolution of the hadron distribution amplitudes, the fundamental gauge-invariant amplitudes which control hadronic process at short distances. More recently, the quark counting rules were shown[18] to be features of a nonperturbative approach based on AdS/QCD and light-front holography, the postulated duality between AdS_5 space and QCD in physical space–time. This analysis is discussed further in Sec. 6.

The pQCD analysis also predicts a number of other phenomenological features of hard exclusive processes. These include:

- **Hadron Helicity Conservation**[19] — The dominant pQCD interactions conserve quark chirality at every quark vertex such as $gq_\uparrow \to q_\uparrow$ and $g \to q_\uparrow \bar{q}_\downarrow$. The quark–chirality interactions bring in powers of the quark mass and thus faster power fall-off. In addition, the leading power contribution to hard exclusive amplitudes are derived from wave-functions with zero quark orbital angular momentum. Thus the total S^z of the incident and final hadrons is conserved at leading twist:

$$\lambda_A + \lambda_B = \lambda_C + \lambda_D \,.$$

  This rule also accounts for the dominance of helicity-conserving form factors, such as the dominance of the Dirac hadron helicity-conserving form factor of the proton $F_1(Q^2) \propto \frac{1}{Q^4}$ relative to the helicity-flip Pauli form factor $F_2(Q^2) \propto \frac{1}{Q^6}$ at a high momentum transfer $Q^2 = -t \gg \Lambda^2_{\rm QCD}$. However, the dominant decay $J/\psi \to \pi\rho$ violates hadronic helicity conservation as well as the OZI rule. The solution[20] to this anomaly may be connected to higher Fock states of the final-state mesons which contain "intrinsic" $c\bar{c}$ pairs. A discussion of this puzzle is given in Ref. 21.
- **Color Transparency**[22–24] — Perturbative QCD predicts that at high moment transfer $p_T^2 \gg \Lambda^2_{\rm QCD}$, the hadronic wave-functions which control hard scattering amplitudes are dominated by configurations where all of the hadron constituents have small impact separation $b_\perp^2 \simeq \frac{1}{p_T^2}$. These configurations interact weakly inside a nuclear environment because of their small color-dipole moment; i.e., the color gauge interactions of the quark and antiquark of a meson tend to cancel when they have small impact separation. Thus a hard-scattering exclusive reaction can occur anywhere in the nucleus in a quasi-elastic reaction such as $pA \to ppX$ without absorption in the initial or final state. These small-size configurations expand to physical size hadrons at a time interval proportional to the hadron's energy. The cross-section for $pA \to ppX$ is thus predicted to be additive in the number of protons in the nuclear target at very large momentum transfer: $E\frac{d\sigma}{d^3p}(pA \to ppX) \simeq ZE\frac{d\sigma}{d^3p}(pp \to pp)$. The expectations of color transparency have been verified in hard exclusive channels such as $\gamma^* p \to \rho^0 p$ at high photon virtuality Q^2 and high momentum transfer quasi-elastic pp scattering. Comprehensive reviews are given in Refs. 24 and 25. The color transparency of hard exclusive reactions is in direct contradiction to traditional Glauber theory,[26] where the large size of hadron–nucleon cross-sections implies that hadronic interactions in

a nuclear target are relegated to the nuclear periphery. For example, in low momentum transfer reactions the cross-section $\sigma(pA \to \pi X)$ behaves $A^{1/3}$ since the proton can only interact on the front surface, and thus the emitted pion can only emerge from the nuclear perimeter.[27]

Color transparency was tested directly in the dijet diffractive reaction $\pi A \to \text{Jet Jet } A'$ in a Fermilab experiment by Ashery *et al.*[25,28] The emerging jets correspond to the dissociation of the pion $\pi \to q\bar{q}$ with closely balanced transverse momenta $\sim \pm\vec{k}_\perp$, so that the nucleus remains intact. See Fig. 3. At large $k_\perp^2$, the dominant configurations of the pion wavefunction have small impact separation $b_\perp^2 \sim 1/k_\perp^2$ and thus small color dipole moments. The measurements of the nuclear A dependence are in agreement with the color transparency predictions of Strikman *et al.*[25,28]

However, in one classic experiment, color transparency was observed to fail dramatically: A strong nuclear suppression is observed in $pA \to ppX$ at

Key Ingredients in E791 Experiment

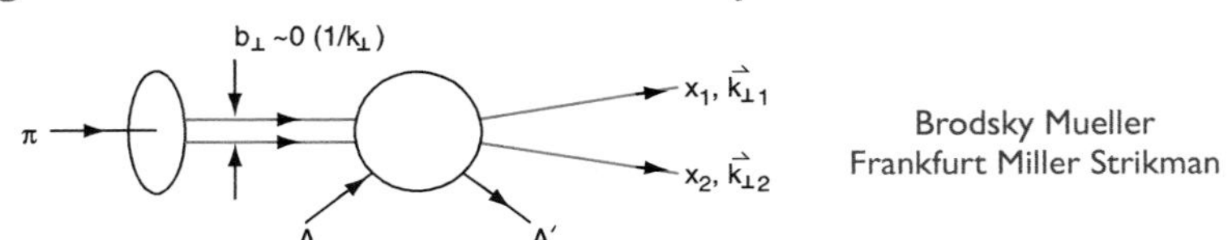

Small color-dipole moment pion not absorbed;
interacts with each nucleon coherently
QCD COLOR Transparency

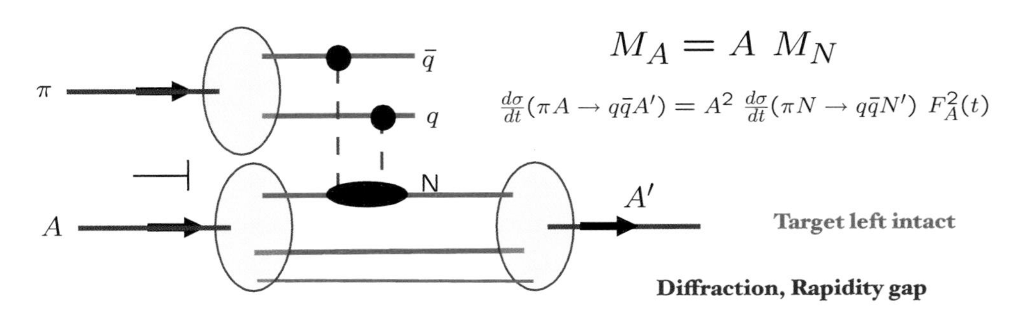

Fig. 3. Illustration of pion diffractive dissociation to dijets in a nucleus $\pi A \to \text{Jet Jet} + A$. The nucleus A remains intact. When the jets are produced at high and opposite transverse momenta $\pm\vec{p}_\perp$, one samples the pion LF wave-function at $b_\perp \simeq \frac{1}{p_\perp}$ where the quark and antiquark have small transverse separation; i.e., the dijet has a small color dipole moment. The color transparency prediction of negligible absorption in the nucleus was verified by the Fermilab experiment of Aitala *et al.*[25,28]

$\sqrt{s} \simeq 5$ GeV.[29] This is the same energy where a surprisingly strong spin–spin correlation $R_{NN} \simeq 4$ is observed.[30] Here R_{NN} is the ratio $\frac{d\sigma}{dt}(p_\uparrow p_\uparrow \to pp)/\frac{d\sigma}{dt}(p_\downarrow p_\downarrow \to pp)$, where the incident protons are polarized normal to the scattering plane. The anomalously strong spin–spin correlation and the associated breakdown of color transparency can be understood as due to the physics of the charm threshold:[31] $\sqrt{s} \simeq 5$ GeV is the energy where a produced $c\bar{c}$ plus the six spectator light quarks are all at the same rapidity and thus can interact strongly. In fact, one can interpret the data for evidence for the production of an "octoquark" $|uuduudc\bar{c}\rangle$ resonance in the $J = 1$, $S = 1$, pp to pp partial wave.[21,31,32] Large-size protons are involved, and thus color transparency breaks down in the energy domain where the octoquark resonance is excited.

- **Nuclear Effects and Hidden Color** — At first approximation, a nucleus can be regarded a composite of protons and nuclei bound by the exchange of mesons. However, quantum chromodynamics predicts that at high resolution, the structure of the nucleus is much more complex: the nucleus itself becomes revealed as a bound-state system of confined quarks and gluons, the fundamental quantum fields of QCD. Even the fundamental nature of the nuclear force is reinterpreted in QCD. In fact, measurements of the elastic scattering of protons at high momentum transfer show that the dominant interaction at short distances between nucleons is the interchange of their constituent quarks.[33]

  An important example where QCD differs from traditional nuclear physics in a fundamental way is presence of the "hidden-color" degrees of freedom in the nuclear eigensolution.[34–37] If one regards the deuteron as a composite of six color-triplet quarks, then there are five different color-singlet states, only one of which matches the usual proton–neutron degrees of freedom. In fact, when one considers color-octet gluon exchange at short inter-nucleon distances, then all five color-singlet components enter the deuteron eigensolution. This can be shown explicitly using rigorous pQCD evolution equations which couple together all of the components.[38] One can prove from the evolution of the deuteron distribution amplitude[38] that at $Q^2 \to \infty$ all five color-singlet components have equal weight.

  Thus QCD predicts the dominance of the hidden-color configurations when a deuteron scatters at very large momentum transfers in elastic electron–deuteron scattering, or when it photo-disintegrates into pairs of baryonic resonances, or when the quark in the nucleus carries a high momentum fraction $x > 1$, beyond the kinematics of a single nucleon.

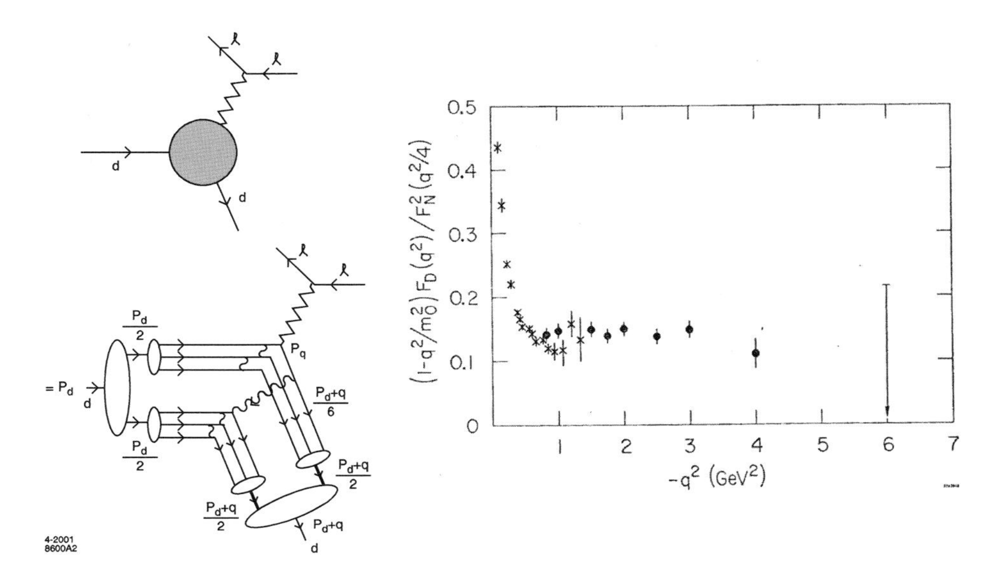

Fig. 4. (a) Physics of of the deuteron reduced form factor $F_d^{\text{reduced}}(Q^2)$. In the weak binding limit, each nucleon scatters in electron–deuteron elastic scattering with half the momentum transfer. (b) Experimental test of the scaling prediction $Q^2 F_d(Q^2) \to$ constant for the reduced deuteron form factor.

The hidden-color degrees of freedom can manifest themselves as the dominant contribution to the elastic form factor of the deuteron at very large Q^2. It is useful to factorize the deuteron helicity-conserving form factor $F_d(Q^2) = A(Q^2)$ in the form[34]

$$F_d(Q^2) \equiv F_p(Q^2/4)F_n(Q^2/4)F_d^{\text{reduced}}(Q^2)$$

since in the weak binding limit each nucleon scatters in electron–deuteron elastic scattering with half the momentum transfer. The quark counting rules predict $F_d^{\text{reduced}}(Q^2) \sim \frac{1}{Q^2}$, the same scaling as a meson. This scaling prediction has been confirmed experimentally.[35] See Fig. 4. The magnitude of $Q^2 F_d^{\text{reduced}}(Q^2)$ is much larger than expected from conventional nuclear physics based on the small deuteron binding, thus indicating a dominant role of hidden color in the hard-scattering domain.[38] The hidden color of the deuteron LFWF also predicts a large coupling of $\gamma d \to \Delta^+\Delta^0$ in exclusive photodisintegration.[39]

3. The Front Form

The light-front (LF) formalism is based on Dirac's "front form,"[40] where the time variable $\tau = t + z/c$ is the time along the light-front. For a review, see

Ref. 41. For example, when when one takes a flash photograph, the resulting photographs is at fixed light-front time τ. In fact physical measurements such as deep inelastic scattering on a proton capture the structure of the proton at fixed LF time within the causal horizon, not at a fixed "instant" time t which would requires acausal information. The form factors are thus a primary measure of hadron structure at fixed LF time $\tau = t + z/c$. Given the QCD Lagrangian, one can derive the LF Hamiltonian $H_{\mathrm{LF}} = -i\partial/\partial\tau$, the LF time evolution operator.

The eigensolutions of the QCD light-front Hamiltonian for each hadron satisfy the LF Heisenberg equation:

$$H_{\mathrm{LF}}|\Psi_I\rangle = M_I^2|\Psi_I\rangle\,,$$

where the eigensolution $\Psi_H\rangle$ for each hadron H can be expanded as a sum over free Fock states:

$$|\Psi_H\rangle = \sum_n \psi_{n/H}(x_i, \vec{k}_{\perp_i}, \lambda_i)|n\rangle\,.$$

The n constituents of the Fock states $|n\rangle$ for a hadron with momentum $P^{\pm} = P^0 \pm P_z$ and $\vec{P}_{\perp}$ have physical momenta $p_i^+ = x_i P^+$ and $\vec{p}_{\perp i} = x_i \vec{P}_{\perp} + \vec{k}_{\perp i}$. The momenta $k^+ = k^0 + k^z$ are conserved and always positive. One also has $M^2 = P^+P^- - P_{\perp}^2$. The Fock state with minimum n ($n = 2$ for mesons and $n = 3$ for baryons) are called the "valence" Fock states. Since the $+$ and transverse momenta are conserved in the front form, $\sum_i^n x_{i-1} = 1$ and $\sum_i^n \vec{k}_{\perp_i} = 0$.

The coefficients in the Fock expansion are the hadronic light-front wave-functions. See Fig. 5. The LFWFs are the hadronic eigensolutions of H_{LF} which in turn is derived from the QCD Lagrangian. They control virtually every observable in QCD (see Fig. 6). A remarkable feature of the light-front formalism is that the LFWFs are independent of the hadron's momentum P^+ and $P_{\perp}$. No Wigner or Melosh boosts are required. In principle, the LF Heisenberg equation can be solved by matrix diagonalization of the LF Hamiltonian using discretized light-cone quantization (DLCQ).[41–44] In fact, this can be done to essentially arbitrary accuracy for QCD[45] and other theories[46] in one space and one LF time. As we shall discuss in Sec. 6, one can obtain predictions for the nonperturbative LFWFs of hadrons with confined quarks from a novel method called "light-front holography."[47]

Form factors and other current matrix elements $\langle p+q|J^{\mu}(0)|p\rangle$ which are measured by virtual photon exchange in elastic or inelastic lepton–hadron scattering can be elegantly written in terms of the overlap of initial-state and final-state light-front wave-functions (LFWFs) of the hadrons.[12,48,49]

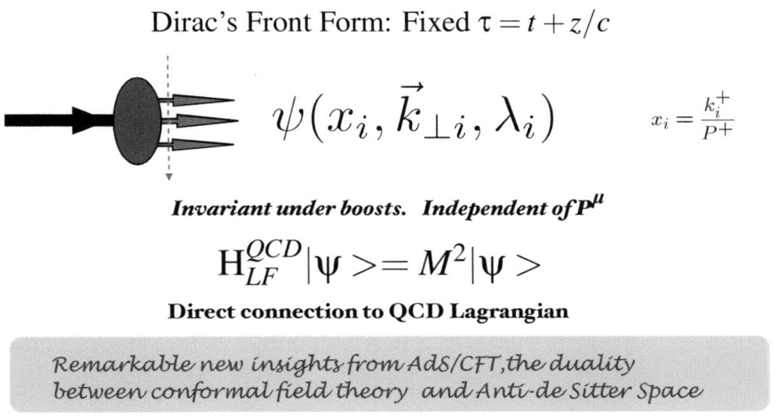

Dirac's Front Form: Fixed $\tau = t + z/c$

$$\psi(x_i, \vec{k}_{\perp i}, \lambda_i) \qquad x_i = \frac{k_i^+}{P^+}$$

Invariant under boosts. Independent of P^μ

$$H_{LF}^{QCD}|\psi> = M^2|\psi>$$

Direct connection to QCD Lagrangian

Remarkable new insights from AdS/CFT, the duality between conformal field theory and Anti-de Sitter Space

Fig. 5. Light-front wave-functions are the frame-independent hadronic eigensolutions of the light-front Hamiltonian H_{LF}, which in turn is derived from the QCD Lagrangian.

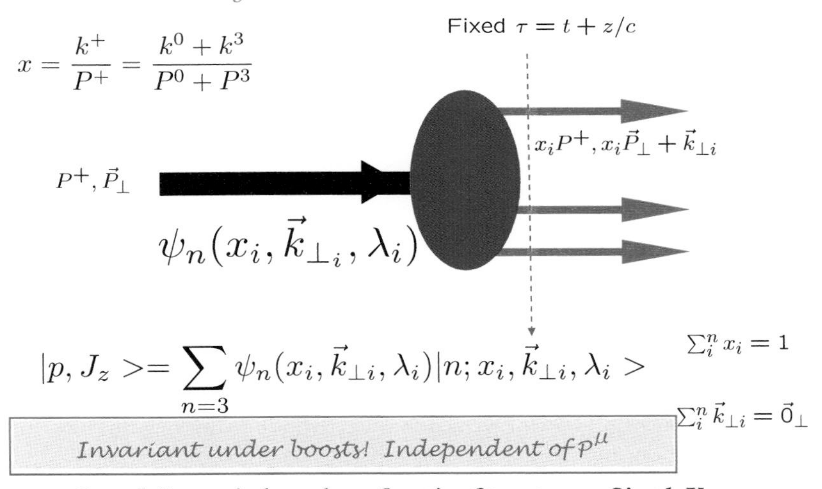

Fig. 6. The LFWF describes the state at a fixed LF time $\tau = t + z/c$ and is off-shell in $P^- = \frac{\mathcal{M}^2 + P_\perp^2}{P^+}$ and thus in invariant mass. The Fock state wave-functions $\psi_n(x_i, \vec{k}_{\perp i}, \lambda_i)$ are independent of the hadron's 4-momentum $P^\mu = (P^+, P^-, \vec{P}_\perp)$. The LFWFs can also be predicted from light-front holography.

The Drell–Yan–West (DYW) formula[12,48,49] for the spacelike ($t < 0$) form factor of a spin-zero meson form factors is given by the overlap of initial and final-state light-front wave-functions:

$$F(q^2) = \int d^2\vec{k}_\perp \int_0^1 dx \psi_H(x, \vec{k}_\perp)\psi_H(x, \vec{k}_\perp + (1-x)\vec{q}_\perp)\,,$$

where $t = -q^2 < 0$. (A sum over the struck quarks weighted by the quark charge and a sum over all hadronic Fock states is understood.) An essential step in the derivation of the DYW formula is to evaluate the current matrix element $F(q^2) = \langle p + q|j^+(0)|p\rangle/P^+$ in the Lorentz frame with $q^+ = 0$ and thus $-t = -q^2 = Q^2 = \vec{q}_\perp^{\,2}$. Thus the kinematical coordinates of the exchanged virtual photon is $q^\mu = (q^+, q^-, \vec{q}_\perp) = (0, q^-, q_\perp)$, where $q_\perp^2 = Q^2 = -q^2$ and $q^- = 2p \cdot q/P^+$. The choice $q^+ = 0$ avoids the need to evaluate contribution from the overlap of LFWFs with different number of constituents. Since $q^+ = 0$, LF coordinate $x = k^+/P^+$ does not change at the quark–photon–quark vertex. See Fig. 7. If the struck quark of the initial state has LF momenta $k_\perp$ and x, then the momenta of the final-state constituents are x and $k_\perp + (1-x)\vec{q}_\perp$ for the struck quark and x and $\vec{k}_{\perp I} + x_i\vec{q}_\perp$ for the spectator constituents. In the case of spin, the LFWFs multiply LF spinors where $\bar{u}\gamma^+ u = 1$. Thus the DYW LF formula holds for the Dirac form factor $F_1(q^2)$ where the initial and final state nucleon have the same helicity. The corresponding formula for the spin-flip Pauli form factor $F_2(q^2)$ is:[49]

$$\begin{aligned}
&(\vec{q}_\perp x + i\vec{q}_\perp y) \times F_2(q^2) \\
&\quad = \langle p + q, S^z = -1/2|j^+(0)|p, S^z = +1/2\rangle/P^+ \\
&\quad = \int d^2\vec{k}_\perp \int_0^1 dx\, \psi_H^{S_z=-1/2}(x, \vec{k}_\perp)\psi_H^{S^z=1/2}(x, \vec{k}_\perp + (1-x)\vec{q}_\perp)\,.
\end{aligned}$$

Since the quark spin is not flipped by the current operator j^+, the Pauli form factor is the overlap of the initial-state LFWF where the spin of struck quark is parallel to the proton spin with the spin-antiparallel LFWF of the final state baryon. Thus the quarks in the baryon must have orbital angular momentum in order to have a nonzero Pauli form factor and nonzero anomalous moment. These frame-independent light-front expressions are the relativistic generalizations of the overlap formula of Schrödinger wave-functions for form factors in nonrelativistic quantum mechanics. One can immediately derive the quark counting rules for form factors by noting that the LFWFs fall-off as $\frac{1}{k_\perp^2}$ at large $k_\perp$ for each pair of internal interactions.

The simplicity of the DYW formula for current matrix elements is remarkable. One can also derive an exact formula in the two-particle Bethe–Salpeter

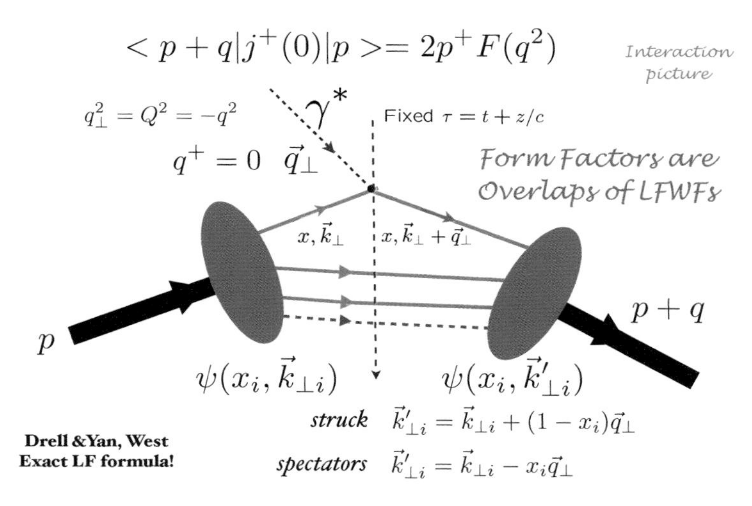

Fig. 7. Illustration of the Drell–Yan–West formula for the calculation of form factors and other current matrix elements as overlaps of boost-invariant LFWFs.

formalism by integrating first over k^-. In fact, the LFWFs are related to the Bethe–Salpeter wave-functions evaluated at fixed LF time τ.[50] However, all cross-graph kernels must appear in the BS formalism in order to maintain crossing symmetry. In fact, one must include all crossed-graph kernels simply to recover the familiar Dirac–Coulomb equation and the form factors of the muonium μ^+e^- atom at infinite muon mass.[51]

If one tries to evaluate form factors using "instant" time, one must include contributions where the virtual photon interacts with the connected currents from $q\bar{q}g$ vertices appearing from the vacuum. One must also boost the instant-form wave-functions,[52] a formidable dynamical problem and also include the overlaps of Fock states with different particle number. The instant form expressions are frame-dependent and acausal. One can formally recover the DYW formula in the instant form by evaluating the current matrix element in a Lorentz frame with $P^z \to -\infty$, the "infinite momentum frame."[53,54] However, it is much more natural to simply quantize at fixed τ.

As discussed by Soper,[55] the exact DYW expression for form factors and other current matrix elements can be written in coordinate space:

$$F(q^2) = \int_0^1 dx \int d^2\vec{\eta}_\perp e^{i\vec{\eta}_\perp\cdot\vec{q}_\perp}\tilde{\rho}(x, \vec{\eta}_\perp)\,,$$

where x is the LF momentum of the struck quark, $q_\perp^2 = Q^2$ and $\vec{\eta}_\perp = \sum_{j=1}^{n-1} x_j\vec{b}_\perp$.

For example, in the case of the valence $q\bar{q}$ Fock state of a meson, $\vec{\eta}_\perp = (1-x)\vec{b}_\perp$, where $\vec{b}_\perp$ is the transverse impact coordinate conjugate to $-\vec{k}_\perp$. At large $Q^2 = -q^2$, $F(q^2)$ is dominated by the LF density $\tilde{\rho}(x, \vec{\eta}_\perp)$ in the domain $\vec{\eta}_\perp^2 \leqq \frac{1}{Q^2}$. This condition is true for all Fock states, independent of the number of constituents. The small $\vec{\eta}_\perp^2$ domain corresponds to small frame-independent separation of the constituents of the hadron; only small color-dipole moments then appear — the underlying physics for color transparency. The region of fixed $k_\perp$ at $x \simeq 1$ has no special role.

4. Dominance of Quark Interchange

One of the prominent features of atom–atom scattering in QED is the dominance of electron exchange:[56] two atoms can scatter by exchanging an electron. This process is called "spin-exchange" since the spins of the exchanged electrons can be interchanged. The interchange amplitude is the physics underlying the covalent bond when the two atoms form a molecule. The analog of electron interchange in atom–atom scattering is quark interchange.[33] For example, in $K^+p \to K^+p$ elastic scattering, the u quark common to both hadrons can be exchanged. This is illustrated in Fig. 8. The interchange amplitude can be expressed as a simple overlap of the four light-front wavefunctions of the interacting hadrons:[33]

$$M^{AB\to CD}_{\text{interchange}}(s,t) = \frac{1}{2(2\pi)^3} \int d^2\vec{k}_\perp \int_0^1 dx \psi_A(x, \vec{k}_\perp + (1-x)\vec{q}_\perp + x r_\perp) \times \psi_C(x, \vec{k}_\perp + x\vec{r}_\perp)\Delta(x, \vec{k}_\perp)\psi_B(x, \vec{k}_\perp)\psi_D(x, \vec{k}_\perp + (1-x)\vec{q}_\perp) .$$

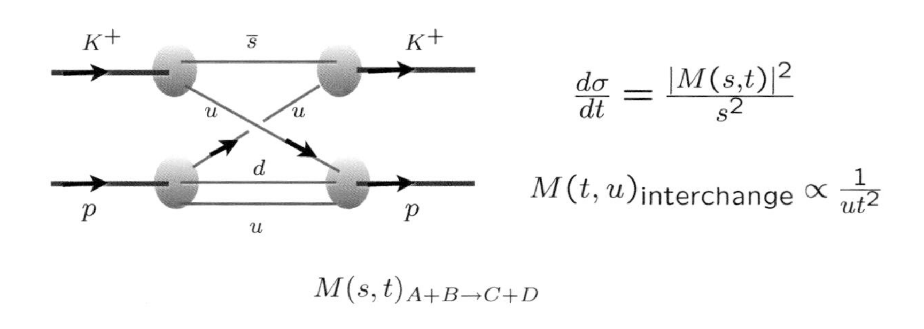

Fig. 8. The quark interchange mechanism for K^+p elastic scattering.

The kinematics require $q_\perp^2 = -t$, $\vec{r}_\perp^{\,2} = u$ and $\vec{r}_\perp \cdot \vec{q}_\perp = 0$. The Mandelstam variables satisfy $s+t+u = \sum_i M_i^2$. The quantity $\Delta = \sum_{I=A,B,C,D}(s - K_a - K_b - K - c - K_d$ is the inverse of the LF energy denominator which appears during the interchange of the quarks, where $K_i = (k_{\perp i}^2 + m_i^2)/x_i$ $(i = a, b, c, d)$ is the LF kinetic energy of each constituent in that intermediate state. The quantity Δ is in effect the sum of the internal binding potentials $\sum_{i=1}^4 V_i$ summed over the four hadrons. The resulting differential cross-section is then

$$\frac{d\sigma}{dt}(AB \to CD)(s,t) = \frac{1}{s^2}\left|M_{\text{interchange}}^{AB\to CD}(s,t)\right|^2 .$$

This relativistic amplitude reduces in the nonrelativistic limit to an overlap of Schrödinger wave-functions.

The interchange amplitude can also be written in impact space in a form similar to that of the Soper equation, where now the exponential involves $\vec{\eta}_\perp \cdot \vec{q}_\perp$ and $\vec{\eta}_\perp \cdot \vec{r}_\perp$. Since they are orthogonal vectors, we can take $\vec{r}_\perp \propto \hat{x}$ and $\vec{q}_\perp \propto \hat{y}$. In this case, the hard-scattering domain of high u and high t (or high $p_t^2 = tu/s$ and fixed θ_{CM}) requires the LF density $\tilde{\rho}$ to be evaluated at $\vec{\eta}_y^{\,2} \leq \frac{1}{|t|}$ and $\vec{\eta}_x^{\,2} \leq \frac{1}{|u|}$. This establishes the short-distance dominance and color transparency of the amplitude. The quark counting rules also emerge. (The argument works in 3+1 space because there are two transverse directions. It would not work for $A + B \to C + D + E$ at fixed angle!) In the case of $K^+p \to K^+p$ elastic scattering, the quark constituent interchange prediction is

$$M_{CIM}^{K^+p\to K^+p}(s,t) \propto \frac{\kappa^6}{ut^2}$$

at high u and t; it agrees well with measurements. See Fig. 9. Here κ represents the mass scale of QCD. The $1/u$ dependence of the amplitude extends Regge theory to the large spacelike momentum transfer; in fact, the Regge trajectory curves over to $\alpha_R(t) \to -1$ at large t; it is no longer linear. The corresponding interchange amplitude for nucleon–nucleon elastic scattering is

$$M_{CIM}^{p+p\to p+p}(s,t) \propto \frac{\kappa^8}{u^2t^2} .$$

One of the striking features of hard two-body exclusive hadronic scattering reactions is the strong dominance of the quark exchange or interchange contributions over gluon exchange. In fact, White *et al.*[57] have analyzed 20 different exclusive processes and found that the quark exchange amplitude strongly dominates over gluon exchange in every reaction. This is particularly surprising since Landshoff[58] has shown that the amplitude where three

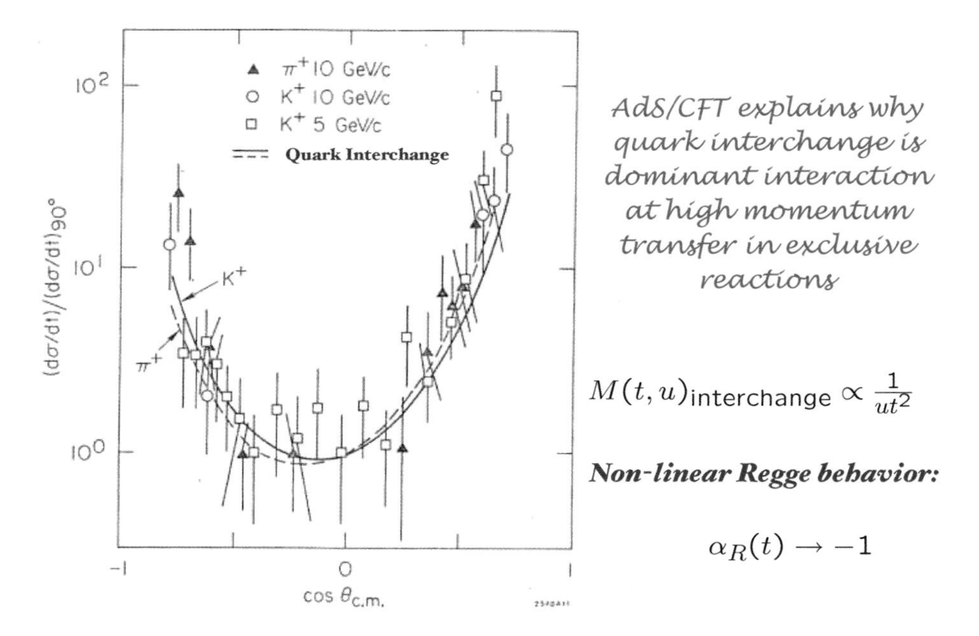

Fig. 9. Data for $K^+p \to K^+p$ elastic scattering at fixed $\theta_{\rm CM}$. The data are consistent with the u-quark interchange prediction $M_{K^+p\to K^+p} \propto 1/ut^2$, Reggeon intercept $\alpha_R(t) \to -1$ at large t, and the $1/s^8$ scaling of the cross-section.

gluons — each exchanging one-third of the momentum transfer between each quark–quark pair — should strongly dominate in large-angle hard $pp \to pp$ elastic scattering if one compares amplitudes. See also Refs. 1 and 59. However, there is no sign of the expected gluon-exchange contribution in the measured cross-section. The strong dominance of the interchange amplitude may be due to the fact that no extra interactions are needed beyond the internal dynamics of confinement when one evaluates the quark interchange amplitude; all of the effects of gluon exchange at the small $t/9$ virtualities involved in these elastic scattering reactions may be encapsulated as a "flux tube"[60] which generates the color confinement potential. If this is the case, hard gluon interactions may become manifest at very large momentum transfers. Other examples where gluon exchange contributions are evidently absent and the OZI rule fails are discussed in Ref. 21.

5. Perturbative QCD Factorization for Hard Exclusive Processes

One of fundamental testing grounds for perturbative QCD are hard exclusive processes. Rigorous theorems based on factorization of the hard and soft dynamics and the operator product expansion (OPE) can be based on first principles in QCD. An introduction to the theory of exclusive reactions and additional references are given in Ref. 11. For a recent review and applications to deeply virtual Compton scattering, see Ref. 59.

At high momentum transfer large compared to the QCD nonperturbative scale, hadronic amplitudes can be factorized at leading power as a convolution[16,17] of a hard scattering amplitude T_H with the product of process-independent hadronic distribution amplitudes $\phi_H(x_i, Q)$. The hard-scattering amplitude T_H is obtained by replacing each hadron by its valence quarks, moving as free partons collinear with the hadron. For example, for meson–meson scattering the factorization takes the form

$$\mathcal{M}_{A+B\to C+D} = \int dx_i \phi_C(x_c, \lambda_c, \tilde{Q})\phi_D(x_d, \lambda_d, \tilde{Q})T_H(x_i, \lambda_i, Q^2, \theta_{cm}) \times \phi_A(x_a, \lambda_a, \tilde{Q})\phi_B(x_b, \lambda_b, \tilde{Q}),$$

where the x_i are the light-front fractions $x_i = k_i^+/P_H^+$ carried by each valence quark in the initial-state and final-state hadrons, and the $\lambda_i = S_i^z = \pm 1/2$ are the respective LF spin projections of the quarks. The distribution amplitude $\phi_H(x, \tilde{Q})$ defined below is the interpolating amplitude between the valence quarks and the full dynamics of the hadron. The pQCD factorization is illustrated for meson–baryon elastic scattering in Fig. 10 and for the proton form factor in Fig. 11. In the case of two-photon annihilation into meson pairs, the incident photons act as elementary fields at leading order in $1/p_T$. One then has

$$\mathcal{M}_{\gamma\gamma\to M\bar{M}} = \int_0^1 dx \int_0^1 dy\, \phi_{\bar{M}}(y, \tilde{Q}_y)T_H(x, y, s, \theta_{cm})\phi_M(x, \tilde{Q}_x),$$

where T_H is the scattering amplitude for $\gamma\gamma \to [q\bar{q}][q\bar{q}]$ computed as if each valence quark and antiquark is collinear to the momentum of its respective ingoing or

outgoing meson. Here $\tilde{Q}_x = \min_{x_1,x_2} Q$. PQCD factorization can also be applied to obtain the leading contribution to hard exclusive scattering amplitudes such as Compton scattering, exclusive two-photon annihilation amplitudes, meson photoproduction, and baryon–baryon scattering, as well as deeply virtual Compton scattering $\gamma^* p \to \gamma p$. In each case, the hard-scattering amplitude T_H can be computed order by order in pQCD, starting with the sum of Born diagrams at first order $\alpha_s(p_T^2)$ connected by a single exchanged gluon.

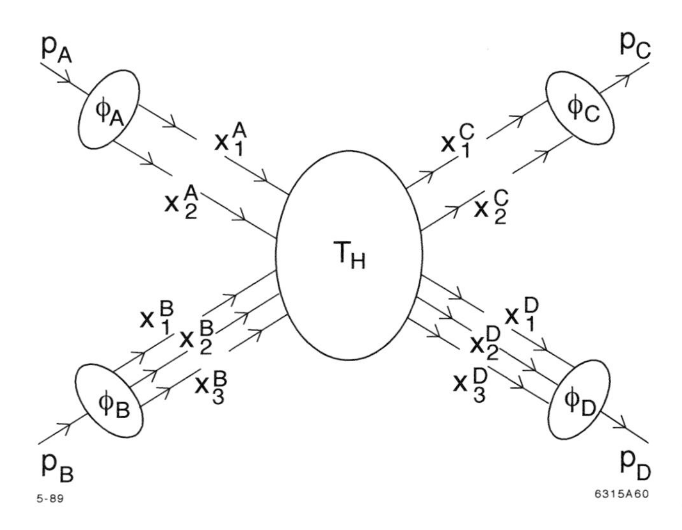

Fig. 10. PQCD factorization of the hard scattering amplitude for elastic meson–baryon scattering at large momentum transfer.

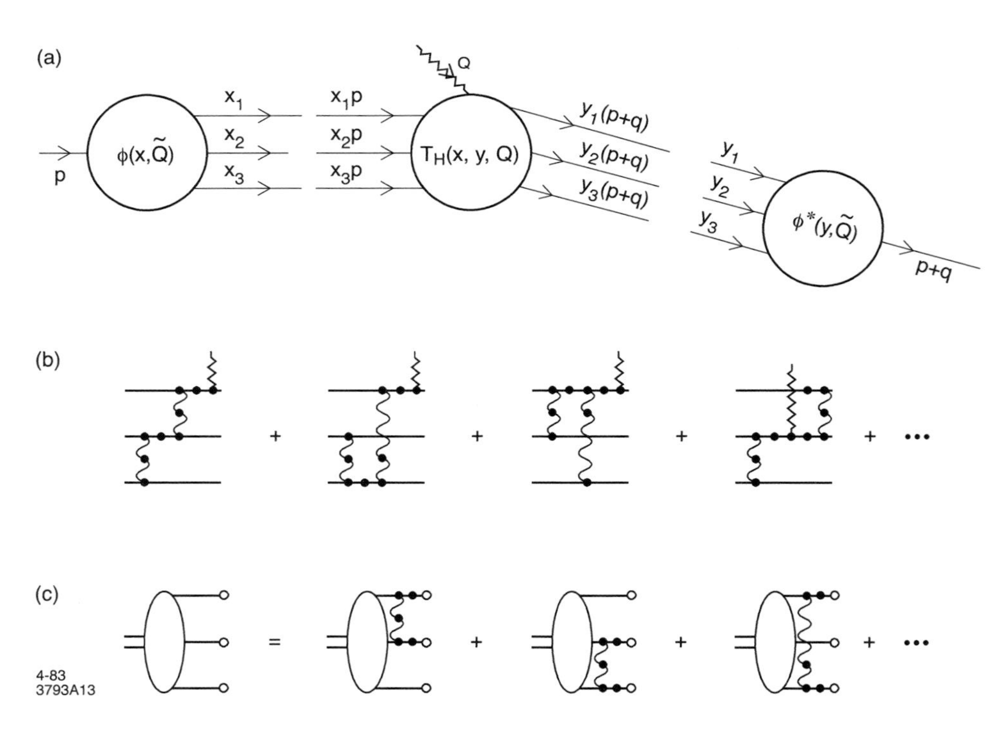

Fig. 11. (a) Factorization of the hard scattering amplitude applied to nucleon form factors at large momentum transfer. (b) The hard scattering amplitude T_H is computed at order α_s^2. (c) Leading order contributions to the evolution kernel for the nucleon distribution amplitude.

The gauge invariant hadron distribution amplitudes $\phi_H(x_i, Q)$ contain the fundamental nonperturbative dynamics of QCD in the short distance regime.[61] The distribution amplitude in light-cone gauge is the integral over transverse momentum of the hadron LFWF $\psi_H(x_i, k_{\perp i})$, the valence Fock state eigensolution of the QCD light-front Hamiltonian:

$$\phi_M(x_i, Q) = d_F^{-1}(Q) \int d^2k_\perp \theta(k_\perp^2 < Q^2) \psi_{q\bar{q}/M}(x_i, \vec{k}_\perp)$$

and

$$\phi_B(x_i, Q) = d_F^{-3/2}(Q) \int [d^2k_\perp] \theta(k_\perp^2 < Q^2) \psi_{qqq/B}(x_i, \vec{k}_\perp) .$$

Here $d_F^{-1/2}$ is the renormalization factor for each fermion line. For simplicity, spin indices have been suppressed. The integration implies that the impact separation between the valence quarks is limited to $b_\perp^2 \le 1/Q^2$. It is convenient to assume the light-cone gauge A^+0 so that the gluon fields have physical polarization; in gauges other than $A^+ = 0$ light-cone gauge, a Wilson line between the quark and antiquark fields is required. The distribution amplitudes $\phi(x_i, Q)$ are universal — they depend specifically on the valence LFWF of the hadron, and they are thus independent of the specific hard scattering process. The distribution amplitude is frame invariant, independent of the hadron's P_H^+ and $P_{\perp H}$. Since it is defined at fixed LF time, the distribution amplitude is also causal. One can also derive[62] the distribution amplitude $\phi_M(x_i, Q)$ from the hadron's Bethe–Salpeter wave-function $\langle H|\overline{\psi}(x)\psi(0)|0\rangle$ evaluated at fixed LF time $\tau = x^+ = 0$.

The renormalization scale of $\alpha_s(\mu_R^2)$ appearing in T_H can be fixed at each order using the "Principle of Maximum Conformality (PMC),"[63] the rigorous extension of the BLM procedure.[64] One shifts the scale α_s so that none of the β_i terms associated with the evolution of the coupling appear. The resulting pQCD prediction is then scheme-independent and, to high accuracy, independent of the choice of the initial scale μ_R. The "renormalon" terms which make the series diverge as $n!$ are absent. An important example is the pion form factor measured in $\ell\pi \to \ell\pi$. In this case the one gluon exchange kernel gives

$$F_\pi(Q^2) = \frac{16\pi\alpha_s(Q^2)}{3Q^2} \int_0^1 dx \int_0^1 dy \frac{\phi(x, \tilde{Q}_x)\phi(y\tilde{Q}_y)}{x(1-x)y(1-y)} .$$

See Fig. 12. An approximate formula for meson form factors was first given by Farrar and Jackson in Ref. 65.

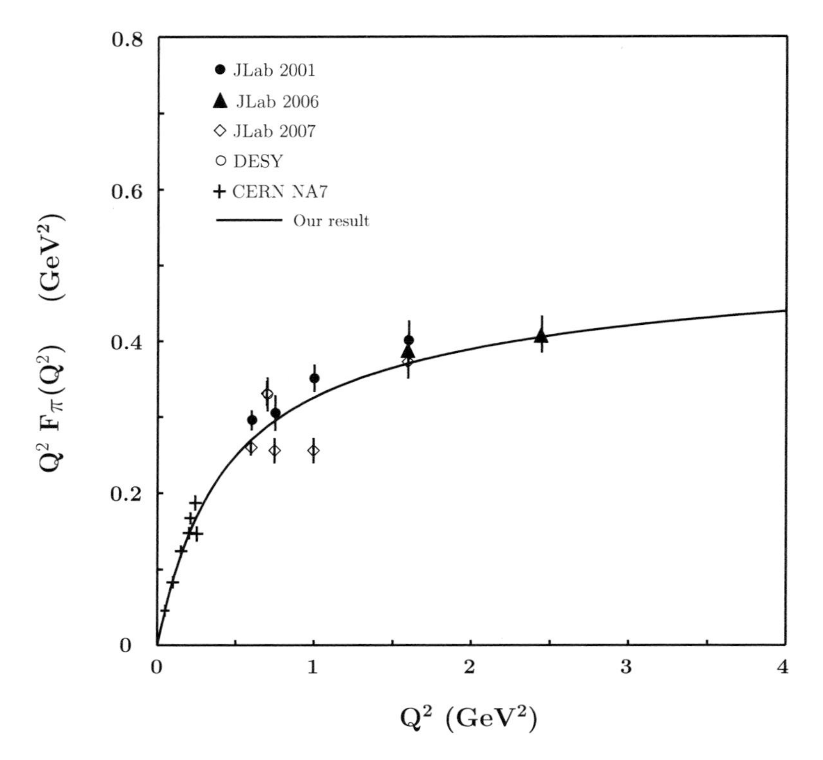

Fig. 12. The pion form factor compared with AdS/QCD predictions and pQCD. From Ref. 66. The asymptotic power $Q^2 F_\pi(Q^2) \simeq$ constant is consistent with quark counting.

At high transverse momentum, the meson LFWF falls off as $\psi_M(x, \vec{k}_\perp) \sim \alpha_s(k_\perp^2)/k_\perp^2$. The distribution amplitude thus evolves as the scale Q^2 of the hard subprocess increases. The resulting "ERBL"[16,17] evolution equation at leading order in α_s is

$$Q\frac{\partial}{\partial Q}\phi(x, Q) = \frac{\alpha_s(Q^2)}{4\pi}\int_0^1 dy \frac{V(x, y)}{y(1-y)}[\phi(y, Q) - 2\phi(x, Q)]\,.$$

The evolution kernel $V(x, y)$ is derived at leading order in pQCD from the one-gluon exchange interaction:

$$V(x, y) = 4C_F\left[x(1-y)\theta(y-x)\delta_{-h,\bar{h}} + \frac{\Delta}{y-x} + (x \to 1-x, y \to 1-y)\right] = V(y, x).$$

The operator Δ in the potential is defined such that

$$\Delta\frac{\phi(y, Q)}{y(1-y)} \equiv \frac{\phi(y, Q)}{y(1-y)} - \frac{\phi(x, Q)}{x(1-x)}$$

which eliminates the singular behavior at $y = x$.

Solutions of the ERBL evolution equation for the distribution amplitude has the general form

$$\phi(x,Q) = x(1-x)\sum_{n=0}^{\infty} \log[Q^2/\Lambda^2]^{-\gamma_n} a_n C_n^{3/2}(2x-1)\,,$$

where the γ_n are the non-singlet anomalous dimensions associated with the operator product $\psi(z/2)\overline{\psi}(-z/2)$ where $z^+ = 0$:

$$\gamma_n = 2C_F\left[1 + 4\sum_{k=2}^{n+1}\frac{1}{k} - \frac{2\delta_{-(h,\bar{h})}}{(n+1)(n+2)}\right] \geq 0\,.$$

This expansion can also be obtained explicitly from the operator-product expansion near the light-cone; i.e., for $z^2 = -z_\perp^2 \sim 1/Q^2$. There is a corresponding evolution equation for the three-quark baryon distribution amplitude.

The coefficients a_n of the Gegenbauer polynomials $C_n^{3/2}$ that appear in the solution to the ERBL equation are determined from the nonperturbative dynamics of the meson; i.e., the form of $\phi_M(x, Q_0)$ at the nonperturbative QCD scale. The Q-independent coefficient a_0 can be determined from the decay meson constant f_M obtained from weak leptonic decays $u\bar{d} \to W^+ \to \bar{\mu}^+\nu$ since the anomalous dimension γ_0 vanishes: $\gamma_0 = 0$. For example, $a_0/6 = \int dx_0^1 \phi_\pi(x.Q) = f_\pi/2\sqrt{n}_C$. The large Q^2 behavior of the meson distribution amplitude at large Q^2 is universal: $\phi_M(x, Q \to \infty)x(1-x)f_\pi$, independent of the nonperturbative input. The underlying hadron LF wave-function must vanish at x, $(1-x) \to 0$ to ensure that the expectation value of the LF kinetic energy $(k_\perp^2 + m^2)/x(1-x)$ is finite.

The LFWFs and their resulting distribution amplitudes can also be used to compute other hadronic observables; for example, the structure functions measured in deep inelastic scattering are effectively the probability distributions defined from the absolute squares of the LFWFs integrated over the transverse momenta.[41] The generalized parton amplitudes, such as E and H, which underly deeply virtual Compton scatting $\gamma * p \to \gamma p$ in the "handbag" approximation, can also be evaluated as convolutions of initial and final proton LFWFs. An example is given in Ref. 67.

6. Light-Front Holography

One of the most exciting recent developments in the analysis of exclusive processes in QCD has been a new approach to nonperturbative QCD based on "light-front holography."[68] Holography can relate a theory in five dimensions

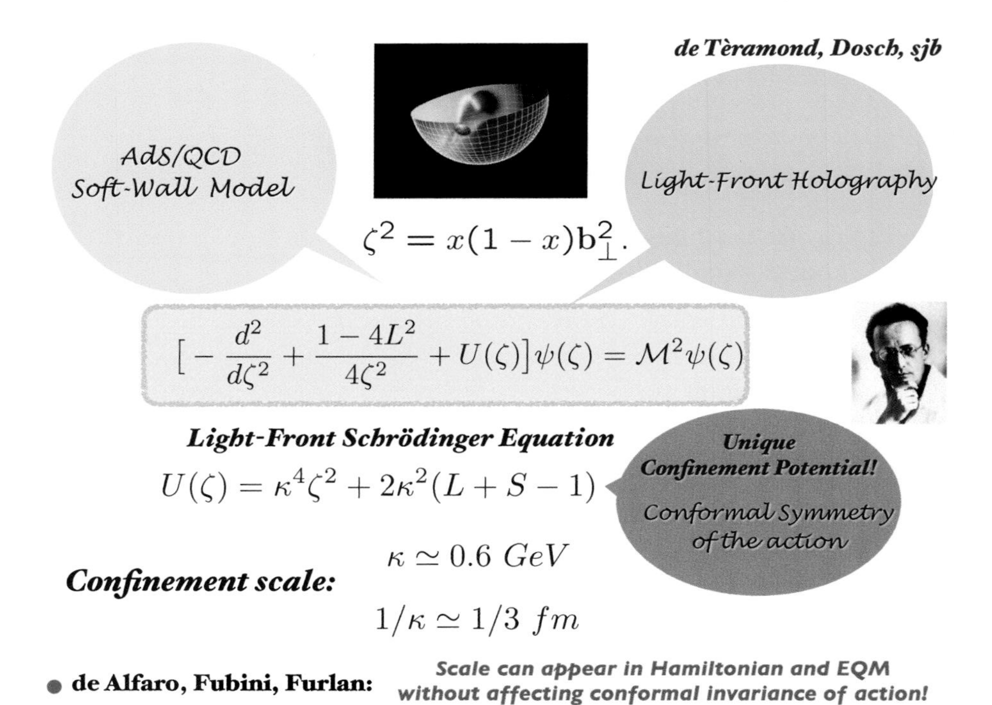

Fig. 13. The LF Schrödinger equation and LF wave-function derived from AdS/QCD and LF holography. The unique form of the confinement potential $U(\zeta^2)$ is a consequence AdS/QCD and the principle of de Alfaro, Fubini, Furlan.[70] The eigenfunction LF Schrödinger equation predicts a zero-mass pion and the same Regge slope in n and L.

to a theory in one less dimension. The LF holographic analysis utilizes the compact AdS_5 space in $4+1$ space–time dimensions, since it provides a geometrical representation of the conformal group and the invariant transverse separation $\vec{\zeta}_\perp^2 \equiv \frac{x}{1-x}\vec{\eta}_\perp^2$ in $3+1$ space at fixed LF time τ is holographically dual[69] to the fifth dimension z coordinate in AdS_5 space. If one includes a factor of $\exp +\kappa^2 z^2$ (the 'dilaton') in the AdS_5 action, a confinement potential with a fundamental scale κ appears in the corresponding LF Schrödinger equations of motion in physical space–time. See Fig. 13 and Ref. 68.

Remarkably, the resulting action in physical space–time remains conformal despite the appearance of the mass scale κ in the LF Hamiltonian and the equations of motion. The LF confinement potential in the meson sector has the unique harmonic oscillator form using the extension of principle of extended conformal invariance developed by de Alfaro, Fubini, Furlan[70] to LF theory. The resulting formalism is frame-independent and causal. Only one parameter, the mass scale that sets the scale of confinement $\kappa = \sqrt{\lambda}$

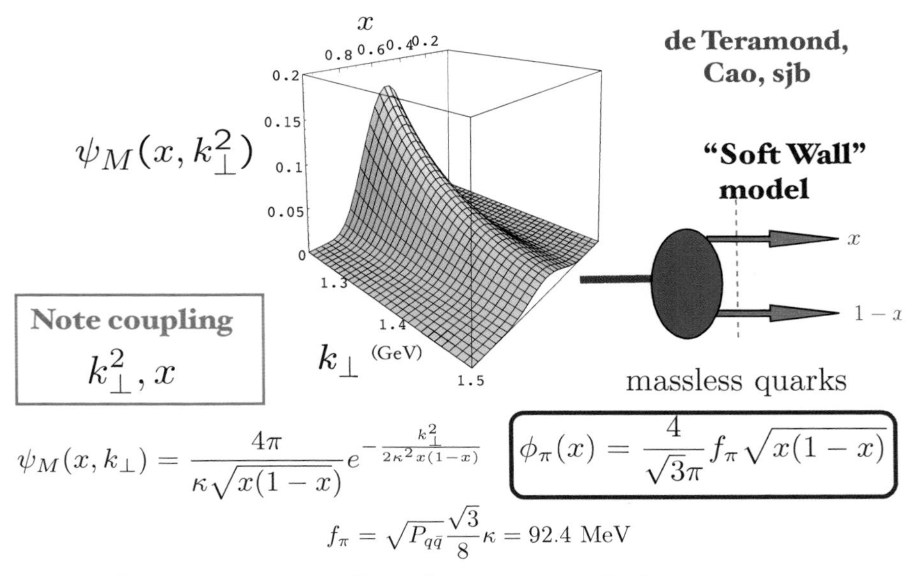

Fig. 14. Predictions for the pion LF wave-function and distribution amplitude from the LF Schrödinger equation.[76]

appears in the AdS/QCD analysis. One can solve the LF Schrödinger equation with the confining kernel $\kappa^4\zeta^2 = \kappa^4 b_\perp^2 x(1-x)$, derived from LF holography and AdS/QCD. The result is a pion with zero mass for $m_q = 0$ and pion excitations which satisfy the Regge form $M_{n,L}^2 = 2\kappa^2(n+L)$ with the same slope in n and L, explaining important features of hadron dynamics.[71] The form of the nonperturbative distribution amplitude from this approach is $\phi_\pi(x) \propto f_\pi\sqrt{x(1-x)}$ at small Q^2. See Fig. 14. This form has also been recently derived using the Bethe–Salpeter formalism in Ref. 72. The distribution then evolves by ERBL evolution to the universal asymptotic form, $\phi_\pi(x,Q) \to \frac{3f_\pi}{\sqrt{n_C}} x(1-x)$ at $Q^2 \to \infty$; i.e., the scale where all hadron structure functions evolve[73] by DGLAP evolution to δ functions at $x = 0$. For a recent review, see Ref. 74. A corresponding LF Dirac equation resulting from light-front holography which determines nucleon spectroscopy and dynamics, can also be derived using superconformal quantum mechanics.[75]

The hard scattering domain for hard exclusive processes is $z^2 = \vec{\zeta}_\perp^2 < x/(1-x)Q^2$, and thus the behavior of the AdS amplitude $\Phi(z)$ at small z controls exclusive amplitudes at large momentum transfer. Since the power-

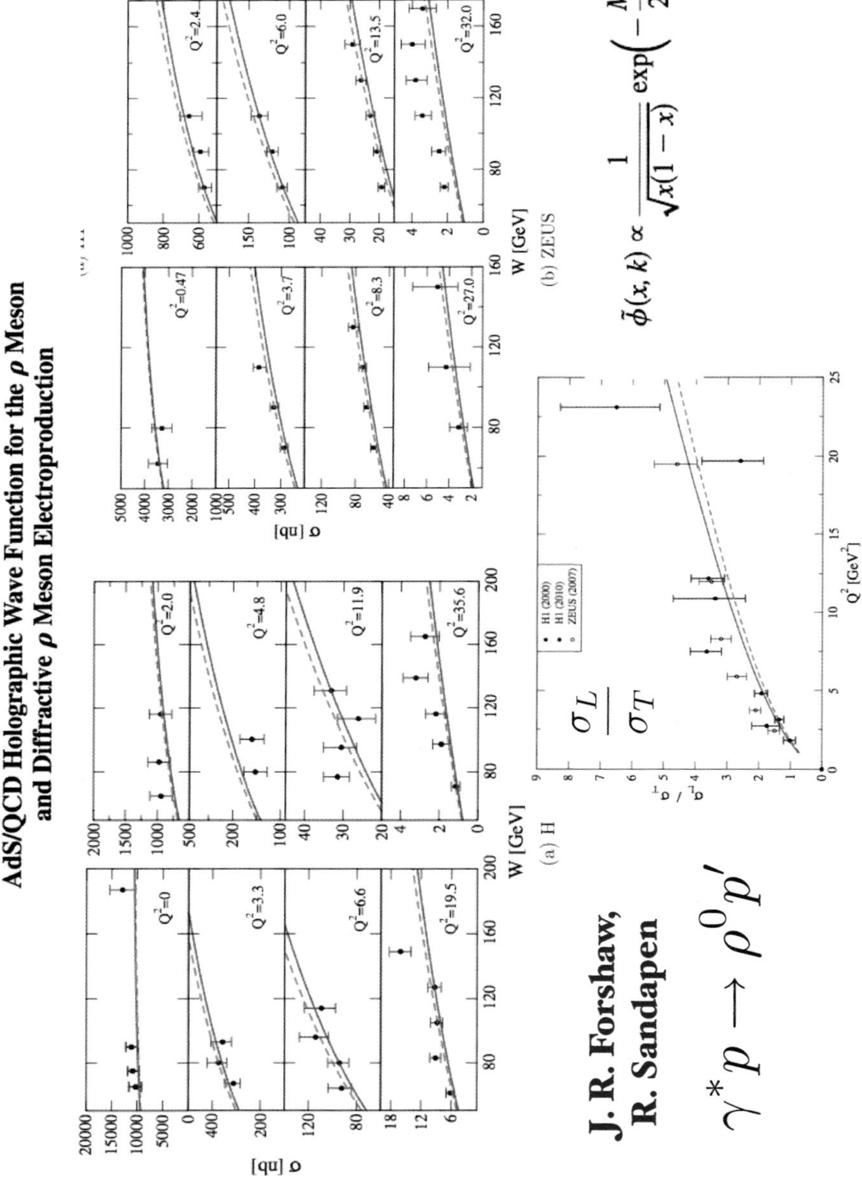

Fig. 15. Prediction[77] for diffractive ρ electroproduction $\gamma * p \to \rho^0 p'$ using the LFWF given by AdS/QCD and LF holography. There are no new parameters.

Using $SU(6)$ flavor symmetry and normalization to static quantities

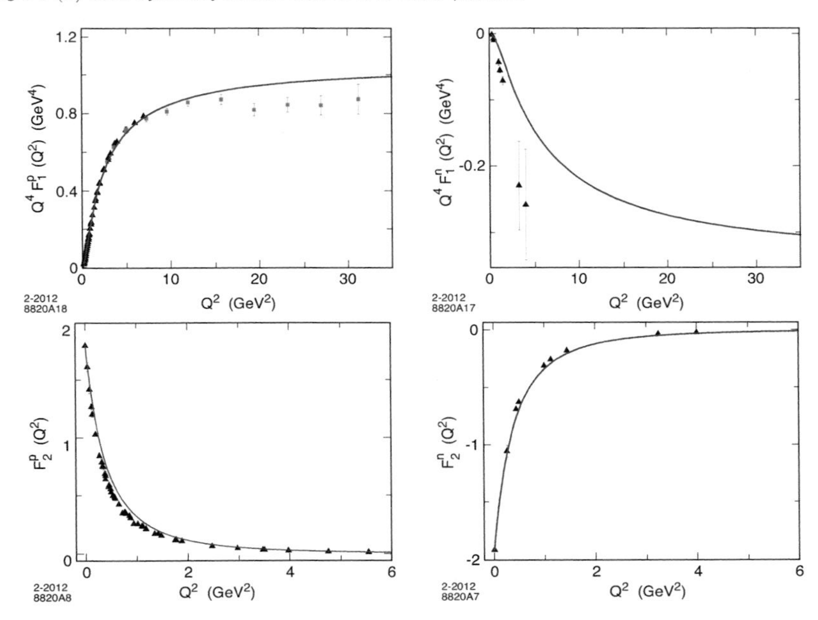

Fig. 16. Light-front holographic predictions[79] for the nucleon form factors normalized to their static values.

law behavior of $\Phi(z)$ at small z is determined by the twist of the short-distance interpolating operator of the hadron, the results agree with the quark counting rules, independent of the choice of the dilaton profile at large z, as was first shown for form factors by Polchinski and Strassler.[18] The LFWF for the ρ^0 meson obtained using same approach gives a remarkably good description of diffractive ρ electroproduction $\gamma * p \to \rho^0 p'$ as shown by Sandapen and Forshaw.[77] See Fig. 15. The meson distribution amplitudes play a central role in QCD phenomenology since they describe *hadronization at the amplitude level* — the transition between the off-shell quark and anti-quark with momentum fractions x and $1-x$, transverse separation $b \perp \sim 1/Q$ and opposite helicities within the meson. Light-Front Holography also predicts the spectroscopy and dynamics[69,78] for hadrons with half-integer spin from a Light-Front Dirac equation. One obtains Regge trajectories with the same slope in n and L, thus reproducing the nucleon entries listed but the PDG. The same results can be also derived using superconformal algebra.[75] This formalism thus leads to dynamical predictions as well as spectroscopy. Predictions for the proton and neutron form factors from the AdS/QCD LF Dirac equation are shown in Fig. 16.

6.1. *The photon-to-pion transition form factor from AdS/QCD light-front holography*

A primary test of the pQCD formalism for hard exclusive reactions is the photon-to-pseudo-scalar meson transition form factor $F_{\gamma\to M}(Q^2)$ since it only involves one hadron. It can be measured at e^+e^- colliders via two-photon reactions where one of the incident leptons scatters with large momentum transfer and the other photon is nearly real: $k^2 \sim 0$. The invariant amplitude for the photon-to-pseudo-scalar meson transition form factor is defined as[16] $M_{\mu,\nu} = \epsilon_{\mu\nu\sigma\tau} p^\sigma_M q^\tau F_{\gamma\to M}(Q^2)$. The transition form factor can be computed from first principles at leading twist and leading order in α_s:

$$Q^2 F_{\gamma\to M}(Q^2) = \frac{4}{\sqrt{3}} \int_0^1 dx \frac{\phi(x,(1-x)Q)}{1-x} \left[1 + \mathcal{O}\left(\alpha_s, \frac{m^2}{Q^2}\right)\right],$$

Since $\phi(x, Q \to \infty) = \sqrt{3} f_\pi x(1-x)$, the asymptotic photon-to-pion form factor is predicted without ambiguity: $Q^2 F_{\pi\gamma}(Q^2 \to \infty) = 2f_\pi$. A complete prediction[80,81] valid at all Q^2 can be made using light-front holography including modifications from the dressed current derived from AdS_5. The AdS/QCD prediction is compared with measurements in Fig. 17. The

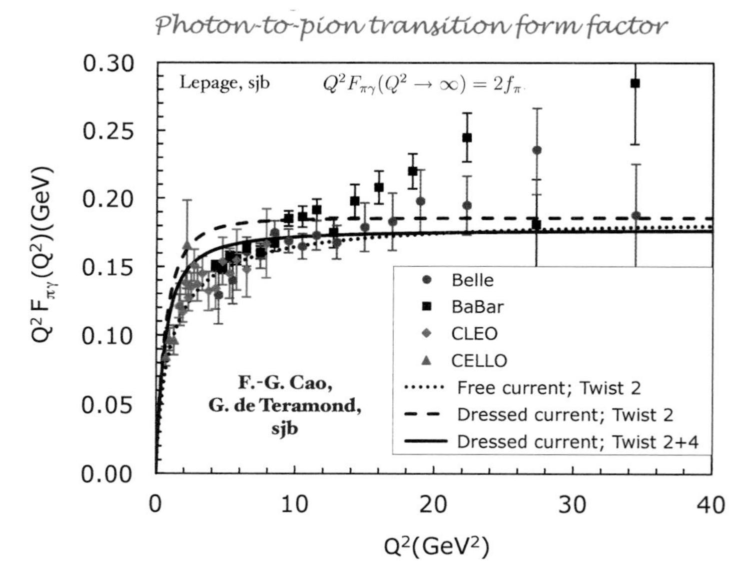

Fig. 17. The $\gamma\gamma^* \to \pi^0$ transition form factor: $Q^2 F_{\pi\gamma}(Q^2)$ as a function of $Q^2 = -q^2$. The dotted curve is the result predicted by AdS/QCD using a Chern–Simons form.[80,81] The dashed and solid curves include the effects of using a confined EM current for twist-2 and twist-2 plus twist-4, respectively. The high Q^2 data are from BaBar, Ref. 83 and Belle, Ref. 84.

result is also consistent with the pQCD analysis. The Belle data[82] agrees well with the unified AdS/QCD predictions. However, the BaBar data[83] appears to diverge at $Q^2 \sim 10$ GeV^2. A possibility is that the BaBar[83] signal, which deviates from the pQCD asymptotic prediction, includes a background contribution to $e^+e^- \to e^{+\prime}e^{-\prime}\pi^0$ events, due to the interference of contributions from the spacelike $\gamma^*\gamma \to \pi^0$ and timelike $\gamma^* \to \pi^0 + \gamma$ transition form factors.

6.2. *Prediction for the QCD running coupling from AdS/QCD light-front holography*

Light-Front Holography also predicts the functional dependence of the QCD running coupling $\alpha_s(Q^2) \propto \exp(-Q^2/4\kappa^2)$ in the nonperturbative region of small Q^2. When one matches the running coupling $\alpha_s^{g1}(Q^2)$ defined from the Bjorken sum rule[87] and its derivative to the pQCD high Q^2 prediction using the $\overline{MS}$ scheme, one obtains a running coupling valid from low to high Q^2. See Refs. 85 and 86. The value of $\Lambda_{\overline{MS}}$ can then be determined from the ρ mass. See Fig. 18. The coupling $\alpha_s^{g1}(Q^2)$ is the effective charge

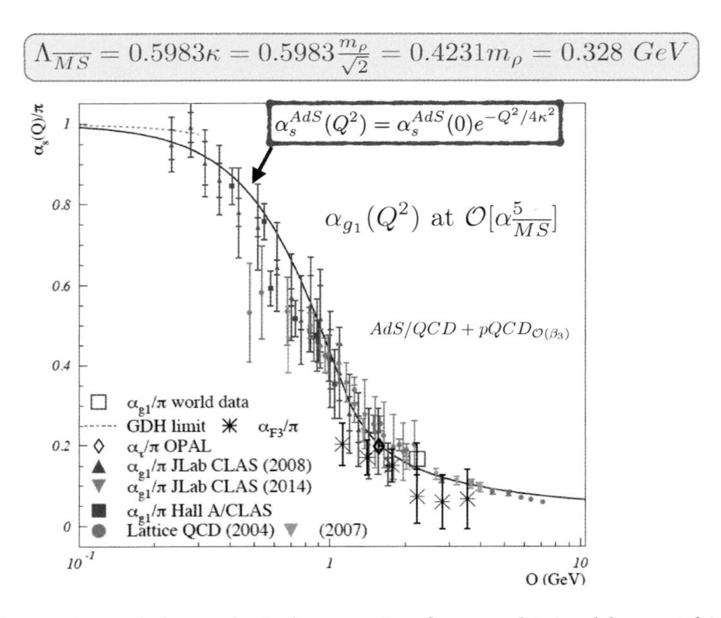

Fig. 18. Comparison of the analytical expression for α_{g_1} obtained by matching the hard pQCD and nonperturbative predictions from AdS/QCD and light-front holography with experimental JLab data for the Bjorken sum rule and, lattice QCD results. The coupling $\alpha_s^{g1}(Q^2)$ is the effective charge defined from the Bjorken sum rule. The constraint at $Q^2 = 0$ is derived from the Drell–Hearn Sum Rule. From Refs. 85 and 86.

defined from the Bjorken sum rule.[88] The AdS/QCD Light-Front Holography approach thus describes both the dynamics of exclusive process and the hadron spectrum in terms of one parameter, the mass scale κ and a unique confining potential.

7. Conclusions

The discovery of Bjorken scaling[89] in deep inelastic lepton–proton scattering[90,91] — and its parton model interpretation by Feynman[92] and by Bjorken and Paschos[93] — was the critical observation showing that hadrons are indeed bound states of quarks. The discovery[94] of the scaling of $R_{e^+e^-}(s)$ and its stepwise character as one passes quark thresholds, counts the number of quark flavors that exist in nature. A third crucial discovery was that the fall-off of the rate for hard hadronic exclusive reactions, such as form factors at large momentum transfer, enumerates[1–3] the number of quarks or anti-quarks in any hadronic bound state — two for mesons, three for baryons, six for the deuteron, etc. Thus even though quarks are confined and cannot emerge separately as asymptotic states, one use the scaling of exclusive amplitudes to rigorously count their degree of compositeness. We have also noted that by duality and the exclusive–inclusive connection, one obtains spectator counting rules[13] for the power-law fall-off of both structure functions at $x \to 1$ and fragmentation functions at $z \to 1$, predictions consistent with the Gribov–Lipatov crossing relations.[95]

The quark counting rules apply not only to elastic and transition form factors, but also to hard exclusive $2 \to N$ processes at fixed angle. The power-law fall-off reflects the twist of the leading interpolating operator for each hadron at short distances $x^2 \to 0$. The pQCD analysis give the leading power-law fall-off; one obtains corrections higher order in $1/Q^2$ and $1/p_T^2$ from higher non-valence Fock states, internal transverse momentum and mass scales. In addition, the pQCD analysis modifies the nominal leading power-law fall-off by anomalous dimensions derived from the operator product expansion, ERBL evolution of the distribution amplitudes,[16,17] and by the evolution of the QCD running coupling.[96–98] The counting rules can also be derived nonperturbatively[18] using light-front holography and the soft-wall modification of the AdS_5 action.

Constituent counting rules also provide an important tool for determining the structure of exotic hadrons.[99] For example, one could measure the power-law fall-off of the rate for the production of pairs of *tetraquarks* such as the

Z_c^+ at large center-of-mass squared s:

$$\frac{\sigma(e^+e^- \to Z_c^+(c\bar{c}u\bar{d}) + \bar{Z}_{\bar{c}}^-(\bar{c}c\bar{u}d))}{\sigma(e^+e^- \to \mu\bar{\mu})} = |F_{Z_c}(s)|^2 \propto \frac{1}{s^N}.$$

If $N = 2(4-1) = 6$, this would prove that the Z_c^+ is indeed a bound state of a two quarks and two antiquarks. One can also determine the constituent content of the Z_c^+ with a significantly larger rate by checking

$$\frac{\sigma(e^+e^- \to Z_c^+(c\bar{c}u\bar{d}) + \pi^-(\bar{u}d))}{\sigma(e^+e^- \to \mu\bar{\mu})} \propto \frac{1}{s^4}$$

since $N = 4+2-2 = 4$. The ratio of e^+e^- exclusive annihilation rates:

$$\frac{\sigma(e^+e^- \to Z_c^+(c\bar{c}u\bar{d}) + \pi^-(\bar{u}d))}{\sigma(e^+e^- \to \Lambda_c(cud)\bar{\Lambda}_c(\bar{c}\bar{u}\bar{d}))}$$

has cancelling heavy-quark mass and scaling corrections; its scaling should be sensitive to the fundamental QCD composition of the Z_c^+; e.g., whether it is primarily a "molecular" ($[c\bar{c}] + [u\bar{d}]$) state, analogous to "nuclear-bound quarkonium" ($J/\psi + A$),[100,101] or whether it is a diquark–diquark ($[c\bar{d}]_{3_C} + [u\bar{c}]_{3_C}$) resonance of QCD color-triplets.[102]

Acknowledgments

This research was supported by the Department of Energy, contract DE–AC02–76SF00515. SLAC-PUB-16162. I am indebted to my collaborators who made these studies possible and to Professor Harald Fritzsch for his invitation to write this contribution to his historical volume, '50 Years of Quarks', to be published by World Scientific.

References

1. S. J. Brodsky and G. R. Farrar, *Phys. Rev. Lett.* **31**, 1153 (1973).
2. S. J. Brodsky and G. R. Farrar, *Phys. Lett. D* **11**, 1309 (1975).
3. V. A. Matveev, R. M. Muradyan and A. N. Tavkhelidze, *Lett. Nuovo Cimento* **5S2**, 907 (1972) [*Lett. Nuovo Cimento* **5**, 907 (1972)].
4. H. Fritzsch, M. Gell-Mann and H. Leutwyler, *Phys. Lett. B* **47**, 365 (1973).
5. S. J. Brodsky, Y. Frishman and G. P. Lepage, *Phys. Lett. B* **167**, 347 (1986).
6. S. J. Brodsky, P. Damgaard, Y. Frishman and G. P. Lepage, *Phys. Lett. D* **33**, 1881 (1986).
7. S. J. Brodsky, Y. Frishman, G. P. Lepage and C. T. Sachrajda, *Phys. Lett. B* **91**, 239 (1980).
8. X. D. Ji, J. P. Ma and F. Yuan, *Phys. Rev. Lett.* **90**, 241601 (2003), arXiv:hep-ph/0301141.

9. P. V. Landshoff and J. C. Polkinghorne, *Phys. Lett. B* **44**, 293 (1973).
10. D. W. Sivers, S. J. Brodsky and R. Blankenbecler, *Phys. Rept.* **23**, 1 (1976).
11. S. J. Brodsky and G. P. Lepage, *Adv. Ser. Direct. High Energy Phys.* **5**, 93 (1989).
12. S. D. Drell and T. M. Yan, *Phys. Rev. Lett.* **24**, 181 (1970).
13. S. J. Brodsky, M. Burkardt and I. Schmidt, *Nucl. Phys. B* **441**, 197 (1995), arXiv:hep-ph/9401328.
14. E. L. Berger and S. J. Brodsky, *Phys. Rev. Lett.* **42**, 940 (1979).
15. G. P. Lepage and S. J. Brodsky, *Phys. Rev. Lett.* **43**, 545 (1979) [Erratum: *ibid.* **43**, 1625 (1979)].
16. G. P. Lepage and S. J. Brodsky, *Phys. Lett. D* **22**, 2157 (1980).
17. A. V. Efremov and A. V. Radyushkin, *Phys. Lett. B* **94**, 245 (1980).
18. J. Polchinski and M. J. Strassler, *Phys. Rev. Lett.* **88**, 031601 (2002), arXiv:hep-th/0109174.
19. S. J. Brodsky and G. P. Lepage, *Phys. Lett. D* **24**, 2848 (1981).
20. S. J. Brodsky and M. Karliner, *Phys. Rev. Lett.* **78**, 4682 (1997), arXiv:hep-ph/9704379.
21. S. Brodsky, G. de Teramond and M. Karliner, *Ann. Rev. Nucl. Part. Sci.* **62**, 1 (2012), arXiv:1302.5684 [hep-ph].
22. S. J. Brodsky and A. H. Mueller, *Phys. Lett. B* **206**, 685 (1988).
23. S. J. Brodsky, L. Frankfurt, J. F. Gunion, A. H. Mueller and M. Strikman, *Phys. Lett. D* **50**, 3134 (1994), arXiv:hep-ph/9402283.
24. D. Dutta, K. Hafidi and M. Strikman, *Prog. Part. Nucl. Phys.* **69**, 1 (2013), arXiv:1211.2826 [nucl-th].
25. D. Ashery, *Prog. Part. Nucl. Phys.* **56**, 279 (2006).
26. R. J. Glauber and G. Matthiae, *Nucl. Phys. B* **21**, 135 (1970).
27. R. H. Dalitz and D. R. Yennie, *Phys. Rev.* **105**, 1598 (1957).
28. E791 Collab. (E. M. Aitala *et al.*), *Phys. Rev. Lett.* **86**, 4773 (2001), arXiv:hep-ex/0010044.
29. J. Aclander *et al.*, *Phys. Lett. C* **70**, 015208 (2004), arXiv:nucl-ex/0405025.
30. E. A. Crosbie *et al.*, *Phys. Lett. D* **23**, 600 (1981).
31. S. J. Brodsky and G. F. de Teramond, *Phys. Rev. Lett.* **60**, 1924 (1988).
32. M. Bashkanov, S. J. Brodsky and H. Clement, *Phys. Lett. B* **727**, 438 (2013), arXiv:1308.6404 [hep-ph].
33. J. F. Gunion, S. J. Brodsky and R. Blankenbecler, *Phys. Lett. D* **8**, 287 (1973).
34. S. J. Brodsky and B. T. Chertok, *Phys. Rev. Lett.* **37**, 269 (1976).
35. S. J. Brodsky and B. T. Chertok, *Phys. Lett. D* **14**, 3003 (1976).
36. V. A. Matveev and P. Sorba, *Nuovo Cimento A* **45**, 257 (1978).
37. V. A. Matveev and P. Sorba, *Lett. Nuovo Cimento* **20**, 435 (1977).
38. S. J. Brodsky, C. R. Ji and G. P. Lepage, *Phys. Rev. Lett.* **51**, 83 (1983).
39. B. L. G. Bakker and C. R. Ji, *Prog. Part. Nucl. Phys.* **74**, 1 (2014).
40. P. A. M. Dirac, *Rev. Mod. Phys.* **21**, 392 (1949).
41. S. J. Brodsky, H. C. Pauli and S. S. Pinsky, *Phys. Rept.* **301**, 299 (1998), arXiv:hep-ph/9705477.
42. H. C. Pauli and S. J. Brodsky, *Phys. Lett. D* **32**, 2001 (1985).
43. S. S. Chabysheva and J. R. Hiller, arXiv:1409.6333 [hep-ph].

44. B. L. G. Bakker *et al.*, *Nucl. Phys. B* (*Proc. Suppl.*) **251-252**, 165 (2014), arXiv:1309.6333 [hep-ph].
45. K. Hornbostel, S. J. Brodsky and H. C. Pauli, *Phys. Lett. D* **41**, 3814 (1990).
46. S. Hellerman and J. Polchinski, *The Many Faces of the Superworld*, ed. M. A. Shifman (World Scientific, Singapore, 1999), pp. 142–155, arXiv:hep-th/9908202.
47. S. J. Brodsky, G. F. de Téramond and H. G. Dosch, *Few Body Syst.* **55**, 407 (2014), arXiv:1310.8648 [hep-ph].
48. G. B. West, *Phys. Rev. Lett.* **24**, 1206 (1970).
49. S. J. Brodsky and S. D. Drell, *Phys. Lett. D* **22**, 2236 (1980).
50. R. Alkofer, A. Holl, M. Kloker, A. Krassnigg and C. D. Roberts, *Few Body Syst.* **37**, 1 (2005), arXiv:nucl-th/0412046.
51. S. J. Brodsky, *Brandeis Univ 1969, Proceedings, Atomic Physics and Astrophysics*, Vol. 1, New York, 1971, pp. 91–169.
52. S. J. Brodsky and J. R. Primack, *Ann. Phys.* **52**, 315 (1969).
53. S. Weinberg, *Phys. Rev.* **150**, 1313 (1966).
54. J. B. Kogut and D. E. Soper, *Phys. Lett. D* **1**, 2901 (1970).
55. D. E. Soper, *Phys. Lett. D* **15**, 1141 (1977).
56. H. T. C. Stoof, J. M. V. A. Koelman and B. J. Verhaar, *Phys. Rev. B* **38**, 4688 (1988).
57. C. White *et al.*, *Phys. Lett. D* **49**, 58 (1994).
58. P. V. Landshoff, *Phys. Lett. D* **10**, 1024 (1974).
59. S. Wallon, arXiv:1403.3110 [hep-ph].
60. N. Isgur and J. E. Paton, *Phys. Lett. D* **31**, 2910 (1985).
61. G. P. Lepage and S. J. Brodsky, *Phys. Lett. B* **87**, 359 (1979).
62. J. Segovia, L. Chang, I. C. Cloet, C. D. Roberts, S. M. Schmidt and H. S. Zong, *Phys. Lett. B* **731**, 13 (2014), arXiv:1311.1390 [nucl-th].
63. S. J. Brodsky, M. Mojaza and X. G. Wu, *Phys. Lett. D* **89**, 014027 (2014), arXiv:1304.4631 [hep-ph].
64. S. J. Brodsky, G. P. Lepage and P. B. Mackenzie, *Phys. Lett. D* **28**, 228 (1983).
65. G. R. Farrar and D. R. Jackson, *Phys. Rev. Lett.* **43**, 246 (1979).
66. T. Gutsche, V. E. Lyubovitskij, I. Schmidt and A. Vega, arXiv:1410.6424 [hep-ph].
67. S. J. Brodsky, M. Diehl and D. S. Hwang, *Nucl. Phys. B* **596**, 99 (2001), arXiv:hep-ph/0009254.
68. G. F. de Teramond and S. J. Brodsky, *Phys. Rev. Lett.* **102**, 081601 (2009), arXiv:0809.4899 [hep-ph].
69. S. J. Brodsky and G. F. de Teramond, *Phys. Rev. Lett.* **96**, 201601 (2006), arXiv:hep-ph/0602252.
70. V. de Alfaro, S. Fubini and G. Furlan, *Nuovo Cimento A* **34**, 569 (1976).
71. S. J. Brodsky, G. F. De Téramond and H. G. Dosch, *Phys. Lett. B* **729**, 3 (2014), arXiv:1302.4105 [hep-th].
72. L. Chang, I. C. Cloet, J. J. Cobos-Martinez, C. D. Roberts, S. M. Schmidt and P. C. Tandy, *Phys. Rev. Lett.* **110**, 132001 (2013), arXiv:1301.0324 [nucl-th].
73. C. G. Callan, Jr. and D. J. Gross, *Phys. Rev. Lett.* **22**, 156 (1969).

74. S. J. Brodsky, G. F. de Teramond, H. G. Dosch and J. Erlich, arXiv:1407.8131 [hep-ph].
75. G. F. de Teramond, H. G. Dosch and S. J. Brodsky, arXiv:1411.5243 [hep-ph].
76. S. J. Brodsky and G. de Teramond, *Int. J. Mod. Phys. Conf. Ser.* **20**, 53 (2012), arXiv:1208.3020 [hep-ph].
77. J. R. Forshaw and R. Sandapen, *Phys. Rev. Lett.* **109**, 081601 (2012), arXiv:1203.6088 [hep-ph].
78. G. F. de Téramond, S. J. Brodsky and H. G. Dosch, *EPJ Web Conf.* **73**, 01014 (2014), arXiv:1401.5531 [hep-ph].
79. G. F. de Teramond and S. J. Brodsky, arXiv:1203.4025 [hep-ph].
80. S. J. Brodsky, F. G. Cao and G. F. de Teramond, *Phys. Lett. D* **84**, 075012 (2011), arXiv:1105.3999 [hep-ph].
81. S. J. Brodsky, F. G. Cao and G. F. de Teramond, *Phys. Lett. D* **84**, 033001 (2011), arXiv:1104.3364 [hep-ph].
82. Belle Collab. (S. Uehara *et al.*), *Phys. Lett. D* **86**, 092007 (2012), arXiv:1205.3249 [hep-ex].
83. BaBar Collab. (B. Aubert *et al.*), *Phys. Lett. D* **80**, 052002 (2009), arXiv:0905.4778 [hep-ex].
84. Belle Collab. (H. Nakazawa), *PoS* **ICHEP2012**, 356 (2013).
85. S. J. Brodsky, G. F. de Téramond, A. Deur and H. G. Dosch, arXiv:1410.0425 [hep-ph].
86. A. Deur, S. J. Brodsky and G. F. de Teramond, arXiv:1409.5488 [hep-ph].
87. J. D. Bjorken, *Phys. Lett. D* **1**, 1376 (1970).
88. S. J. Brodsky, G. F. de Teramond and A. Deur, *Phys. Lett. D* **81**, 096010 (2010), arXiv:1002.3948 [hep-ph].
89. J. D. Bjorken, *Phys. Rev.* **148**, 1467 (1966).
90. E. D. Bloom *et al.*, *Phys. Rev. Lett.* **23**, 930 (1969).
91. M. Breidenbach *et al.*, *Phys. Rev. Lett.* **23**, 935 (1969).
92. R. P. Feynman, *Selected Papers of Richard Feynman*, ed. L. M. Brown, pp. 560–655.
93. J. D. Bjorken and E. A. Paschos, *Phys. Rev.* **185**, 1975 (1969).
94. B. Richter, *Conf. Proc.* **C751125**, 327 (1975).
95. V. N. Gribov and L. N. Lipatov, *Sov. J. Nucl. Phys.* **15**, 438 (1972) [*Yad. Fiz.* **15**, 781 (1972)].
96. H. D. Politzer, *Phys. Rept.* **14**, 129 (1974).
97. D. J. Gross and F. Wilczek, *Phys. Lett. D* **9**, 980 (1974).
98. A. J. Buras, *Rev. Mod. Phys.* **52**, 199 (1980).
99. H. Kawamura, S. Kumano and T. Sekihara, arXiv:1410.0494 [hep-ph].
100. S. J. Brodsky, I. A. Schmidt and G. F. de Teramond, *Phys. Rev. Lett.* **64**, 1011 (1990).
101. M. E. Luke, A. V. Manohar and M. J. Savage, *Phys. Lett. B* **288**, 355 (1992), arXiv:hep-ph/9204219.
102. S. J. Brodsky, D. S. Hwang and R. F. Lebed, *Phys. Rev. Lett.* **113**, 112001 (2014), arXiv:1406.7281 [hep-ph].

Quark–Gluon Soup — The Perfectly Liquid Phase of QCD

Ulrich Heinz*

Physics Department, The Ohio State University,
Columbus, Ohio 43210-1117, USA
heinz.9@osu.edu

At temperatures above about 150 MeV and energy densities exceeding 500 MeV/fm^3, quarks and gluons exist in the form of a plasma of free color charges that is about 1000 times hotter and a billion times denser than any other plasma ever created in the laboratory. This quark–gluon plasma (QGP) turns out to be strongly coupled, flowing like a liquid. About 35 years ago, the nuclear physics community started a program of relativistic heavy-ion collisions with the goal of producing and studying QGP under controlled laboratory conditions. This article recounts the story of its successful creation in collider experiments at Brookhaven National Laboratory and CERN and the subsequent discovery of its almost perfectly liquid nature, and reports on the recent quantitatively precise determination of its thermodynamic and transport properties.

1. Emergent Phenomena in QCD Matter

Like other condensed matter systems, sufficiently large ensembles of quarks and gluons exhibit emergent phenomena that are difficult to predict directly from the fundamental quantum chromodynamic (QCD) interactions. Quark–gluon plasma (QGP), the type of matter that initially filled the entire universe during the first 10 microseconds of its life, has entirely different properties from cold nuclear matter, made of gluon-bound three-quark clusters known as protons and neutrons, which forms the core of our existence today. Unraveling the complex phenomena of nuclear and hadronic structure in terms of fundamental quark degrees of freedom, and identifying SU(3) gauge field theory, with gluons as force mediators, as the correct description of their fundamental interactions have been, without doubt, remarkable theoretical achievements. But the successful prediction of a phase transition to a much more homogeneous phase of color-deconfined quarks and gluons at ~ 3 times larger energy density, and its subsequent experimental

*This work is supported by the Department of Energy, Office of Science, Office of Nuclear Physics under Award No. DE-SC0004286.

discovery followed by detailed investigations of its properties in heavy-ion collision experiments have had a similarly strong impact on our understanding of the many facets of the strong interaction. The complexity of the phase diagram of strongly interacting matter, especially if one allows oneself to vary the quark masses, is now known to rival that of water, featuring first and second order phase transitions, continuous crossover transitions and critical points (Fig. 1). The impacts of these insights on other fields of physics where strongly coupled multi-particle systems are of interest (the study of ultracold atoms, condensed matter physics, plasma physics and string theory) has been profound, creating a multitude of new interdisciplinary connections.

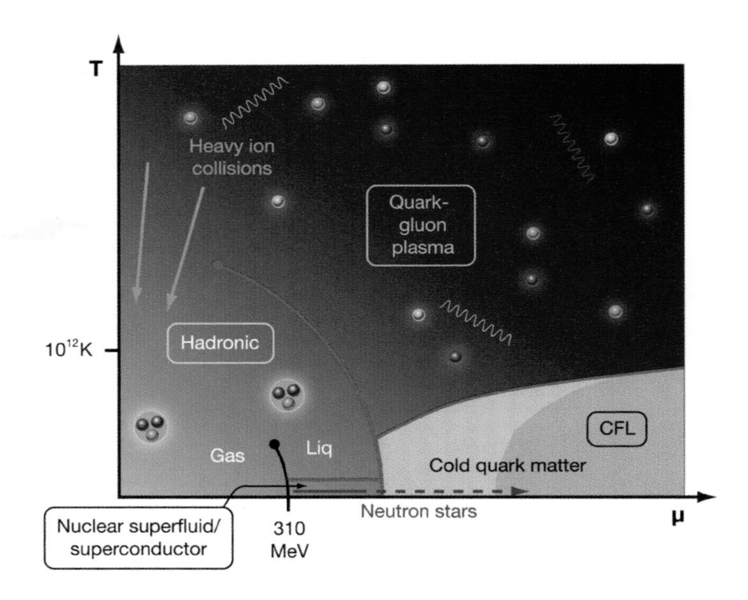

Fig. 1. The QCD phase diagram.[1]

This contribution discusses the discovery of quark–gluon plasma in laboratory experiments using heavy-ion collisions at ultra-relativistic energies, and our recent dramatic progress in quantitatively characterizing its thermodynamic and transport properties. It also points out striking similarities between the Big Bang that spawned our universe and the "Little Bangs" created in heavy-ion collisions[2] and reports on the recent convergence towards the Little Bang Standard Model (LBSM),[3] a standard theoretical framework for the dynamical evolution of relativistic heavy-ion collisions that provides a precise quantitative explanation of the multitude of phenomena observed in such collisions.

2. The Road to Quark–Gluon Matter in the Laboratory: The Relativistic Heavy-Ion Program

Less than a year after the discoveries of asymptotic freedom[4,5] and QCD,[6] adventurous theorists[7,8] advanced the idea that strongly interacting matter should undergo a phase transition at high densities to color-deconfined, weakly interacting quark matter, possibly to be found in the core of neutron stars. This speculation soon received support from lattice QCD,[9,10] where a similar transition was indeed found[11] when the conditions of high density were caused by heating the system to high temperatures, at zero net baryon density, as approximately realized in the early universe.[a] The creation of nuclear matter at several times its equilibrium density of 0.16 nucleons/fm^3 (corresponding to an equilibrium energy density of about 150 MeV/fm^3) in the laboratory, using high-energy nuclear collisions, had been pursued since the mid-1970s at the BEVALAC in Berkeley. To reach temperatures sufficient for a transition to quark matter[b] was thought, however, to require much higher beam energies.

In the mid-1980s both CERN and Brookhaven National Laboratory therefore started a fixed-target program of ultra-relativistic heavy-ion collisions, at 15 to 200 times the beam energies available at the BEVALAC, by injecting initially small nuclei (^{16}O and ^{32}S at the CERN, ^{28}Si at BNL) into their existing accelerators (SPS at CERN, AGS at BNL) and putting together a number of make-shift detectors, with heavy use of components left behind by earlier particle physics experiments. The program operated in "parasitic mode," with most of the beam time allocated to the particle physics experiments that constituted the primary research programs at these accelerators, and was mounted with very limited funding. Still, these experiments yielded a number of highly interesting phenomena,[14] including strong indications for thermalization and collective transverse flow[15] and several that matched predictions made by proponents from the "quark matter camp," most prominently among them an enhanced generation of strangeness[16] and a suppression of J/ψ production.[17] The limitations in acceptance and capabilities of the detectors and the relative smallness of the projectile nuclei made it, however, difficult to arrive at unambiguous interpretations of these data and to convincingly exclude standard hadronic interpretations.

---

[a]Since then our knowledge from lattice QCD of the properties of this phase transition has reached impressive precision.[12,13]

[b]At that time its critical temperature was estimated to exceed 200 MeV, but we now know[12,13] that at zero net baryon density the transition is a rapid but continuous crossover with an inflection temperature in the range $145 < T < 163$ MeV, corresponding to a critical energy density $180 < e_{cr} < 500$ MeV/fm^3, i.e. just a few times the energy density of equilibrium nuclear matter.

A second round of experiments, with significantly upgraded and several newly constructed dedicated heavy-ion detectors, and using truly heavy ion beams (2–11A GeV breams of ^{197}Au in the AGS, 40–160A GeV beams of ^{207}Pb extracted from a newly constructed ion source at the SPS), took place from 1995 to 2003. These collisions literally generated thousands of particles per event, making it obvious even to the skeptic that the community had succeeded in creating small samples of a new form of "condensed matter." Contrary to standard matter condensed matter, the properties of this new matter were controlled by the strong interaction of QCD. The new and old data together led to much increased clarity that culminated in a CERN press release on 10 Feb. 2000 on a "New State of Matter created at CERN"[18] that declared the incompatibility of the observations with interpretations based on standard hadronic degrees of freedom while noting the consistency of the experimental data with theoretical expectations based on the hypothesis of quark–gluon plasma formation. It shied away, however, from claiming discovery of the QGP, mainly because of the absence of a simultaneous theoretical interpretation of all the observations within a single compelling dynamical framework. This left open the possibility of being able to explain the data without QGP, and indeed the literature teemed with papers that claimed successful "conventional" interpretations of individual experimental results. In other words, heavy-ion theory was in a sorry state. The happiness and world-wide excitement over the achievements of the Pb-beam program at CERN were tempered by a significant amount of flak that Maurice Jacob and I received for our "New State of Matter" white paper,[18] especially from the other side of the atlantic where scientists were gearing up for the first collider experiments at the RHIC facility at BNL and accused us of neglecting the standards of scientific rigor.

Experiments at the Relativistic Heavy-Ion Collider (RHIC) at BNL began taking data in the summer of 2000, with Au + Au collisions at 10 times higher energies than achievable at the SPS. A suite of dedicated state-of-the-art heavy-ion detectors, with both overlapping and complementary capabilities, and the easier tracking environment in collider experiment which avoid the strong forward collimation of produced particles caused by the center-of-mass motion in fixed target experiments, produced a qualitatively superior set of new data that included, for the first time, measurements of jets in heavy-ion collisions. Within less than half a year, at the *Quark Matter 2001* conference in Stony Brook,[19] it was clear that the RHIC data confirmed the qualitative picture[18] extracted from the data collected at lower energies in almost every aspect and complemented it with qualitatively new insights on "jet quenching" (the predicted[20] suppression of high-p_T particles resulting from energy loss of fast colored particles (quarks or gluons) moving through a dense medium of unconfined color charges). The evidence for the creation of matter in local thermal equilibrium that expanded collectively according to the laws of relativistic fluid dynamics[21] became cleaner and stronger, and the much larger multiplicity densities of particles created at RHIC energies now required unambiguously initial densities for the energy deposited in the collision that exceeded the critical value for deconfinement by one to two orders of magnitude. Energy densities of this magnitude, combined with evidence for a large degree of thermalization already less than about 1 fm/c after nuclear impact,[22] made the conclusion unavoidable that QGP had been created in these collisions. The confirmation in 2003 that the hallmarks of QGP formation (anisotropic hydrodynamic flow and jet quenching) were absent[23] (or at least so weak that, with the data precision available at the time, no sign of them could be found — see remarks below) convinced the RHIC community that they had indeed created QGP in 200A GeV Au + Au collisions.[24] The inability to make these signals go away[25] by lowering the collision energy in the 2010/11 RHIC Beam Energy Scan to below the top SPS energy strongly suggests that the "New States of Matter" created at RHIC and in Pb + Pb collisions at the SPS are fundamentally the same, i.e. QGP was indeed created already in those earlier CERN experiments, and possibly even in the Au + Au collisions studied at the AGS.

The RHIC data posed a fundamental theoretical problem: How can matter made of deconfined quarks and gluons with "strong" interactions that become weak at high energies thermalize so quickly? As discussed in later sections, the bulk of the matter produced in the collision can be described very well using relativistic fluid dynamics with almost vanishing viscosity. Since viscosity is proportional to the mean free path of the medium constituents, small viscosity requires strong coupling. Theorists took a long time to wrap their heads around this: wasn't the deconfinement transition at high temperature triggered by the weakness of the strong interactions at high energies? The experimental data told us that this question had to be answered in the negative: It is impossible to understand the collective flow data by assuming a weakly interacting, gaseous plasma.[22] The QGP is a strongly coupled liquid. In fact, it is the most perfect liquid ever created in the lab![c] It is also

[c]As a medium, QGP is also more strongly coupled than cold nuclear matter and other forms of matter made of hadrons: In hadronic matter, the colored fundamental degrees of freedom of QCD are clustered into color neutral objects, neutralizing their strong primary color interactions and only leaving behind much weaker residual meson exchange interactions that fall off exponentially over distances beyond the confinement range. For this reason the hadron resonance gas behaves much more like a "gas" while the quark–gluon plasma acts like a liquid.

highly opaque to colored probes, as witnessed by the observed large parton energy loss.[26] These two phenomena are fundamentally related,[27] and they mutually reinforce the conclusion of a strongly coupled plasma.[24]

The last decade of RHIC experiments, joined in 2010 by heavy-ion collision experiments at the Large Hadron Collider (LHC) at another order of magnitude higher energies, has been dedicated to determine, with quantitative precision, the thermodynamic and transport properties of the QGP. This also requires, of course, a quantitatively precise theoretical framework for the description of the dynamical evolution of the collision fireballs (Little Bangs) and its effect on the various experimental probes used to determine these properties. The development of such a framework and the nature of the recent studies will be discussed below. First, however, I will comment on a deep conceptual connection between the Little Bangs and the Big Bang which will help put heavy-ion collisions in a larger context and help our thinking about them.

3. The Big Bang and the Little Bangs

As mentioned already, ultra-relativistic heavy-ion collisions at RHIC and the LHC produce fireballs made of extraordinarily hot matter, at initial energy densities (at the time when the matter reaches approximate local thermal equilibrium) that exceed the energy density of atomic nuclei in their ground states by two to three orders of magnitude. Due to enormous pressure gradients between the fireball center and the surrounding vacuum, these fireballs undergo explosive collective expansion, cooling down rapidly through several different states of matter, finally fragmenting into thousands of free-streaming hadrons whose energy and momentum distributions can be detected in the detectors set up around the collider rings. The evolution history of these "Little Bangs" has much similarity with the Big Bang that created our Universe (Fig. 2): Both undergo Hubble-like expansion,[d] feature a hierarchy of decoupling processes that are driven by the expansion dynamics (rather than by the finite geometric size of the Little Bang), with chemical decoupling of the finally observed particle abundances[e] preceding

[d]The relative velocity between two matter elements increases roughly linearly with their relative distance.

[e]In the Big Bang the chemical composition is frozen during the process of primordial nucleosynthesis at an age of about 3 minutes, in the Little Bang this happens during hadrosynthesis at the quark–hadron transition at an age of about 20–30 ioctoseconds.[2]

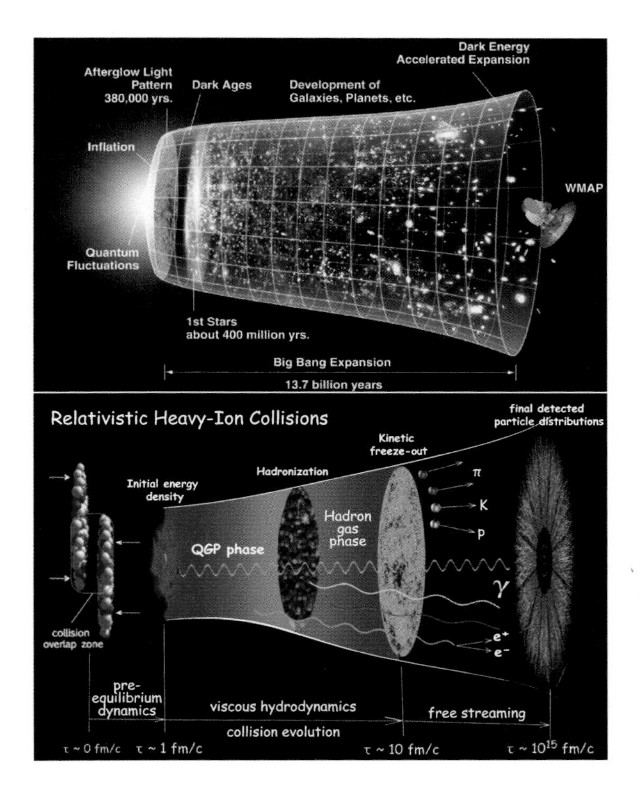

Fig. 2. Artist's conception of the evolution of the Big Bang (top — credit: NASA) and the Little Bang (bottom — credit: Paul Sorensen and Chun Shen).

kinetic decoupling,[f] and with initial-state quantum fluctuations[g] (shown in Figs. 3 and 4) imprinting themselves onto the experimentally observed final state.[h]

[f]In the Big Bang, the formation of neutral atoms by charge recombination at an age of about 380,000 yr makes the universe transparent to light and freezes the thermal Bose–Einstein energy distribution of the Cosmic Microwave Background radiation. When the Little Bang reaches an age of about 40–50 ioctoseconds, the final-stage hadron gas becomes so dilute that strong interactions between the hadrons cease and their energies and momenta are "frozen out."

[g]The initial wave function of the universe, stretched by cosmic inflation, seeds density fluctuations in the Big Bang. In the Little Bang, the initial quark and gluon wave functions inside the nucleons within the colliding nuclei control its initial density distribution.

[h]In the Big Bang the initial density fluctuations evolve under the action of gravity, described by Einstein's general theory of relativity, into the observed Cosmic Microwave temperature fluctuation spectrum shown in Figs. 3 (left) and 4 (top), and ultimately into today's distributions of stars, galaxies, and galaxy clusters and superclusters (see top panel of Fig. 2). In the Little Bang, they evolve through viscous hydrodynamics into the measured anisotropic collective flow patterns and their event-by-event fluctuations. By making the QGP such a perfect liquid, Nature has given us a huge break: If the QGP were significantly more viscous, all initial state fluctuations would be wiped out by dissipation during the life of the Little Bang, and the window onto the initial state provided by the observation of finalist flow fluctuations would be closed.

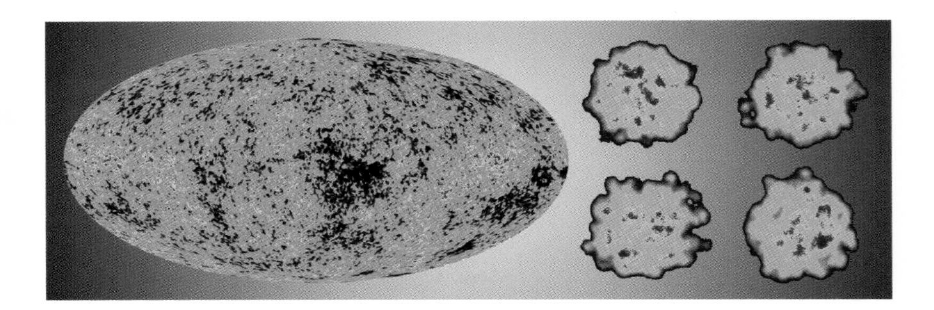

Fig. 3. Temperature fluctuation spectrum of the Big Bang at age 380,000 yr (left), as measured through the Cosmic Microwave Background radiation by WMAP, and of four typical Little Bangs created in central Pb + Pb collisions at the LHC (right), at age 0.2 fm/c, as calculated[28] from the IP-Glasma model.[29]

Of course, the Big and Little Bangs are quite different in other aspects: Their expansion rates differ by about 18 orders of magnitude; the Little Bang's expansion is 3-dimensional and driven by pressure gradients, not 4-dimensional and controlled by gravity; Little Bangs evolve on time scales of ioctoseconds, not billions of years; distances are measured in femtometers rather than light years. Most importantly, the Little Bang Standard Model is still under construction. This overview discusses recent progress of the edifice.

4. Eccentricity Fluctuations, Anisotropic Flows, and Flow Fluctuations

We can observe only one Big Bang (the one that produced our universe), but at RHIC and LHC we have experimentally created and studied billions of Little Bangs. Each Little Bang is different: Highly successful phenomenology based on hydrodynamic evolution models[21,31] has taught us that the initially very dense quark–gluon matter created in heavy-ion collisions reaches approximate local thermal equilibrium on a very short time scale[22] of order 1 fm/c (3 ioctoseconds), after which it evolves according to the macroscopic laws of relativistic viscous fluid dynamics. Pressure gradients in the fluctuating density profile (see Fig. 5 for a specific example) are the hydrodynamic forces that accelerate the fluid and cause it to expand and dilute. Spatial anisotropies and inhomogeneities in the initial density profile transverse to the beam direction lead to corresponding anisotropies in the final transverse expansion flow velocity profile that are imprinted on the momenta of the experimentally observed particles emitted from the collision. Each Little Bang features its own final flow velocity profile which (within hydrodynamics) is a deterministic classical response to the initial conditions that fluctuate from collision to collision due to quantum fluctuations in the initial nuclear wave function. We now know that the quark–gluon liquid that makes up the matter of the Little Bang during the first half of its life has very small viscosity, behaving like an almost ideal fluid.[21] This is a fantastic gift of Nature since it allows us to study experimentally the spectrum of initial-state quantum fluctuations through the final-state anisotropic flow fluctuations. Had the quark–gluon plasma (QGP) turned out to be highly viscous, all initial-state fluctuations would have been wiped out by dissipation before final decoupling of the emitted particles, thereby closing this observation window on the initial state of the Little Bang and on the quantum nature of the initial energy deposition process.

The Little Bang pressure p is related to its energy density e by its equation of state (EOS), and the anisotropy of the initial pressure gradients can be characterized by a series of harmonic eccentricity coefficients

$$\begin{aligned} \varepsilon_1 e^{i\Phi_1} &\equiv -\frac{\int r\,dr\,d\varphi\, r^3 e^{i\varphi}\, e(r,\varphi)}{\int r\,dr\,d\varphi\, r^3 e(r,\varphi)}\,, \\ \varepsilon_n e^{in\Phi_n} &\equiv -\frac{\int r\,dr\,d\varphi\, r^n e^{in\varphi}\, e(r,\varphi)}{\int r\,dr\,d\varphi\, r^n e(r,\varphi)} \quad (n>1)\,, \end{aligned} \tag{1}$$

where $e(r,\varphi)$ is the initial energy density distribution in the plane transverse to the beam direction. ε_n characterizes the magnitude of the nth harmonic deformation coefficient, and the angle Φ_n gives the direction of the corresponding deformation component in the lab frame. Both the magnitudes ε_n and their orientations Φ_n fluctuate from event to event. The mean values $\langle\varepsilon_n\rangle$ averaged over many events from three popular initial energy deposition models[31] are shown as a function of harmonic index n in the left bottom panel of Fig. 4. They represent the primordial temperature fluctuation spectrum of the Little Bang. We see that each collision centrality generates a different class of Little Bangs, with its own $\langle\varepsilon_n\rangle$ power spectrum, and that the three models shown give quite different results for these power spectra. If we can somehow measure the centrality dependence of the initial $\langle\varepsilon_n\rangle$ power spectrum, we have a powerful constraint on the initial nuclear quark and gluon wave functions on our hands. This is reminiscent of the crucial role the CMB temperature power spectrum (shown in the top panel of Fig. 4) has played in nailing down the parameters of the Standard Model of Big Bang Cosmology.

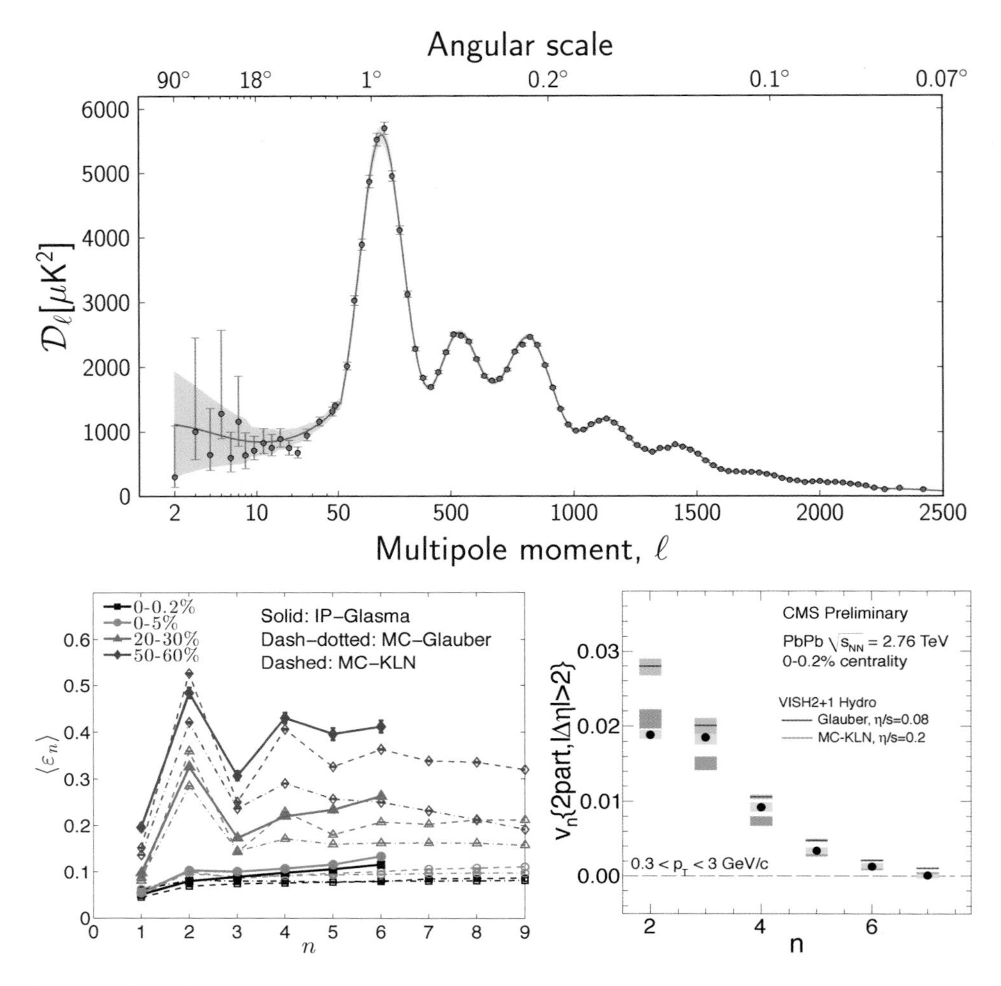

Fig. 4. *Top:* Temperature power spectrum of the Big Bang at age 380,000 yr, as measured by the Planck satellite.[30] *Bottom left*: Primordial eccentricity power spectrum of the Little Bangs created in 2.76A TeV Pb + Pb collisions of different centralities, at $\tau = 0$, from the energy density distributions of three different initial-state models (IP-Glasma, MC-Glauber, MC-KLN).[31] *Bottom right*: The final Little Bang flow power spectrum for ultracentral (0–0.2% centrality) Pb + Pb collisions at $\sqrt{s} = 2.76A$ TeV at the LHC, measured by the CMS Collaboration[32] (Wei Li, *Quark Matter 2012*) and calculated with viscous hydrodynamics `VISH2+1`[33] using MC-Glauber and MC-KLN initial conditions with the indicated specific shear viscosities.

The hydrodynamic response to the initial ε_n-spectrum of the Little Bang is reflected in the observed transverse momentum distributions $dN_i/(dy\, p_T\, dp_T\, d\phi)$ of the various emitted hadronic species i, and in correlations among them. The transverse momentum spectra can again be

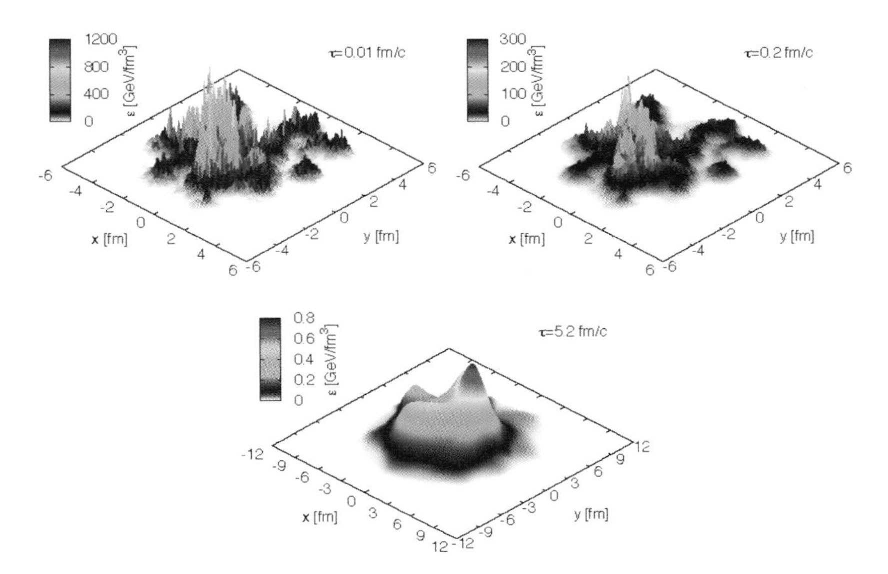

Fig. 5. Typical transverse energy density profiles $e(x, y)$ from the IP-Glasma model[28] for a semiperipheral ($b = 8$ fm) Au + Au collision at $\sqrt{s} = 200A$ GeV, at times $\tau = 0.01$,0.2, and 5.2 fm/c. From $\tau = 0.01$ fm/c to 0.2 fm/c the fireball evolves out of equilibrium according to the Glasma model;[34] at $\tau = 0.2$ fm/c the energy momentum tensor from the IP-Glasma evolution is Landau-matched to ideal fluid form (for technical reasons[35] the viscous pressure components are set to zero at the matching time) and henceforth evolved with viscous Israel–Stewart fluid dynamics, assuming $\eta/s = 0.12$ for the specific shear viscosity. The pre-equilibrium Glasma evolution is seen to somewhat wash out the large initial energy density fluctuations. The subsequent dissipative hydrodynamic evolution further smoothes these fluctuations. The asymmetric pressure gradients due to the prominent dipole asymmetry in the initial state of this particular event (visible as a left–right asymmetry of the density profile in the left upper panel) is seen to generate a dipole ("directed flow") component in the hydrodynamic flow pattern that pushes matter towards the right during the later evolution stages.

characterized by a set of complex harmonic coefficients

$$V_n = v_n e^{in\Psi_n} := \frac{\int p_T\, dp_T\, d\phi\, e^{in\phi} \frac{dN}{dy\, p_T\, dp_T\, d\phi}}{\int p_T\, dp_T\, d\phi\, \frac{dN}{dy\, p_T\, dp_T\, d\phi}} \equiv \left\{e^{in\phi}\right\}, \tag{2}$$

$$V_n(p_T) = v_n(p_T) e^{in\Psi_n(p_T)} := \frac{\int d\phi\, e^{in\phi} \frac{dN}{dy\, p_T\, dp_T\, d\phi}}{\int d\phi\, \frac{dN}{dy\, p_T\, dp_T\, d\phi}} \equiv \left\{e^{in\phi}\right\}_{p_T}. \tag{3}$$

Here ϕ is the azimuthal angle around the beam direction of the particle's transverse momentum $\boldsymbol{p}_\mathrm{T}$, and the curly brackets denote the average over particles from a single collision. Each particle species i has its own set of anisotropic flow coefficients V_n. Eq. (2) defines the flow coefficients and associated flow angles for the entire event, whereas Eq. (3) is the analogous definition for the subset of particles in the event with a given magnitude of the transverse momentum p_T. I suppress the dependence of both types of flow coefficients on the rapidity y. v_n are known as the "integrated" anisotropic flows, $v_n(p_T)$ are called "differential" flows. By definition, both v_n and $v_n(p_T)$ are positive definite. Hydrodynamic simulations show that in general the flow angles Ψ_n depend on p_T, and that, as a function of p_T, $\Psi_n(p_T)$ wanders around the "average angle" Ψ_n that characterizes the integrated flow v_n of the entire event.[36]

For each collision system, centrality class, and collision energy, the fluctuating initial state of the Little Bang is characterized by a distinct probability distribution $P(\varepsilon_n, \Phi_n)$, one particular moment of which is the $\langle\varepsilon_n\rangle$ power spectrum shown in Fig. 4. For different n, the angles Φ_n are more or less correlated with the direction of the impact parameter $\boldsymbol{b}$, but this direction cannot be measured experimentally. The viscous hydrodynamic evolution relates the probability distribution for the initial complex eccentricity coefficients deterministically to probability distributions $P(V_n)$ and $P(V_n(p_T))$ for the integrated and differential final complex harmonic flow coefficients. These distributions are particle species specific; for unidentified charged particles, they were recently measured by the ATLAS Collaboration.[37] As will be discussed in the next section, the relation between the final complex flow coefficients V_n and the initial eccentricity coefficients $\varepsilon_n e^{in\Phi_n}$ (and thus between the corresponding flow and eccentricity probability distributions characterizing each class of collisions) depends on the viscosity of the Little Bang matter. One goal of the relativistic heavy-ion program is to both constrain the QGP viscosity and identify the correct theory for the initial-state quantum fluctuations by performing a complete experimental reconstruction of the final multi-dimensional distributions $P(V_n)$ and $P(V_n(p_T))$.

Due to limited statistics arising from the finite number of particles emitted by each Little Bang, neither the magnitudes v_n nor the flow angles Ψ_n can be accurately determined for a single event. Experimental flow measures therefore involve angle correlations between two or more particles; for example, $\langle \{e^{in\phi_1}\}_{p_{T1}} \{e^{-in\phi_2}\}_{p_{T2}} \rangle = \langle v_n(p_{T1}) v_n(p_{T2}) \cos[n(\Psi_n(p_{T1}) - \Psi_n(p_{T2}))] \rangle$ where the first particle from the event has transverse momentum $\boldsymbol{p}_{T1} = (p_{T1}, \phi_1)$ and the second particle has $\boldsymbol{p}_{T2} = (p_{T2}, \phi_2)$, the average $\{\cdots\}$ is over all such particles in the event and $\langle\cdots\rangle$ indicates the average of the result over many Little Bangs of the selected class. One sees that the event-by-event fluctuations and transverse momentum dependences of both the anisotropic flow magnitudes v_n and their associated flow angles Ψ_n affect these experimental observables. Different such observables correspond to different correlation functions between the v_n's and Ψ_n's all of which can, for a given initial energy deposition model with probability distribution $P(\varepsilon_n, \Phi_n)$, be computed from the hydrodynamically predicted probability distributions $P(V_n)$ and $P(V_n(p_T))$. One can define experimental observables that separate the fluctuations of the anisotropic flow magnitudes $v_n(p_T)$ from those in the flow angles $\Psi_n(p_T)$.[36] These should be powerful discriminators between different $P(V_n)$ and $P(V_n(p_T))$ distributions, and thus between different initial-state fluctuation models. Experimental studies of these v_n and Ψ_n fluctuations and their p_T-dependences have just gotten under way; the results will be interesting and should significantly advance the construction of the Little Bang Standard Model.

The angles Φ_n associated with the initial deformation parameters ε_n are not only correlated with the (unmeasurable) direction of the impact parameter $\boldsymbol{b}$, but also with each other, in ways that are predicted by (and thus depend on) the initial energy deposition model.[38,39] These so-called participant-plane correlations in the initial state are translated by hydrodynamic response into final-state flow angle correlations. These flow angle correlations were measured in Pb + Pb collisions at the LHC by the ALICE[40] and ATLAS[41] Collaborations. All measured two- and three-plane flow angle correlations[41] are qualitatively reproduced by viscous hydrodynamic calculations, both in their magnitudes and centrality dependences[39] (bottom panels in Fig. 6). They differ, however, qualitatively from the initial participant-plane correlations,[39] often even in sign (top panels in Fig. 6). Hydrodynamic evolution is nonlinear and leads to mode-coupling between different harmonics.[42] For example, elliptic and triangular deformations ε_2 and ε_3 provide a nonlinear contribution to pentangular flow V_5 which, for large impact parameters where the elliptic deformation ε_2 is big, overwhelms the linear response to ε_5 and completely decorrelates the pentangular flow plane Ψ_5 from the angle Φ_5 of the initial pentangular density deformation.[42] Only by accounting for mode-coupling, either directly by following the non-linear hydrodynamic evolution[39] or through a non-linear response analysis that keeps at least second-order terms,[43] can the experimental data be reproduced. These flow-plane correlations thus represent an *experimentum crucis* in support of the hydrodynamic paradigm for the Little Bang;[39] dynamical models without a large degree of local thermalization and hydrodynamic collective flow will not be able to describe the dynamical change of character between the initial-state participant-plane and final-state flow-plane correlations.[i]

[i]The strength of the flow-plane correlations depends on the QGP shear viscosity.[39] A careful quantitative study of these correlations by more precise experiments and more systematic theoretical analyses should be able to separate the effects arising from the initial-state fluctuation spectrum and from dissipative transport.

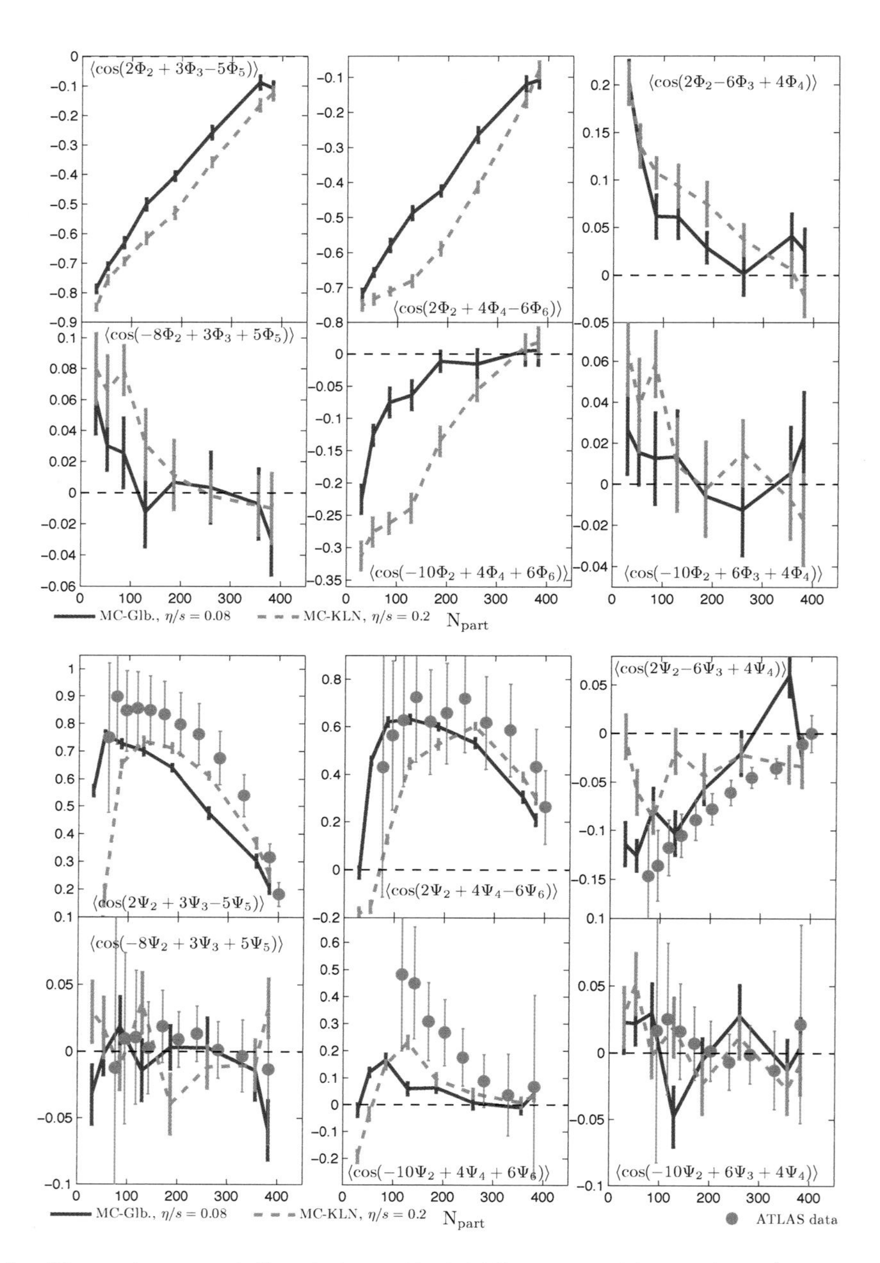

Fig. 6. Three-plane correlations between the initial-state participant planes (top 6 panels) and the final-state flow planes (bottom 6 panels), for $2.76A$ TeV Pb + Pb collisions as functions of collision centrality, indicated by the number of participant nucleons N_{part} on the horizontal axes. The solid (dashed) lines are for MC-Glauber (MC-KLN) initial conditions evolved event-by-event with viscous hydrodynamics using $\eta/s = 0.08$ (0.2) for the specific shear viscosity.[39] Filled circles show the experimental values measured by the ATLAS Collaboration.[41]

5. QGP Shear Viscosity from Anisotropic Flow Measurements

The efficiency of the fluid to convert spatial anisotropies in the pressure gradients into anisotropic flows is degraded by shear viscosity. The "conversion efficiency" v_n/ε_n is therefore a measure for the specific shear viscosity η/s of the expanding fluid. This is shown in Fig. 7. Higher harmonics, reflecting variations on smaller spatial scales, are suppressed more strongly by shear viscosity than lower flow harmonics.

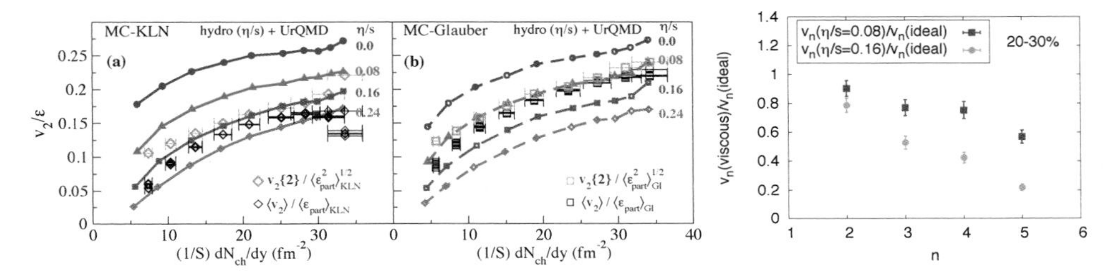

Fig. 7. *Left*: The eccentricity-scaled integrated elliptic flow of all charged hadrons, $v_2^{\rm ch}(\eta/s)/\varepsilon$, as a function of total charged hadron multiplicity density per unit overlap area, $(1/S)(dN_{\rm ch}/dy)$. The experimental data points show two measures for the elliptic flow ($\langle v_2\rangle$ (Ref. 44) and $v_2\{2\}$ (Ref. 45)) from 200A GeV Au–Au collisions at RHIC, measured by the STAR Collaboration. Both panels use the same sets of data, but use different average initial eccentricities $\langle\varepsilon\rangle$ and overlap areas $\langle S\rangle$ (obtained from the MC-Glauber and MC-KLN models) to normalize the vertical and horizontal axes. The theoretical curves were computed with the `VISHNU` model,[46] for different (temperature-independent) choices of the specific QGP shear viscosity $(\eta/s)_{\rm QGP}$. *Right*: Viscous suppression of v_n for $\eta/s = 0.08$ (squares) and 0.16 (circles), as function of harmonic index n, for Au + Au collisions at 20–30% centrality.[47]

The azimuthally symmetric part of the collective transverse flow, called radial flow, boosts particles from low to high transverse momentum in proportion to their masses; it is thus responsible for the distribution of the hydrodynamically generated momentum anisotropies over the various particle species and in p_T.[49] To correctly describe the differential anisotropic flows $v_n(p_T)$, for all charged hadrons or for specific identified hadron species, thus requires that the total hydrodynamic momentum anisotropy (which "measures" η/s), the radial flow (which is also sensitive to bulk viscosity that suppresses radial flow[50]), and the chemical composition of the system at final decoupling (which reflects the kinetics of chemical freeze-out) are *all* correctly described by the model. By analyzing instead the total flow anisotropies $v_n^{\rm ch}$ of *all* charged hadrons, integrated over p_T, one can strongly reduce the model sensitivity to bulk viscosity and final chemical composition and obtain a much more robust estimate of η/s.[46,49] The left panel in Fig. 7 shows such an extraction of η/s from elliptic flow data. The `VISHNU` model[46,51] used for this extraction couples the viscous hydrodynamic evolution of the QGP phase to a microscopic kinetic evolution of the hadronic phase after hadronization, thereby eliminating many uncertainties in earlier simplified calculations that described the hadronic phase macroscopically, too. The different theoretical lines show that, for a given initial energy deposition model, $v_2^{\rm ch}/\varepsilon_2$ is *only* sensitive to the specific QGP shear viscosity. Once $(\eta/s)_{\rm QGP}$ has been adjusted to the measured total charged hadron elliptic flow as in Fig. 7 (resulting in $(\eta/s)_{\rm QGP} \simeq 0.2$ for MC-KLN and $(\eta/s)_{\rm QGP} \simeq 0.08$ for MC-Glauber initial conditions), one finds that the model also correctly describes the p_T-spectra and differential elliptic flows $v_2(p_T)$ of all charged hadrons together, as well as for specific identified hadron species (pions, kaons, protons), for all collision centralities.[49] Furthermore, it correctly *predicted* the analogous observables for Pb + Pb collisions at the LHC,[33,52,53] in particular an increased mass splitting between the differential elliptic flows for light and heavy hadrons.[33,54]

Unfortunately, the two left panels in Fig. 7 also show that different initial energy deposition models lead to very different estimates for $(\eta/s)_{\rm QGP}$, due to their different initial eccentricities ε_2 (see Fig. 4). This model ambiguity can be resolved by analyzing simultaneously elliptic and triangular flow (v_2 and v_3) data. Such an analysis eliminates the MC-KLN model with $(\eta/s)_{\rm QGP} = 0.2$ as a viable candidate,[55] demonstrating the power of a comprehensive set of anisotropic flow data to (over)constrain the dynamical evolution model. Unfortunately, the MC-Glauber model did not enjoy a much longer life: The v_n data ($n = 2, \ldots, 7$) from ultra-central Pb + Pb collisions shown by CMS at *Quark Matter 2012* (see the right bottom plot in Fig. 4) and the event-by-event v_n probability distributions ($n = 2, \ldots, 4$) measured by ATLAS and shown at the same conference[37] (see the right panel in Fig. 8) can not simultaneously be described by viscous hydrodynamics for *any* choice of $(\eta/s)_{\rm QGP}$ if initial fluctuation spectra from either of these two models (MC-KLN and MC-Glauber) are used.

Fortunately, a new and much more successful initial-state model saved the day: the IP-Glasma model,[28,56] based on the Color Glass Condensate idea,[34] implements gluonic field fluctuations inside nucleons[28,58] as well as gluon saturation effects.[34,57] As shown in Fig. 8, these IP-Glasma initial conditions (represented by the solid lines in Fig. 4, lower left panel), when (after a short initial pre-equilibrium stage modeled by classical Yang–Mills evolution) evolved with viscous fluid dynamics, reproduce the entire measured

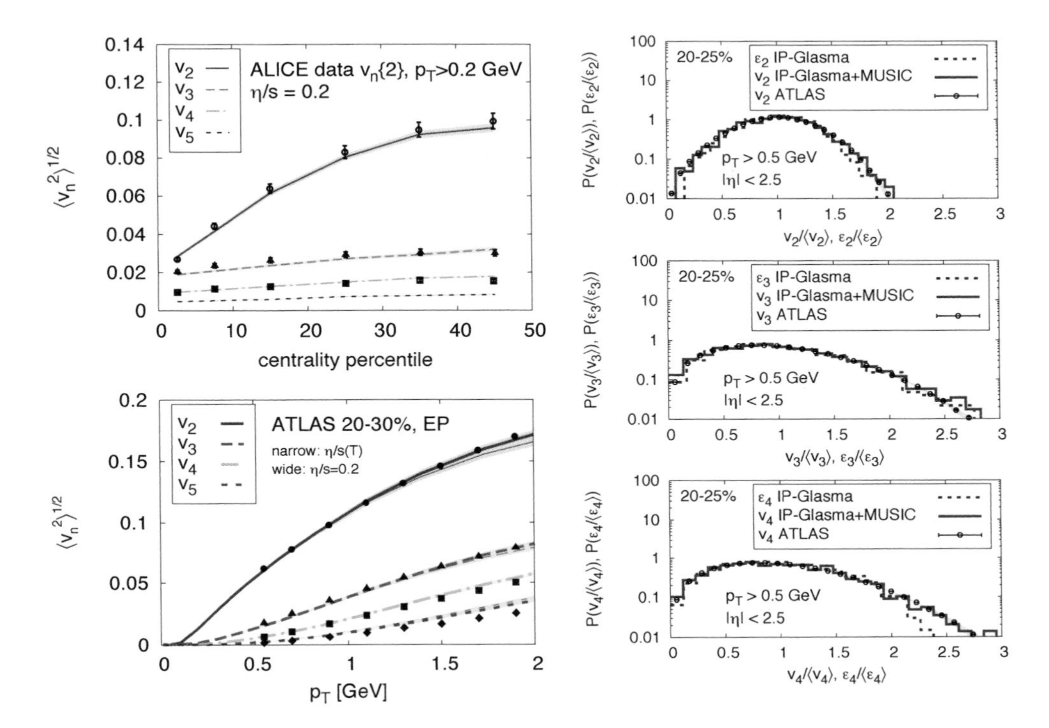

Fig. 8. *Top left*: The centrality dependence of $v_n\{2\}$ from $2.76A$ TeV Pb + Pb collisions measured by ALICE[40] compared to viscous hydrodynamic model calculations.[35] *Bottom left*: Comparison of $v_n(p_T)$ for the same collision system at 20–30% centrality from ATLAS[48] with hydrodynamical calculations, using both a constant average and a temperature dependent η/s.[35] *Right*: Scaled distributions of $v_{2,3,4}$ (from top to bottom) from viscous hydrodynamics with IP-Glasma initial conditions[35] compared with experimental data from ATLAS[37] and with the scaled distributions of the corresponding initial eccentricities $\varepsilon_{2,3,4}$. Nonlinear hydrodynamic evolution causes slightly larger variances for the v_n distributions compared to those of ε_n. The data in both panels are from Pb + Pb collisions at the LHC.

spectrum of charged hadron anisotropic flow coefficients v_n, both integrated over and differential in p_T, for all collision centralities, as well as the measured[37] event-by-event distribution of v_2, v_3, and v_4, again for a range of collision centralities.[35] This is shown in Fig. 8 for Pb + Pb collisions at the LHC but also holds for Au + Au collisions at RHIC.[35] The only difference is that for LHC energies the effective specific shear viscosity of the QGP fluid must be chosen somewhat larger ($(\eta/s)_{\mathrm{LHC}} \simeq 0.2$) than at top RHIC energy ($(\eta/s)_{\mathrm{RHIC}} \simeq 0.12$). Both data sets are consistent with a temperature dependent specific shear viscosity $(\eta/s)(T)$ proposed in Ref. 59 (the thin lines in the lower left panel in Fig. 8) that has a minimum value around $1/(4\pi) = 0.08$ at the pseudocritical temperature T_{cr} where the QGP hadronizes and then rises to about 5 times that minimal value at $2T_{\rm cr}$. While this temperature-dependence of η/s is not yet tightly constrained by the available analyses of RHIC and LHC data, the need for a somewhat larger effective shear viscosity at the higher temperatures probed at the LHC, already noted in earlier work,[59,60] appears to solidify, and the early evidence from RHIC data for a very low specific shear viscosity near $T_{\rm cr}$,[61] perhaps as low as the KSS bound[62] of $1/(4\pi)$, is now strongly supported.[35]

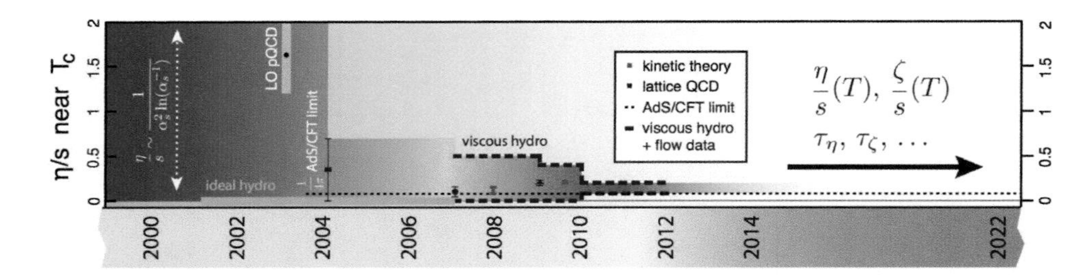

Fig. 9. Progress in the uncertainty of the QGP shear viscosity $(\eta/s)_{\rm QGP}$ over the last decade.[29]

Figure 9 illustrates the evolution of our understanding of the transport properties "quark–gluon soup" over the last decade or so. Initial estimates[63] based on leading order perturbative QCD gave results that were uncertain by at least an order of magnitude. The work by Son *et al.* on strong coupled plasmas[62] first gave an indication that the QGP shear viscosity could be quite small, as originally suggested by the unexpected semi-quantitative success of ideal fluid dynamics.[21,22] Significant progress was made with the development of viscous relativistic fluid dynamics in 2007 and the viscous hydro+cascade hybrid models in 2010 which reduced the uncertainty in $(\eta/s)_{\rm QGP}$ to about a factor of 3. As of today we know that at RHIC and LHC energies $(\eta/s)_{\rm QGP} \simeq 2/(4\pi)$, with an uncertainty of about 50% in both directions. High precision data for the entire v_n spectrum from pp, pA and AA collisions and a careful analysis of the influence of event-by-event initial state quantum fluctuations has now opened the possibility to constrain $(\eta/s)_{\rm QGP}$ to within a few percent relative precision. This opens the window to establishing meaningful constraints also for the bulk viscosity ζ/s and its temperature dependence, as well as the shear and bulk relaxation times τ_η, τ_ζ. Recent additional improvements in the dynamical framework[65] will help to further decrease systematic theoretical uncertainties.

6. Summary: Status of and Prospects for the Little Bang Standard Model

The Little Bang Standard Model is still under construction, but its key features are showing through the scaffolding: (1) Every heavy-ion collision system, centrality class, and collision energy generates a different class of Little Bangs, each with its own characteristic initial density fluctuation and final anisotropic flow fluctuation spectrum. (2) The initial fluctuation spectrum can be computed from the Color Glass Condensate (CGC) theory using e.g. the IP-Glasma model; gluon field fluctuations inside the nucleons within the colliding nuclei play an essential role in this spectrum. (3) After a very short pre-equilibrium evolution stage, not lasting much longer than 1 fm/c, which is best described using classical Yang–Mills field dynamics corrected for quantum fluctuations, the Little Bangs are in the QGP phase which undergoes viscous hydrodynamic evolution until it has cooled down to the pseudo-critical temperature for hadronization, $T_{\rm cr} \simeq 160$ MeV. The specific QGP shear viscosity is small, around $1/(4\pi)$ near $T_{\rm cr}$, rising modestly at higher temperatures; its effective value for $200A$ GeV Au + Au collisions at RHIC is around $0.12 = 1.5/(4\pi)$, for $2.76A$ TeV Pb + Pb collisions it is around $0.2 = 2.5/(4\pi)$. This small shear viscosity allows flow anisotropies to build in response to the initial fluctuations and anisotropies in the pressure gradients, to values that are large enough to be experimentally observed in the final state but still visibly attenuated by shear viscous effects, following an η/s-dependent pattern where higher harmonic flow coefficients are suppressed more strongly than lower harmonics. The spectrum of flow anisotropies, their magnitudes, directions, particle species and p_T-dependences, and event-by-event fluctuations provide a rich menu of experimental observables from which the QGP shear viscosity and initial fluctuation spectra can be reconstructed. (4) After hadronization, the Little Bang continues to evolve as a dilute, highly dissipative hadron resonance gas that is highly inefficient in generating any additional momentum anisotropies but also doesn't erase those established earlier during the QGP phase. This makes it possible to extract from hadronic final state observables quantitative information about the earlier QGP and CGC phases.

In this relatively simple form, the Little Bang Standard Model applies to Little Bangs created at top RHIC and LHC energies. At lower collision energies, our understanding of the initial conditions becomes less reliable and the lifetime of the hydrodynamic QGP stage shrinks, making a quantitative theory of the Little Bang more challenging, but not less interesting.

References

1. M. Alford, *Physics* **3**, 44 (2010).
2. U. Heinz, Hunting down the quark gluon plasma in relativistic heavy ion collisions, in *Strong and Electroweak Matter '98 (SEWM98)*, eds. J. Ambjorn *et al.* (World Scientific, Singapore, 1999), pp. 81–100, arXiv:hep-ph/9902424; U. Heinz, *Nucl. Phys. A* **661**, 140 (1999); U. Heinz, *Nucl. Phys. A* **685**, 414 (2001).
3. U. Heinz, *J. Phys. Conf. Ser.* **455**, 012044 (2013).
4. D. J. Gross and F. Wilczek, *Phys. Rev. Lett.* **30**, 1343 (1973).
5. H. D. Politzer, *Phys. Rev. Lett.* **30**, 1346 (1973).
6. H. Fritzsch, M. Gell-Mann and H. Leutwyler, *Phys. Lett. B* **47**, 365 (1973).
7. J. C. Collins and M. J. Perry, *Phys. Rev. Lett.* **34**, 1353 (1975).
8. G. Baym and S. A. Chin, *Phys. Lett. B* **62**, 241 (1976).
9. K. G. Wilson, *Phys. Rev. D* **10**, 2445 (1974).
10. M. Creutz, *Phys. Rev. D* **21**, 2308 (1980).
11. L. D. McLerran and B. Svetitsky, *Phys. Lett. B* **98**, 195 (1981); J. Kuti, J. Polonyi and K. Szlachanyi, *Phys. Lett. B* **98**, 199 (1981); J. Engels, F. Karsch, H. Satz and I. Montvay, *Phys. Lett. B* **101**, 89 (1981).
12. S. Borsanyi, G. Endrodi, Z. Fodor, A. Jakovac, S. D. Katz, S. Krieg, C. Ratti and K. K. Szabo, *J. High Energy Phys.* **1011**, 077 (2010).
13. HotQCD Collab. (A. Bazavov *et al.*), The equation of state in (2+1)-flavor QCD, arXiv:1407.6387 [hep-lat].
14. J. W. Harris and B. Müller, *Annu. Rev. Nucl. Part. Sci.* **46**, 71 (1996).
15. E. Schnedermann, J. Sollfrank and U. Heinz, *Phys. Rev. C* **48**, 2462 (1993).
16. J. Rafelski and B. Müller, *Phys. Rev. Lett.* **48**, 1066 (1982) [Erratum: *ibid.* **56**, 2334 (1986)]; P. Koch, B. Müller and J. Rafelski, *Phys. Rept.* **142**, 167 (1986).
17. T. Matsui and H. Satz, *Phys. Lett. B* **178**, 416 (1986).
18. U. Heinz and M. Jacob, Evidence for a new state of matter: An assessment of the results from the CERN lead beam program, arXiv:nucl-th/0002042; CERN Press Release, Feb. 10, 2000: New State of Matter created at CERN, `http://press.web.cern.ch/press-releases/2000/02/new-state-matter-created-cern`.
19. T. J. Hallman, D. E. Kharzeev, J. T. Mitchell and T. Ullrich (eds.), *Nucl. Phys. A* **698**, 1–708 (2002).
20. M. Gyulassy and M. Plümer, *Phys. Lett. B* **243**, 432 (1990); X. N. Wang and M. Gyulassy, *Phys. Rev. Lett.* **68**, 1480 (1992).
21. P. F. Kolb and U. Heinz, Hydrodynamic description of ultrarelativistic heavy-ion collisions, in *Quark-Gluon Plasma 3*, eds. R. C. Hwa *et al.* (World Scientific, Singapore, 2004), pp. 634–714, arXiv:nucl-th/0305084.
22. U. Heinz and P. F. Kolb, *Nucl. Phys. A* **702**, 269 (2002).
23. PHOBOS Collab. (B. B. Back *et al.*), *Phys. Rev. Lett.* **91**, 072302 (2003), arXiv:nucl-ex/0306025; PHENIX Collab. (S. S. Adler *et al.*), *Phys. Rev. Lett.* **91**, 072303 (2003); STAR Collab. (J. Adams *et al.*), *Phys. Rev. Lett.* **91**, 072304 (2003); BRAHMS Collab. (I. Arsene *et al.*), *Phys. Rev. Lett.* **91**, 072305 (2003).
24. M. Gyulassy, The QGP discovered at RHIC, in *Structure and Dynamics of*

Elementary Matter, NATO ASI Science Series II: Mathematics, Physics and Chemistry, Vol. 166, eds. W. Greiner *et al.* (Kluwer Academic, Dordrecht, 2004), pp. 159–182, arXiv:nucl-th/0403032; M. Gyulassy and L. McLerran, *Nucl. Phys. A* **750**, 30 (2005).
25. STAR Collab. (L. Adamczyk *et al.*), *Phys. Rev. C* **88**, 014902 (2013).
26. K. M. Burke *et al.*, *Phys. Rev. C* **90**, 014909 (2014).
27. A. Majumder, B. Müller and X. N. Wang, *Phys. Rev. Lett.* **99**, 192301 (2007).
28. B. Schenke, P. Tribedy and R. Venugopalan. *Phys. Rev. Lett.* **108**, 252301 (2012); B. Schenke, P. Tribedy and R. Venugopalan, *Phys. Rev. C* **86** 034908.
29. R. Tribble (chair), A. Burrows *et al.*, Implementing the 2007 Long Range Plan. Report to the Nuclear Science Advisory Committee, January 31, 2013. Available at `http://science.energy.gov/np/nsac/reports/`.
30. Planck Collab. (P. A. R. Ade *et al.*), Planck 2013 results. XV. CMB power spectra and likelihood, arXiv:1303.5075 [astro-ph.CO].
31. U. Heinz and R. Snellings, *Annu. Rev. Nucl. Part. Sci.* **63**, 123 (2013).
32. J. W. Harris, D. Kharzeev and T. Ullrich, *CERN Courier* **52**(9), 17 (2012).
33. C. Shen, U. Heinz, P. Huovinen and H. Song, *Phys. Rev. C* **84**, 044903 (2011).
34. A. Kovner, L. D. McLerran and H. Weigert, *Phys. Rev. D* **52**, 6231 (1995); Y. V. Kovchegov and D. H. Rischke, *Phys. Rev. C* **56**, 1084 (1997); A. Krasnitz and R. Venugopalan, *Nucl. Phys. B* **557**, 237 (1999); A. Krasnitz and R. Venugopalan, *Phys. Rev. Lett.* **84**, 4309 (2000); A. Krasnitz and R. Venugopalan, *Phys. Rev. Lett.* **86**, 1717 (2001); T. Lappi, *Phys. Rev. C* **67**, 054903 (2003); T. Lappi and L. D. McLerran, *Nucl. Phys. A* **772**, 200 (2006).
35. C. Gale, S. Jeon, B. Schenke, P. Tribedy and R. Venugopalan, *Phys. Rev. Lett.* **110**, 012302 (2013); C. Gale, S. Jeon, B. Schenke, P. Tribedy and R. Venugopalan, *Nucl. Phys. A* **904-905**, 409c (2013).
36. U. Heinz, Z. Qiu and C. Shen, *Phys. Rev. C* **87**, 034913 (2013).
37. ATLAS Collab. (J. Jia *et al.*), *Nucl. Phys. A* **904-905**, 421c (2013); ATLAS Collab. (G. Aad *et al.*), *J. High Energy Phys.* **1311**, 183 (2013).
38. J. Jia and S. Mohapatra, *Eur. Phys. J. C* **73**, 2510 (2013), arXiv:1203.5095 [nucl-th]; J. Jia and D. Teaney, *Eur. Phys. J. C* **73**, 2558 (2013).
39. Z. Qiu and U. Heinz, *Phys. Lett. B* **717**, 261 (2012).
40. ALICE Collab. (K. Aamodt *et al.*), *Phys. Rev. Lett.* **107**, 032301 (2011); ALICE Collab. (A. Bilandzic *et al.*), *Nucl. Phys. A* **904-905**, 515c (2013).
41. ATLAS Collab. (J. Jia *et al.*), *Nucl. Phys. A* **910-911**, 276 (2013); ATLAS Collab. (G. Aad *et al.*), *Phys. Rev. C* **90**, 024905 (2014).
42. Z. Qiu and U. Heinz, *Phys. Rev. C* **84**, 024911 (2011).
43. D. Teaney and L. Yan, *Phys. Rev. C* **86**, 044908 (2012); D. Teaney and L. Yan, *Nucl. Phys. A* **904-905**, 365c (2013).
44. J.-Y. Ollitrault, A. M. Poskanzer and S. A. Voloshin, *Phys. Rev. C* **80**, 014904 (2009).
45. STAR Collab. (J. Adams *et al.*), *Phys. Rev. C* **72**, 014904 (2005).
46. H. Song, S. A. Bass and U. Heinz, *Phys. Rev. C* **83**, 024912 (2011); H. Song, S. A. Bass, U. Heinz, T. Hirano and C. Shen, *Phys. Rev. Lett.* **106**, 192301 (2011).
47. B. Schenke, S. Jeon and C. Gale, *Phys. Rev. C* **85**, 024901 (2012).

48. ATLAS Collab. (G. Aad *et al.*), *Phys. Rev. C* **86**, 014907 (2012).
49. H. Song, S. A. Bass, U. Heinz, T. Hirano and C. Shen, *Phys. Rev. C* **83**, 054910 (2011).
50. H. Song and U. Heinz, *Phys. Rev. C* **81**, 024905 (2010).
51. C. Shen, Z. Qiu, H. Song, J. Bernhard, S. Bass and U. Heinz, The iEBE-VISHNU code package for relativistic heavy-ion collisions, arXiv:1409.8164 [nucl-th].
52. U. Heinz, C. Shen and H. Song, *AIP Conf. Proc.* **1441**, 766 (2012).
53. ALICE Collab. (B. Abelev *et al.*), *Phys. Rev. Lett.* **109**, 252301 (2012); ALICE Collab. (B. Abelev *et al.*), *Phys. Rev. C* **88**, 044910 (2013).
54. B. Müller, J. Schukraft and B. Wyslouch, *Annu. Rev. Nucl. Part. Sci.* **62**, 361 (2012).
55. Z. Qiu, C. Shen and U. Heinz, *Phys. Lett. B* **707**, 151 (2012).
56. B. Schenke, P. Tribedy and R. Venugopalan, *Phys. Rev. C* **86**, 034908 (2012).
57. J. Bartels, K. J. Golec-Biernat and H. Kowalski, *Phys. Rev. D* **66**, 014001 (2002); H. Kowalski and D. Teaney, *Phys. Rev. D* **68**, 114005 (2003).
58. P. Tribedy and R. Venugopalan, *Nucl. Phys. A* **850**, 136 (2011).
59. H. Niemi, G. S. Denicol, P. Huovinen, E. Molnar and D. H. Rischke, *Phys. Rev. Lett.* **106**, 212302 (2011); H. Niemi, G. S. Denicol, P. Huovinen, E. Molnar and D. H. Rischke, *Phys. Rev. C* **86**, 014909 (2012).
60. M. Luzum, *Phys. Rev. C* **83**, 044911 (2011); H. Song, S. A. Bass and U. Heinz, *Phys. Rev. C* **83**, 054912 (2011).
61. R. A. Lacey and A. Taranenko, *Proceedings of Science* **CFRNC2006**, 021 (2006).
62. G. Policastro, D. T. Son and A. O. Starinets, *Phys. Rev. Lett.* **87**, 081601 (2001); P. Kovtun, D. T. Son and A. O. Starinets, *Phys. Rev. Lett.* **94**, 111601 (2005).
63. P. B. Arnold, G. D. Moore and L. G. Yaffe, *J. High Energy Phys.* **0011**, 001 (2000); P. B. Arnold, G. D. Moore and L. G. Yaffe, *J. High Energy Phys.* **0305**, 051 (2003).
64. B. Schenke and R. Venugopalan, *Phys. Rev. Lett.* **113**, 102301 (2014).
65. D. Bazow, U. Heinz and M. Strickland, Second-order $(2+1)$-dimensional anisotropic hydrodynamics, arXiv:1311.6720 [nucl-th]; M. Strickland, Anisotropic Hydrodynamics: Three lectures, arXiv:1410.5786 [nucl-th].

Quarks and Anomalies

R. J. Crewther
Department of Physics, University of Adelaide,
Adelaide SA 5005, Australia
rodney.crewther@adelaide.edu.au

A nonperturbative understanding of neutral pion decay was an essential step towards the idea that strong interactions are governed by a color gauge theory for quarks. Some aspects of this work and related problems are still important.

1. Quarks Before QCD

Any Caltech theory student in the late 1960's, particularly if Murray Gell-Mann was their supervisor, had to be good at distinguishing various "quark models." Were we talking about "constituent" or "current" quarks, and within those categories, was model dependence an issue? Quarks were somehow fundamental, but it was not even clear that their dynamics should be governed by a local field theory. The main tactic was to "abstract" rules which seemed to be model independent and led to physical consequences which could be compared with existing data.

By that time, the quark idea was several years old, dating from work completed independently by the end of 1963: Gell-Mann's quarks,[1] Zweig's aces,[2] and (a reference I have just heard of) Petermann's "spineurs (avec)... des valeurs non entières de la charge."[3] These papers had in common

(1) structures $q\bar{q}$ for mesons and qqq for baryons built from nonrelativistic *constituent* quarks q and anti-quarks $\bar{q}$,
(2) the idea that $SU(3)$ mass formulas[5,6] are due to the strange quark s being heavier than the up and down quarks u, d, and
(3) concerns about whether the fractional charges would be observable.

Gell-Mann and Zweig were led to (1) by the need to explain the absence of exotic $SU(3)$ multiplets in the Eightfold Way.[5,7] Zweig analyzed the constituent quark model in detail, deriving properties such as spins, parities and masses for various $SU(3)$ multiplets. Gell-Mann had a separate aim: to reproduce current algebra, a set of equal-time commutators for $SU(3) \times SU(3)$

currents[4] which he had previously managed to abstract without using quarks. For this, he needed *current quarks*, i.e. relativistic fields $q(x)$ and $\bar{q}(x)$ for each flavor $q = u, d, s$, from which electromagnetic, weak and other $SU(3) \times SU(3)$ currents could be constructed.

Immediately, there were concerns about constituent quark statistics. How can a baryon like Δ^{++} exist as an S-wave spin-flavor symmetric state $|u \uparrow u \uparrow u \uparrow\rangle$ if quarks are spin-$\frac{1}{2}$ fermions? It is hard to imagine ground states being P-wave, so instead, it was proposed that quarks are either[9] para-fermions[10] of order 3 or[11–13] fermions with an extra quantum number taking three values, which we now know as *color*.[14–16] The observed fermionic baryons $|qqq\rangle$ are then symmetric in space-spin-flavor (a) for paraquarks automatically, or (b) for fermion quarks antisymmetrized in a color $SU(3)$ singlet state[12,13] (but not $SO(3)$, because that would allow colorless diquark states $|qq\rangle$).

Whether para-particle or colored multiplets would appear at higher energies or be banned completely (quark confinement) was not clear. In an attempt to make these extra states appear less weird, colored quarks were initially given integer charges[12,13] which, however, depended on the color index. Then photons could excite color from hadrons and perhaps induce transitions to a deconfined (weird) sector.

In the model eventually adopted in 1972,[15,16] quarks became colored fermions with fractional charges, with 3 colors for each charge or flavor. As a result, the electromagnetic and weak currents became color $SU(3)$ singlets, like the observed hadronic spectrum. Confinement was as unclear for this model as the others. If confinement were not absolute, the model could have degenerate color multiplets and fractionally charged states above some threshold energy. Comparing all of these models, it was concluded that, as models of *constituent* quarks, they were hard to distinguish below thresholds for deconfinement.

However the 1972 model was also designed to take into account color for *current* quarks. The rest of this article describes how studies of short-distance behavior[17] and the reaction[18] $\pi^0 \to \gamma\gamma$ led to this.

2. Scale Invariance at High Energies

I started life at Caltech as a graduate student in the fall of 1968. The very first seminar, on Tuesday October 1, was "Partons" by Richard P. Feynman, with Murray Gell-Mann sitting near the front. Feynman had just returned from a summer in SLAC hearing about Bjorken's work[19] on scaling in deep inelastic lepton-nucleon scattering and developing a model of point scatterers (partons) to give the same results. Murray kept asking "but Richard, what are their quantum numbers? Are they quarks?" but Richard's sole concern was scaling due to scattering by "grains of sand inside the nucleon." (A year or so later, my fellow student Finn Ravndal got him interested in quarks.)

Murray began supervising me two months later and in due course asked me, as an initial research exercise, to try using the Cutkosky bootstrap model[20] to generate higher symmetries like $SU(6)$. That produced hundreds of equations. Fortunately, just a few of them could be used to show that there could be no consistent solution. Murray commented that he hadn't intended the exercise "to be so vigorous" and suggested that I take a trip while he thought of a suitable PhD topic. My fellow student Chris Hamer and I had already planned to drive around the US that summer (1969), so we left immediately and on the way back, stopped at Aspen.

Murray had just started working on scale invariance as an approximate symmetry of hadrons, and suggested that I do the same. This would involve the energy–momentum tensor $\theta_{\mu\nu}$ as well as the $SU(3)\times SU(3)$ currents. Did I know about the Belinfante[21] tensor? Fortunately, I did (from Geoff Opat, supervisor of Chris and myself as Masters students in Melbourne, 1966–68). In that case, the next step was to understand all 14 pages of Wilson's paper on operator product expansions.[17]

Wilson generalized current algebra, replacing equal-time limits of commutators by short-distance limits of products of currents and other observables such as $\theta_{\mu\nu}$. Instead of a single term on the right-hand side, he obtained an asymptotic expansion $\sum_n \mathcal{C}_n O_n$ with coefficient functions

$$\mathcal{C}_1 \gg \mathcal{C}_2 \gg \mathcal{C}_3 \gg \cdots \tag{1}$$

in order of decreasing singularity times observable operators O_1, O_2, $O_3, \ldots$ of increasing operator dimensionality (in mass units). Equal-time commutators, such as in Gell-Mann's current algebra and Bjorken's work on scaling, could be recovered by noting that, since commutators vanish for space-like separations, their equal-time limits are controlled by the short-distance behavior of the relevant operator product. Checks in renormalized perturbation theories or for free current quarks indicated that, apart from quantum number constraints, the same set of operators $\{O_n\}$ tended to appear in the expansion, whatever the operator product used to generate them: "a limited set of licensed operators," as Murray put it.

A key feature of Wilson's work was his critique[22] of canonical field theory: operators usually *cannot* be multiplied at the same point, equal-time commutators may be singular, and T-ordering with step functions $\theta(t - t')$

can fail. These *anomalies* arise wherever renormalization is necessary. In particular, renormalized perturbation theory produces $\log^p\big(\mu^2(x-y)^2\big)$ factors at short distances, where μ is the renormalization scale. When summed up *à la* Gell-Mann and Low,[23] anomalous powers may be produced. If the ultraviolet limit is controlled by a nontrivial Gell-Mann–Low fixed point, scale invariance becomes exact at short distances, with anomalous dimensions for all operators O_n except those which are conserved or partially conserved. I was happy to abstract these rules and learn the renormalization group later.

Wilson's paper[17] also featured a Sec. VII "Applications" with five subsections, each equivalent to a separate publication. Subsection D "$\pi^0 \to \gamma\gamma$ Problem" drew my attention because (a) it explained how short-distance singularities determine contact terms in low-energy Ward identities and (b) I had seen the papers of Bell and Jackiw[24] and Adler[25] on the axial anomaly. Could the three-point function $T\langle \text{vac}|J_\alpha J_\beta J_{\mu 5}|\text{vac}\rangle$ of the electromagnetic and axial-vector currents J_α and $J_{\mu 5}$ be determined at short distances without using perturbation theory? Noting Wilson's comment (Sec. VIII) that "the prospects for obtaining such a solution seem dim at present," I filed the problem away as a challenge for the future.

At that time, the main question was whether Bjorken scaling is exact or not. Bjorken[19] obtained scaling by assuming that an infinite set of equal-time commutators of J_α with its derivatives is finite, i.e. not zero. It was quickly established that this was equivalent to assuming canonical or free-field (parton) behavior for the coefficient functions (1). Towers of these short-distance singularities could be summed to form terms in an operator product expansion near the light cone $(x-y)^2 \to 0$, the limit in position space conjugate to Bjorken's limit.[26,27] By then, quarks were widely believed to be responsible for scaling, so the proposal of Fritzsch and Gell-Mann[28] to abstract the light-cone expansion from free-quark theory was logical.

The argument against exact Bjorken scaling was led by Wilson.[29] Interactions tend to increase the dimensions of composite-field operators O_n which are not conserved exactly or partially, making higher-n functions (1) less singular on the light cone. The difficulty for this point of view was explaining why these anomalous corrections were *all* so small. Nevertheless, I tended to belong to this school of thought. My concern was that any tensor operator $O(x)$ lacking an anomalous dimension would be at least partially conserved, because the leading singularity of $\langle \text{vac}|O(x)O(y)|\text{vac}\rangle$ would be canonical and hence divergenceless. Therefore, my view was that the *only* operators allowed to have canonical dimension were $\theta_{\mu\nu}$ and the $SU(3)\times SU(3)$ currents.

In particular, there was the $U(1)$ problem, which I knew from Gell-Mann's 1969 Hawaii lectures.[30] If we abstract from the free-quark model, the isoscalar current

$$J^0_{\mu 5} = \bar{u}\gamma_\mu\gamma_5 u + \bar{d}\gamma_\mu\gamma_5 d \tag{2}$$

is conserved in the $SU(2) \times SU(2)$ limit. This is a disaster[31] because then the $SU(2) \times SU(2)$ condensate

$$\langle \text{vac}|\bar{u}u + \bar{d}d|\text{vac}\rangle \neq 0 \tag{3}$$

also acts as an axial $U(1)$ condensate. In addition to π^+, π^0, π^-, there would have to be a fourth Nambu–Goldstone boson, an isoscalar 0^- meson of mass $O(m_\pi)$. If just one extra conserved current could cause so much trouble, we certainly did not want an infinite tower of them.

The choice between canonical and anomalous dimensions would be cleared up by asymptotic freedom[32,33] four years later. Only $\theta_{\mu\nu}$ and the $SU(N_f) \times SU(N_f)$ currents behave canonically. Coefficients $\mathcal{C}_n$ of other operators O_n have their canonical behavior modified by inverse logarithmic powers, corresponding to a very weak violation of Bjorken scaling. The $U(1)$ problem was not so easily dismissed, and while majority opinion is that it is understood, they all miss the reference[34] where problems yet to be resolved are analyzed.

3. Approximate Scale Invariance at Low Energies

At the same time (1969–71), I was supposed to be working on my PhD research project. Clearly the hadronic ground state $|\text{vac}\rangle$ breaks scale invariance very strongly, given the 1 GeV scale set by baryons. This could be simply due to scale invariance being badly broken explicitly, with the trace θ^μ_μ large as an operator. The alternative is that scale invariance is approximately conserved in the Nambu–Goldstone mode with a massless 0^{++} *dilaton* in the limit $\theta^\mu_\mu \to 0$.

The analogy with chiral symmetry was obvious. Both chiral $SU(3) \times SU(3)$ and scale symmetry would be manifest at short distances and hidden elsewhere by the effects of their Goldstone bosons, the 0^- octet π, K, η and the 0^+ singlet dilaton σ (not to be confused with the field "σ" of the sigma model). I chose the simplest case where scale invariance was the result of taking the chiral $SU(3) \times SU(3)$ limit, so the 3-flavor version of the chiral condensate (3) could also act as a scale condensate.

This picture is no longer entirely accurate, given that QCD renormalization effects break scale symmetry everywhere, including short distances.

However if 3-flavor QCD has an infrared fixed point $\alpha_{\rm IR}$, where θ^μ_μ vanishes apart from $O(m_{u,d,s})$ corrections, the essential features of the original scheme can be reproduced by a double expansion in the running gluon coupling α_s about $\alpha_{\rm IR}$ and the light quark masses $O(m_{u,d,s})$ about zero.[35]

The dilaton idea is contained in footnote 38 of the 1962 current algebra paper.[4] A "resonance or quasi-resonance" which dominates a dispersion relation for

$$\langle \text{particle}|\theta^\mu_\mu|\text{particle}\rangle = \text{particle mass} \tag{4}$$

yields "a relation of the Goldberger–Treiman type" where "the coupling of the resonant state to different particles is roughly proportional to their masses." In the scale-invariant limit,[30] the vacuum would become degenerate, as for exact chiral symmetry, except for the degeneracy being noncompact. Physical predictions are then the result of expanding in m_σ^2 about zero.

The term "dilaton" is often used in a manner which is distinct from the scheme above or even contradicts it. The earliest variant was Fujii's proposal[37] of a finite-range scalar component of gravity. Gell-Mann called it a "Brans–Dickeon" after the well-known proponents of the scalar-tensor theory of gravity,[38] but the name did not stick. In modern times, "dilaton" is often used for a scalar particle which has zero mass classically but becomes massive due to quantum corrections, such as Higgs bosons which acquire mass due to dimensional transmutation.[39] Since there is no way of "turning off" such a mass, this has nothing to do with dilatons in the original sense.

In my student days, the main candidate for σ was $\epsilon(700)$, whose existence was not clear. Final state pions interact very strongly in the 0^{++} channel, so there was good reason to assume the presence of a resonance far off shell. However that meant that it was very hard to pin down in phase-shift analyses. It was declared dead in the 1976 particle data tables, but in recent years, has been resurrected as the broad but clearly defined resonance[36] $f_0(500)$.

If dilatons couple to mass, why is its coupling to pions so large? In leading order, one would expect $F_\sigma g_{\sigma\pi\pi}$ to be $2m_\pi^2$ for the coupling $g_{\sigma\pi\pi}\sigma\boldsymbol{\pi}\cdot\boldsymbol{\pi}$, where F_σ is the analogue of the pion decay constant $F_\pi \simeq 93$ MeV and has a similar order of magnitude:

$$\langle\sigma(q)|\theta_{\mu\nu}|\text{vac}\rangle = (F_\sigma/3)\big(q_\mu q_\nu - g_{\mu\nu}q^2\big)\,. \tag{5}$$

The solution, on which I based my PhD thesis, was to note that the result is really

$$F_\sigma g_{\sigma\pi\pi} = 2m_\pi^2 + O\big(m_\sigma^2\big) \tag{6}$$

and use approximate chiral symmetry to deduce the coefficient of m_σ^2:

$$F_\sigma g_{\sigma\pi\pi} = -m_\sigma^2 + O(m_\pi^2)\,. \tag{7}$$

This implies a width of a few hundred MeV, as required. I did it the hard way, using basic current algebra, and so took too long to obtain a mass formula for $m_\sigma^2 F_\sigma^2$. In the meantime, John Ellis was working on his PhD in Cambridge (UK), and obtained both Eq. (7) and the mass formula by efficient use of a chiral-scale effective Lagrangian. A few months later, we met and were able to compare notes at the 1971 Coral Gables conference.[40,41]

No account of these times would be complete without mentioning the episode in 1970 when Feynman became excited about Bose statistics for quarks. He hoped to explain the $\Delta I = 1/2$ rule for nonleptonic decays of strange particles. How quarks could possibly be bosons was a matter for future study; perhaps their bad statistics would not matter if they were confined. Almost immediately, we heard that the idea had already been suggested,[43,44] but the interest generated by Feynman[45] in this key problem was good for particle physics. A few months later, the correct version of the idea was proposed[46] (also anticipated in Japan[47]): for fermion quarks with color (and even for paraquarks[48]), the color antisymmetrization of qqq states plus current algebra implies the $\Delta I = 1/2$ rule for nonleptonic hyperon decays, but says nothing about $\Delta I = 1/2$ for $K \to \pi\pi$.

Since nonleptonic strange particle decays had been a problem for so long,[42] my interest was piqued. I told Murray of this, carefully avoiding any suggestion that quarks could be bosons (which I didn't believe anyway), and drew the response "watch out, it's a can of worms!" I was too busy finishing my PhD to pursue it; otherwise, I may have drawn Fig. 1, which is required by approximate chiral-scale invariance. It shows that the $\Delta I = 1/2$ rule for kaons is due to a large contribution from the dilaton pole. Only after 40-odd years, with help from my young colleague Lewis Tunstall, can I report a solution to that problem.[35] For hyperon decays, the $\Delta I = 1/2$ rule

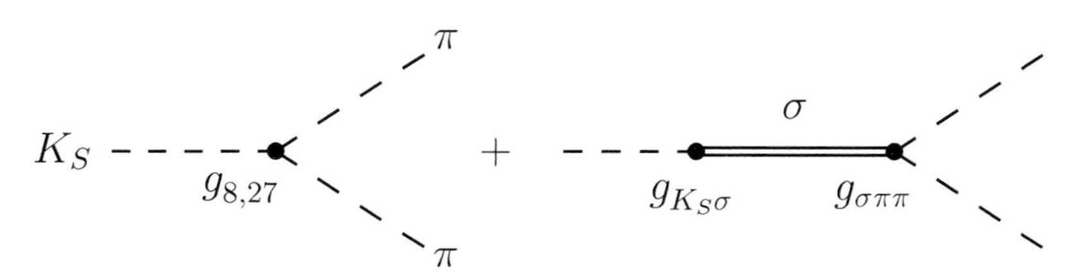

Fig. 1. Tree diagrams in chiral-scale perturbation theory[35] for $K_S \to \pi\pi$. The vertex amplitudes due to **8** and **27** contact couplings g_8 and g_{27} are dominated by the σ/f_0 pole amplitude. The magnitude of $g_{K_S\sigma}$ can be deduced from $K_S \to \gamma\gamma$ and $\gamma\gamma \to \pi\pi$.

is understood, but current algebra does not seem to work: that part of the problem is still a can of worms.

After Coral Gables, there was a thesis to be written, and a suitable way of ending it had to be found. What else could dilatons do?

From the literature on axial anomalies, I knew about Schwinger's 1951 paper on gauge invariance.[49] In Sec. V, he obtained unique results for both $\pi^0 \to \gamma\gamma$ and $\sigma \to \gamma\gamma$ in one-loop Yukawa theory by imposing gauge invariance on the renormalization procedure. In terms of the electromagnetic field tensor $F_{\mu\nu}$, fermion mass M, Yukawa coupling g, and fine-structure constant α, the answer for $\sigma \to \gamma\gamma$ is

$$\mathcal{L}^{\text{Yukawa}}_{\sigma\gamma\gamma} = -\frac{\alpha g}{6\pi M}\sigma F^{\mu\nu}F_{\mu\nu}\,. \tag{8}$$

In the second-last paragraph of my thesis, I noted that this breaks scale invariance (operator dimension $\neq 4$), so if M plays the role of F_σ as well as F_π, perhaps both $F_\pi g_{\pi\gamma\gamma}$ and $F_\sigma g_{\sigma\gamma\gamma}$ are anomalous. Already, Wilson had shown[22] that one-loop corrections in $\lambda\phi^4$ theory break scale invariance, which he interpreted as an anomaly in the trace of $\theta_{\mu\nu}$. Perhaps there is an electromagnetic trace anomaly due to strong interactions? I was moving to a post-doctoral job at Cornell; as soon as I arrived, I would try to extend Wilson's method for $\pi^0 \to \gamma\gamma$ to $\sigma \to \gamma\gamma$.

4. Derivation of $\pi^0 \to \gamma\gamma$ for Nonperturbative Pions

When Schwinger analyzed $\pi^0 \to \gamma\gamma$, chiral invariance and PCAC (partially conserved axial current) were unknown. At issue was the equivalence

$$\phi\bar{\psi}\gamma_5\psi \leftrightarrow -(i/2M)\bar{\psi}\gamma_\mu\gamma_5\psi\partial^\mu\phi\,, \quad \phi = \pi^0 \text{ field} \tag{9}$$

between pseudoscalar and pseudovector couplings for the one-fermion-loop triangle diagram. The trouble was that the product of the fermion fields at the same point is singular. The solution was to consider ψ and $\bar{\psi}$ at different points x' and x'' and make the analysis gauge invariant: then the limit $x' \to x''$ becomes finite. Rephrased in terms of chiral symmetry, the problem is that the Noether construction fails because (a) it requires

$$\frac{\partial\mathcal{L}}{\partial\partial^\mu\psi} = \bar{\psi}\gamma_\mu \quad \text{and} \quad \delta_{\text{axial}}\psi = \gamma_5\psi \tag{10}$$

to be multiplied at the same point and (b) it does not work for nonlocal expressions produced by point splitting. The axial anomaly[24,25] is responsible for this failure: it is the finite counterterm mismatch between gauge invariant and chiral invariant renormalization prescriptions for axial-vector operators.

Wilson's version of this was designed to avoid perturbation theory. In particular, if pions are $q\bar{q}$ states which become Nambu–Goldstone bosons in the chiral limit, they are certainly not perturbative and so should not be represented by a perturbative field ϕ.

The other key feature of his approach was the use of short distance analysis. The connection between axial and trace anomalies and short distance behavior is best illustrated by considering first how equal-time commutators produce contact terms $\sim \delta^4(x-y)$ in ordinary Ward identities.

Given a free massive boson field φ, let $\partial_\mu\varphi$ play the role of a current. Canonically, the divergence of $T\{\varphi\partial_\mu\varphi\}$ is found by writing the T-product in terms of step functions $\theta(\pm x_0)$ and unordered field products, differentiating the step functions

$$\frac{\partial}{\partial x^\mu}\theta(\pm x_0) = \pm\delta(x_0)g_{0\mu} \tag{11}$$

and substituting $\partial^2\varphi = -m^2\varphi$:

$$\partial^\mu T\{\varphi(0)\partial_\mu\varphi(x)\} = [\partial_0\varphi(x), \varphi(0)]\delta(x_0) - m^2 T\{\varphi(0)\varphi(x)\}\,. \tag{12}$$

In this case, the contact term can be found by substituting a canonical commutator:

$$[\partial_0\varphi(x), \varphi(0)]\delta(x_0) = -i\delta^4(x)I\,, \quad I = \text{identity operator}\,. \tag{13}$$

The short-distance method is to note that a term $\sim \delta^4(x)$ can arise only if ∂^μ acts on a singularity $\sim 1/x^3$ at $x \sim 0$. The leading term of the operator product expansion for $T\{\varphi(0)\partial_\mu\varphi(x)\}$ is given by the propagator of the massless theory

$$T\{\varphi(0)\partial_\mu\varphi(x)\} \to \frac{x_\mu}{2\pi^2(x^2 - i\epsilon)^2}I\,, \quad I = \text{identity operator}\,. \tag{14}$$

Substituting $\partial^\mu(x_\mu/x^4) = -2i\pi^2\delta^4(x)$ and $\partial^2\varphi = -m^2\varphi$, we find

$$\partial^\mu T\{\varphi(0)\partial_\mu\varphi(x)\} = -i\delta^4(x)I - m^2 T\{\varphi(0)\varphi(x)\} \tag{15}$$

in agreement with Eqs. (12) and (13).

For the axial anomaly, the problem is to evaluate the quantity

$$S = -\frac{\pi^2}{12}\epsilon^{\mu\nu\alpha\beta}\iint d^4x\, d^4y\, x_\mu y_\nu T\langle\text{vac}|J_\alpha(x)J_\beta(0)\partial^\gamma J_{\gamma5}(y)|\text{vac}\rangle\,, \tag{16}$$

where data for $\pi^0 \to \gamma\gamma$ and approximate $SU(2) \times SU(2)$ symmetry imply $S \simeq +0.5$. The constant S normalizes the contact term in an anomalous

Ward identity of the form

$$\partial_y^\nu \text{“}T\text{”}\langle\text{vac}|J_\alpha(x)J_\beta(0)J_{\nu5}(y)|\text{vac}\rangle$$
$$= \frac{S}{2\pi^2}\epsilon_{\alpha\beta\mu\nu}\partial_x^\mu\partial_y^\nu\delta^4(x)\delta^4(y) + T\langle\text{vac}|J_\alpha(x)J_\beta(0)\partial^\nu J_{\nu5}(y)|\text{vac}\rangle\,. \quad (17)$$

The contact term scales as $1/\{\text{length}\}^{10}$, so it must be generated by a short distance singularity

$$J_\alpha(x)J_\beta(0)J_{\nu5}(y) \sim 1/\{\text{length}\}^9 \quad (18)$$

as *both* x_μ and y_μ tend to zero. (Do not confuse this with the short-distance properties of $J_\alpha J_\beta \partial^\gamma J_{\gamma5}$ in Eq. (16), where the condition dim $\partial^\gamma J_{\gamma5} < 4$ ensures convergence of the integral.)

In Eq. (17), a *single* derivative ∂_y^ν produces *a product of two delta functions*, so it is clear that θ-functions in time *cannot* be used to construct “T”. This example exposes the limitations of canonical field theory very effectively.

In perturbation theory, it has long been known[50] but not often noted that time ordering is part of the renormalization procedure. In general, “T” must be regarded as an operation which depends on the renormalization prescription. The difference between two time-ordering procedures for a given operator product is a set of contact terms at coinciding points. In the case of the triangle diagram coupled to photons, electromagnetic gauge invariance specifies the renormalization procedure completely:

$$\text{“}T\text{”} \to T_{\text{e'mag}}\,. \quad (19)$$

Wilson[17] circumvented the “T” problem by excising a small neighbourhood around lines of coinciding points in the integral (16). Let the region of integration be restricted to the region

$$\mathcal{R} = \{|x_0| > \epsilon, |y_0| > \epsilon', |x_0 - y_0| > \epsilon''\} \quad (20)$$

shown in Fig. 2, so that Eq. (16) becomes

$$S = -\frac{\pi^2}{12}\epsilon^{\mu\nu\alpha\beta}\iint_{\mathcal{R}} d^4x\, d^4y\, x_\mu y_\nu T\langle\text{vac}|J_\alpha(x)J_\beta(0)\partial^\gamma J_{\gamma5}(y)|\text{vac}\rangle + O(\epsilon,\epsilon',\epsilon'')\,. \quad (21)$$

Of course, S does not depend on ϵ, ϵ', or ϵ''. Within $\mathcal{R}$, define

$$\begin{aligned} X_\gamma &= \epsilon^{\mu\nu\alpha\beta}x_\mu y_\nu T\langle\text{vac}|J_\alpha(x)J_\gamma(0)J_{\beta5}(y) + J_\gamma(x)J_\alpha(0)J_{\beta5}(y)|\text{vac}\rangle\,, \\ Y_\gamma &= \epsilon^{\mu\nu\alpha\beta}x_\mu y_\nu T\langle\text{vac}|J_\alpha(x)J_\beta(0)J_{\gamma5}(y) + J_\alpha(x)J_\gamma(0)J_{\beta5}(y)|\text{vac}\rangle\,, \end{aligned} \quad (22)$$

where now time ordering with θ-functions is allowed because $\mathcal{R}$ excludes coinciding points. This also means that derivatives commute with the T-operation, so we can obtain

$$S = -\frac{\pi^2}{12} \iint_{\mathcal{R}} d^4x\, d^4y \big(\partial_x^\gamma X_\gamma + \partial_y^\gamma Y_\gamma\big) + O(\epsilon, \epsilon', \epsilon'') \tag{23}$$

by using current conservation $\partial^\gamma J_\gamma = 0$, translation invariance of $|\text{vac}\rangle$ and symmetry $x \leftrightarrow y$ of the integral to $O(\epsilon, \epsilon', \epsilon'')$. If Σ is the surface in 8-dimensional space which bounds $\mathcal{R}$, we have

$$S = -\frac{\pi^2}{12} \int_\Sigma d\vec{\Sigma} \cdot \vec{Z} + O(\epsilon, \epsilon', \epsilon'')\,, \tag{24}$$

where $\vec{Z} = (X_\gamma, Y_\gamma)$ is an 8-dimensional vector formed from the components of X_γ and Y_γ.

So S is given by the result of taking ϵ, ϵ' and ϵ'' to zero in Eq. (24). Since the current operators commute at space-like separations, their products at short distances are all that we need. If we consider (say) $\epsilon \to 0$ and exclude the x, $y \sim 0$ neighbourhood where the axes in Fig. 2 meet, we have $x \sim 0$ for fixed y, which means expanding in $J_\alpha(x)J_\beta(0)$ to produce an equal-time commutator. There could be three commutators in principle, one for each axis, but explicit checks confirm the conclusion[51] that they all vanish. Therefore S is entirely determined by the leading VVA short-distance singularity

$$T\{J_\alpha(x)J_\beta(0)J_{\gamma5}(y)\} \sim G_{\alpha\beta\gamma}(x,y)I\,, \quad x,\, y \sim 0\,, \tag{25}$$

so it can be calculated if the three-point function $G_{\alpha\beta\gamma}$ is known.

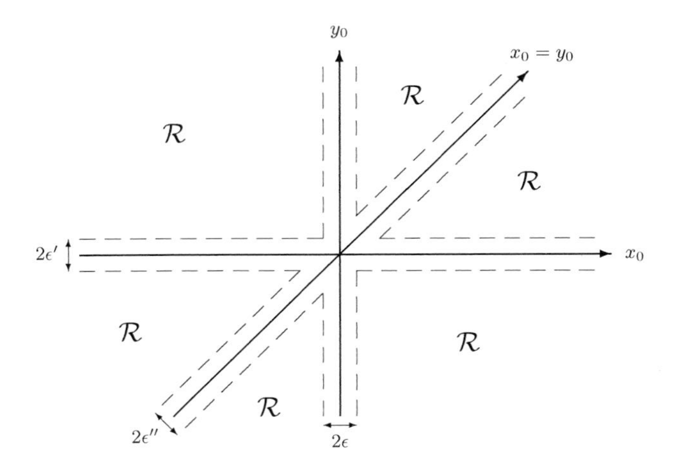

Fig. 2. Integration region $\mathcal{R}$ defined by Eq. (20).

At this point, I tried the same analysis for the trace anomaly. Let the amplitude for photons to couple to the hadronic energy–momentum tensor be

$$\langle\gamma(\epsilon_1, k_1)\gamma(\epsilon_2, k_2)|\theta^\mu_\mu(0)|\mathrm{vac}\rangle = (\epsilon_1\cdot\epsilon_2 k_1\cdot k_2 - \epsilon_1\cdot k_2\epsilon_2\cdot k_1)F\big((k_1+k_2)^2\big)\,. \tag{26}$$

As in Eq. (16) for S, the trace anomaly corresponds to the low-energy limit $k_1,\, k_2 \sim 0$:

$$F(0) = -\frac{\pi\alpha}{3}\iint d^4x\, d^4y\, x\cdot y\langle\mathrm{vac}|J^\alpha(x)J_\alpha(0)\theta^\mu_\mu(y)|\mathrm{vac}\rangle\,. \tag{27}$$

The aim was to substitute the formula for the divergence of the conformal current

$$\partial^\mu_y\{(2y_\lambda y^\nu - \delta^\nu_\lambda y^2)\theta_{\mu\nu}(y)\} = 2y_\lambda\theta^\mu_\mu(y) \tag{28}$$

and integrate by parts. To my surprise, I found that it was not necessary to exclude coinciding points as in Fig. 2. Instead, I found that an answer could be found directly by restricting just the x integration to $|x_0| > \eta$ for small $\eta > 0$ to keep the $x, y \sim 0$ singularity

$$T\{J_\alpha(x)J_\beta(0)\theta_{\mu\nu}(y)\} \sim K_{\alpha\beta\mu\nu}(x,y)I \tag{29}$$

under control. Then integration by parts with respect to y produced known equal-time commutators, so the y integral could be done, with the result

$$F(0) = -\frac{i\pi\alpha}{6}\int_{|x_0|>\eta} d^4x\,\partial^\nu_x\{x^2x_\nu T\langle\mathrm{vac}|J^\alpha(x)J_\alpha(0)|\mathrm{vac}\rangle\} + O(\eta)\,. \tag{30}$$

We have $J_\alpha J_\beta \sim R/x^6$ at short distances, where R is the asymptotic Drell–Yan ratio

$$R = \big\{(\sigma(e^+e^- \to \mathrm{hadrons})/(\sigma(e^+e^- \to \mu^+\mu^-)\big\}_{\mathrm{energy}\to\infty} \tag{31}$$

so the x integral can also be done, yielding an exact result:

$$F(0) = 2R\alpha/3\pi\,. \tag{32}$$

In effect, an anomalous term[a] $(R\alpha/6\pi)F_{\mu\nu}F^{\mu\nu}$ is induced in the trace of the energy–momentum tensor by electromagnetism.[18] The same result was found independently by Mike Chanowitz and John Ellis[52] via a momentum-space analysis. It was the immediate precursor of the gluonic trace anomaly $\beta(\alpha_s)/(4\alpha_s)G^a_{\mu\nu}G^{a\mu\nu}$ found a few years later.[53]

[a] The extrapolation in $(k_1+k_2)^2$ from zero to m^2_σ used to estimate $F_\sigma g_{\sigma\gamma\gamma}$ from $F(0)$ has had to be modified,[35] because π, K loop diagrams compete with the σ-pole amplitude.

The unexpected feature of the analysis leading to Eq. (32) was that, although the 3-point singularity $K_{\alpha\beta\mu\nu}$ is responsible for the presence of the trace anomaly, its full functional form is not needed: only the subregion $x - y \ll x,\, y$ *within* the $x,\, y \sim 0$ region is needed. In Fig. 2, this subregion connects the central area $x,\, y \sim 0$ to other $x \sim y$ regions along the diagonal axis $x_0 = y_0$.

That led me to consider *nested* operator product expansions, where an expansion such as

$$T\{A(x)B(0)\} \sim \sum_m \mathcal{C}_m(x) O'_m(0) \quad \text{for } x \sim 0 \tag{33}$$

is substituted into a larger expansion, e.g.

$$T\{A(x)B(0)C(y)\} \sim \sum_n f_n(x,y) O_n(0) \quad \text{for} \quad x, y \sim 0\,. \tag{34}$$

This is legitimate provided that y is independent of the limit $x \to 0$, i.e. $x \ll y$. Then a subsequent limit $y \to 0$ can be taken:

$$T\{O'_m(0)C(y)\} \sim \sum_n \mathcal{C}_{mn}(y) O_n(0)\,. \tag{35}$$

The result is a set of consistency conditions[18]

$$f_n(x,y) \sim \sum_m \mathcal{C}_m(x) \mathcal{C}_{mn}(y)\,. \tag{36}$$

The idea works at short distances (and not on other parts of light cones) provided the limits are *nested*. For the example above, fix $\hat{x}$ and $\hat{y}$ in

$$x = \rho_1 \rho_2 \hat{x} \quad \text{and} \quad y = \rho_2 \hat{y} \tag{37}$$

and take the limits $\rho_1 \to 0$ and $\rho_2 \to 0$ independently. This is the position-space version of Weinberg's limiting procedure[54] used to classify the asymptotic behavior of amplitudes and hence justify power counting methods for renormalization.

An obvious next step was to apply this procedure to the short-distance VVA function $G_{\alpha\beta\gamma}$ of Eq. (25). Let

$$u = x^2 - i\epsilon\,, \quad v = y^2 - i\epsilon\,, \quad w = (x-y)^2 - i\epsilon\,. \tag{38}$$

The relevant two-point expansions are

$$\begin{aligned} T\{J_\alpha(x)J_\beta(0)\} &\sim R(g_{\alpha\beta}x^2 - 2x_\alpha x_\beta)I/(\pi u)^4 + K\epsilon_{\alpha\beta\lambda\mu}x^\lambda J^\mu_5(0)/(3\pi^2 u^2)\,, \\ T\{J^\mu_5(0)J_{\gamma 5}(y)\} &\sim R'(\delta^\mu_\gamma y^2 - 2y^\mu y_\gamma)I/(\pi v)^4\,, \end{aligned} \tag{39}$$

where R' is the isovector part of R, and K is measurable in polarized deep-inelastic electroproduction or in $e^+ + e^- \to \mu^+ + \mu^- + \pi^0$. The result

$$G_{\alpha\beta\gamma}(x,y) \underset{x \ll y}{\longrightarrow} \{K\epsilon_{\alpha\beta\lambda\mu}x^\lambda/(3\pi^2 u^2)\}R'(\delta^\mu_\gamma y^2 - 2y^\mu y_\gamma)I/(\pi v)^4 \tag{40}$$

said something about the normalization of $G_{\alpha\beta\gamma}$, but without a formula valid for the whole x, $y \sim 0$ region, the calculation of S could not be completed.

In the meantime, I was checking products of $\theta_{\mu\nu}$ with other currents to see if the absence of the soft trace at short distances would imply asymptotic[b] conformal invariance, as indicated by Eq. (28). Satisfied that it did, I required conformal invariance for $G_{\alpha\beta\gamma}$, found that it had to be proportional to the triangle diagram, and then found that this result had already been published by Schreier.[55]

So the evaluation of the VVA singular function was complete:

$$G_{\alpha\beta\gamma}(x,y) = \frac{KR'}{12\pi^6 u^2 v^2 w^2}\mathrm{Tr}\{\gamma_\alpha\gamma\cdot x\gamma_\beta\gamma\cdot y\gamma_\gamma(\gamma\cdot x - \gamma\cdot y)\gamma_5\}\,. \tag{41}$$

When $G_{\alpha\beta\gamma}$ is substituted in Eq. (24), the equal-time commutator regions give no contribution (as before), so as long as (say) ϵ' is held fixed, the limits $\epsilon \to 0$ and $\epsilon'' \to 0$ can be taken without intruding on the short-distance region. Thus[c]

$$S = \frac{\pi^2}{12}\int d^3y \int d^4x\{\tilde{Y}_0(y_0 = \epsilon') - \tilde{Y}_0(y_0 = -\epsilon')\} + O(\epsilon')\,, \tag{42}$$

where $\tilde{Y}_\gamma$ is the $x, y \sim 0$ part of Y_γ:

$$\begin{aligned}\tilde{Y}_\gamma &= \epsilon^{\mu\nu\alpha\beta}x_\mu y_\nu\{G_{\alpha\beta\gamma}(x,y) + G_{\alpha\gamma\beta}(x,y)\}\\ &= -\frac{4KR'}{3\pi^6 u^2 v^2 w^2}y_\gamma\{x^2y^2 - (x\cdot y)^2\}\,.\end{aligned} \tag{43}$$

Do the x-integral

$$\int d^4x\{x^2y^2 - (x\cdot y)^2\}/(uw)^2 = -3\pi^2 i/2 \tag{44}$$

and then the y-integral

$$\int d^3y/v^2 = -\pi^2 i/|y_0| \tag{45}$$

[b]This has *nothing* to do with the properties of the vacuum state. As noted at the beginning of Sec. 3, $|\mathrm{vac}\rangle$ breaks scale and hence conformal invariance very strongly. This may be due to explicit symmetry breaking or to the symmetry being realized in the Nambu–Goldstone mode.

[c]These details, taken from a letter I wrote to Fritzsch and Gell-Mann at the time,[56] should have been part of Ref. 10 of my paper[18] but it was never finished.

to obtain the desired formula[18]

$$3S = KR' . \tag{46}$$

It relates the low-energy amplitude S to high-energy amplitudes R' and K.

Being anxious to avoid model dependence, I allowed for the possibility that the electromagnetic current is not a pure $SU(3)$ octet,

$$4R' \leqslant 3R . \tag{47}$$

At that time, we did not know R, R' or K. In particular, data showing scaling behavior for $e^+ + e^- \to$ hadrons was not available until the 1974 London Conference, a year after asymptotic freedom. The value of S could well be exactly 0.5, so Adler's nonrenormalization theorem[25,57] for S suggested a theory with three species of quark, but I could not see how to deal with colored electromagnetic currents or paraquark operators.

So when I received by return mail a letter from Gell-Mann proposing colored fractional quarks with color-neutral currents, it seemed to me that this clarified matters from the point of view of Adler's theorem, but I felt (for reasons discussed above) that having free quarks on the light cone was going too far. However, sometimes an oversimplification can lead to correct answers — in this case, QCD[58] and asymptotic freedom.[32,33]

Initially, the QCD proposal looked good as a model of constituent quarks, but not for current quarks and the $\pi^0 \to \gamma\gamma$ analysis. Having found the electromagnetic trace anomaly, we knew already that θ^μ_μ would have anomalous gluonic terms proportional to $G^a_{\mu\nu}G^{a\mu\nu}$ which would break scale and conformal invariance at short distances.

What asymptotic freedom did was to turn QCD into a good theory of current quarks as well as constituent quarks. The breaking of scale invariance at short distances was minimal, being associated with operators which are not conserved exactly or partially. The analysis of $\pi^0 \to \gamma\gamma$ can be still be carried through since all of the equations remain valid: they can be derived by using asymptotic freedom instead of asymptotic conformal invariance. The results for three colors are

$$R = 2 , \quad R' = 1.5 , \quad K = 1 , \quad \text{and} \quad S = 0.5 , \tag{48}$$

where the nonrenormalization theorem for S is *not* used. All of this goes through *without* treating pions perturbatively.

The method of nested operator product expansions is now not needed for the $\pi^0 \to \gamma\gamma$ derivation, but it is generally valid in renormalized field theory. As a result, coupling constant dependence of the form

$$3S = K(g)R'(g) \tag{49}$$

can be investigated.[59] This program has been extensively pursued by Andrei Kataev, Stan Brodsky and their collaborators.[60]

In retrospect, there came a time when abstracting physics had to give way to guessing the correct model. I remember that time well.

References

1. M. Gell-Mann, *Phys. Lett.* **8**, 214 (1964).
2. G. Zweig, CERN Reports 8182/TH.401 and 8419/TH.412 (1964).
3. A. Petermann, *Nucl. Phys.* **63**, 349 (1965).
4. M. Gell-Mann, *Phys. Rev.* **125**, 1067 (1962).
5. M. Gell-Mann, Caltech report CTSL-20 (1961), reprinted in *The Eightfold Way*, ed. M. Gell-Mann and Y. Ne'eman (Benjamin, New York, 1964).
6. S. Okubo, *Prog. Theor. Phys.* **27**, 949 (1962).
7. Y. Ne'eman, *Nucl. Phys.* **26**, 222 (1961).
8. K. G. Wilson, *Phys. Rev.* **179**, 1499 (1969).
9. O. W. Greenberg, *Phys. Rev. Lett.* **13**, 598 (1964).
10. H. S. Green, *Phys. Rev.* **90**, 270 (1953).
11. B. V. Struminsky, Joint Institute of Nuclear Research publication P-1939, Dubna, January 1965; translation http://arxiv.org/abs/0904.0343.
12. M. Y. Han and Y. Nambu, *Phys. Rev. B* **139**, 1006 (1965).
13. Y. Miyamoto, *Commemoration Issue for the Thirtieth Anniversary of the Meson Theory by Dr. H. Yukawa*, Progress of Theoretical Physics (Supplement) (1965), p. 187.
14. D. B. Lichtenberg, *Unitary Symmetry and Elementary Particles* (Academic Press, New York, 1970), pp. 227–228.
15. M. Gell-Mann, *Acta Phys. Austriaca, Suppl.* **IX**, 733 (1972).
16. W. A. Bardeen, H. Fritzsch and M. Gell-Mann, in *Scale and Conformal Symmetry in Hadron Physics*, Frascati, May 1972, ed. R. Gatto (Wiley, New York, 1973), p. 139.
17. K. G. Wilson, *Phys. Rev.* **179**, 1499 (1969).
18. R. J. Crewther, *Phys. Rev. Lett.* **28**, 1421 (1972).
19. J. D. Bjorken, *Phys. Rev.* **179**, 1547 (1969).
20. R. E. Cutkosky, *Phys. Rev.* **131**, 1888 (1963).
21. F. J. Belinfante, *Physica* **7**, 449 (1940).
22. K. G. Wilson, *Phys. Rev. D* **2**, 1478 (1970).
23. M. Gell-Mann and F. E. Low, *Phys. Rev.* **95**, 1300 (1954).
24. J. S. Bell and R. Jackiw, *Nuovo Cimento A* **60**, 47 (1969).
25. S. L. Adler, *Phys. Rev.* **177**, 2426 (1969); *Lectures on Elementary Particles and Quantum Field Theory*, Vol. 1, Brandeis University Summer Institute (MIT Press, Cambridge, Mass., 1970).
26. Y. Frishman, *Phys. Rev. Lett.* **25**, 966 (1970); *Broken Scale Invariance and the Light Cone, Vol. 2, Coral Gables Conference*, January 1971 (Gordon and Breach, New York, 1971), p. 61; *Ann. Phys. (N.Y.)* **66**, 373 (1971).

27. R. Brandt and G. Preparata, *Nucl. Phys. B* **27**, 541 (1971); *Broken Scale Invariance and the Light Cone, Vol. 2, Coral Gables Conference*, January 1971 (Gordon and Breach, New York, 1971), p. 43.
28. H. Fritzsch and M. Gell-Mann, *Broken Scale Invariance and the Light Cone, Vol. 2, Coral Gables Conference*, January 1971 (Gordon and Breach, New York, 1971), p. 1.
29. K. Wilson, *Broken Scale Invariance and the Light Cone, Vol. 2, Coral Gables Conference*, January 1971 (Gordon and Breach, New York, 1971), p. 122.
30. M. Gell-Mann, *1969 Hawaii Topical Conference on Particle Physics* (Western Periodicals Co., Los Angeles, 1970).
31. S. L. Glashow, *Hadrons and their Interactions*, Erice 1967, ed. A. Zichichi (Academic Press, New York, 1968), p. 83.
32. H. D. Politzer, *Phys. Rev. Lett.* **30**, 1346 (1973).
33. D. J. Gross and F. Wilczek, *Phys. Rev. Lett.* **30**, 1343 (1973).
34. R. J. Crewther, Chiral properties of quantum chromodynamics, in *Field Theoretical Methods in Particle Physics*, NATO Advanced Study Institutes Series, Vol. 55B, Kaiserslautern, 1979, ed. W. Rühl (Plenum, New York, 1980), p. 529, www.physics.adelaide.edu.au/theory/staff/crewther/chiral.pdf.
35. R. J. Crewther and L. C. Tunstall, arXiv:1203.1321; *Mod. Phys. Lett. A* **28**, 1360010 (2013); arXiv:1312.3319; *EPJ Web Conf.* **73**, 03006 (2014).
36. I. Caprini, G. Colangelo and H. Leutwyler, *Phys. Rev. Lett.* **96**, 132001 (2006).
37. Y. Fujii, *Nature Phys. Sci.* **234**, 5 (1971); *Ann. Phys. (N.Y.)* **69**, 494 (1972).
38. C. H. Brans and R. H. Dicke, *Phys. Rev.* **124**, 925 (1961).
39. S. Coleman and E. Weinberg, *Phys. Rev. D* **7**, 1888 (1973).
40. J. Ellis, *Nucl. Phys. B* **22**, 478 (1970); *Broken Scale Invariance and the Light Cone, Vol. 2, Coral Gables Conference*, January 1971 (Gordon and Breach, New York, 1971), p. 77.
41. R. J. Crewther, *Phys. Lett. B* **33**, 305 (1970); *Broken Scale Invariance and the Light Cone, Vol. 2, Coral Gables Conference*, January 1971 (Gordon and Breach, New York, 1971), p. 136.
42. M. Gell-Mann and A. Pais, *Proceedings of the 1954 Glashow Conference on Nuclear and Meson Physics*, eds. E. H. Bellamy and R. G. Moorhouse (Pergamon Press, London, New York, 1955).
43. T. Goto, O. Hara and S. Ishida, *Prog. Theor. Phys.* **43**, 849 (1970), and references therein to articles in Japanese.
44. C. H. Llewellyn Smith, *Ann. Phys. (N.Y.)* **53**, 521 (1969); *Phys. Rev. D* **1**, 3194 (1970).
45. R. P. Feynman, M. Kislinger and F. Ravndal, *Phys. Rev. D* **3**, 2706 (1971).
46. J. C. Pati and C. H. Woo, *Phys. Rev. D* **3**, 2920 (1971).
47. K. Miura and T. Minamikawa, *Prog. Theor. Phys.* **38**, 954 (1967).
48. R. L. Kingsley, *Phys. Lett. B* **40**, 387 (1972).
49. J. Schwinger, *Phys. Rev.* **82**, 664 (1951).
50. E. C. G. Stueckelberg and A. Petermann, *Helv. Phys. Acta* **26**, 499 (1953).
51. D. G. Sutherland, *Phys. Lett.* **23**, 384 (1966).
52. M. S. Chanowitz and J. Ellis, *Phys. Lett. B* **40**, 397 (1972).

53. S. L. Adler, J. C. Collins and A. Duncan, *Phys. Rev. D* **15**, 1712 (1977); P. Minkowski, Berne PRINT-76-0813, September 1976; N. K. Nielsen, *Nucl. Phys. B* **120**, 212 (1977); J. C. Collins, A. Duncan and S. D. Joglekar, *Phys. Rev. D* **16**, 438 (1977).
54. S. Weinberg, *Phys. Rev.* **118**, 838 (1960).
55. E. J. Schreier, *Phys. Rev. D* **3**, 980 (1971).
56. T. Y. Cao, *From Current Algebra to Quantum Chromodynamics: A Case for Structural Realism* (Cambridge University Press, Cambridge, UK, 2010).
57. S. L. Adler and W. A. Bardeen, *Phys. Rev.* **182**, 1517 (1969).
58. H. Fritzsch and M. Gell-Mann, *Proc. XVI Int. Conf. on High Energy Physics*, Vol. 2, Chicago, 1972, p. 135.
59. S. L. Adler, C. G. Callan, D. J. Gross and R. Jackiw, *Phys. Rev. D* **6**, 2982 (1972).
60. D. J. Broadhurst and A. Kataev, *Phys. Lett. B* **315**, 179 (1993); S. J. Brodsky, G. T. Gabadadze, A. L. Kataev, H. J. Lu, *Phys. Lett. B* **372**, 133 (1996); A. Kataev, *J. High Energy Phys.* **1402**, 092 (2014); S. J. Brodsky, M. Mojaza and X. G. Wu, *Phys. Rev. D* **89**, 014027 (2014).

Lessons from Supersymmetry: "Instead-of-Confinement" Mechanism

M. Shifman[*,‡] and A. Yung[*,†]

[*] *William I. Fine Theoretical Physics Institute, University of Minnesota, Minneapolis, MN 55455, USA*
[†] *Petersburg Nuclear Physics Institute, Gatchina, St. Petersburg, 188300, Russia*
[‡] *shifman@umn.edu*

We review physical scenarios in different vacua of $\mathcal{N} = 2$ supersymmetric QCD deformed by the mass term μ for the adjoint matter. This deformation breaks supersymmetry down to $\mathcal{N} = 1$ and, at large μ, the theory flows to $\mathcal{N} = 1$ QCD. We focus on dynamical scenarios which can serve as prototypes of what we observe in real-world QCD. The so-called $r = N$ vacuum is especially promising in this perspective. In this vacuum an "instead-of-confinement" phase was identified previously, which is qualitatively close to the conventional QCD confinement: the quarks and gauge bosons screened at weak coupling, at strong coupling evolve into monopole–antimonopole pairs confined by non-Abelian strings. We review genesis of this picture.

1. Introduction

Fifty years ago quarks introduced by Gell-Mann and Zweig,[1,2] which shortly after became three-colored[3,4] opened the gate to modern high-energy physics. Discovery of asymptotic freedom in Yang–Mills theories[5,6] and creation of quantum chromodynamics (QCD)[4–6] completed this process. Gradually it became clear that a peculiar phenomenon in QCD — quark (or, more generally, color) confinement presented a novel dynamical phase in field theory with which one had never encountered before. This phase is inherent to Yang–Mills theories at strong coupling. A possible physical explanation of the confinement phase — the so-called dual Meissner effect — was suggested in the mid-1970s[7–10] but nobody managed to demonstrate it analytically under controllable conditions until much later.

The advent of supersymmetry changed that. In the mid-1990s Seiberg and Witten considered $\mathcal{N} = 2$ super-Yang–Mills theory with the $SU(2)$ gauge group. In this model a vacuum manifold exists, i.e. a (complex) flat direction. Seiberg and Witten identified[11,12] two points on this manifold in which a monopole or a dyon become massless. A crucial fact was that supersymmetry allowed them to use analytical extrapolations to give a precise meaning to monopoles and dyons in non-Abelian field theory. Then they showed that a weak deformation of the above model forces the monopoles/dyons to condense, thus triggering the dual Meissner effect in super-Yang–Mills and ensuing color confinement. The Seiberg–Witten vacua[11,12] are referred to as the monopole vacua.

The Seiberg–Witten solution is based on a cascade gauge symmetry breaking. At a high energy scale the non-Abelian gauge group is broken down to $U(1)$ (or a generic Abelian subgroup in more general models) by condensation of the adjoint scalar field $\mathcal{A}$. An effective low-energy theory near the monopole vacuum includes the Abelian gauge fields and magnetically charged matter represented by light monopoles or dyons.

The latter condense upon a small $\mathcal{N} = 2$ breaking deformation $\mu\mathcal{A}^2$ at a much lower scale μ. Their vacuum expectation values (VEVs) are of the order[a] of $\mu\Lambda_2$. Simultaneously, formation of confining color-electric flux tubes (strings) occurs. Their tension is minute being proportional to $\mu\Lambda_2$.

The Seiberg–Witten solution gave a strong impetus for studies of Yang–Mills dynamics at strong coupling. Another inspiration came from the so-called Seiberg duality.[13,14] Pairs of completely different $\mathcal{N} = 1$ super-Yang–Mills theories (supersymmetric QCD, or SQCD for short) were identified which were related as follows: one theory from the pair in the ultraviolet domain (UV), flows to the second theory from the pair in the infrared (IR) limit. Of course, this can happen only if the second theory is at strong coupling. Formally the Seiberg duality is an extension of the notion of the 't Hooft anomaly matching.[15] Its roots are deeper, however, and shortly after its discovery were found to ascend to string theory.

Before the Seiberg–Witten solution theorists were aware of three basic phases in gauge theories: the Higgs, Coulomb and confining regimes. It turned out that the phase structure of super-Yang–Mills theory is much richer. For instance, contrived matter sectors were shown[16] to lead to a number of exotic phases unheard of previously.

The Seiberg duality refers to $\mathcal{N} = 1$ massless SQCD. Several years ago we started a project of detailing Seiberg's duality in various vacua by using the exact Seiberg–Witten solution as an intermediate step in our derivations, see e.g. Refs. 17–20. Detailing means that we isolated discrete vacua with specific properties (to be discussed below) by adjusting quark mass parameters which are present in our $\mathcal{N} = 2$ deformed theory. To pass from $\mathcal{N} = 2$ to

[a] We introduce a shorthand notation for the dynamical scale parameters Λ_2 and Λ_1, for the scale parameters $\Lambda_{\mathcal{N}=2}$ and $\Lambda_{\mathcal{N}=1}$ appearing in $\mathcal{N} = 2$ and $\mathcal{N} = 1$ theories, respectively.

$\mathcal{N} = 1$ we had to consider large rather than small values of the parameter μ in front of the deformation term. In these studies we discovered yet a novel phase of super-Yang–Mills which we called "instead of confinement."[21] A review of this phase is one of the main goals of this article.

Another goal — discussion of confining strings which are as close as possible to the conjectured QCD strings — also requires the large-μ limit. Indeed, the Seiberg–Witten confinement at small μ is essentially Abelian,[22–25] with strings of the Abrikosov type. We wanted to construct non-Abelian strings, i.e. those with non-Abelian moduli on their worldsheet. At large μ adjoint scalars of the Seiberg–Witten model decouple and we then approach $\mathcal{N} = 1$ limit in which the confinement mechanism is expected to be non-Abelian. Moreover, it is believed that confinement in real-world QCD is also non-Abelian.

Thus, there were two reasons why we wanted, starting from the small-μ limit to pass to large values of μ (or, equivalently, the $\mathcal{N} = 1$ limit). We hoped that in passing from $\mathcal{N} = 2$ SQCD via the decoupling of the adjoint scalars, we would be able to see how the Abelian Seiberg–Witten confinement transforms itself into a regime with non-Abelian confinement.

This program turned out to be challenging. The effective (dual) theory which describes low-energy physics of μ-deformed $\mathcal{N} = 2$ SQCD below the scale of the adjoint VEVs is in fact the Abelian gauge theory with light monopoles. It is infrared-free and stays at weak coupling as long as VEVs of the monopoles (not to be confused with the adjoint VEVs) are small enough. As was mentioned above, these VEVs are proportional to $\mu\Lambda_2$; thus, at small μ the condition for weak coupling is met. However, moving toward the desired large-μ limit breaks this condition. The effective theory goes into a strong coupling regime. The analytic control is lost. This was a genuine obstacle.

Recent breakthrough developments allowed us to resolve this problem. First, we developed a benchmark model which was slightly different from that of Seiberg and Witten. In our benchmark model the gauge group is $U(N)$ (rather than $SU(N)$) and the number of quark flavors is N_f where $(N+1) < N_f < \frac{3}{2}N$. In this version of $\mathcal{N} = 2$ SQCD with the μ deformation switched on one can identify a number of the so-called r vacua which are characterized by a parameter r, the number of the condensed (s)quarks in the classical domain of large and generic quark mass parameters m_A (here $A = 1, \ldots, N_f$). Clearly, r cannot exceed N, the rank of the gauge group. The monopole vacua (there are N of them) corresponds to $r = 0$. Quarks do not condense in the monopole vacua.

The key observation is as follows.[19,20] In this model there is a subset of r vacua (to be referred to as *zero vacua*) which can be found at $r < (N_f - N)$. In these vacua the gaugino condensate is parametrically small provided the quark mass parameters are small too. This subset includes a part of the monopole vacua. The smallness of the gaugino condensate ensures that quantum effects are small in the zero vacua, hence physics can be described in terms of weakly coupled dual theory.

In Refs. 19 and 20 we demonstrated that the Seiberg–Witten confinement, present in the zero vacua at small μ, disappears at large μ. This happens because the scale of the gaugino condensate controls the confinement radius (tensions of the confining strings) in a certain sector of the dual theory. As we increase μ, confinement becomes weaker and weaker in this sector, and eventually quarks are "liberated." Effective dual theory has $\mathrm{U}(N_f - N)$ gauge group. This theory is infrared-free and stays at weak coupling at low energies.

It appears to be in the mixed Coulomb-Higgs phase for quarks. Namely, r quarks condense, while the $\mathrm{U}(N_f - N - r)$ subgroup is in the Coulomb phase, see Refs. 19 and 20 for a more rigorous and detailed discussion. Therefore, the zero vacua (in particular, the monopole zero vacua of strongly coupled μ-deformed SQCD) are not good prototypes for physics in real-world QCD.

Nevertheless, μ-deformed SQCD is a rich theory with a variety of r vacua with different infrared behaviors. We discovered that certain other vacua in this theory are more promising in providing us with a prototype for real-world QCD dynamics. This will be described below.

As was mentioned above, the zero vacua support a weakly coupled dual description at large μ due to the smallness of the gaugino condensate. Another exceptional vacuum is the $r = N$ vacuum in which the maximal number of quarks condense (in the weakly coupled domain of the large quark masses). In this vacuum the gaugino condensate is *identically zero*. This vacuum also has a weakly coupled — the so-called r-dual — description, in the large-μ limit.[18,19,21]

In this article we review infrared dynamics in μ-deformed SQCD focusing mostly on the $r = N$ vacuum. It is just this vacuum which supports a new phase, namely, the "instead-of-confinement" phase.[17–19,21] In this phase quarks and gauge bosons screened at weak coupling evolve at strong coupling into monopole–antimonopole pairs confined by non-Abelian strings.

These monopole–antimonopole stringy mesons have "correct" (adjoint or singlet) quantum numbers with respect to the global group, in much the same way as mesons in real-world QCD. Moreover, they lie on the Regge trajectories. Thus, this phase is qualitatively rather similar to what we observe in real-world QCD. The role of QCD constituent quarks is played by monopoles.

The paper is organized as follows. In Sec. 2 we review the $r = N$ vacuum at weak coupling, i.e. at large ξ (the parameter ξ is defined in Eq. (4)). In Sec. 3 we turn to r-duality in this vacuum and the "instead-of-confinement" mechanism in the small-μ limit. In Sec. 4 we discuss the large μ-limit. Section 5 briefly summarizes our results and conclusions.

2. The $r = N$ Vacuum at Large ξ

Our benchmark model reduces to $\mathcal{N} = 2$ SQCD with the $U(N)$ gauge group in the absence of μ-deformation. The matter sector consists of N_f massive quark hypermultiplets. We assume that $N_f > N + 1$ but $N_f < \frac{3}{2}N$. The latter inequality ensures that the dual theory is infrared free.

This theory is described in detail in Refs. 26 and 27, see also the reviews in Ref. 28. The field content is as follows. The $\mathcal{N} = 2$ vector multiplet consists of the U(1) gauge field A_μ and the SU(N) gauge field A^a_μ, where $a = 1, \ldots, N^2 - 1$, as well as their Weyl fermion superpartners plus complex scalar fields a, and a^a and their Weyl superpartners, respectively. These complex scalar fields present the bosonic sector of the *adjoint scalars*.

The matter sector of the U(N) theory contains N_f quark multiplets which consist of the complex scalar fields q^{kA} and $\tilde{q}_{Ak}$ (squarks) and their fermion superpartners — all in the fundamental representation of the SU(N) gauge group. Here $k = 1, \ldots, N$ is the color index while A is the flavor index, $A = 1, \ldots, N_f$.

In addition, as was mentioned, we add the mass term μ for the adjoint scalar superfield breaking $\mathcal{N} = 2$ supersymmetry down to $\mathcal{N} = 1$. This deformation term

$$\mathcal{W}_{\text{def}} = \mu \,\text{Tr}\, \Phi^2 \,, \quad \Phi \equiv \frac{1}{2}\mathcal{A} + T^a \mathcal{A}^a \tag{1}$$

does not break $\mathcal{N} = 2$ supersymmetry in the small-μ limit, see Refs. 23, 25, 26, however, at large μ this theory flows to $\mathcal{N} = 1$ SQCD. The fields $\mathcal{A}$ and $\mathcal{A}^a$ in Eq. (1) are chiral superfields, the $\mathcal{N} = 2$ superpartners of the U(1) and SU(N) gauge bosons.

In this theory we can find a set of r vacua, where r is the number of condensed (s)quarks in the classical domain of large generic quark masses m_A ($A = 1, \ldots, N_f$, and $r \leq N$). In this review we will focus on the $r = N$ vacua. Dynamical scenarios in the $r < N$ vacua are considered in Refs. 19 and 20.

These vacua have the maximal possible number of condensed quarks, namely, $r = N$. Moreover, the gauge group $U(N)$ is completely Higgsed in these vacua, and, as a result, they support non-Abelian strings.[26,29–31] These strings result in confinement of monopoles.

First, we will assume that μ is small, much smaller than the quark masses

$$|\mu| \ll |m_A|, \quad A = 1, \ldots, N_f. \tag{2}$$

In the quasiclassical domain of large quark masses the squark fields develop VEVs triggered by the deformation parameter μ,

$$\langle q^{kA} \rangle = \langle \bar{\tilde{q}}^{kA} \rangle = \frac{1}{\sqrt{2}} \begin{pmatrix} \sqrt{\xi_1} & \cdots & 0 & 0 & \cdots & 0 \\ \cdots & \cdots & \cdots & \cdots & \cdots & \cdots \\ 0 & \cdots & \sqrt{\xi_N} & 0 & \cdots & 0 \end{pmatrix}, \tag{3}$$
$$k = 1, \ldots, N, \quad A = 1, \ldots, N_f,$$

where we present the squark fields as matrices in the color (k) and flavor (A) indices, while new parameters ξ are given (in the quasiclassical approximation) by

$$\xi_P \approx 2\mu m_P, \quad P = 1, \ldots, N. \tag{4}$$

The quark condensate (3) implies the spontaneous breaking of both gauge and flavor symmetries. A diagonal global $SU(N)$ combining the gauge $SU(N)$ and an $SU(N)$ subgroup of the flavor $SU(N_f)$ survives in the limit of equal (or almost equal) quark masses. This is the color-flavor locking.

Thus, the unbroken global symmetry we are left with is

$$SU(N)_{C+F} \times SU(\tilde{N}) \times U(1), \quad \tilde{N} \equiv N_f - N. \tag{5}$$

Here $SU(N)_{C+F}$ is an unbroken global color-flavor rotation, which involves only the first N flavors, while the $SU(\tilde{N})$ factor refers to the flavor rotation of the remaining $\tilde{N}$ quarks.

The presence of the global $SU(N)_{C+F}$ symmetry is the reason for formation of the non-Abelian strings.[26,27,29–31] At small μ these strings are BPS-saturated,[23,25] and their tensions are determined[27] by the parameters ξ_P introduced in (4),

$$T_P = 2\pi|\xi_P|, \quad P = 1, \ldots, N. \tag{6}$$

The above non-Abelian strings, with non-Abelian moduli in the coset

$$SU(N)_{C+F}/SU(N-1) \times U(1)$$

on their worldsheet, confine monopoles. In fact, in the U(N) theories confined elementary monopoles are junctions of two "neighboring" strings with labels P and $(P+1)$, see Ref. 28 for a more detailed review.

Now, let us briefly discuss the elementary excitation spectrum in the bulk. Since both U(1) and SU(N) gauge groups are broken by the squark condensation, all gauge bosons become massive. To the leading order in μ, $\mathcal{N}=2$ supersymmetry is not broken. In fact, with nonvanishing ξ_P's (see Eq. (4)), both the quarks and adjoint scalars combine with the gauge bosons to form long $\mathcal{N}=2$ supermultiplets.[25] In the equal quark mass limit $\xi_P \equiv \xi$, and all states come in representations of the unbroken global group (5), namely, in the singlet and adjoint representations of SU$(N)_{C+F}$,

$$(1,1)\,, \quad (N^2-1,1)\,, \tag{7}$$

and in the bifundamental representations

$$(\bar{N},\tilde{N})\,, \quad (N,\bar{\tilde{N}})\,. \tag{8}$$

The representations in (7) and (8) are marked with respect to two non-Abelian factors in (5). The singlet and adjoint fields are (i) the gauge bosons, and (ii) the first N flavors of squarks q^{kP} ($P=1,\ldots,N$), together with their fermion superpartners. The bifundamental fields are the quarks q^{kK} with $K=N+1,\ldots,N_f$. Quarks transform in the two-index representations of the global group (5) due to the color-flavor locking.

The above quasiclassical analysis is valid if the theory is at weak coupling. From (3) we see that the weak coupling condition is

$$\sqrt{\xi} \sim \sqrt{\mu m} \gg \Lambda_2\,, \tag{9}$$

where we assume all quark masses to be of the same order $m_A \sim m$. This condition means that the quark masses are large enough to compensate for the smallness of μ.

3. r-Dual Theory

In this section we review non-Abelian r duality in the $r=N$ vacua first established in Refs. 17 and 32 at small μ. This is an important part of our consideration on which we base further analysis, in particular the conclusion of the instead-of-confinement phase.

Let us relax the condition (9) and pass to the strong coupling domain at

$$|\sqrt{\xi_P}| \ll \Lambda_2\,, \quad |m_A| \ll \Lambda_2\,, \tag{10}$$

while keeping μ small.

In nonsupersymmetric theories such as QCD this step cannot be carried out analytically. This is the point where supersymmetry becomes important. More exactly, we exploit the exact Seiberg–Witten solution on the Coulomb branch[11,12] in our theory. We start at large $\xi \sim \mu m$ (in the equal quark mass limit) and then go to the equal mass small-ξ limit via the domain of large $\Delta m \sim \Delta m_{AB} \equiv (m_A - m_B)$.

At $\Delta m \sim \Lambda_2$ the theory enters a strong coupling regime and undergoes a crossover. We use the Seiberg–Witten curve to find the dual gauge group.[17,18] The domain (10) can be described in terms of weakly coupled (infrared free) r-dual theory with the gauge group

$$\mathrm{U}(\tilde{N}) \times \mathrm{U}(1)^{N-\tilde{N}}\,, \tag{11}$$

and N_f flavors of light quark-like dyons.[b] Note, that we refer to our dual theory as the "r dual" because $\mathcal{N} = 2$ duality described here can be generalized to other r vacua with $r > N_f/2$.

This leads to a theory with the dual gauge group $\mathrm{U}(N_f - r) \times \mathrm{U}(1)^{N-N_f+r}$.[21] However, the $\mathcal{N} = 1$ deformation of these r dual theories at larger μ can be performed at weak coupling only in the $r = N$ vacuum,[20] which we will discuss below.

The light dyons D^{lA} ($l = 1, \ldots, \tilde{N}$ and $A = 1, \ldots, N_f$) are in the fundamental representation of the gauge group $\mathrm{SU}(\tilde{N})$ and are charged under the Abelian factors indicated in Eq. (11). In addition, there are $(N - \tilde{N})$ light dyons D^J ($J = \tilde{N} + 1, \ldots, N$), neutral under the $\mathrm{SU}(\tilde{N})$ group, but charged under the U(1) factors.

The color charges of all these dyons are identical to those of quarks.[c] This is the reason why we call them quark-like dyons. However, these dyons are not quarks.[17] As we will show below they belong to a different representation of the global color-flavor locked group. Most importantly, condensation of these dyons still leads to confinement of *monopoles*.

The dyon condensates have the form:[18,27]

$$\langle D^{lA} \rangle = \langle \bar{\tilde{D}}^{lA} \rangle = \frac{1}{\sqrt{2}} \begin{pmatrix} 0 & \cdots & 0 & \sqrt{\xi_1} & \cdots & 0 \\ \cdots & \cdots & \cdots & \cdots & \cdots & \cdots \\ 0 & \cdots & 0 & 0 & \cdots & \sqrt{\xi_{\tilde{N}}} \end{pmatrix}, \tag{12}$$

$$\langle D^J \rangle = \langle \bar{\tilde{D}}^J \rangle = \sqrt{\frac{\xi_J}{2}}, \quad J = (\tilde{N} + 1), \ldots, N\,. \tag{13}$$

---

[b] The $\mathrm{SU}(\tilde{N})$ gauge group was first identified at the root of the baryonic Higgs branch in $\mathcal{N} = 2$ $\mathrm{SU}(N)$ SQCD with massless quarks and vanishing ξ parameters using the Seiberg–Witten curve in Ref. 33.

[c] Because of monodromies[11,12,34] the quarks pick up at strong coupling root-like color-magnetic charges in addition to their weight-like color-electric charges.[17]

The important feature apparent in (12), as compared to the squark VEVs in the original theory (3), is a "vacuum leap."[17] Namely, if we pick up the vacuum with nonvanishing VEVs of the first N quark flavors in the original theory at large ξ, and then reduce ξ below Λ_2, the system undergoes a crossover transition and ends up in the vacuum of the r-dual theory with the dual gauge group (11) and nonvanishing VEVs of $\tilde{N}$ last dyons (plus VEVs of $(N - \tilde{N})$ dyons that are $\mathrm{SU}(\tilde{N})$ singlets).

The parameters ξ_P in (12) and (13) are determined[27] by the quantum version of the classical expressions (4). They can be written in terms of roots of the Seiberg–Witten curve.

The first $\tilde{N}$ parameters ξ_P which determine VEVs of the non-Abelian dyons in (12) are small,

$$\xi_P = 2\mu m_{P+N} \sim \xi^{\mathrm{small}} \sim \mu m\,, \quad P = 1, \ldots, \tilde{N}\,. \tag{14}$$

This is a reflection of the fact that the non-Abelian sector of the dual theory is infrared free and is at weak coupling in the domain (10). Other ξ's which determine VEVs of the Abelian dyons in (13) are large,

$$\xi_P \sim \xi^{\mathrm{large}} \sim \mu \Lambda_{\mathcal{N}=2}\,, \quad P = \tilde{N}+1, \ldots, N\,. \tag{15}$$

As long as we keep ξ_P and masses small enough (i.e. in the domain (10)) the coupling constants of the infrared-free r-dual theory (frozen at the scale of the dyon VEVs) are small: the r-dual theory is at weak coupling.

3.1. *"Instead-of-confinement" mechanism*

Now, we are ready to explain the regime which we called "instead-of-confinement." Let us consider the limit of almost equal quark masses. Both, the gauge group and the global flavor $\mathrm{SU}(N_f)$ group, are broken in the vacuum. The form of the dyon VEVs in (12) shows that the r-dual theory is also in the color-flavor locked phase. Namely, the unbroken global group of the dual theory is

$$\mathrm{SU}(N) \times \mathrm{SU}(\tilde{N})_{C+F} \times \mathrm{U}(1)\,, \tag{16}$$

where this time the $\mathrm{SU}(\tilde{N})$ global group arises from color-flavor locking.

In much the same way as in the original theory, the presence of the global $\mathrm{SU}(\tilde{N})_{C+F}$ symmetry is the reason behind formation of the non-Abelian strings. Their tensions are still given by Eq. (6), where the parameters ξ_P are determined by (14).[18,27] These strings still confine monopoles.[17,d]

In the equal-mass limit the global unbroken symmetry (16) of the dual theory at small ξ coincides with the global group (5) of the original theory in the $r = N$ vacuum at large ξ. However, this global symmetry is realized in two very distinct ways in the dual pair at hand. As was already mentioned, the quarks and U(N) gauge bosons of the original theory at large ξ come in the following representations of the global group (5):

$$(1, 1), \ (N^2 - 1, 1), \ (\bar{N}, \tilde{N}) \quad \text{and} \quad (N, \bar{\tilde{N}}).$$

At the same time, the dyons and U($\tilde{N}$) gauge bosons of the r-dual theory form

$$(1, 1), \ (1, \tilde{N}^2 - 1), \ (N, \bar{\tilde{N}}) \quad \text{and} \quad (\bar{N}, \tilde{N}) \tag{17}$$

representations of (16). We see that the adjoint representations of the color-flavor locked subgroup are different in two theories.

The quarks and gauge bosons which form the adjoint $(N^2 - 1)$ representation of SU(N) at large ξ and the quark-like dyons and dual gauge bosons which form the adjoint $(\tilde{N}^2 - 1)$ representation of SU($\tilde{N}$) at small ξ are, in fact, *distinct* states.[17]

Thus, the quark-like dyons are not quarks. At large ξ they are heavy solitonic states. However below the crossover at small ξ they become light and form fundamental "elementary" states D^{lA} of the r-dual theory. And *vice versa*, quarks are light at large ξ but become heavy below the crossover.

This raises the question: what exactly happens to quarks when we reduce ξ?

They are in the "instead-of-confinement" phase. The Higgs-screened quarks and gauge bosons at small ξ decay into the monopole–antimonopole pairs on the curves of marginal stability (the so-called wall crossing).[17,32] The general rule is that the only states that exist at strong coupling inside the curves of marginal stability are those which can become massless on the Coulomb branch.[11,12,34] For the r-dual theory these are light dyons shown in Eq. (12), gauge bosons of the dual gauge group and monopoles.

The monopoles and antimonopoles produced at small nonvanishing values of ξ in the decay process of the adjoint $(N^2 - 1, 1)$ states cannot escape

[d] An explanatory remark regarding our terminology is in order. Strictly speaking, the dyons carrying root-like electric charges are confined as well. We refer to all such states collectively as to "monopoles." This is to avoid confusion with the quark-like dyons which appear in Eqs. (12) and (13). The latter dyons carry weight-like electric charges. As was already mentioned, their color charges are identical to those of quarks, see Ref. 17 for further details.

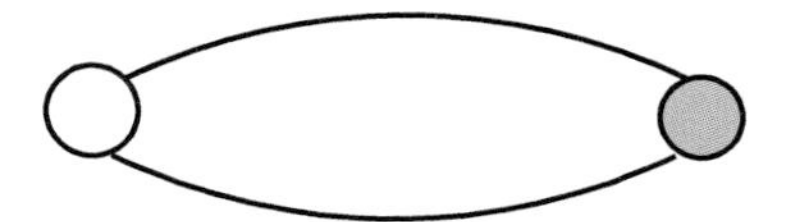

Fig. 1. Meson formed by a monopole–antimonopole pair connected by two strings. Open and closed circles denote the monopole and antimonopole, respectively.

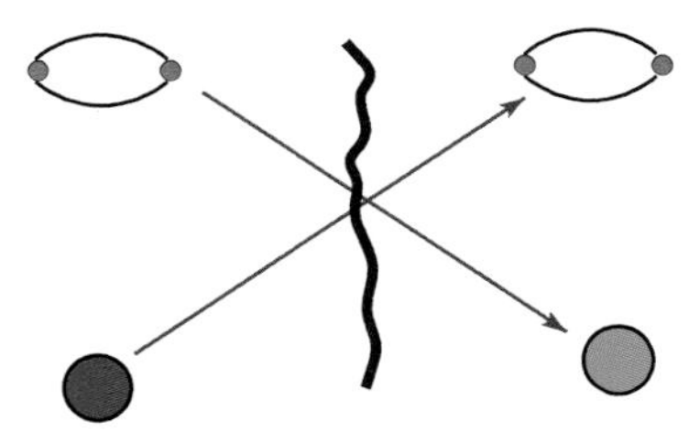

Fig. 2. Schematic picture of the crossover from large to small ξ. Blue circles denote light quarks, while green circles denote light quark-like dyons. Monopole–antimonopole mesons are shown as in Fig. 1.

from each other and fly to opposite infinities because they are confined. Therefore, the (screened) quarks and gauge bosons evolve into stringy mesons (in the strong coupling domain of small ξ) shown in Fig. 1, namely monopole–antimonopole pairs connected by two strings.[17,18]

The flavor quantum numbers of stringy monopole–antimonopole mesons were studied in Ref. 32 in the framework of an appropriate two-dimensional $CP(N-1)$ model which describes worldsheet dynamics of the non-Abelian strings.[26,29–31] In particular, confined monopoles are seen as kinks in this worldsheet theory. If two strings in Fig. 1 are "neighboring" strings P and $P+1$ ($P = 1,\ldots,(N-1)$), each meson is in the two-index representation $M_A^B(P,P+1)$ of the flavor group, where the flavor indices are $A,B = 1,\ldots,N_f$. It splits into a singlet, adjoint and bifundamental representations of the global unbroken group (16). In particular, at small ξ the adjoint representation of SU(N) contains former (screened) quarks and gauge bosons of the original theory.

The picture of the crossover is schematically shown in Fig. 2. The left and right sides of this figure correspond to large and small values of ξ, respectively. Quarks are light at large ξ. They evolve into monopole–antimonopole stringy mesons at small ξ. Moreover, heavy monopole–antimonopole stringy mesons present at large ξ become light at small ξ and form "fundamental" charged matter of the r-dual theory, namely, quark-like dyons.

We see that the monopole–antimonopole stringy mesons have "correct" (adjoint or singlet) quantum numbers with respect to the global group, in much the same way as mesons in real-world QCD. Let us explain this. For example, in the actual world a $U(1)$ subgroup of the global flavor group is gauged with respect to electromagnetic interactions. The correct (adjoint or singlet) global quantum numbers of the monopole–antimonopole stringy mesons mean, in particular, that they have integer rather than fractional (like 2/3 or $-1/3$) electric charges.

Moreover, because these mesons are formed by strings, they lie on the Regge trajectories. Thus, the monopole–antimonopole mesons of the instead-of-confinement phase are qualitatively similar to real-world QCD mesons. The role of QCD constituent quarks is played by monopoles.

4. Flowing to $\mathcal{N} = 1$ QCD

In this section we will discuss what happens to the r-dual theory in the $r = N$ vacuum once we increase μ, see Refs. 18, 19, 21. We also discuss the relation of our dual theory to the Seiberg's duality.

4.1. *Intermediate μ: Emergence of the $U(\tilde{N})_{\rm gauge}$*

Combining Eqs. (12), (13), (14) and (15) we see that in the domain (10) the VEVs of the non-Abelian dyons D^{lA} are much smaller than those of the Abelian dyons D^J. This circumstance is the most crucial. It allows us to increase μ and decouple the adjoint fields without violating the weak coupling condition in the dual theory.[18]

First let us consider intermediate values of μ which are large enough to decouple the adjoint matter.[18,21] We uplift μ to the intermediate domain

$$|\mu| \gg |m_A|\,, \quad A = 1, \ldots, N_f\,, \quad \mu \ll \Lambda_2\,. \tag{18}$$

The VEVs of the Abelian dyons (13) are large. This makes the U(1) gauge fields of the dual group (11) heavy. Decoupling these gauge factors, together with the adjoint matter and the Abelian dyons themselves, we obtain the low-energy theory with the $U(\tilde{N})$ gauge fields and a set of non-Abelian dyons

$$D^{lA}\,, \quad l = 1, \ldots, \tilde{N}\,, \quad A = 1, \ldots, N_f\,. \tag{19}$$

The superpotential for D^{lA} has the form[18]

$$\mathcal{W} = -\frac{1}{2\mu}\,(\tilde{D}_A D^B)(\tilde{D}_B D^A) + m_A(\tilde{D}_A D^A)\,, \tag{20}$$

where the color indices are contracted inside each parentheses. Minimization of this superpotential leads to the VEVs (12) for the non-Abelian dyons determined by $\xi^{\rm small}$, see (14).

Below the scale μ, our theory becomes dual to $\mathcal{N}=1$ SQCD. This r-dual theory has the scale

$$\tilde{\Lambda}_1^{N-2\tilde{N}} = \frac{\Lambda_2^{N-\tilde{N}}}{\mu^{\tilde{N}}} \,. \tag{21}$$

In order to keep this infrared-free theory in the weak coupling regime we impose the constraint

$$|\sqrt{\mu m}| \ll \tilde{\Lambda}_1 \,. \tag{22}$$

This means that at large μ we must keep the quark masses sufficiently small.

Note that for the intermediate μ we assume that $\mu \ll \Lambda_2$. This condition guarantees that the heavy Abelian $U(1)^{N-\tilde{N}}$ sector is at weak coupling too, and is indeed heavy. If we relax the condition $\mu \ll \Lambda_2$ this sector enters a strong coupling regime, and certain states could in principle become light and show up in our low-energy $U(\tilde{N})$ theory.

4.2. *Connection to Seiberg's duality*

The gauge group of our r-dual theory is $\mathrm{U}(\tilde{N})$, the same as the gauge group of the Seiberg's dual theory.[13,14] This suggests that there should be a close relation between two duals. For intermediate values of μ this relation was found in Refs. 20 and 35.

Originally Seiberg's duality was formulated for $\mathcal{N}=1$ SQCD which in our setup corresponds to the limit $\mu \to \infty$. Therefore, in the original formulation Seiberg's duality referred to the monopole vacua with $r=0$. Other vacua, with $r \neq 0$, have condensates of r quark flavors $\langle \tilde{q}q \rangle_A \sim \mu m_A$ and, therefore, disappear in the limit $\mu \to \infty$: they become runaway vacua.

However, Seiberg's duality can be (and in fact, was) generalized to the case of μ-deformed $\mathcal{N}=2$ SQCD.[36,37] If the mass term μ is large then μ-deformed $\mathcal{N}=2$ SQCD flows to $\mathcal{N}=1$ SQCD with an additional quartic quark superpotential. This theory has all r vacua which were present in the original $\mathcal{N}=2$ theory in the small-μ limit.

The generalized Seiberg dual theory in the case of μ-deformed $\mathrm{U}(N)$ $\mathcal{N}=2$ SQCD at large but finite μ has the $\mathrm{U}(\tilde{N})$ gauge group, N_f flavors of Seiberg's "dual quarks" h^{lA} (here $l=1,\ldots,\tilde{N}$ and $A=1,\ldots,N_f$) and the following superpotential:

$$\mathcal{W}_S = -\frac{\kappa^2}{2\mu}\,\mathrm{Tr}(M^2) + \kappa m_A M_A^A + \tilde{h}_{Al} h^{lB} M_B^A \,, \tag{23}$$

where M_A^B is the Seiberg neutral mesonic field defined as

$$(\tilde{q}_A q^B) = \kappa M_A^B \,. \tag{24}$$

The parameter κ above has dimension of mass and is needed to formulate Seiberg's duality.[13,14] Two last terms in (23) were originally suggested by Seiberg, while the first term is a generalization to finite μ. This generalization originates from the quartic quark superpotential.[36,37]

Now, let us assume the fields M_A^B to be heavy and integrate them out. This implies that κ is large. Integrating out the M fields in (23) we arrive at

$$\mathcal{W}_S^{\rm LE} = \frac{\mu}{2\kappa^2}\,(\tilde{h}_A h^B)(\tilde{h}_B h^A) + \frac{\mu}{\kappa}\,m_A(\tilde{h}_A h^A)\,. \tag{25}$$

The change of variables

$$D^{lA} = \sqrt{-\frac{\mu}{\kappa}}\,h^{lA}, \quad l = 1,\dots,\tilde{N}\,, \quad A = 1,\dots,N_f \tag{26}$$

brings this superpotential to the form

$$\mathcal{W}_S^{\rm LE} = \frac{1}{2\mu}\,(\tilde{D}_A D^B)(\tilde{D}_B D^A) - m_A(\tilde{D}_A D^A)\,. \tag{27}$$

We see that the r-dual and Seiberg's dual theories match each other. At intermediate μ the Seiberg M meson is heavy and should be integrated out implying the superpotential (27) which agrees with the superpotential (20) obtained in the r-dual theory.

This match, together with the identification (26), reveals the physical nature of Seiberg's "dual quarks." They are not monopoles as one could naively think. Instead, they are quark-like dyons appearing in the r-dual theory below the crossover. Their condensation leads to confinement of monopoles and the "instead-of-confinement" phase[21] for quarks and gauge bosons of the original theory.

4.3. *Large μ*

Finally, we pass to the large-μ domain. Increasing μ we simultaneously reduce m keeping $\xi^{\rm small}$ sufficiently small, see (22). Namely, we assume

$$\xi^{\rm small} \sim \mu m \ll \tilde{\Lambda}_1\,, \quad \mu \gg \Lambda_1\,, \tag{28}$$

where Λ_1 is the scale of the original $\mathcal{N} = 1$ SQCD,

$$\Lambda_1^{2N-\tilde{N}} = \mu^N \Lambda_2^{N-\tilde{N}}\,. \tag{29}$$

This ensures that our low-energy $U(\tilde{N})$ r-dual theory is at weak coupling. However, the Abelian $U(1)^{N-\tilde{N}}$ sector ultimately enters the strong coupling regime. As was already mentioned, we loose analytic control over this sector and, in particular, certain states can become light and show up in our low-energy $U(\tilde{N})$ theory.

This is exactly what happens at large values of μ and is, in fact, required by the 't Hooft anomaly matching.[15] Large values of μ require a chiral limit of small m due to the condition (28). In this limit we need to match global anomalies in terms of original and dual theories. In fact, without light Seiberg M meson the anomalies do not match. This was checked initially in Ref. 13 for the limit $\mu \to \infty$ and presented a basis for the discovery of Seiberg's duality. Moreover, recently it was confirmed[19] for our μ-deformed theory at large but finite μ and massive quarks in the domain (28).

Natural candidates for the Seiberg M mesons in the r-dual theory are the stringy mesons $M_A^B(P, P+1)$ (with $P = \tilde{N}, \ldots, (N-1)$) from the Abelian $U(1)^{N-\tilde{N}}$ sector. This sector is at strong coupling at large μ; therefore, certain states from this sector can become light. Perturbative states from this sector (quark-like dyons and Abelian gauge fields) are singlets with respect to the global group (16) and cannot play the role of the M mesons. Stringy mesons $M_A^B(P, P+1)$ (where $P = 1, \ldots, (\tilde{N}-1)$) from the $U(\tilde{N})$ low-energy theory also cannot play the role of the M mesons. First, they are represented in the $U(\tilde{N})$ low-energy theory as nonperturbative solitonic states and cannot be added to this theory as new "fundamental" or "elementary" fields. Second, they are too heavy, with masses of the order of $\sqrt{\xi^{\rm small}}$, determined by the tensions of the non-Abelian strings, which can be calculated at weak coupling.

Thus, we proposed in Ref. 19 that the Seiberg M_A^B mesons come from a multitude of the monopole–antimonopole stringy mesons $M_A^B(P, P+1)$ (where $P = \tilde{N}, \ldots, (N-1)$) from the Abelian $U(1)^{N-\tilde{N}}$ sector. At large μ the M meson should become light, with mass of the order of m. It should be incorporated in the $U(\tilde{N})$ low-energy theory as a new "fundamental" or "elementary" field. Note, that other states from the Abelian sector are still heavy and decouple.

Since our $U(\tilde{N})$ r-dual theory is at weak coupling we can write down its effective action. Using the procedure described in Sec. 4.2 in the opposite direction we integrate the M-meson in the superpotential (20). In this way we arrive at

$$\mathcal{W} = \frac{\kappa^2}{2\mu}\,\mathrm{Tr}(M^2) - \kappa m_A M_A^A + \frac{\kappa}{\mu}\,\tilde{D}_{Al} D^{lB}\, M_B^A\,, \tag{30}$$

where

$$\kappa \sim \begin{cases} \mu^{\frac{3}{4}}\Lambda_{\mathcal{N}=2}^{\frac{1}{4}}, & \mu \ll \Lambda_{\mathcal{N}=2}, \\ \sqrt{\mu m}, & \mu \gg \Lambda_{\mathcal{N}=2}. \end{cases} \tag{31}$$

This dependence guarantees that the M meson is heavy, with mass of the order of $\sqrt{\xi^{\rm large}}$ at intermediate μ, and becomes light, with mass of the order of m at large μ, see Ref. 19 for details.

5. Conclusions

Quarks, gluons, and other notions of which M. Gell-Mann was a pioneer got a new life in the era of supersymmetry, when supersymmetry-based methods became powerful — and quite often, unique — tools in the studies of confinement and other nontrivial features of gauge dynamics at strong coupling.

In this brief article we summarized some applications of non-Abelian strings and reviewed phases of $\mathcal{N} = 1$ SQCD obtained from μ-deformed $\mathcal{N} = 2$ SQCD in the limit of large μ. We identified "promising" vacua among all r vacua — promising in the quest of confinement similar to that inherent to QCD. The number of r vacua as a function of r is shown in Fig. 3.

The zero vacua represent a subset of vacua at $r < \tilde{N}$ with parametrically small gaugino condensate in the limit of the small quark masses. In this limit the zero vacua are described in terms of a weakly coupled dual infrared-free

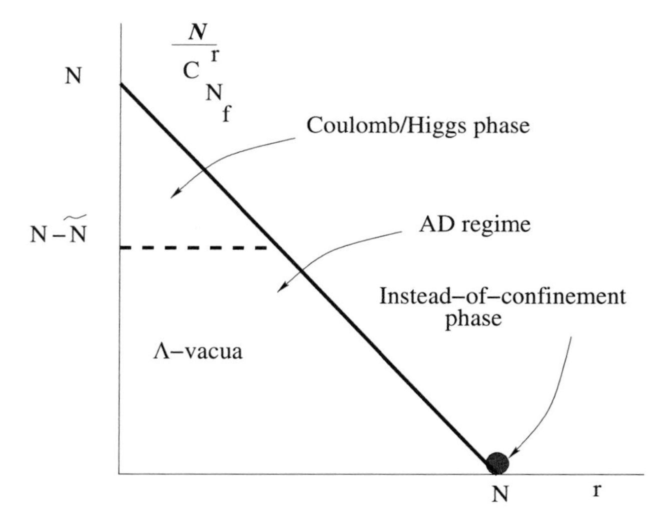

Fig. 3. Phases of r vacua in $\mathcal{N} = 1$ SQCD. Upper triangle represents the zero vacua, while black circle denotes the $r = N$ vacuum.

$U(\tilde{N})$ gauge theory with r condensed quarks. This theory is in the mixed Coulomb-Higgs phase.[19,20]

The Λ vacua (of which we said little, if at all) have no weak coupling description at large μ and small m.[20] In a certain limit they flow into a conformal Argyres–Douglas-like strongly coupled regime.[38,39]

The regime closest to what we observe in real-world QCD is represented by the instead-of-confinement phase which occurs in the $r = N$ vacuum at strong coupling. The monopole–antimonopole stringy mesons formed in this phase are qualitatively similar to mesons in QCD. They have "correct" quantum numbers with respect to the global group (singlet plus adjoint) and lie on the Regge trajectories.

Acknowledgments

This work is supported in part by DOE grant DE-SC0011842. The work of A. Yung was supported by FTPI, University of Minnesota, by RFBR Grant No. 13-02-00042a and by Russian State Grant for Scientific Schools RSGSS-657512010.2. The work of A. Yung was supported by RSCF Grant No. 14-22-00281.

References

1. M. Gell-Mann, *Phys. Lett.* **8**, 214 (1964).
2. G. Zweig, An SU(3) model for strong interaction symmetry and its breaking, CERN-TH-401 (1964) [also in *Developments in the Quark Theory of Hadrons*, Vol. 1, eds. D. Lichtenberg and S. Rosen (Hadronic Press, Nonantum, MA, 1980), pp. 22–101].
3. O. W. Greenberg, *Phys. Rev. Lett.* **13**, 598 (1964).
4. H. Fritzsch and M. Gell-Mann, Current algebra: Quarks and what else?, in *Proc. XVI Int. Conf. on High Energy Physics*, Vol. 2, Chicago, 1972, eds. J. D. Jackson and A. Roberts (FNAL, Batavia, Illinois, 1972), p. 135, reprinted in arXiv:hep-ph/0208010.
5. D. J. Gross and F. Wilczek, *Phys. Rev. Lett.* **30**, 1343 (1973).
6. H. D. Politzer, *Phys. Rev. Lett.* **30**, 1346 (1973).
7. Y. Nambu, *Phys. Rev. D* **10**, 4262 (1974).
8. G. 't Hooft, Gauge theories with unified weak, electromagnetic and strong interactions, in *Proc. E.P.S. Int. Conf. on High Energy Physics*, Palermo, 23–28 June 1975, ed. A. Zichichi (Editrice Compositori, Bologna, 1976).
9. G. 't Hooft, *Nucl. Phys. B* **190**, 455 (1981).
10. S. Mandelstam, *Phys. Rep.* **23**, 245 (1976).
11. N. Seiberg and E. Witten, *Nucl. Phys. B* **426**, 19 (1994) [Erratum: *ibid.* **430**, 485 (1994)], arXiv:hep-th/9407087.

12. N. Seiberg and E. Witten, *Nucl. Phys. B* **431**, 484 (1994), arXiv:hep-th/9408099.
13. N. Seiberg, *Nucl. Phys. B* **435**, 129 (1995), arXiv:hep-th/9411149.
14. K. A. Intriligator and N. Seiberg, *Nucl. Phys. B* (*Proc. Suppl.*) **45C**, 1 (1996), arXiv:hep-th/9509066.
15. G. 't Hooft, Naturalness, chiral symmetry, and spontaneous chiral symmetry breaking, in *Recent Developments in Gauge Theories*, eds. G. 't Hooft, C. Itzykson, A. Jaffe, H. Lehmann, P. K. Mitter, I. M. Singer and R. Stora (Plenum Press, New York, 1980) [Reprinted in *Dynamical Symmetry Breaking*, eds. E. Farhi *et al.* (World Scientific, Singapore, 1982), p. 345 and in G. 't Hooft, *Under the Spell of the Gauge Principle* (World Scientific, Singapore, 1994), p. 352].
16. F. Cachazo, N. Seiberg and E. Witten, *J. High Energy Phys.* **0304**, 018 (2003), arXiv:hep-th/0303207.
17. M. Shifman and A. Yung, *Phys. Rev. D* **79**, 125012 (2009), arXiv:0904.1035 [hep-th].
18. M. Shifman and A. Yung, *Phys. Rev. D* **83**, 105021 (2011), arXiv:1103.3471 [hep-th].
19. M. Shifman and A. Yung, *Phys. Rev. D* **90**, 065014 (2014), arXiv:1403.6086 [hep-th].
20. M. Shifman and A. Yung, *Phys. Rev. D* **87**, 106009 (2013), arXiv:1304.0822 [hep-th].
21. M. Shifman and A. Yung, *Phys. Rev. D* **86**, 025001 (2012), arXiv:1204.4165 [hep-th].
22. M. R. Douglas and S. H. Shenker, *Nucl. Phys. B* **447**, 271 (1995), arXiv:hep-th/9503163.
23. A. Hanany, M. Strassler and A. Zaffaroni, *Nucl. Phys. B* **513**, 87 (1998), arXiv:hep-th/9707244.
24. M. Strassler, *Prog. Theor. Phys. Suppl.* **131**, 439 (1998), arXiv:hep-lat/9803009.
25. A. I. Vainshtein and A. Yung, *Nucl. Phys. B* **614**, 3 (2001), arXiv:hep-th/0012250.
26. M. Shifman and A. Yung, *Phys. Rev. D* **70**, 045004 (2004), arXiv:hep-th/0403149.
27. M. Shifman and A. Yung, *Phys. Rev. D* **82**, 066006 (2010), arXiv:1005.5264 [hep-th].
28. M. Shifman and A. Yung, *Rev. Mod. Phys.* **79**, 1139 (2007), arXiv:hep-th/0703267 [an expanded version in *Supersymmetric Solitons* (Cambridge University Press, 2009)].
29. A. Hanany and D. Tong, *J. High Energy Phys.* **0307**, 037 (2003), arXiv:hep-th/0306150.
30. R. Auzzi, S. Bolognesi, J. Evslin, K. Konishi and A. Yung, *Nucl. Phys. B* **673**, 187 (2003), arXiv:hep-th/0307287.
31. A. Hanany and D. Tong, *J. High Energy Phys.* **0404**, 066 (2004), arXiv:hep-th/0403158.
32. M. Shifman and A. Yung, *Phys. Rev. D* **81**, 085009 (2010), arXiv:1002.0322 [hep-th].

33. P. Argyres, M. Plesser and N. Seiberg, *Nucl. Phys. B* **471**, 159 (1996), arXiv:hep-th/9603042.
34. A. Bilal and F. Ferrari, *Nucl. Phys. B* **516**, 175 (1998), arXiv:hep-th/9706145.
35. M. Shifman and A. Yung, *Phys. Rev. D* **86**, 065003 (2012), arXiv:1204.4164 [hep-th].
36. G. Carlino, K. Konishi and H. Murayama, *Nucl. Phys. B* **590**, 37 (2000), arXiv:hep-th/0005076.
37. A. Giveon and D. Kutasov, *Nucl. Phys. B* **796**, 25 (2008), arXiv:0710.0894 [hep-th].
38. P. C. Argyres and M. R. Douglas, *Nucl. Phys. B* **448**, 93 (1995), arXiv:hep-th/9505062.
39. P. C. Argyres, M. R. Plesser, N. Seiberg and E. Witten, *Nucl. Phys. B* **461**, 71 (1996), arXiv:hep-th/9511154.

Quarks and a Unified Theory of Nature Fundamental Forces

I. Antoniadis

Albert Einstein Center for Fundamental Physics,
Institute for Theoretical Physics, Bern University,
Sidlerstrasse 5 CH-3012 Bern, Switzerland
Ecole Polytechnique, 91128 Palaiseau, France
antoniadis@itp.unibe.ch

Quarks were introduced 50 years ago opening the road towards our understanding of the elementary constituents of matter and their fundamental interactions. Since then, a spectacular progress has been made with important discoveries that led to the establishment of the Standard Theory that describes accurately the basic constituents of the observable matter, namely quarks and leptons, interacting with the exchange of three fundamental forces, the weak, electromagnetic and strong force. Particle physics is now entering a new era driven by the quest of understanding of the composition of our Universe such as the unobservable (dark) matter, the hierarchy of masses and forces, the unification of all fundamental interactions with gravity in a consistent quantum framework, and several other important questions. A candidate theory providing answers to many of these questions is string theory that replaces the notion of point particles by extended objects, such as closed and open strings. In this short note, I will give a brief overview of string unification, describe in particular how quarks and leptons can emerge and discuss what are possible predictions for particle physics and cosmology that could test these ideas.

1. Introduction

During the last few decades, physics beyond the Standard Model (SM) was guided from the problem of mass hierarchy. This can be formulated as the question of why gravity appears to us so weak compared to the other three known fundamental interactions corresponding to the electromagnetic, weak and strong nuclear forces. Indeed, gravitational interactions are suppressed by a very high energy scale, the Planck mass $M_P \sim 10^{19}$ GeV, associated with a length $l_P \sim 10^{-35}$ m, where they are expected to become important. In a quantum theory, the hierarchy implies a severe fine tuning of the fundamental parameters in more than 30 decimal places in order to keep the masses of elementary particles at their observed values. The reason is that quantum radiative corrections to all masses generated by the Higgs vacuum expectation value (VEV) are proportional to the ultraviolet cutoff which in the presence of gravity is fixed by the Planck mass. As a result, all masses are "attracted" to become about 10^{16} times heavier than their observed values.

Besides compositeness, there are two main ideas that have been proposed and studied extensively during the last decades, corresponding to different approaches of dealing with the mass hierarchy problem. (1) Low energy supersymmetry with all superparticle masses in the TeV region. Indeed, in the limit of exact supersymmetry, quadratically divergent corrections to the Higgs self-energy are exactly cancelled, while in the softly broken case, they are cutoff by the supersymmetry breaking mass splittings. (2) TeV scale strings, in which quadratic divergences are cutoff by the string scale and low energy supersymmetry is not needed. Both ideas are experimentally testable at high-energy particle colliders and in particular at LHC. Below, I discuss their implementation in string theory.

The appropriate and most convenient framework for low energy supersymmetry and grand unification is the perturbative heterotic string. Indeed, in this theory, gravity and gauge interactions have the same origin, as massless modes of the closed heterotic string, and they are unified at the string scale M_s. As a result, the Planck mass M_P is predicted to be proportional to M_s:

$$M_P = M_s/g\,, \tag{1}$$

where g is the gauge coupling. In the simplest constructions all gauge couplings are the same at the string scale, given by the four-dimensional (4d) string coupling, and thus no grand unified group is needed for unification. In our conventions $\alpha_{\rm GUT} = g^2 \simeq 0.04$, leading to a discrepancy between the string and grand unification scale $M_{\rm GUT}$ by almost two orders of magnitude. Explaining this gap introduces in general new parameters or a new scale, and the predictive power is essentially lost. This is the main defect of this framework, which remains though an open and interesting possibility.[1]

The other idea has as natural framework of realization type I string theory with D-branes. Unlike in the heterotic string, gauge and gravitational interactions have now different origin. The latter are described again by closed strings, while the former emerge as excitations of open strings with endpoints confined on D-branes.[2] This leads to a braneworld description of our universe, which should be localized on a hypersurface, i.e. a membrane extended in p spatial dimensions, called p-brane (see Fig. 1). Closed strings

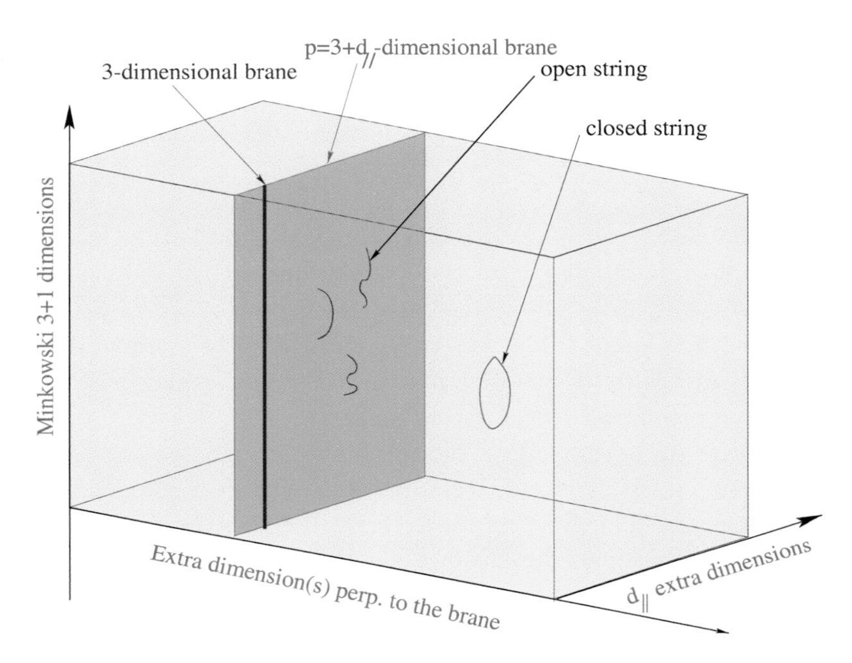

Fig. 1. In the type I string framework, our Universe contains, besides the three known spatial dimensions (denoted by a single blue line), some extra dimensions ($d_{\parallel} = p - 3$) parallel to our world p-brane (green plane) where endpoints of open strings are confined, as well as some transverse dimensions (yellow space) where only gravity described by closed strings can propagate.

propagate in all nine dimensions of string theory: in those extended along the p-brane, called parallel, as well as in the transverse ones. On the contrary, open strings are attached on the p-brane. Obviously, our p-brane world must have at least the three known dimensions of space. But it may contain more: the extra $d_{\parallel} = p-3$ parallel dimensions must have a finite size, in order to be unobservable at present energies, and can be as large as $\text{TeV}^{-1} \sim 10^{-18}$ m.[3] On the other hand, transverse dimensions interact with us only gravitationally and experimental bounds are much weaker: their size should be less than about 0.1 mm.[4]

2. Framework of Low Scale Strings

In type I theory, the different origin of gauge and gravitational interactions implies that the relation between the Planck and string scales is not linear as (1) of the heterotic string. The requirement that string theory should be weakly coupled, constrain the size of all parallel dimensions to be of order of the string length, while transverse dimensions remain unrestricted. Assuming an isotropic transverse space of $n = 9 - p$ compact dimensions of common radius $R_\perp$, one finds:

$$M_P^2 = \frac{1}{g^4} M_s^{2+n} R_\perp^n\,, \quad g_s \simeq g^2\,, \tag{2}$$

where g_s is the string coupling. It follows that the type I string scale can be chosen hierarchically smaller than the Planck mass at the expense of introducing extra large transverse dimensions felt only by gravity, while keeping the string coupling small.[5] The weakness of 4d gravity compared to gauge interactions (ratio M_W/M_P) is then attributed to the largeness of the transverse space $R_\perp$ compared to the string length $l_s = M_s^{-1}$.

An important property of these models is that gravity becomes effectively $(4+n)$-dimensional with a strength comparable to those of gauge interactions at the string scale. The first relation of Eq. (2) can be understood as a consequence of the $(4+n)$-dimensional Gauss law for gravity, with

$$M_*^{(4+n)} = M_s^{2+n}/g^4 \tag{3}$$

the effective scale of gravity in $4+n$ dimensions. Taking $M_s \simeq 1$ TeV, one finds a size for the extra dimensions $R_\perp$ varying from 10^8 km, .1 mm, down to a Fermi for $n = 1, 2$, or 6 large dimensions, respectively. This shows that while $n = 1$ is excluded, $n \geq 2$ is allowed by present experimental bounds on gravitational forces.[4,6] Thus, in these models, gravity appears to us very weak at macroscopic scales because its intensity is spread in the "hidden" extra dimensions. At distances shorter than $R_\perp$, it should deviate from Newton's law, which may be possible to explore in laboratory experiments (see Fig. 2).

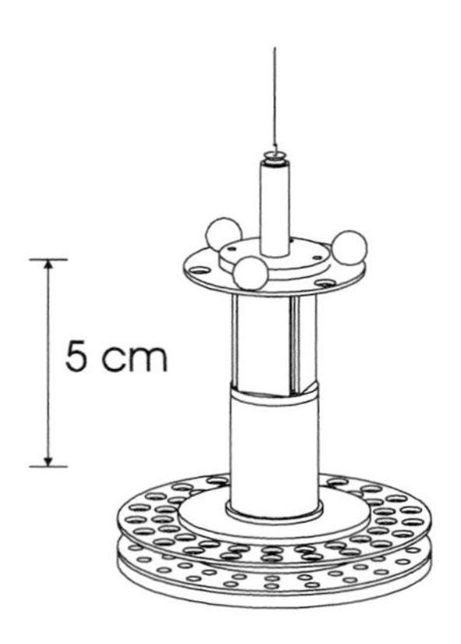

Fig. 2. Torsion pendulum that tested Newton's law at 55 μm.

The main experimental implications of TeV scale strings in particle accelerators are of three types, in correspondence with the three different sectors that are generally present: (i) new compactified parallel dimensions, (ii) new extra large transverse dimensions and low scale quantum gravity, and (iii) genuine string and quantum gravity effects. On the other hand, there exist interesting implications in non accelerator table-top experiments due to the exchange of gravitons or other possible states living in the bulk.

3. Large Number of Species

Here, we point out that low scale gravity with large extra dimensions is actually a particular case of a more general framework, where the UV cutoff is lower than the Planck scale due to the existence of a large number of particle species coupled to gravity.[7] Indeed, it was shown that the effective UV cutoff M_* is given by

$$M_*^2 = M_P^2/N\,, \tag{4}$$

where the counting of independent species N takes into account all particles which are not broad resonances, having a width less than their mass. The derivation is based on black hole evaporation but here we present a shorter argument using quantum information storage.[8] Consider a pixel of size L containing N species storing information. The energy required to localize N wave functions is then given by N/L, associated with a Schwarzschild radius $R_s = N/LM_P^2$. The latter must be less than the pixel size in order to avoid the collapse of such a system to a black hole, $R_s \leq L$, implying a minimum size $L \geq L_{\min}$ with $L_{\min} = \sqrt{N}/M_P$ associated precisely with the effective UV cutoff $M_* = L_{\min}$ given in Eq. (4). Imposing $M_* \simeq 1$ TeV, one should then have $N \sim 10^{32}$ particle species below about the TeV scale!

In the string theory context, there are two ways of realizing such a large number of particle species by lowering the string scale at a TeV:

(1) In large volume compactifications with the SM localized on D-brane stacks, as described in the previous section. The particle species are then the Kaluza–Klein (KK) excitations of the graviton (and other possible bulk modes) associated with the large extra dimensions, given by $N = R_\perp^n l_s^n$, up to energies of order $M_* \simeq M_s$.
(2) By introducing an infinitesimal string coupling $g_s \simeq 10^{-16}$ with the SM localized on Neveu–Schwarz NS5-branes in the framework of little strings.[9] In this case, the particle species are the effective number of string modes that contribute to the black hole bound:[10] $N = 1/g_s^2$ and gravity does not become strong at $M_s \sim \mathcal{O}(\text{TeV})$.

Note that both TeV string realizations above are compatible with the general expression (2), but in the second case there is no relation between the string and gauge couplings.

4. Standard Model on D-branes

The gauge group closest to the Standard Model one can easily obtain with D-branes is $U(3) \times U(2) \times U(1)$. The first factor arises from three coincident "color" D-branes. An open string with one end on them is a triplet under $SU(3)$ and carries the same $U(1)$ charge for all three components. Thus, the $U(1)$ factor of $U(3)$ has to be identified with *gauged* baryon number. Similarly, $U(2)$ arises from two coincident "weak" D-branes and the corresponding Abelian factor is identified with *gauged* weak-doublet number. Finally, an extra $U(1)$ D-brane is necessary in order to accommodate the Standard Model without breaking the baryon number.[11] In principle this $U(1)$ brane can be chosen to be independent of the other two collections with its own gauge coupling. To improve the predictability of the model, we choose to put it on top of either the color or the weak D-branes.[12] In either case, the model has two independent gauge couplings g_3 and g_2 corresponding, respectively, to the gauge groups $U(3)$ and $U(2)$. The $U(1)$ gauge coupling g_1 is equal to either g_3 or g_2.

Let us denote by Q_3, Q_2 and Q_1 the three $U(1)$ charges of $U(3) \times U(2) \times U(1)$, in a self explanatory notation. Under $SU(3) \times SU(2) \times U(1)_3 \times U(1)_2 \times U(1)_1$, the members of a family of quarks and leptons have the following quantum numbers:

$$\begin{aligned}
&Q(\mathbf{3}, \mathbf{2}; 1, w, 0)_{1/6}\,, \\
&u^c(\bar{\mathbf{3}}, \mathbf{1}; -1, 0, x)_{-2/3}\,, \\
&d^c(\bar{\mathbf{3}}, \mathbf{1}; -1, 0, y)_{1/3}\,, \\
&L(\mathbf{1}, \mathbf{2}; 0, 1, z)_{-1/2}\,, \\
&l^c(\mathbf{1}, \mathbf{1}; 0, 0, 1)_{1}\,.
\end{aligned} \tag{5}$$

The values of the $U(1)$ charges x, y, z, w will be fixed below so that they lead to the right hypercharges, shown for completeness as subscripts.

It turns out that there are two possible ways of embedding the Standard Model particle spectrum on these stacks of branes,[11] which are shown pictorially in Fig. 3. The quark doublet Q corresponds necessarily to a massless excitation of an open string with its two ends on the two different collections of branes (color and weak). As seen from the figure, a fourth brane stack is

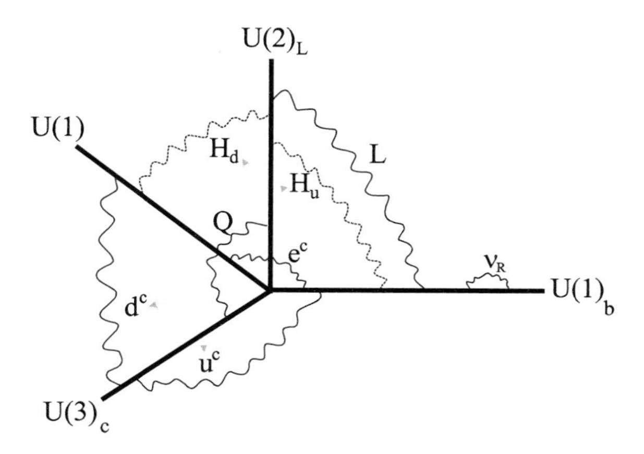

Fig. 3. A minimal Standard Model embedding on D-branes.

needed for a complete embedding, which is chosen to be a $U(1)_b$ extended in the bulk. This is welcome since one can accommodate right handed neutrinos as open string states on the bulk with sufficiently small Yukawa couplings suppressed by the large volume of the bulk.[13] The two models are obtained by an exchange of the up and down antiquarks, u^c and d^c, which correspond to open strings with one end on the color branes and the other either on the $U(1)$ brane, or on the $U(1)_b$ in the bulk. The lepton doublet L arises from an open string stretched between the weak branes and $U(1)_b$, while the antilepton l^c corresponds to a string with one end on the $U(1)$ brane and the other in the bulk. For completeness, we also show the two possible Higgs states H_u and H_d that are both necessary in order to give tree-level masses to all quarks and leptons of the heaviest generation.

4.1. *Hypercharge embedding and the weak angle*

The weak hypercharge Y is a linear combination of the three $U(1)$'s:

$$Y = Q_1 + \frac{1}{2}Q_2 + c_3 Q_3 ; \quad c_3 = -1/3 \text{ or } 2/3 , \tag{6}$$

where Q_N denotes the $U(1)$ generator of $U(N)$ normalized so that the fundamental representation of $SU(N)$ has unit charge. The corresponding $U(1)$ charges appearing in Eq. (5) are $x = -1$ or 0, $y = 0$ or 1, $z = -1$, and $w = 1$ or -1, for $c_3 = -1/3$ or $2/3$, respectively. The hypercharge coupling g_Y is

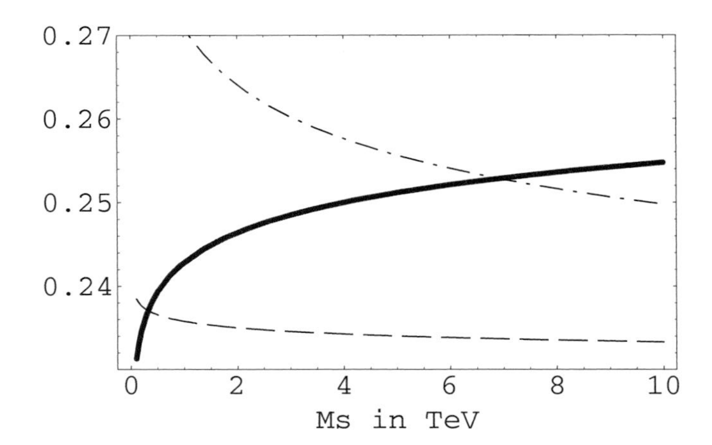

Fig. 4. The experimental value of $\sin^2\theta_W$ (thick curve), and the theoretical predictions (8).

given by:[a]

$$\frac{1}{g_Y^2} = \frac{2}{g_1^2} + \frac{4c_2^2}{g_2^2} + \frac{6c_3^2}{g_3^2} . \tag{7}$$

It follows that the weak angle $\sin^2\theta_W$, is given by:

$$\sin^2\theta_W \equiv \frac{g_Y^2}{g_2^2 + g_Y^2} = \frac{1}{2 + 2g_2^2/g_1^2 + 6c_3^2 g_2^2/g_3^2} , \tag{8}$$

where g_N is the gauge coupling of $SU(N)$ and $g_1 = g_2$ or $g_1 = g_3$ at the string scale. In order to compare the theoretical predictions with the experimental value of $\sin^2\theta_W$ at M_s, we plot in Fig. 4 the corresponding curves as functions of M_s. The solid line is the experimental curve. The dashed line is the plot of the function (8) for $g_1 = g_2$ with $c_3 = -1/3$ while the dotted-dashed line corresponds to $g_1 = g_3$ with $c_3 = 2/3$. The other two possibilities are not shown because they lead to a value of M_s which is too high to protect the hierarchy. Thus, the second case, where the $U(1)$ brane is on top of the color branes, is compatible with low energy data for $M_s \sim 6-8$ TeV and $g_s \simeq 0.9$.

From Eq. (8) and Fig. 4, we find the ratio of the $SU(2)$ and $SU(3)$ gauge couplings at the string scale to be $\alpha_2/\alpha_3 \sim 0.4$. This ratio can be arranged by an appropriate choice of the relevant moduli. For instance, one may choose the color and $U(1)$ branes to be D3 branes while the weak branes to be D7 branes. Then, the ratio of couplings above can be explained by choosing the volume of the four compact dimensions of the seven branes to be $V_4 = 2.5$ in

[a]The gauge couplings $g_{2,3}$ are determined at the tree-level by the string coupling and other moduli, like radii of longitudinal dimensions. In higher orders, they also receive string threshold corrections.

string units. This being larger than one is consistent with the picture above. Moreover it predicts an interesting spectrum of KK states for the Standard model, different from the naive choices that have appeared hitherto: the only Standard Model particles that have KK descendants are the W bosons as well as the hypercharge gauge boson. However, since the hypercharge is a linear combination of the three $U(1)$'s, the massive $U(1)$ KK gauge bosons do not couple to the hypercharge but to the weak doublet number.

4.2. *The fate of $U(1)$'s, proton stability and neutrino masses*

It is easy to see that the remaining three $U(1)$ combinations orthogonal to Y are anomalous. In particular there are mixed anomalies with the $SU(2)$ and $SU(3)$ gauge groups of the Standard Model. These anomalies are cancelled by three axions coming from the closed string RR (Ramond) sector, via the standard Green–Schwarz mechanism.[14] The mixed anomalies with the non-anomalous hypercharge are also cancelled by dimension five Chern–Simmons type of interactions.[11] An important property of the above Green–Schwarz anomaly cancellation mechanism is that the anomalous $U(1)$ gauge bosons acquire masses leaving behind the corresponding global symmetries. This is in contrast to what would had happened in the case of an ordinary Higgs mechanism. These global symmetries remain exact to all orders in type I string perturbation theory around the orientifold vacuum. This follows from the topological nature of Chan–Paton charges in all string amplitudes. On the other hand, one expects non-perturbative violation of global symmetries and consequently exponentially small in the string coupling, as long as the vacuum stays at the orientifold point. Thus, all $U(1)$ charges are conserved and since Q_3 is the baryon number, proton stability is guaranteed.

Another linear combination of the $U(1)$'s is the lepton number. Lepton number conservation is important for the extra dimensional neutrino mass suppression mechanism described above, that can be destabilized by the presence of a large Majorana neutrino mass term. Such a term can be generated by the lepton-number violating dimension five effective operator $LLHH$ that leads, in the case of TeV string scale models, to a Majorana mass of the order of a few GeV. Even if we manage to eliminate this operator in some particular model, higher order operators would also give unacceptably large contributions, as we focus on models in which the ratio between the Higgs VEV and the string scale is just of order $\mathcal{O}(1/10)$. The best way to protect tiny neutrino masses from such contributions is to impose lepton number conservation.

A bulk neutrino propagating in $4+n$ dimensions can be decomposed in a series of 4d KK excitations denoted collectively by $\{m\}$:

$$S_{\rm kin} = R_\perp^n \int d^4x \sum_{\{m\}} \left\{ \bar\nu_{Rm} \partial\!\!\!/ \nu_{Rm} + \bar\nu_{Rm}^c \partial\!\!\!/ \nu_{Rm}^c + \frac{m}{R_\perp} \nu_{Rm}\nu_{Rm}^c + {\rm c.c.} \right\}, \quad (9)$$

where ν_R and ν_R^c are the two Weyl components of the Dirac spinor and for simplicity we considered a common compactification radius $R_\perp$. On the other hand, there is a localized interaction of ν_R with the Higgs field and the lepton doublet, which leads to mass terms between the left-handed neutrino and the KK states ν_{Rm}, upon the Higgs VEV v:

$$S_{\rm int} = g_s \int d^4x\, H(x)L(x)\nu_R(x, y=0) \quad \rightarrow \quad \frac{g_s v}{R_\perp^{n/2}} \sum_m \nu_L \nu_{Rm}\,, \quad (10)$$

in strings units. Since the mass mixing $g_s v/R_\perp^{n/2}$ is much smaller than the KK mass $1/R_\perp$, it can be neglected for all the excitations except for the zero-mode ν_{R0}, which gets a Dirac mass with the left-handed neutrino

$$m_\nu \simeq \frac{g_s v}{R_\perp^{n/2}} \simeq v\frac{M_s}{M_p} \simeq 10^{-3} - 10^{-2}\ {\rm eV}\,, \quad (11)$$

for $M_s \simeq 1$–10 TeV, where the relation (2) was used. In principle, with one bulk neutrino, one could try to explain both solar and atmospheric neutrino oscillations using also its first KK excitation. However, the later behaves like a sterile neutrino which is now excluded experimentally. Therefore, one has to introduce three bulk species (at least two) ν_R^i in order to explain neutrino oscillations in a "traditional way," using their zero-modes ν_{R0}^i.[15] The main difference with the usual seesaw mechanism is the Dirac nature of neutrino masses, which remains an open possibility to be tested experimentally.

5. Minimal Standard Model Embedding

In this section, we perform a general study of SM embedding in three brane stacks with gauge group $U(3)\times U(2)\times U(1)$,[16] and present an explicit example having realistic particle content and satisfying gauge coupling unification.[17] We consider in general non-oriented strings because of the presence of the orientifold plane that gives rise to mirror branes. An open string stretched between a brane stack $U(N)$ and its mirror transforms in the symmetric or antisymmetric representation, while the multiplicity of chiral fermions is given by their intersection number.

The quark and lepton doublets (Q and L) correspond to open strings stretched between the weak and the color or $U(1)$ branes, respectively. On

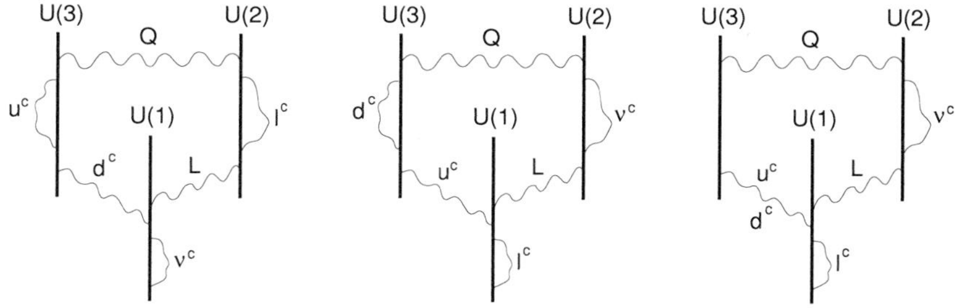

Fig. 5. Pictorial representation of models A, B and C.

the other hand, the u^c and d^c antiquarks can come from strings that are either stretched between the color and $U(1)$ branes, or that have both ends on the color branes (stretched between the brane stack and its orientifold image) and transform in the antisymmetric representation of $U(3)$ (which is an anti-triplet). There are therefore three possible models, depending on whether it is the u^c (model A), or the d^c (model B), or none of them (model C), the state coming from the antisymmetric representation of color branes. It follows that the antilepton l^c comes in a similar way from open strings with both ends either on the weak brane stack and transforming in the antisymmetric representation of $U(2)$ which is an $SU(2)$ singlet (in model A), or on the Abelian brane and transforming in the "symmetric" representation of $U(1)$ (in models B and C). The three models are presented pictorially in Fig. 5.

Thus, the members of a family of quarks and leptons have the following quantum numbers:

$$
\begin{array}{lll}
\text{Model A} & \text{Model B} & \text{Model C} \\
\hline
Q(\mathbf{3},\mathbf{2};1,1,0)_{1/6} & (\mathbf{3},\mathbf{2};1,\varepsilon_Q,0)_{1/6} & (\mathbf{3},\mathbf{2};1,\varepsilon_Q,0)_{1/6} \\
u^c(\bar{\mathbf{3}},\mathbf{1};2,0,0)_{-2/3} & (\bar{\mathbf{3}},\mathbf{1};-1,0,1)_{-2/3} & (\bar{\mathbf{3}},\mathbf{1};-1,0,1)_{-2/3} \\
d^c(\bar{\mathbf{3}},\mathbf{1};-1,0,\varepsilon_d)_{1/3} & (\bar{\mathbf{3}},\mathbf{1};2,0,0)_{1/3} & (\bar{\mathbf{3}},\mathbf{1};-1,0,-1)_{1/3} \\
L(\mathbf{1},\mathbf{2};0,-1,\varepsilon_L)_{-1/2} & (\mathbf{1},\mathbf{2};0,\varepsilon_L,1)_{-1/2} & (\mathbf{1},\mathbf{2};0,\varepsilon_L,1)_{-1/2} \\
l^c(\mathbf{1},\mathbf{1};0,2,0)_1 & (\mathbf{1},\mathbf{1};0,0,-2)_1 & (\mathbf{1},\mathbf{1};0,0,-2)_1 \\
\nu^c(\mathbf{1},\mathbf{1};0,0,2\varepsilon_\nu)_0 & (\mathbf{1},\mathbf{1};0,2\varepsilon_\nu,0)_0 & (\mathbf{1},\mathbf{1};0,2\varepsilon_\nu,0)_0
\end{array}
\tag{12}
$$

where the last three digits after the semi-column in the brackets are the charges under the three Abelian factors $U(1)_3 \times U(1)_2 \times U(1)$, that we will call Q_3, Q_2 and Q_1 in the following, while the subscripts denote the corresponding hypercharges. The various sign ambiguities $\varepsilon_i = \pm 1$ are due to the fact that the corresponding Abelian factor does not participate in the hypercharge combination (see below). In the last lines, we also give the quantum numbers of a possible right-handed neutrino in each of the three models. These are in fact all possible ways of embedding the SM spectrum in three sets of branes.

The hypercharge combination is:

$$\begin{aligned} \text{Model A}: \quad & Y = -\frac{1}{3}Q_3 + \frac{1}{2}Q_2\,, \\ \text{Model B, C}: \quad & Y = \frac{1}{6}Q_3 - \frac{1}{2}Q_1 \end{aligned} \tag{13}$$

leading to the following expressions for the weak angle:

$$\begin{aligned} \text{Model A}: \quad & \sin^2\theta_W = \frac{1}{2 + 2\alpha_2/3\alpha_3} = \frac{3}{8}\bigg|_{\alpha_2=\alpha_3}\,, \\ \text{Model B, C}: \quad & \sin^2\theta_W = \frac{1}{1 + \alpha_2/2\alpha_1 + \alpha_2/6\alpha_3} \\ & \qquad = \frac{6}{7 + 3\alpha_2/\alpha_1}\bigg|_{\alpha_2=\alpha_3}\,. \end{aligned} \tag{14}$$

In the second part of the above equalities, we used the unification relation $\alpha_2 = \alpha_3$, that can be imposed if for instance $U(3)$ and $U(2)$ branes are coincident, leading to a $U(5)$ unified group. Alternatively, this condition can be generally imposed under mild assumptions.[17] It follows that model A admits natural gauge coupling unification of strong and weak interactions, and predicts the correct value for $\sin^2\theta_W = 3/8$ at the unification scale M_{GUT}. On the other hand, model B corresponds to the flipped $SU(5)$ where the role of u^c and d^c is interchanged together with l^c and ν^c between the **10** and $\bar{\mathbf{5}}$ representations.[18]

Besides the hypercharge combination, there are two additional $U(1)$'s. It is easy to check that one of the two can be identified with $B - L$. For instance, in model A choosing the signs $\varepsilon_d = \varepsilon_L = -\varepsilon_\nu = -\varepsilon_H = \varepsilon_{H'}$, it is given by:

$$B - L = -\frac{1}{6}Q_3 + \frac{1}{2}Q_2 - \frac{\varepsilon_d}{2}Q_1\,. \tag{15}$$

Finally, the above spectrum can be easily implemented with a Higgs sector, since the Higgs field H has the same quantum numbers as the lepton doublet or its complex conjugate.

6. Conclusions

In this note, dedicated to 50 years after the proposal of quarks as elementary constituents of protons and neutrons, I gave a short overview of how they can emerge in string theory that provides a consistent quantum framework of unification of all fundamental forces of Nature, including gravity. String theory introduces a new fundamental energy scale associated with the string tension, or equivalently with the inverse string size. Its value can be high, near the four-dimensional Planck mass, compatible with traditional (supersymmetric) grand unification, or lower, up to the TeV scale providing an answer alternative to supersymmetry for solving the so-called hierarchy problem. The appropriate framework for such a realization is the (weakly coupled) type I theory of closed and open strings with D-branes. I have shown how the Standard Model can be embedded in such a framework.

References

1. For a review, see e.g. K. R. Dienes, *Phys. Rept.* **287**, 447 (1997), arXiv:hep-th/9602045, and references therein.
2. C. Angelantonj and A. Sagnotti, *Phys. Rept.* **371**, 1 (2002) [Erratum: *ibid.* **376**, 339 (2003)], arXiv:hep-th/0204089.
3. I. Antoniadis, *Phys. Lett. B* **246**, 377 (1990).
4. D. J. Kapner, T. S. Cook, E. G. Adelberger, J. H. Gundlach, B. R. Heckel, C. D. Hoyle and H. E. Swanson, *Phys. Rev. Lett.* **98**, 021101 (2007).
5. N. Arkani-Hamed, S. Dimopoulos and G. R. Dvali, *Phys. Lett. B* **429**, 263 (1998), arXiv:hep-ph/9803315; I. Antoniadis, N. Arkani-Hamed, S. Dimopoulos and G. R. Dvali, *Phys. Lett. B* **436**, 257 (1998), arXiv:hep-ph/9804398.
6. J. C. Long and J. C. Price, *Comptes Rendus Physique* **4**, 337 (2003); R. S. Decca, D. Lopez, H. B. Chan, E. Fischbach, D. E. Krause and C. R. Jamell, *Phys. Rev. Lett.* **94**, 240401 (2005); R. S. Decca *et al.*, arXiv:0706.3283 [hep-ph]; S. J. Smullin, A. A. Geraci, D. M. Weld, J. Chiaverini, S. Holmes and A. Kapitulnik, arXiv:hep-ph/0508204; H. Abele, S. Haeßler and A. Westphal, in *271th WE-Heraeus-Seminar*, Bad Honnef (2002).
7. G. Dvali, arXiv:0706.2050 [hep-th]; *Int. J. Mod. Phys. A* **25**, 602 (2010), arXiv:0806.3801 [hep-th]; G. Dvali and M. Redi, *Phys. Rev. D* **77**, 045027 (2008), arXiv:0710.4344 [hep-th]; R. Brustein, G. Dvali and G. Veneziano, *J. High Energy Phys.* **0910**, 085 (2009), arXiv:0907.5516 [hep-th].
8. G. Dvali and C. Gomez, *Phys. Lett. B* **674**, 303 (2009).
9. I. Antoniadis and B. Pioline, *Nucl. Phys. B* **550**, 41 (1999); I. Antoniadis, S. Dimopoulos and A. Giveon, *J. High Energy Phys.* **0105**, 055 (2001), arXiv: hep-th/0103033; I. Antoniadis, A. Arvanitaki, S. Dimopoulos and A. Giveon, arXiv:1102.4043.
10. G. Dvali and D. Lust, arXiv:0912.3167 [hep-th]; G. Dvali and C. Gomez, arXiv:1004.3744 [hep-th].

11. I. Antoniadis, E. Kiritsis and T. N. Tomaras, *Phys. Lett. B* **486**, 186 (2000); I. Antoniadis, E. Kiritsis, J. Rizos and T. N. Tomaras, *Nucl. Phys. B* **660**, 81 (2003).
12. G. Shiu and S.-H. H. Tye, *Phys. Rev. D* **58**, 106007 (1998); Z. Kakushadze and S.-H. H. Tye, *Nucl. Phys. B* **548**, 180 (1999); L. E. Ibáñez, C. Muñoz and S. Rigolin, *Nucl. Phys. B* **553**, 43 (1999).
13. K. R. Dienes, E. Dudas and T. Gherghetta, *Nucl. Phys. B* **557**, 25 (1999), arXiv:hep-ph/9811428; N. Arkani-Hamed, S. Dimopoulos, G. R. Dvali and J. March-Russell, *Phys. Rev. D* **65**, 024032 (2002), arXiv:hep-ph/9811448; G. R. Dvali and A. Y. Smirnov, *Nucl. Phys. B* **563**, 63 (1999).
14. A. Sagnotti, *Phys. Lett. B* **294**, 196 (1992); L. E. Ibáñez, R. Rabadán and A. M. Uranga, *Nucl. Phys. B* **542**, 112 (1999); E. Poppitz, *Nucl. Phys. B* **542**, 31 (1999).
15. H. Davoudiasl, P. Langacker and M. Perelstein, *Phys. Rev. D* **65**, 105015 (2002), arXiv:hep-ph/0201128.
16. I. Antoniadis, E. Kiritsis and T. N. Tomaras, *Phys. Lett. B* **486**, 186 (2000), arXiv:hep-ph/0004214; I. Antoniadis, E. Kiritsis, J. Rizos and T. N. Tomaras, *Nucl. Phys. B* **660**, 81 (2003), arXiv:hep-th/0210263; R. Blumenhagen, B. Kors, D. Lust and T. Ott, *Nucl. Phys. B* **616**, 3 (2001), arXiv:hep-th/0107138; M. Cvetic, G. Shiu and A. M. Uranga, *Nucl. Phys. B* **615**, 3 (2001), arXiv: hep-th/0107166; I. Antoniadis and J. Rizos, 2003 unpublished work.
17. I. Antoniadis and S. Dimopoulos, *Nucl. Phys. B* **715**, 120 (2005), arXiv:hep-th/0411032.
18. S. M. Barr, *Phys. Lett. B* **112**, 219 (1982); J. P. Derendinger, J. E. Kim and D. V. Nanopoulos, *Phys. Lett. B* **139**, 170 (1984); I. Antoniadis, J. R. Ellis, J. S. Hagelin and D. V. Nanopoulos, *Phys. Lett. B* **194**, 231 (1987).

$SU(8)$ Family Unification with Boson–Fermion Balance

Stephen L. Adler
Institute for Advanced Study,
Einstein Drive, Princeton, NJ 08540, USA
adler@ias.edu

We formulate an $SU(8)$ family unification model motivated by requiring that the theory should incorporate the graviton, gravitinos, and the fermions and gauge fields of the standard model, with boson–fermion balance. Gauge field $SU(8)$ anomalies cancel between the gravitinos and spin $\frac{1}{2}$ fermions. The 56 of scalars breaks $SU(8)$ to $SU(3)_{\text{family}} \times SU(5) \times U(1)/Z_5$, with the fermion representation content needed for "flipped" $SU(5)$ with three families, and with residual scalars in the 10 and $\overline{10}$ representations that break flipped $SU(5)$ to the standard model. Dynamical symmetry breaking can account for the generation of 5 representation scalars needed to break the electroweak group. Yukawa couplings of the 56 scalars to the fermions are forbidden by chiral and gauge symmetries, so in the first stage of $SU(8)$ breaking fermions remain massless. In the limit of vanishing gauge coupling, there are $N = 1$ and $N = 8$ supersymmetries relating the scalars to the fermions, which restrict the form of scalar self-couplings and should improve the convergence of perturbation theory, if not making the theory finite and "calculable." In an Appendix we give an analysis of symmetry breaking by a Higgs component, such as the $(1, 1)(-15)$ of the $SU(8)$ 56 under $SU(8) \supset SU(3) \times SU(5) \times U(1)$, which has nonzero $U(1)$ generator.

1. Introduction

The presence of bosons and fermions in Nature makes the idea of a fundamental boson–fermion balance appealing, and this has motivated an extensive search for supersymmetric extensions of the standard model. However, since the observed particle mass spectrum is not supersymmetric, supersymmetry breaking must be invoked, and despite much effort a definitive model, and a definitive symmetry breaking mechanism, have yet to emerge. We turn in this paper to another possibility, that boson–fermion balance without full supersymmetry is the relevant property of the unification theory, and construct a model based on this philosophy motivated by $SU(8)$ unification and supergravity.

2. Counting States

The model we study is inspired by the state structure of maximal $SO(8)$ supergravity. The usual counting of on-shell states for $N = 8$ supergravity is one graviton with 2 helicity states, 8 Majorana gravitinos with 16 helicity states, 28 vectors with 56 helicity states, 56 Majorana fermions with 112 helicity states, and 70 scalars with 70 helicity states. Thus there are $2+56+70 = 128$ boson states, and $16+112 = 128$ fermion states, giving the required boson–fermion balance, and interacting models with this field content exist. Unfortunately, however, these models do not contain the full particle and gauge group content needed for the standard model.

In a fascinating comment in his magisterial work on "Group theory for unified model building," Richard Slansky wrote:[1] "One may wish to speculate about a future unified theory of all interactions and all elementary particles that would resemble SO_8 supergravity but involve sacrificing some principle now held sacred, so that the notion of extended supergravity could be generalized. In such a hypothetical theory, an internal symmetry group G larger than SO_8 would be gauged by spin 1 bosons, and both the spin $\frac{3}{2}$ and spin $\frac{1}{2}$ fermions would be assigned to representations of G. It is then very natural to suppose that the spin $\frac{3}{2}$ fermions would belong to some basic representation of G and would include only color singlets, triplets and antitriplets. The spin $\frac{1}{2}$ particles would then presumably be assigned to a more complicated representation. These speculations are a major motivation for this review, as they were for Ref. 6."(Slansky's Ref. 6 is the paper by Gell-Mann, Ramond and Slansky.[2])

The rest of this paper proceeds in the spirit of Slansky's remarks (which as we shall see, describe the model that we construct). We begin by noting that if the 70 scalars are eliminated from the counting, and their degrees of freedom are redistributed to the two helicities of 35 vectors, we are left with $28 + 35 = 63$ vectors in all, which can be assigned to the adjoint representation of an $SU(8)$ group. The remaining representations in the counting, the 8 and 56, can be interpreted as the fundamental and rank three antisymmetric tensor representations of $SU(8)$, giving an "$SU(8)$ graviton" multiplet consisting of the graviton, the 8 gravitinos, the 63 vectors, and the 56 fermions. There are still 128 boson and 128 fermion helicities in this model, but the state structure is no longer the one corresponding to unitary supersymmetry representations in Hilbert space. Since we are working in four dimensions, and the model is not supersymmetric, we switch at this point from Majorana fermions to the usual left chiral (L) Weyl fermions used in grand unification, but the state counting is the same.

There is a long history of $SU(8)$ unification models in the literature; see Refs. 3–14. Of particular interest are the papers of Curtright and Freund,[3] C. Kim and Roiesnel,[7] and J. Kim and Song,[9] which incorporate spin $\frac{1}{2}$ fermions through single left chiral $\bar{8}$, $\overline{28}$, and 56 representations of $SU(8)$. Under breaking to $SU(5)$, the $\overline{28}$ of $SU(8)$ contains three copies of the $\bar{5}$ of $SU(5)$, and the 56 of $SU(8)$ contains three copies of the 10 of $SU(5)$, so this representation content incorporates the three standard model families. Additionally, the paper of Curtright and Freund explicitly ties the representation numbers 8, 28, and 56 to those appearing in $N = 8$ supergravity, with the suggestion that the $SU(8)$ gauge bosons may appear as bound states, as suggested by Cremmer and Julia.[15]

Returning to the "$SU(8)$ graviton" multiplet, the 56_L of fermions contains three families in the $SU(5)$ 10_L representation. In order to incorporate three $SU(5)$ $\bar{5}_L$ families into a model with boson–fermion balance, we adjoin to the "$SU(8)$ graviton" multiplet a "$SU(8)$ matter" multiplet consisting of a complex scalar field in the 56 representation of $SU(8)$, and *two* copies of a fermion spin $\frac{1}{2}$ field in the $\overline{28}_L$ representation of $SU(8)$. Use of a complex scalar is necessary since the 56 is a complex representation, and so cannot be assigned to a real scalar multiplet. Boson–fermion balance then requires that we double the number of $\overline{28}_L$ representations, so that the number of spin $\frac{1}{2}$ helicity states is $2 \times 2 \times 28 = 112$, equal to the number of helicity states in a complex 56 scalar. (Although boson–fermion balance could be achieved with a single 28_L of fermions and a complex 28 of scalars, $SU(8)$ anomalies would not cancel, and $SU(8)$ could not be broken to $SU(3) \times SU(5)$.) The $SU(8)$ fermion and boson content of the model is summarized in Table 1.

3. Anomaly Cancelation

To have a consistent $SU(8)$ gauge theory, anomalies must cancel. In the papers of Curtright and Freund,[3] Kim and Roiesnel,[7] and Kim and Song,[9] this is achieved through

$$\begin{aligned} \text{anomaly}(\bar{8}_L) &= -1\,, \\ \text{anomaly}(\overline{28}_L) &= -4\,, \\ \text{anomaly}(56_L) &= 5\,, \\ \text{total anomaly} &= -1 - 4 + 5 = 0\,. \end{aligned} \tag{1}$$

In our model anomaly cancelation involves the same representations, up to conjugation, but different counting. Instead of a spin $\frac{1}{2}$ $\bar{8}_L$, our "$SU(8)$ graviton" multiplet contains a spin $\frac{3}{2}$ 8_L. Since the chiral anomaly of a spin $\frac{3}{2}$

Table 1. Field content of the model, with the top part of the table showing the "$SU(8)$ graviton" multiplet, and the bottom part of the table showing the "$SU(8)$ matter" multiplet. The linearized graviton $h_{\mu\nu}$ is defined by $g_{\mu\nu} = \eta_{\mu\nu} + \kappa h_{\mu\nu}$, with $\eta_{\mu\nu}$ the Minkowski metric and κ the gravitational coupling. Branching rules are from Slansky[1] with $U(1)$ generators (or charges) in parentheses, followed in curly brackets by equivalent $U(1)$ generators modulo 5. (The modulo 5 ambiguities in these assignments have been used to give the assignments needed for flipped $SU(5)$, plus states that can be paired into condensates after family symmetry breaking or are neutral with respect to the $SU(3) \times SU(5) \times U(1)/Z_5$ force.) Square brackets on the field subscripts and superscripts indicate complete antisymmetrization of the enclosed indices. The indices α, β, γ range from 1 to 8, the index A runs from 1 to 63, and μ, ν are Lorentz indices.

Field	Spin	$SU(8)$ rep.	Helicities	Branching to $SU(3) \times SU(5) \times U(1)$
$h_{\mu\nu}$	2	1	2	1
ψ_μ^α	Weyl $\frac{3}{2}$	8_L	16	$(3,1)(-5)\{0\} + (1,5)(3)\{-2\}$
A_μ^A	1	63	126	$(1,1)(0)\{0\} + (8,1)(0)\{0\} + (3,\bar{5})(-8)\{2\}$ $+ (\bar{3},5)(8)\{-2\} + (1,24)(0)\{0\}$
$\chi^{[\alpha\beta\gamma]}$	Weyl $\frac{1}{2}$	56_L	112	$(1,1)(-15)\{0\} + (1,\overline{10})(9)\{-1\}$ $+ (\bar{3},5)(-7)\{3\} + (3,10)(1)\{1\}$
$\lambda_{1[\alpha\beta]}$	Weyl $\frac{1}{2}$	$\overline{28}_L$	56	$(3,1)(10)\{5\} + (1,\overline{10})(-6)\{-1\} + (\bar{3},\bar{5})(2)\{-3\}$
$\lambda_{2[\alpha\beta]}$	Weyl $\frac{1}{2}$	$\overline{28}_L$	56	$(3,1)(10)\{5\} + (1,\overline{10})(-6)\{-1\} + (\bar{3},\bar{5})(2)\{-3\}$
$\phi^{[\alpha\beta\gamma]}$	complex 0	56	112	$(1,1)(-15)\{0\} + (1,\overline{10})(9)\{-1\}$ $+ (\bar{3},5)(-7)\{-2\} + (3,10)(1)\{1\}$

particle is three times that of the corresponding spin $\frac{1}{2}$ particle,[16,17,a] the 8_L of gravitinos contributes 3 to the anomaly count. The 56_L of spin $\frac{1}{2}$ fermions contributes 5 as before, while the two $\overline{28}_L$ of spin $\frac{1}{2}$ fermions contribute -8, giving

$$\begin{aligned} 3 \times \text{anomaly } (8_L) &= 3\,, \\ 2 \times \text{anomaly } (\overline{28}_L) &= -8\,, \\ \text{anomaly } (56_L) &= 5\,, \\ \text{total anomaly in our model} &= 3 - 8 + 5 = 0\,. \end{aligned} \tag{2}$$

So anomalies cancel, but by a different mechanism than in Refs. 3, 7, 9. Anomaly cancellation with the counting of Eq. (2) (using the conjugate

[a]For the spin $\frac{3}{2}$ anomaly, see Eq. (4.6) and for the spin $\frac{3}{2}$ beta function, see Eq. (3.48) of Ref. 17.

representations $\bar{8}$, 28, and $\overline{56}$) was noted by Marcus[18] in a study of dynamical gauging of $SU(8)$ in $N = 8$ supergravity.

4. Gauge Symmetry Breaking and State Content

We turn to the issue of gauge symmetry breaking. Symmetry breaking in our model is initiated by a Brout–Englert–Higgs (BEH) mechanism using the complex scalar field in the "$SU(8)$ matter" multiplet, which is in the 56 of $SU(8)$. (This can be accomplished by either an explicit negative mass for the scalar in the action, or by an alternative that we favor, the Coleman–Weinberg[19] mechanism induced by radiative corrections starting from a massless scalar.) Since the 56 representation of $SU(8)$ branches to the 56_v of $SO(8)$, not to a singlet of $SO(8)$, the symmetry breaking pathway of our model cannot pass through $SO(8) \times U(1)$. Referring to Table 1, which gives the branching of the 56 of $SU(8)$ to $SU(3) \times SU(5) \times U(1)$, we see that there is a singlet (1,1) of $SU(3) \times SU(5)$ with a nonzero $U(1)$ generator of -15. Hence there are two interesting symmetry breaking pathways. In the first, the BEH mechanism breaks $SU(8)$ to $SU(3) \times SU(5)$, with the $U(1)$ gauge symmetry either completely broken or, as discussed in App. A, broken to $U(1)/Z$. It is then natural to identify the $SU(3)$ factor as a family symmetry group, and the $SU(5)$ factor and fermion content as the usual minimal grand unification group.[20] In the second, the $U(1)$ gauge symmetry breaks only to $U(1)/Z_5$, that is, after symmetry breaking there is an equivalence between values of $U(1)$ generators that differ by multiples of 5, as a result of a periodicity in the $U(1)$ generator of the broken symmetry ground state, which is discussed in detail in App. A. It is again natural to identify the unbroken $SU(3)$ factor as a family symmetry group. An inspection of the $U(1)$ generators modulo 5, given in curly brackets in Table 1, shows that the fermion content in this breaking pathway contains all the representations needed for flipped $SU(5)$ grand unification.[21–23,b]

To elaborate on this, the basic flipped $SU(5)$ model[c] consists of a $10\{1\}$ for the quark doublet Q, the down quark d^c, and the right handed neutrino N; a $\bar{5}\{-3\}$ for the lepton doublet L and the up quark u^c; and a $1\{5\}$ for the charged lepton e^c. Referring to Table 1, we see that χ contains a

[b] There are many more recent papers on flipped $SU(5)$, and two that we found helpful are Refs. 24 and 25.

[c] Wikipedia article on "Flipped $SU(5)$," whose notation we follow here. This article includes the Z_5 factor, which does not appear in the original flipped $SU(5)$ articles.[21–23] It is not clear whether this was intended to indicate a ground state modulo 5, or was just a shorthand for the division by 5 in the flipped $SU(5)$ hypercharge formula.

$(3, 10)\{1\}$, while λ_2 contains a $(\bar{3}, \bar{5})\{-3\}$ and a $(3, 1)\{5\}$. This gives three 3 or $\bar{3}$ families of the states needed for basic flipped $SU(5)$. Note that we have chosen the $U(1)$ charge assignments modulo 5 needed to make this correspondence possible. This guarantees that the correct particle charge assignments are obtained after further breaking to the standard model, and also implies that $SU(5)$ anomalies cancel within the set of spin $\frac{1}{2}$ states assigned to flipped $SU(5)$, without invoking the spin $\frac{3}{2}$ states. The remaining states are the $(\bar{3}, 5)\{3\}$ in χ and the $(\bar{3}, \bar{5})\{-3\}$ in λ_1, which after family symmetry breaking can pair to form a condensate; the $(1, 1)\{0\}$ in χ, which does not feel the $SU(3) \times SU(5) \times U(1)/Z_5$ force and could be a dark matter candidate; and the $(3, 1)(10)\{5 \equiv 0\}$ in $\lambda_{1,2}$, which together with the $(\bar{3}, 5)(-7)\{3\}$ in χ can form a condensate which leads to the standard model Higgs through dynamical chiral symmetry breaking (see below). There are also three $(1, \overline{10})\{-1\}$, one in each of the fermions χ, λ_1, and λ_2, which after family symmetry $SU(3)$ breaking can form condensates with the $(3, 10)\{1\}$ in χ to affect the particle mass spectrum. We note finally that an extended version of flipped $SU(5)$, proposed recently by Barr,[25] introduces a vector-like pair $5\{-2 \equiv 3\} + \bar{5}\{2 \equiv -3\}$ in each family and uses them to argue that proton decay can be rotated away.

There are residual boson states left after the 56 representation boson ϕ breaks the group $SU(8)$, with 63 generators, to $SU(3) \times SU(5)$, with $8 + 24 = 32$ generators, plus the single additional generator of the discrete group $U(1)/Z_5$ when $U(1)$ is not completely broken. Since $63 - 33 = 30$ real components of ϕ are absorbed to form longitudinal components of the broken $SU(8)$ generators in the $(3, \bar{5})(-8)\{2\}$ and $(\bar{3}, 5)(8)\{-2\}$ representations, these components can only come from the $(\bar{3}, 5)(-7)\{-2\}$ representation in the branching expansion for ϕ in Table 1. So the residual boson states necessarily are the representations $(1, \overline{10})(9)\{-1\}$ and $(3, 10)(1)\{1\}$, plus the $(1, 1)(-15)\{0\}$ when $U(1)$ is only broken to $U(1)/Z_5$. Since breaking minimal $SU(5)$ to the standard model requires a scalar in the 24 representation, the symmetry breaking pathway to $SU(3) \times SU(5)$ with minimal $SU(5)$ requires dynamical generation of this 24, to be further discussed below.

On the other hand, the residual boson states after $SU(8)$ breaking contain the Higgs boson representations needed to break flipped $SU(5)$ to the standard model.[21–25] Elaborating on this, the basic flipped $SU(5)$ model uses a $\overline{10}\{-1\}$ and a $10\{1\}$ of scalars to break flipped $SU(5)$ to the standard model, and these representations are residual components of the scalar ϕ. To further break the electroweak group of the standard model to the electromagnetic $U(1)$ group, flipped $SU(5)$ requires a $5\{-2\}$ of scalars, which contains the standard model Higgs. This is not present as a residual scalar component of ϕ, but as shown below can by generated in our model by dynamical symmetry breaking.

5. Asymptotic Freedom and Global Symmetries

The $SU(8)$ representation content of the model has a small enough spin 0, spin $\frac{1}{2}$, and spin $\frac{3}{2}$ content to keep the theory asymptotically free,

$$\frac{1}{3}[11c(1) - 26c(\text{Weyl } 3/2) - 2c(\text{Weyl } 1/2) - c(\text{complex } 0)]$$
$$= \frac{1}{3}[11 \times 16 - 26 \times 1 - 2 \times (15 + 2 \times 6) - 15] = 27 > 0\,, \quad (3)$$

with $c(s)$ the index of the $SU(8)$ representation with spin s. (For the spin $\frac{3}{2}$ beta function see Curtright,[26] Duff,[17] and Fradkin and Tseytlin;[27,d] the index c is tabulated as ℓ in the tables of Slansky.[1]) Thus the $SU(8)$ coupling increases as the energy decreases, which can trigger dynamical symmetry breaking in addition to the symmetry breaking provided by the elementary Higgs fields. In addition to a locally gauged $SU(8)$ symmetry, our model admits a number of global chiral symmetries associated with the fermion fields.[28] The first is an overall chiral $U(1)$ symmetry associated with an overall $U(1)$ rephasing of all of the fermion fields, spin $\frac{3}{2}$ as well as spin $\frac{1}{2}$. It will be convenient to regard this phase as associated with the 8_L of spin $\frac{3}{2}$ fermions, labeled ψ_μ in Table 1. We expect this global symmetry to be broken by the usual instanton and anomaly mechanism that is invoked to solve the "$U(1)$ problem" in QCD.[29–31] The second global symmetry is an overall $U(1)$ rephasing of the spin $\frac{1}{2}$ 56_L fermion fields χ relative to ψ_μ. Finally, since the kinetic Lagrangian contains the doubled spin $\frac{1}{2}$ $\overline{28}_L$ representation spanned by the fermion basis $\lambda_{1,2}$, there is a global $U(2)$ symmetry associated with mixing of these basis states, relative to the phase of ψ_μ.

6. Dynamical versus Elementary Higgs Symmetry Breaking

As already noted, in the $SU(8) \supset SU(3) \times SU(5)$ symmetry breaking pathway, the $SU(5)$ 24 representation needed for breaking to the standard model must be generated dynamically. A quick review of the theory of dynamical symmetry breaking is given in App. B. A strategy for getting a 24, following Ref. 32, would be to generate a 24 condensate at the unification scale, which

[d]Fradkin and Tseytlin used instanton counting to reproduce the spin $\frac{3}{2}$ beta function coefficient given by Curtright.[26] I wish to thank Arkady Tseytlin for email correspondence about this calculation.

violates the chiral symmetries of the theory, and so leads to a 24 Goldstone boson, which could then serve as the 24 Higgs. There are two problems with this scenario. The first is that the only way to generate a 24 representation of $SU(5)$ from the representations in Table 1 is through either $\bar{5} \times 5$ or $\overline{10} \times 10$, both of which contain an $SU(5)$ singlet in addition to a 24. Since the singlet is always the most attractive channel (see Eq. (B.6)), dynamical generation of a 24 seems unlikely.[28,33,34] The second problem is that if gauge couplings were strong enough for a 24 condensate to be formed at the unification scale, then one would expect that unification scale condensates involving the wanted fermions in the $(3,10)$ representation in Table 1 would also form, removing these states from the low energy spectrum. So getting the standard model from our theory through an $SU(8) \supset SU(3) \times SU(5)$ symmetry breaking pathway is not plausible.

The situation is more favorable for the $SU(8) \supset SU(3) \times SU(5) \times U(1)/Z_5$ symmetry breaking pathway, which as explained in App. A involves a ground state that is periodic in the $U(1)$ generator. This pathway does not require dynamical condensates to break $SU(3) \times SU(5) \times U(1)/Z_5$ to the standard model; the residual elementary scalar states in ϕ can do this, as well as breaking the $SU(3)$ family symmetry. (Before $SU(5)$ breaking, the $(3,10)(1)\{1\}$ scalar can break family $SU(3)$ to $SU(2) \times U(1)$, accommodating two light families and one heavy one, while after $SU(5)$ breaking family symmetry can be completely broken.[35]) However, we have seen that the residual components of the 56 scalar, after $SU(8)$ symmetry breaking, do not contain the $5\{-2\}$ needed in flipped $SU(5)$ to break the electroweak symmetry of the standard model, so here the dynamical symmetry breaking mechanism of Ref. 32 is needed. Referring to Table 1, we see that the $(\overline{3},5)(-7)$ component of χ can pair with the $(3,1)(10)$ component of the doublet λ_a, to form a doublet condensate which is in the representation $(1,5)(3)\{3 \equiv -2\}$, with the family $SU(3)$ group acting as the hypercolor or "technicolor" force in binding the condensate in the most attractive singlet channel. Since this condensate breaks the $U(2)\big(= U(1) \times SU(2)\big) \times U(1)$ global chiral symmetry of the doublet λ_a and of χ to a diagonal $U(1)$, there will be a $U(1)$ singlet and a $SU(2)$ triplet of Goldstone bosons with the needed flipped $SU(5)$ quantum numbers $5\{-2\}$. These Goldstone bosons are still gauged under the $SU(5)$ group, so the Coleman–Weinberg mechanism will generate symmetry breaking potentials for them, leading to the electroweak symmetry breaking of the standard model. Our model thus suggests that in addition to the observed (presumably singlet) Higgs boson, there should also be an $SU(2)$ triplet of Higgs bosons with the same standard model quantum numbers.

7. The Gauge Sector Action

We turn now to writing down the action for the gauge sector of our model. Since all fermion representations are antisymmetrized direct products of fundamental 8 representations, we need only use generators t_A for the fundamental 8 of $SU(8)$ to construct covariant derivatives of the fermion fields. We follow here the conventions of Ref. 36, and take the t_A to be anti-self-adjoint, with commutators and trace normalization given by

$$\begin{aligned}[][t_A, t_B] &= f_{ABC} t_C \,, \\ \mathrm{Tr}(t_A t_B) &= -\frac{1}{2}\delta_{AB} \,,\end{aligned} \tag{4}$$

with implicit summation on repeated indices.

Defining the gauge variation of the gauge potential by

$$\delta_G A_\mu^A = \frac{1}{g}\partial_\mu \Theta^A + f_{ABC} A_\mu^B \Theta^C \,, \tag{5}$$

the gauge covariant field strength $F_{\mu\nu}^A$ is defined as

$$F_{\mu\nu}^A = \partial_\mu A_\nu^A - \partial_\nu A_\mu^A + g f_{ABC} A_\mu^B A_\nu^C \,, \tag{6}$$

and has the gauge variation

$$\delta_G F_{\mu\nu}^A = f_{ABC} F_{\mu\nu}^B \Theta^C \,. \tag{7}$$

We can now define covariant derivatives of the fermion fields of the model. Writing

$$A_{\mu\beta}^{\alpha} = A_\mu^A (t_A)^\alpha{}_\beta \,, \tag{8}$$

the covariant derivatives of the fermion fields ψ_μ^α, $\chi^{[\alpha\beta\gamma]}$ and $\lambda_{a\,[\alpha\beta]}$, $a = 1, 2$ of Table 1 (in the 8, 56 and $\overline{28}$ representations, respectively) are defined by

$$\begin{aligned} D_\nu \psi_\mu^\alpha &= \partial_\nu \psi_\mu^\alpha + g A_{\nu\delta}^\alpha \psi_\mu^\delta \,, \\ D_\nu \chi^{[\alpha\beta\gamma]} &= \partial_\nu \chi^{[\alpha\beta\gamma]} + g\Big(A_{\nu\delta}^\alpha \chi^{[\delta\beta\gamma]} + A_{\nu\delta}^\beta \chi^{[\alpha\delta\gamma]} + A_{\nu\delta}^\gamma \chi^{[\alpha\beta\delta]}\Big) \,, \\ D_\nu \lambda_{a[\alpha\beta]} &= \partial_\nu \lambda_{a[\alpha\beta]} + g\big(A_{\nu\alpha}^\delta \lambda_{a[\delta\beta]} + A_{\nu\beta}^\delta \lambda_{a[\alpha\delta]}\big) \,, \quad a = 1, 2 \,. \end{aligned} \tag{9}$$

Similarly, for the scalar field $\phi^{[\alpha\beta\gamma]}$, the covariant derivative is defined by

$$D_\nu \phi^{[\alpha\beta\gamma]} = \partial_\nu \phi^{[\alpha\beta\gamma]} + g\Big(A_{\nu\delta}^\alpha \phi^{[\delta\beta\gamma]} + A_{\nu\delta}^\beta \phi^{[\alpha\delta\gamma]} + A_{\nu\delta}^\gamma \phi^{[\alpha\beta\delta]}\Big) \,. \tag{10}$$

These give

$$\delta_G \psi_\mu^\alpha = -\theta^A t_{A\delta}^\alpha \psi_\mu^\delta \,, \quad \delta_G D_\nu \psi_\mu^\alpha = -\theta^A t_{A\delta}^\alpha D_\nu \psi_\mu^\delta \,, \tag{11}$$

and similarly for the gauge variations of the other fields and their covariant derivatives.

With the $SU(8)$ covariant derivatives of the fields defined, we can now write down the gauge sector action of the model, with gravity treated in the linearized approximation, as follows. The total action is

$$S(\text{total}) = S(h_{\mu\nu}) + S(\psi_\mu) + S(A_\mu) + S(\chi) + S(\lambda_{1,2}) + S_{\text{kinetic}}(\phi) + S_{\text{self-coupling}}(\phi) + S_{\text{fermion-coupling}}(\phi, \psi_\mu, \chi, \lambda)\,. \tag{12}$$

For $S(h_{\mu\nu})$ we have the usual linearized gravitational action,

$$\begin{aligned} S(h_{\mu\nu}) &= \frac{1}{8}\int d^4x\, h^{\mu\nu} H_{\mu\nu}\,, \\ H_{\mu\nu} &= \partial_\mu\partial_\nu h^\lambda_\lambda + \Box h_{\mu\nu} - \partial_\mu\partial^\lambda h_{\lambda\nu} \\ &\quad - \partial_\nu\partial^\lambda h_{\lambda\mu} - \eta_{\mu\nu}\Box h^\lambda_\lambda + \eta_{\mu\nu}\partial^\lambda\partial^\rho h_{\lambda\rho}\,. \end{aligned} \tag{13}$$

For the gravitino action we have the $SU(8)$ gauged extension of the usual expression,

$$\begin{aligned} S(\psi_\mu) &= \frac{1}{2}\int d^4x\, \bar{\psi}_{\mu\alpha} R^{\mu\alpha}\,, \\ R^{\mu\alpha} &= i\epsilon^{\mu\eta\nu\rho}\gamma_5\gamma_\eta D_\nu\psi^\alpha_\rho \\ &= R^{\mu\alpha}_{\text{free}} + R^{\mu\alpha}_{\text{interaction}}\,, \\ R^{\mu\alpha}_{\text{free}} &= i\epsilon^{\mu\eta\nu\rho}\gamma_5\gamma_\eta\partial_\nu\psi^\alpha_\rho\,, \\ R^{\mu\alpha}_{\text{interaction}} &= ig\epsilon^{\mu\eta\nu\rho}\gamma_5\gamma_\eta A^\alpha_{\nu\delta}\psi^\delta_\rho\,. \end{aligned} \tag{14}$$

Since the free gravitino action is invariant under the gravitino gauge transformation $\psi^\alpha_\rho \to \psi^\alpha_\rho + \partial_\rho\epsilon^\alpha$, a gauge fixing condition is needed to quantize, which can be taken in the covariant form $\gamma^\rho\psi^\alpha_\rho = 0$. The associated ghost fields then play a role in the spin $\frac{3}{2}$ anomaly calculation.[37]

The $SU(8)$ gauge field action has the standard form

$$S(A_\mu) = -\frac{1}{4}\int d^4x F^A_{\mu\nu} F^{A\mu\nu}\,, \tag{15}$$

and the spin $\frac{1}{2}$ fermion actions are

$$\begin{aligned} S(\chi) &= -\frac{1}{2}\int d^4x\, \bar{\chi}_{[\alpha\beta\gamma]}\gamma^\nu D_\nu\chi^{[\alpha\beta\gamma]}\,, \\ S(\lambda_{1,2}) &= -\frac{1}{2}\int d^4x \sum_a \overline{\lambda_a}^{[\alpha\beta]}\gamma^\nu D_\nu\lambda_{a[\alpha\beta]}\,, \end{aligned} \tag{16}$$

where we have written the second line in a form which exhibits its global $U(2)$ invariance. Finally, for the scalar field kinetic action we have

$$S_{\rm kinetic}(\phi) = -\frac{1}{2}\int d^4x (D^\nu\phi)^*_{[\alpha\beta\gamma]} D_\nu \phi^{[\alpha\beta\gamma]} \,. \tag{17}$$

For later use, we note that the equations of motion of the spin $\frac{1}{2}$ fermions and the spin 0 boson, following from these gauged kinetic actions but ignoring for the moment possible additional scalar interaction terms, are

$$\begin{aligned} \gamma^\nu D_\nu \chi^{[\alpha\beta\gamma]} &= 0\,, \\ \gamma^\nu D_\nu \lambda_{a[\alpha\beta]} &= 0\,, \quad a = 1,2\,, \\ D^\nu D_\nu \phi^{[\alpha\beta\gamma]} &= 0\,. \end{aligned} \tag{18}$$

8. Absence of Scalar–Fermion Yukawa Couplings

We turn next to possible Yukawa couplings $S_{\text{fermion-coupling}}(\phi, \psi_\mu, \chi, \lambda)$, which we show must all vanish. The chirality requirements for forming nonzero Yukawa couplings are the same as those for forming condensates discussed in App. B. Thus, chirality requires that Yukawa couplings of the spin $\frac{1}{2}$ fermions must be of the form $\Psi_{L1}^T i\gamma 0 \Psi_{L2}\Phi$, with $\Psi_{1,2}$ any of the spin $\frac{1}{2}$ fermion fields, and Φ either ϕ or ϕ^*, with $SU(8)$ indices contracted to form a singlet. But this is not possible, since the product of two χ has six upper $SU(8)$ indices, the product of two λ has four lower $SU(8)$ indices, and the product of χ with a λ has three upper $SU(8)$ indices and two lower $SU(8)$ indices, none of which can be contracted with a ϕ, with three upper indices, or a ϕ^*, with three lower indices, to form an $SU(8)$ singlet. Hence there are no Yukawa couplings involving the spin $\frac{1}{2}$ fermions by themselves. This implies that after $SU(8)$ symmetry breaking, the spin $\frac{1}{2}$ fermions remain massless, which is essential for getting a three family flipped $SU(5)$ model.

Yukawa couplings of a spin $\frac{3}{2}$ field to a spin $\frac{3}{2}$ field and the scalar are forbidden by a chirality and $SU(8)$ index contraction argument similar to that used in the case of two spin $\frac{1}{2}$ fields. This argument does not forbid couplings of a spin $\frac{3}{2}$ field to a spin $\frac{1}{2}$ field and the scalar of the form $\overline{\lambda_a}^{[\alpha\beta]}\gamma^\nu\psi_\nu^\gamma\phi^*_{[\alpha\beta\gamma]}$ and its conjugate, but these vanish when the gravitino gauge fixing condition $\gamma^\nu\psi_\nu^\gamma = 0$ is imposed.

9. Supersymmetries in the Limit of Zero Gauge Coupling

Let us now consider the free limit of the theory in which the gauge coupling g vanishes, so that the covariant derivatives D_ν become ordinary partial derivatives ∂_ν, and the equations of motion of Eq. (18) simplify to

$$\begin{aligned} \gamma^\nu \partial_\nu \chi^{[\alpha\beta\gamma]} &= 0\,, \\ \gamma^\nu \partial_\nu \lambda_{a\,[\alpha\beta]} &= 0\,, \quad a = 1, 2\,, \\ \partial^\nu \partial_\nu \phi^{[\alpha\beta\gamma]} &= 0\,. \end{aligned} \tag{19}$$

One can then form two conserved $SU(8)$ representation 8 supercurrents,

$$\begin{aligned} J_a^{\mu\alpha} &= \gamma^\nu \big(\partial_\nu \phi^{[\alpha\beta\gamma]}\big) \gamma^\mu \lambda_{a[\beta\gamma]}\,, \\ \partial_\mu J_a^{\mu\,\alpha} &= 0\,, \quad \alpha = 1, \ldots, 8 \quad \text{and} \quad a = 1, 2\,, \end{aligned} \tag{20}$$

and an $SU(8)$ singlet conserved supercurrent,

$$J^\mu = \gamma^\nu (\partial_\nu \phi^*_{[\alpha\beta\gamma]}) \gamma^\mu \chi^{[\alpha\beta\gamma]}\,, \quad \partial_\mu J^\mu = 0\,. \tag{21}$$

In deriving supercurrent conservation we have used the equations of motion together with

$$\gamma^\nu \gamma^\mu \partial_\nu \partial_\mu \Phi = \eta^{\mu\nu} \partial_\nu \partial_\mu \Phi\,, \tag{22}$$

which is a consequence of the commutativity of partial derivatives. The invariance transformation of the free action for which $J_a^{\mu\alpha}$ (with $a = 1$ or 2) is the Noether current is

$$\begin{aligned} \delta\phi^{[\alpha\beta\gamma]} &= \bar{\lambda}_a^{[[\beta\gamma]} \epsilon^{\alpha]}\,, \\ \delta\phi^*_{[\alpha\beta\gamma]} &= \bar{\epsilon}_{[\alpha} \lambda_{a[\beta\gamma]]}\,, \\ \delta\lambda_{a[\alpha\beta]} &= \gamma^\nu \partial_\nu \phi^*_{[\alpha\beta\delta]} \epsilon^\delta\,, \\ \delta\bar{\lambda}_a^{[\alpha\beta]} &= -\bar{\epsilon}_\delta \gamma^\nu \partial_\nu \phi^{[\alpha\beta\delta]}\,, \end{aligned} \tag{23}$$

and the transformation for which J^μ is the Noether current is

$$\begin{aligned} \delta\phi^{[\alpha\beta\gamma]} &= \bar{\epsilon} \chi^{[\alpha\beta\gamma]}\,, \\ \delta\phi^*_{[\alpha\beta\gamma]} &= \bar{\chi}_{[\alpha\beta\gamma]} \epsilon\,, \\ \delta\chi^{[\alpha\beta\gamma]} &= \gamma^\nu \partial_\nu \phi^{[\alpha\beta\gamma]} \epsilon\,, \\ \delta\bar{\chi}_{[\alpha\beta\delta]} &= -\bar{\epsilon} \gamma^\nu \partial_\nu \phi^*_{[\alpha\beta\gamma]}\,. \end{aligned} \tag{24}$$

Since covariant derivatives do not commute, when gauge interactions are included there are no longer conserved supercurrents. For example, if we redefine the singlet current as

$$J^\mu = \gamma^\nu \big(D_\nu \phi^*_{[\alpha\beta\gamma]}\big) \gamma^\mu \chi^{[\alpha\beta\gamma]}\,, \tag{25}$$

then we find

$$\begin{aligned}\partial_\mu J^\mu &= \Big(D_\mu\gamma^\nu\big(D_\nu\phi^*_{[\alpha\beta\gamma]}\big)\Big)\gamma^\mu\chi^{[\alpha\beta\gamma]} + \gamma^\nu\big(D_\nu\phi^*_{[\alpha\beta\gamma]}\big)\gamma^\mu D_\mu\chi^{[\alpha\beta\gamma]} \\ &= \frac{1}{2}\Big([D_\mu, D_\nu]\gamma^{\nu\mu}\phi^*_{[\alpha\beta\gamma]}\Big)\chi^{[\alpha\beta\gamma]} \\ &= \frac{1}{2}g\gamma^{\nu\mu}\Big(F^\delta_{\mu\nu\alpha}\phi^*_{[\delta\beta\gamma]} + F^\delta_{\mu\nu\beta}\phi^*_{[\alpha\delta\gamma]} + F^\delta_{\mu\nu\gamma}\phi^*_{[\alpha\beta\delta]}\Big)\chi^{[\alpha\beta\gamma]}\,, \end{aligned} \tag{26}$$

with $\gamma^{\nu\mu} = \frac{1}{2}[\gamma^\nu, \gamma^\mu]$.

10. Scalar Sector Self-couplings

We consider finally the action terms involving the scalar field without gauging. For the scalar field self-coupling action, taking index permutation possibilities into account, we have

$$S_{\text{self-coupling}}(\phi) = \phi^*_{[\rho\kappa\tau]}\phi^*_{[\alpha\beta\gamma]}\Big(g_1\phi^{[\rho\kappa\tau]}\phi^{[\alpha\beta\gamma]} + g_2\phi^{[\alpha\kappa\tau]}\phi^{[\rho\beta\gamma]}\Big)\,, \tag{27}$$

which is a straightforward generalization of the usual real scalar field ϕ^4 coupling. However, when the gauge coupling g is zero, the kinetic action is invariant under the supersymmetry transformations of Eqs. (23) and (24), which are not invariances of the self-coupling action of Eq. (27). Hence the couplings g_1 and g_2 must be of order g^2 or higher order in the gauge coupling. An important question to be answered is how the invariances of the action affect the renormalization of $g_{1,2}$: in what order of g^2 do they contain logarithms of the ultraviolet cutoff, or are they finite and calculable to all orders? Since there are no Yukawa couplings, it is possible that the theory is calculable in the sense suggested by Weinberg.[38]

11. Discussion

Grand unification has been intensively investigated for over forty years, and many different approaches have been tried. The model proposed here involves three ingredients that do not appear in the usual constructions: (1) boson–fermion balance without full supersymmetry, (2) canceling the spin $\frac{1}{2}$ fermion gauge anomalies against the anomaly from a gauged spin $\frac{3}{2}$ gravitino, and (3) using a scalar field representation with nonzero $U(1)$ generator to break the gauge symmetry, through a ground state with periodic $U(1)$ generator structure. The model has a number of promising features: (1) natural incorporation of three families, (2) incorporation of the experimentally viable flipped $SU(5)$ model, (3) a symmetry breaking pathway to the standard model using the scalar field required by boson–fermion balance, together with a stage of most attractive channel dynamical symmetry breaking, without postulating additional Higgs fields, and (4) vanishing of bare Yukawa couplings and zero gauge coupling supersymmetries, which keeps the spin $\frac{1}{2}$ fermions massless after $SU(8)$ symmetry breaking, and may improve the predictive power of the theory.

This investigation started from an attempt to base a supersymmetric theory on the state counting of Sec. 2. In the free limit of zero $SU(8)$ couplings, we saw that the supercurrents of Eqs. (20) and (21) are conserved, but that the analogous construction does not give a conserved supercurrent when $SU(8)$ gauge interactions are included. Moreover, even in the free limit, there is no corresponding conserved representation 8 supercurrent for the "$SU(8)$ gravity" multiplet, since in $SU(8)$, 8×63 does not contain the totally antisymmetric 56 representation. If one instead looks for an $SO(8)$ representation 8 supercurrent, a similar problem arises, since the direct product of 8 with the symmetric 35 of $SO(8)$ again does not contain the totally antisymmetric 56 representation. So for these reasons we abandoned the search for a supersymmetric model, and instead turned to the weaker condition of boson–fermion balance. (Group representation considerations leave open the possibility of constructing 8 $N = 1$ Lorentz and gauge noncovariant supercurrents in the free limit, by stacking the two helicity components of the symmetric 35 of the gauge field into an artificial 70 component "scalar" $\tilde{\phi}^{[\alpha\beta\gamma\delta]}$.)

Many open issues remain. In Table 1, we used the modulo 5 freedom of the $U(1)/Z_5$ charges to assign these charges so that the representations needed for flipped $SU(5)$ have the usual $U(1)$ charge assignments for that model. This recipe is *ad hoc*, and needs further justification from a detailed study of the dynamics of symmetry breaking with a modular ground state prior to dynamical symmetry breaking. (For example, it would suffice to show that after dynamical symmetry breaking, charge states differing from the wanted ones are separated by a large mass gap, or are absent from the asymptotic spectrum altogether, from anomaly considerations and/or a discrete analog of the familiar vacuum alignment condition.[39]) As is clear, our analysis is focussed solely on boson–fermion balance, Lorentz structures, and group theory, and does not address further dynamical issues such as running couplings, proton decay, generating the standard model mass and mixing parameters, CP violation, and flavor changing neutral current constraints on a multiple Higgs structure. Nonetheless, the issues examined are an essential first step in trying to set up a realistic unification model, and the results look promising; the pieces appear to fit together in a jigsaw puzzle-like fashion reminiscent of what one finds in the standard model.

The model presented here should have distinctive experimental signatures. First, in common with generic flipped $SU(5)$ models, it will have three families of sterile neutrinos. Second, as noted above, in addition to a singlet Higgs boson, it should have an $SU(2)$ triplet of Higgs bosons with the same quantum numbers.

If the model we propose turns out to be the path that Nature follows, there will remain the further question of how the "$SU(8)$ gravity" multiplet and the "$SU(8)$ matter" multiplet of the model are unified in a more fundamental structure, for example, as arising from involutions of a large finite group or from periodic or aperiodic tilings of a large lattice. We note that the "$SU(8)$ gravity" multiplet has 128 boson and fermion helicity states, and the "$SU(8)$ matter" multiplet has 112 boson and fermion helicity states. These numbers respectively match the numbers of half-integer and integer roots of the exceptional group $E(8)$. Is this a numerical coincidence, or a hint of a deep connection with the $E(8)$ root lattice?

Acknowledgments

I wish to acknowledge the hospitality during the summers of 2013 and 2014 of the Aspen Center for Physics, which is supported by the National Science Foundation under Grant No. PHYS-1066293. I wish to thank Edward Witten for instructive comments on the Z_5 and Z_6 factors in the Wikipedia article on flipped $SU(5)$, Paul Langacker for reading through the paper and a helpful email correspondence, Michael Peskin for an informative survey lecture on composite Higgs models at the Princeton Center for Theoretical Science, and Michael Duff for email correspondence about the spin $\frac{3}{2}$ anomaly and beta function values, which brought Refs. 17 and 18 to my attention. I also wish to thank Graham Kribs for giving me the opportunity to talk at the Aspen workshop that he co-organized, and David Curtin for asking a question during my talk about the masses of the scalars.

Appendix A. Higgs Mechanism using a Representation with Nonzero $U(1)$ Charge

In the usual application of the Higgs mechanism to grand unification, such as in the breaking of minimal $SU(5)$ to the standard model $SU(3) \times SU(2) \times U(1)_Y$, a Higgs representation is chosen which contains a component that is a singlet under all three factors of the standard model symmetry group.

Thus, the 24 of SU(5) can be used, since it branches according to $24 = (1,1)(0) + (3,1)(0) + (2,3)(-5) + (2,\bar{3})(5) + (1,8)(0)$, which contains the overall singlet $(1,1)(0)$. This singlet can attain a nonzero expectation in a ground state (the "vacuum") that has a definite value 0 of the unbroken $U(1)$ generator.

In the $SU(8)$ model studied in this paper, only the 56 representation is available as a scalar to break the symmetry to $SU(3) \times SU(5) \times U(1)$, and the component $\phi_{(1,1)(-15)}$ that is an $SU(3) \times SU(5)$ singlet has nonzero $U(1)$ charge -15. By the generalized Wigner–Eckart theorem, this component cannot acquire a nonzero expectation in a ground state $|\Omega\rangle$ that is a $U(1)$ eigenstate with a definite generator value. To get a nonzero expectation, we must take $|\Omega\rangle$ to be a superposition of at least two $U(1)$ eigenstates that differ in their $U(1)$ generators by 15. Anticipating that we want the final result to have a modulo 5 (and not a modulo 15 or modulo 3) structure, we write the ground state as a superposition of $U(1)$ eigenstates displaced from one another by 5. Let G be the $U(1)$ generator, and $|n\rangle$ a $SU(3) \times SU(5)$ singlet that is a $U(1)$ eigenstate with eigenvalue (or $U(1)$ charge) n, so that $G|n\rangle = n|n\rangle$. Then we write the ground state $|\Omega\rangle$ in the form

$$|\Omega\rangle = \sum_n f(n)|5n\rangle\,, \tag{A.1}$$

which for generic $f(n)$ completely breaks the $U(1)$ invariance,

$$\langle\Omega|\phi_{(1,1)(-15)}|\Omega\rangle \neq 0\,. \tag{A.2}$$

As in the similar analysis of the ground state structure of quantum chromodynamics, let us now impose the requirement of clustering. In order for the ground state of a tensor product composite system

$$|\Omega_{A+B}\rangle = \sum_{n_A,n_B} f(n_A + n_B)|A;5n_A\rangle|B;5n_B\rangle \tag{A.3}$$

to factor when the subsystems A, B are widely separated,

$$\begin{aligned} |\Omega_{A+B}\rangle &= |\Omega_A\rangle|\Omega_B\rangle\,, \\ |\Omega_A\rangle &= \sum_{n_A} f(n_A)|A;5n_A\rangle\,, \\ |\Omega_B\rangle &= \sum_{n_B} f(n_B)|B;5n_B\rangle\,, \end{aligned} \tag{A.4}$$

we must require $f(n)$ to obey

$$f(n_A + n_B) = f(n_A)f(n_B)\,. \tag{A.5}$$

This requires that $f(n)$ must have the functional form

$$f(n) = e^{nz} \tag{A.6}$$

for some complex number z. Boundedness as $|n| \to \infty$ requires that $|e^z| = 1$, so e^z is a phase $e^{i\omega}$. The ground state then has the form

$$|\Omega\rangle = \sum_{n=-\infty}^{\infty} e^{in\omega}|5n\rangle\,, \tag{A.7}$$

and $U(1)$ charges are only conserved modulo 5. This ground state corresponds to breaking $SU(8)$ to $SU(3) \times SU(5) \times U(1)/Z_5$, which is the "second symmetry breaking pathway" and the one chosen for our analysis. (The "first symmetry breaking pathway" corresponds either to breaking $U(1)$ completely by using a nonexponential $f(n)$ that violates clustering, or to breaking $U(1)$ to $U(1)/Z$ by choosing the ground state $|\Omega\rangle = \sum_{n=-\infty}^{\infty} \exp{(in\omega)}|n\rangle$, which equivalences the integer $U(1)$ charges all to zero.)

The full basis of states for the second pathway has the form

$$|k\rangle = \sum_{n=-\infty}^{\infty} e^{in\omega}|5n+k\rangle\,, \tag{A.8}$$

with $|k=0\rangle = |\Omega\rangle$. Under a modulo 5 shift we have

$$\begin{aligned}|k+5s\rangle &= \sum_{n=-\infty}^{\infty} e^{in\omega}|5n+k+5s\rangle \\ &= e^{-is\omega}\sum_{n=-\infty}^{\infty} e^{in\omega}|5n+k\rangle = e^{-is\omega}|k\rangle\,,\end{aligned} \tag{A.9}$$

and so the the state basis has a modulo 5 structure up to overall phases. Denoting by $G_\pm$ the raising and lowering operators on the original basis states $|n\rangle$,

$$G_+|n\rangle = |n+1\rangle\,, \quad G_-|n\rangle = |n-1\rangle\,, \tag{A.10}$$

we can rewrite Eq. (A.9) as

$$G_+^{5s}|k\rangle = e^{-is\omega}|k\rangle\,, \quad G_-^{5s}|k\rangle = e^{is\omega}|k\rangle\,. \tag{A.11}$$

Using the generalized Wigner–Eckart theorem, we can relate the ground state expectation of $\phi_{(1,1)(-15)}$ to a constant K times the expectation of G_-^{15},

$$\langle\Omega|\phi_{(1,1)(-15)}|\Omega\rangle = K\langle\Omega|G_-^{15}|\Omega\rangle = Ke^{3i\omega}\langle\Omega|\Omega\rangle \neq 0\,. \tag{A.12}$$

So within the modulo 5 state structure, $\phi_{(1,1)(-15)}$ can attain a nonzero ground state expectation.

Appendix B. Review of Condensate Formation

We first review the Lorentz kinematics of forming condensates from Dirac spinors, and then turn to the dynamics of condensate formation. For any two Dirac spinors Ψ_1 and Ψ_2, both $\overline{\Psi_1}\Psi_2$ and $\overline{\Psi_1^c}\Psi_2$ are Lorentz scalars, with c denoting charge conjugation and with $\bar{\Psi} = \Psi^\dagger i\gamma 0$. In analyzing condensate formation, it is convenient to use real Majorana representation γ_μ matrices, with γ_5 self-adjoint and skew symmetric, and $\gamma 0$ skew symmetric. The chiral projectors P_L, P_R defined by

$$P_L = \frac{1}{2}(1 + \gamma_5)\,, \quad P_R = \frac{1}{2}(1 - \gamma_5)\,, \tag{B.1}$$

then obey

$$\begin{aligned} P_L^\dagger &= P_L\,, \quad P_R^\dagger = P_R\,, \\ P_L^T &= P_R\,, \quad P_R^T = P_L\,, \end{aligned} \tag{B.2}$$

with $\dagger$ the adjoint and T the Dirac transpose. Charge conjugation now reduces to complex conjugation, and so we have

$$\Psi^c = \Psi^*\,, \quad \overline{\Psi^c} = (\Psi^c)^\dagger i\gamma 0 = \Psi^T i\gamma 0\,. \tag{B.3}$$

For left chiral spinors, $\overline{\Psi_{L1}}\Psi_{L2} = 0$, while Eqs. (B.2) and (B.3) imply that

$$\overline{\Psi_{L1}^c}\Psi_{L2} = \Psi_{L1}^T i\gamma 0 \Psi_{L2} \neq 0\,. \tag{B.4}$$

Thus Eq. (B.4) gives the general Lorentz structure of scalar condensates constructed from left chiral spinors. Since γ_0 is skew symmetric, and since spinors anticommute, Eq. (B.4) has the same form when state labels $1, 2$ are interchanged,

$$\Psi_{L1}^T i\gamma 0 \Psi_{L2} = \Psi_{L2}^T i\gamma 0 \Psi_{L1}\,. \tag{B.5}$$

Because this equation involves no complex conjugation, the group representation content of the condensate is simply the direct product of the representation content of Ψ_1 and Ψ_2.

The only way to rigorously determine if condensates form in a theory is to calculate the effective action[40] governing condensate formation, and this is generally not feasible. So to study the dynamics of condensate formation, one falls back on simple rules of thumb, such as determining whether the leading order perturbation theory force between the constituents is attractive. The single gluon exchange potential[28,34] produced when a vector gluon mediates the reaction $A + B \to A + B$ is

$$V = \frac{g^2 K(A+B; A, B)}{2r},$$
$$K(A+B; A, B) = C_2(A+B) - C_2(A) - C_2(B), \tag{B.6}$$

with g the gauge coupling and the C_2 the relevant Casimirs. (The Casimir for a representation R is calculated from the index $\ell(R)$, the dimension $N(R)$, and the dimension of the adjoint representation $N(\text{adjoint})$ by $C_2(R) = \ell(R)N(\text{adjoint})/N(R)$; see Slansky.[1]) When more than one non-Abelian group acts on the fermions forming the condensate, the one gluon exchange potentials associated with each are added.

References

1. R. Slansky, *Phys. Rep.* **79**, 1 (1981).
2. M. Gell-Mann, P. Ramond and R. Slansky, *Rev. Mod. Phys.* **50**, 721 (1978).
3. T. L. Curtright and P. G. O. Freund, *SU*(8) unification and supergravity, in *Supergravity*, eds. P. van Nieuwenhuizen and D. Z. Freedman (North-Holland, 1979).
4. P. Ramond, The family group in grand unified theories, invited talk at the *Sannibel Symposia*, Feb. 1979, arXiv:hep-ph/9809459.
5. P. H. Frampton, *Phys. Lett. B* **89**, 352 (1980).
6. J. Chakrabarti, M. Popović and R. N. Mohapatra, *Phys. Rev. D* **21**, 3212 (1980).
7. C. W. Kim and C. Roiesnel, *Phys. Lett. B* **93**, 343 (1980).
8. Dzh. L. Chkareuli, *Pis'ma Zh. Eksp. Teor. Fiz.* **32**, 684 (1980).
9. J. E. Kim and H. S. Song, *Phys. Rev. D* **25**, 2996 (1982).
10. S. K. Yun, *Phys. Rev. D* **29**, 1494 (1984).
11. S. K. Yun, *Phys. Rev. D* **30**, 1598 (1984).
12. J. L. Chkareuli, *Phys. Lett. B* **300**, 361 (1993).
13. S. M. Barr, *Phys. Rev. D* **78**, 075001 (2008).
14. R. Martinez, F. Ochoa and P. Fonseca, A 3-3-1 model with *SU*(8) unification, arXiv:1105.4623.
15. E. Cremmer and B. Julia, *Phys. Lett. B* **80**, 48 (1978).
16. N. K. Nielsen and H. Römer, *Phys. Lett. B* **154**, 141 (1985).
17. M. J. Duff, Ultraviolet divergences in extended supergravity, in *Supergravity 81*, eds. S. Ferrara and J. G. Taylor (Cambridge University Press, 1982), arXiv:1201.0386.
18. N. Marcus, *Phys. Lett. B* **157**, 383 (1985).
19. S. Coleman and E. Weinberg, *Phys. Rev. D* **7**, 1888 (1973).
20. H. Georgi and S. L. Glashow, *Phys. Rev. Lett.* **32**, 438 (1974).
21. S. M. Barr, *Phys. Lett. B* **112**, 219 (1982).

22. J.-P. Derendinger, J. E. Kim and D. V. Nanopoulos, *Phys. Lett. B* **139**, 170 (1984).
23. I. Antoniadis, J. Ellis, J. S. Hagelin and D. V. Nanopoulos, *Phys. Lett. B* **194**, 231 (1987).
24. A.-C. Davis and N. F. Lepora, *Phys. Rev. D* **52**, 7265 (1995).
25. S. M. Barr, Rotating away proton decay in flipped unification, arXiv:1307.5770.
26. T. L. Curtright, *Phys. Lett. B* **102**, 17 (1981).
27. E. S. Fradkin and A. A. Tseytlin, *Phys. Lett. B* **134**, 301 (1984).
28. M. Peskin, Chiral symmetry and chiral symmetry breaking, in *Les Houches Summer School in Theoretical Physics: Recent Advances in Field Theory and Statistical Mechanics*, eds. J. B. Zuber and R. Stora (North-Holland, 1984).
29. G. 't Hooft, *Phys. Rev. Lett.* **37**, 8 (1976).
30. G. 't Hooft, *Phys. Rev. D* **14**, 3432 (1976).
31. S. Coleman, The uses of instantons, in *Aspects of Symmetry* (Cambridge University Press, Cambridge, 1985), Chap. 7.
32. E. Farhi and L. Susskind, *Phys. Rep.* **74**, 279 (1981).
33. J. M. Cornwall, *Phys. Rev. D* **10**, 500 (1974).
34. S. Raby, S. Dimopoulos and L. Susskind, *Nucl. Phys. B* **169**, 373 (1980).
35. L.-F. Li, *Phys. Rev. D* **9**, 1723 (1974).
36. D. Z. Freedman and A. Van Proeyen, *Supergravity* (Cambridge University Press, Cambridge, 2012).
37. L. Alvarez-Gaumé and E. Witten, *Nucl. Phys. B* **234**, 269 (1983).
38. S. Weinberg, *Phys. Rev. Lett.* **29**, 388 (1972).
39. S. Weinberg, *The Quantum Theory of Fields, Volume II: Modern Applications* (Cambridge University Press, Cambridge, 1996), Sec. 19.3.
40. J. M. Cornwall, R. Jackiw and E. Tomboulis, *Phys. Rev. D* **10**, 2428 (1974).